Y0-BSD-770

Structure of Water and Aqueous Solutions

Structure of Water and Aqueous Solutions

Proceedings of the International Symposium
held at Marburg in July 1973

edited by
Werner A.P. Luck

Verlag Chemie   Physik Verlag

Editor: Prof. Dr. Werner A. P. Luck
Department of Physical Chemistry
Marburg University
D-3550 Marburg
Biegenstraße 12

Copy Editing: Dr. Hans F. Ebel

ISBN 3-527-25588-5 (Verlag Chemie)
ISBN 3-87664-028-8 (Physik Verlag)

Printer and Bookbinder: Hans Richarz Publikationsservice, St. Augustin
Printed in Germany

THE BEST OF HUMANITY IS LIKE WATER;
WATER SERVES ALL.

LAOTSE.

Foreword

There would be no life on earth without water, and in particular there would be no life on earth without the unusual properties of water. Water would melt at about -90 ^{o}C and boil at 1 atm at about -80 ^{o}C if the influence of the H-bonds did not dominate the properties of water. There are organisms with extreme content of 96-97% water, there are organs of mammals which are 97% water (eyes). The fundamental molecule of life DNA, is denaturated by dehydration. There are no growing mechanism without water, there are no life mechanism without water.

Much industrial production would not exist without water, like emulsion-polymerisation, production of Cl_2 from NaCl, flotation, dyeing in textile chemistry and so on. Colloid-chemistry would be only a very small field without water. The solubility properties of water are very important. And there are other anomalous behaviours of water. For example water becomes a gel if one adds the dyestuff pseudo-iso-cyanine to water with a concentration of 1 dye molecule in 2000 water molecules. But until now the enormous progress of science is delayed by this fundamental question of exact knowledge of the water structure.

The modern history of the structure of water starts in 1892 with a paper by Röntgen. "The molecules of type 1, - which we will call ice-molecules, because we will give them certain properties of ice, - are transformed in molecules of type 2 by heat input; however, if the heat content of water is reduced, so will return a corresponding part of ice-molecules. We suppose the non-supercooled water is at every temperature a saturated solution of ice-molecules, which is the more concentrated the lower the temperature is".

In the following 25 years (1892-1917) (I in Fig.1) different authors tried to analyse the properties of liquid water by the assumption of a coupled equilibrium of different aggregates. Under the influence of Röntgen's paper and the point of view of the time that in ice the aggregation number is 10-12 molecules of H_2O the courage to assume large associates was only growing slowly. Therefore the maximum of supposed aggregation number of liquid water was nearly a linear function of the publications year of the different papers till 1915 (Fig. 1).

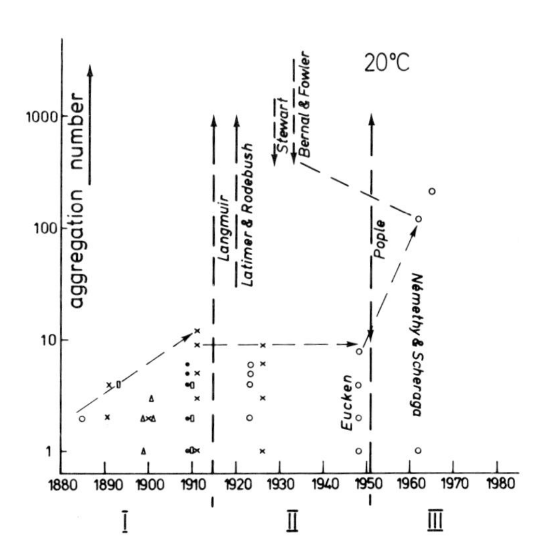

Historic view on water research

In this year Langmuir stressed for liquids that: "every atom is combined chemically (or adsorbed) to all adjacent ones. The molecular weight is, therefore, a term that has very little significance in the case of a liquid".

Latimer and Rodebush in 1920 pointed out: "Indeed the liquid may be made up of large aggregates of molecules continually breaking up and reforming under the influence of the termal agitation". In the next three decades (II Fig.1) water research branched in two different directions. One group following Langmuir's, Latimer's, and Rodebush's ideas, received support from Stewart's x-ray scattering results, which indicated a lattice-like situation in liquids at low temperatures. Stewart (1929, 1930) described liquids as mixture of large aggregates (clusters) and monomers. This assumption of large crystal-like aggregates was also stressed qualitatively by Bernal and Fowler (1933). A second group retained the older model of equilibria of different small aggregates (Tamann, Eucken). This model of small aggregates was inspired by the research of association in solution and was elaborated by Eucken 1948. He could describe the temperature function of specific heat quantitatively with the model of equilibria of monomers, dimers, tetramers and octamers.

At the beginning of the last decade (III Fig.1) Nêmethy and Scheraga (N.S.) gave 1962 a model with quantitative statements of cluster sizes of about 125 molecules at room temperature. This paper also gave quantitative agreement with specific heat data and represents a compromise between the two different viewpoints of small or very large aggregates. The properties of water, for example the specific heat, can be demonstrated quantitavely by very different assumptions. For example, see Eucken's paper (1948) and the paper of Némethy and Scheraga. But the reason is only a mathematical one. Both theories adjust a number of constants (5). Therefore the physical meaning of both theories is that water can only be described by 5 adjusted constants.

In this situation experiments were necessary to decide in which directions theories can be made and if the constants can be determined experimentally.

Earlier in 1952, Haggis Hasted and Buchanan (H.H.B.) had tried to understand the dielectric constant of water to the critical temperature T_C. With a simple formula they tried to estimate the content of broken H-bonds in water from the vaporisation energy and the adjustment of 9% open OH groups in the liquid at the melting point.

The best method to study the H-bonded state is infrared spectroscopy especially in the overtone region. For example the author started in the beginning of the sixties experiments to study I.R. spectra of water up to T_C. From this method one finds a similar function of non-bonded OH-groups. But this experimental method is founded only of the adjustment of zero OH bonds at T_C. This assumption is established by spectroscopic measurements. This experimental function of non-bonded OH groups is similar to the function founded by H.H.B. and with the temperature dependence of the nmr chemical shift and is in agreement with Raman data (Walrafen). It allows the calculation of some properties of water without any other adjusted constants in a two species model.

These experimental data have had the result that the attempts to assume water as a coupled equilibrium of different small species cannot be true and the discussion of monomers below 200 $^{\circ}$C should be neglected. Eucken's model would give at the melting point about 65% non-bonded OH groups. Therefore this second viewpoint of water structure was disproved by experiments.

But in the meantime a new second group was founded. It starts 1951 (Pople) to stress that liquid water should be understood as a continuum of different states of distances and angles. In the last years some papers tried to elucidate these distribution functions (Stillinger 1972, nmr data, x-ray- or neutron scattering experiments). The development of water research in the next years (III Fig.1) will be to improve our knowledge of such distribution functions with the goal to have an exact description of water. Meanwhile approximation methods - how the determination of non-bonded OH groups - can be applied to help practical problems. From this point of view the last years have seen considerable in our understanding of the most important and most anomalous liquid of our planet. The Marburg meeting will attempt to look for the present standpoint.

The point of view of some experts in water research is that we should now stop thinking of simple models of water and should wait till one genius will discover this complicated distribution function. I wonder why this group has no scruple to need the idealized model of first approximation of ideal gas to describe the gaseous state and discuss more complicated models of the real gaseous state if necessary. I think since water is one of the most important materials for society and for science, we cannot wait on this exact description of the water structure. In the meantime for a lot of application problems of water structure in industry, in chemistry or in biochemistry laboratories we need simple models like the efficient model of ideal gas. If we remember that these idealized models are approximations, the scientists, who look for the exact distribution function of water structures, may forgive this simple attempt.

We thus see that the history of water structure shows clearly that scientists are influenced by fashion trends in science (Fig.1), a statement which is verified in an alarming situation by the story of "polywater".

Deryagin, Fedyakin and collaborators studied the properties of water in quartz capillaries and found that in some capillaries (at about 5%) water has some anomalous properties (no freezing point at 0 °C, high viscosity, higher density etc.). Several newspapers and popular publications featured this new "form" of water. The most interesting point of view of this story seems to be that many serious scientists lost their trained constructive criticism by the excitement of this novel field of work. They published many papers on this so - called polywater. Even calculations of quantum chemistry showed that water has to have this modification of "polywater". The spending of some million of dollars of "polywater"-research the knowledge only led to showing that impurities simulated the "new modification" of water and the fashion of "polywater" vanished as suddenly as the fashion of mini-skirts.

Naturally the Marburg Meeting papers were free from such fashion's influence. They give a good and short review on the present status of water research.

Although there is now a comprehensive treatise of water research in the five volumes of the excellent work edited by Felix Franks there is a need for a shorter review of current water research. This review should be of interest to experts, to non-specialists who need to be informed on the modern point of view of water structure, adn to students, who can better appreciate most of the techniques of Physical Chemistry by following one central problem with all modern methods.

The main goal of the meeting has been the structure of pure water, secondly the structure of aqueous solutions. The structure of aqueous solutions is very complicated. There are 6 main parameters of anomalous properties of solutions: amount of H-bonds, distribution function of different H-bonds (angles, distances etc), the life-time of the H-bonds, proton exchange mechanism (especially in acids and bases), special gas hydrate-like structures in presence of hydrophobic groups, hydrate structure around ions or hydrophilic organic molecules. Whilst our knowledge of the water structure has made some progress in the last two decades our knowledge of the structure of solutions has only just begun.

Chapter I gives an introduction to the restrictions for an acceptable model of liquid water (I.2. H.Frank) and an general view on the properties of liquid water (I.3. E.U.Franck). This chapter is introduced by a general view on the interpretation of water properties given by Onsager. Chapter II provides a review of the different theoretical attempts to describe water and solutions and some new aspects of the theory. Chapter III - VIII are organized according to the different experimental methods to study water and solutions. Experimental techniques in the field of work of water research are much more important than in other areas because the water structure and the solution structures are so complicated. The I.R. methods are given in chapter III. In chapter III.3. an attempt is made to describe the water properties by a very simple model. The I.R. matrix technique (chap. III.4.) gives the interesting result that under matrix conditions there is not a continuum of different aggregation states of water but only a small number of about 5 defined species. The Raman methods chapter IV are in good agreement with the I.R. methods and can give now together a first feeling for the structure of water. Scattering methods are given in chapter V. The x-ray and neutron scattering methods show that the distances between first next neighbours in liquid water do not change much with temperature and are of the size of ice-like arrangements. The dielectric methods for studying water, solutions and ice are reported in chapter VI and a review on the NMR methods is given in chapter VII. The Ultrasonic methods (chapter VIII) complete the experimental methods. This method gives interesting information and shows that the results can be described with a minimum of three species. But this method needs many parameters to give a consistent picture of the water properties. Chapter IX serves the second goal of our meeting to introduce scientists to the problems of sea-water desalination. Young collagues who today wish to give their work social significance will find in this chapter one important application of the water research. The last chapter X gives the titles of the short communications of the meeting and limitations some references of the work of these contributors. It is regretted that the volume limitations of this book precludes the publication of the new results presented in these short contributions.

We are fortunate that the main experts in the different field of works were able to accept our invitation to Marburg and that they provided written manuscripts of the major contributions. I hope that our work will succeed in enlarging our knowledge of the most important material of our planet and that one or the other may be stimulated to help in the complicated development of sea-water desalination.

It gives me great pleasure to express my thanks for all the help in organisation of the meeting, especially to G.E.Walrafen and A.P.Zukovskij and to our staff Drs. B.Mann, P.Dechant, U.Schoeler, L.v.Szentpaly, and M.Kuballa; and Messrs. Th.Neikes and W.Lattrich and Frau A.Krumey.

I also had much help for the editorial work of this book. I thank Dr.H. Hallam for guidance with translationational difficulties and G.Andratschke, S.Bamberger, G.Dahm, U.Olias and A.Weinert for much editonial assistance and also their contributions to the smooth running of the meeting.

Marburg, April 1974 W.A.P.Luck

CONTENTS

CHAPTER II THEORY

II.4. THEORY OF HYDROGEN BONDING IN WATER AND ION HYDRATION

P.Schuster

II. 5. SOLUTE-SOLUTE INTERACTIONS AND THE EXCESS PROPERTIES OF AQUEOUS SOLUTIONS

H.L.Friedman; C.V.Krishnan and L.P.Hwang

II. 6. THERMODYNAMICS OF MIXTURES OF AQUEOUS ELECTROLYTE SOLUTIONS

Y.Ch. Wu

CHAPTER III INFRARED METHODS

III. 1. THE FAR-INFRARED SPECTRUM OF WATER

B. Curnutte and D. Williams

III. 2. THE FUNDAMENTAL SPECTRUM OF WATER AND SOLUTIONS

W.A.P. Luck

III. 3. THE OVERTONE REGION

W.A.P. Luck

III. 4. INFRARED MATRIX ISOLATION STUDIES OF WATER

H.E. Hallam

CHAPTER IV RAMAN METHODS

IV. 1. SPONTANEOUS AND STIMULATED RAMAN SPECTRA FROM WATER AND AQUEOUS SOLUTIONS

G.E. Walrafen

IV. 2. STIMULATED RAMAN SCATTERING IN LIQUID WATER

M.J. Colles

IV. 3. RAMAN SPECTRAL STUDIES OF ION-ION AND ION-SOLVENT INTERACTIONS

D.E. Irish

CHAPTER V SCATTERING METHODS

V. 1. X-RAY AND NEUTRON DIFFRACTION FROM WATER AND AQUEOUS SOLUTIONS

A.H. Narten

V. 2. ELECTRON DIFFRACTION STUDIES OF LIQUID WATER

E. Kálmán, E.Lengyel, G.Pálinkás, L.Haklik and E. Eke

CHAPTER VI DIELECTRIC METHODS

VI. 1. THE DIELECTRIC PROPERTIES OF WATER

J.B. Hasted

VI. 2. DIELECTRIC RELAXATION OF WATER IN AQUEOUS SOLUTIONS

R. Pottel, K. Giese and U. Katze

VI. 3. STRUCTURAL INFORMATION FROM DIELECTRIC PROPERTIES OF ICE

C. Jaccard

CHAPTER VII NMR - METHODS

VII. 1. NMR-STUDIES OF PURE WATER

J.A. Glasel

VII. 2. NMR STUDIES OF AQUEOUS ELECTROLYTE SOLUTIONS

H.G. Hertz

VII. 3. NMR STUDIES OF AQUEOUS NON-ELECTROLYTE SOLUTIONS

M.D. Zeidler

CHAPTER VIII ULTRASONIC METHODS

VIII. 1. STRUCTURAL RELAXATION IN WATER AND IONIC SOLUTIONS

K.G. Breitschwerdt

CHAPTER IX SEA-WATER DESALINATION

Opening address

It gives me great pleasure, as dean of the "Fachbereich für Physikalische Chemie der Universität Marburg", to have the opportunity of opening this international meeting concerned with the "Structures of Water and Aqueous Solutions".

In 1529, the Duke of this region invited all of the church reformers of the time to a council concerned with the founding of a new church. Luther, Zwingli, Melanchton and other well-known men assembled at Marburg. They were successful in framing 14 joint statements, but they disagreed on a 15th point dealing with the Lord's supper. Hence, they left Marburg in failure. Our work here, fortunately, is much easier. If we should agree on 14 points, but not on the 15th, our meeting would be a great success!

The history of our university, which was founded in 1527, has aspects related to our present field of interest. The first European Professor of Chemistry was Professor Hartmann in Marburg in 1609. We also have a thermodynamic tradition; Denis Papin was Professor of Physics from 1687 to 1707. And, we have some spectroscopic history since Bunsen was Professor of Chemistry at Marburg in the last century. Further, we have an excellent tradition in Quantum Chemistry. Professor Erich Hückel noted for his contributions to electrolyte solution theory and molecular orbital theory, was Professor of Physics here, and still lives in retirement in Marburg.

This year, Walrafen from the United States, Zukovskij from the Soviet Union, and Miss Ghosh from India have been working in Marburg, and since these workers constituted an international nucleus for research in water structure, the idea of expanding this nucleus into an international meeting of many workers was conceived. Also at that time I became chairman of a research group of the Dechema for sea-water desalination. My assignment was to activate water research and development in Germany, and thus the German Ministry for Research and Technology agreed to inaugurate this German activation with an International Meeting. To this end, it was requested that this meeting bring people involved in fundamental research, together with people concerned with practical applications. Financial help was given by the "German ministry for Research and Technology"by the "Hessisches Kultusministerium" and the "Dechema, Frankfurt".

Finally, there is one other serious matter that I should like to mention, that deals with the German university situation today, and more specifically, with the University of Marburg. Today our young students are being compelled to develop a greater interest in university politics, than in science. If the student's union organizes any aggressive action this lecture hall is overcrowded with students. This meeting was also organized to give our students the possibility to be trained in modern physical chemistry. This meeting gives them the opportunity to learn the latest point of view in water research, one of the most important subjects of science; and on the other hand to learn all modern methods of Physical Chemistry on one research-example by the leading international experts. However if we look around in this lecture hall we can count on the fingers of one hand the number of students of the University of Marburg who participate in our meeting. This catastrophic result may be an objective criteria on the new constitution and on the new spirit of our university-mode by politicians.

I personally believe that physical scientists have been more successful in improving human conditions in the last 200 years than political scientists. Now, however, our politicians are compelling us to work under political conditions, and to ignore those conditions which led to the healthy and vigorous development of science. Indeed, the German universities are actually being renovated along liberal arts and political lines. Therefore, I hope that this meeting will concentrate on the goal of helping society, by helping it through well-established, rigorous, responsible scientific methods. Every scientist has a responsibility to society, of course, but society must have some responsibility to its scientists. With this in mind I would now like to welcome the lecturers and scientists from 20 countries to our meeting.

W.A.P.Luck

I. INTRODUCTORY SURVEY

I. 1. INTERPRETATION OF KINETIC AND EQUILIBRIUM PROPERTIES

L. Onsager
Center for Theoretical Studies
University of Miami
Coral Gables, Florida 33124

ABSTRACT

All solid forms of water exhibit complete networks of hydrogen bonds in more or less regular tetrahedral coordination. The heat of fusion suggests that about 40% of the molecules in the liquid might well be 3-coordinated at the freezing point. These could conceivably perform functions analogous to those of D and L defects, perforce equal in number. Transfers of such defects rearrange the topology of the hydrogen-bonded network, whereby the correlation between viscous and dielectric relaxations seems reasonably explained.

1.1 APPROACH

While we entertain the ambition to explain all the remarkable properties of liquid water in terms of its electronic structure, it is helpful to examine the gaseous and solid phases.

1.2 THE VAPOR

The free molecule in the vapor has a symmetrical bent structure; the O-H distance is a little less than 1 A.U., and the apical angle is about 105°. An electric dipole moment of 1.85 Debye does not quite account for the magnitude of the second virial coefficient, which has been taken as evidence for a significant quadrupole moment /1/, as might be reasonably expected on first principles.

1.3 THE SOLID STRUCTURES

Ordinary ice contracts about 8% on melting; but at moderate pressures in the range 2.1-21.5 kilobar a succession of phases quite a bit denser than the liquid are stable. In the low pressure forms Ih and Ic each molecule is surrounded by four neighbors, disposed at the corners of a regular tetrahedron, at a distance of about 2.75 A.U.; accordingly, there are 12 second nearest neighbors all at the same distance of 4.5 A.U. The high pressure forms II-VI display somewhat distorted tetrahedra of neighbors at distances ranging from 2.75 to 2.9 A.U., with no other molecule approaching much closer than the van der Waals radius of about 3.5 A.U.; but in the densest form VII the molecular arrangement is body-centered cubic with a distance of 2.95 A.U. from corner to center. The interpretation of these findings in terms of hydrogen bonding is amply confirmed by studies of neutron diffraction, infrared and Raman spectra and, particularly for the disordered phases Ih, Ic, III, V, VI and VII, by their electrical and acoustic properties and by their residual entropies. Clearly there is a strong preference for a complete set of hydrogen bonds connecting every molecule. The preferred angle between proton bonds is evidently not very different from 109°; but the variations encountered among ices II-VI entail rather modest increments of energy |2|.

1.4 THE LIQUID

Considering the variety of configurations observed among the solid phases, it stands to reason that a great many disordered structures could be produced by assembling molecules in largely four-coordinated arrangements of tetrahedra not much more distorted than those which occur in the various solid forms. As regards the radial distribution function of pairwise distances in water, B. Kamb /2/ produced a pretty good imitation by combining reasonably smeared r.d.fs. of 50% ice I, 17% ice II and 33% ice III. However, as Kamb rightly emphasizes, this does not at all imply any significant abundance of molecular aggregates large and regular enough to be identified with any of the crystal structures, and it stands to reason that there must be many more irregular configurations of energies compatible with the heat of fusion, which contain very few recognizable microcrystals. One further constraint on speculations about microcrystals is imposed by the fact that water is about as readily undercooled as are most metals or fused salts. Crystals generally contain small (at times large) proportions of point defects like interstitial molecules or ions, or vacancies of either, which are important for transport processes. In the disordered forms of ice the polarization current is carried by the misbonding defects named for Bjerrum /3/. In the solid abnormal co-ordination of a single molecule does not occur isolated, because it depends on the presence of an adjacent interstitial or vacancy. Similarly, a Bjerrum defect must involve a pair of adjacent molecules in such orientations that each one presents either an unbonded hydrogen or an unbonded lone electron pair towards the other, or at least in that general direction. Analogous defects in the liquid phase are not subject to such constraints; quite particularly, a three-coordinated molecule here or there imposes no additional discomfort on its neighborhood. For the purposes of dielectric relaxation etc., such a molecule may be regarded as half a Bjerrum defect; it is $\frac{1}{2}$ D or $\frac{1}{2}$ L, depending on whether its proton or its lone pair is disengaged. Jaccard /4/ estimated in effect 0.68 e.v. for the energy needed to produce a pair of D and L defects in ice; if we divide this by 4, allowing that the self-energy is proportional to the square of the charge, we arrive at an estimate of 0.17 e.v. = 3.9 k cal/mole for the energy required to produce a pair of half Bjerrum defects in water. This estimate will be reduced if we allow that a high frequency dielectric constant of water greater than that of ice ought to be assumed for the purpose in hand.

As analyzed by B. Kamb /2/, the heat of fusion suggests that quite a significant proportion of the hydrogen bonds are broken. In infrared and Raman spectra both the location and the width of the OH stretching bonds are important; a comparison with the frequencies observed in the vapor and in the various solid modifications allows a rough estimate of the proportion of broken bonds. This proportion is seen to vary with the temperature, and the order of magnitude of the temperature coefficient is compatible with estimates of the bond energy. For a naive interpretation we might try to invoke the law of mass-action

$$\alpha^2/(1-\alpha) = K(T) \quad (1)$$

where $2\underline{N}(1-\alpha)$ is the number of intact bonds among $\underline{N}$ molecules, while the numbers of hydrogens and lone electron pairs not involved in bonding are both equal to $2N\alpha$. For a refinement we might recognize electrostatic interaction between these defects and apply a correction of the Debye-Hückel form

$$\Delta \log K = C\ \alpha^{1/2}\ (\varepsilon_\infty T)^{-3/2} \tag{2}$$

where ε_∞ denotes the high frequency value of the dielectric constant -- if that can be defined. However, that leaves unanswered the important question how readily the random thermal motion of the whole connected network will strain a bond to the breaking point, and how often is a bond restored by the recombination of complementary defects? In other words, how much will one added pair of defects contribute to the configurational entropy? This is likely to depend on the defect concentration itself, and we end up with the rather vague formulation

$$\begin{aligned} \alpha^2/(1-\alpha) &= K(T,\alpha) < \infty \\ K(T,\alpha) &\sim K_1(T)\ F(\alpha)? \end{aligned} \tag{3}$$

It is safe to assume that K is bounded; but more detailed predictions may call for difficult analysis. This whole discussion has no meaning unless at least the topology of the network is reasonably well defined. However, we have good reason to hope for that much, considering how the r.d.f. of water, quite unusual among liquids, reminds us particularly of ice I and to a lesser extent of the denser solid forms.

In his attempt to explain the large heat of fusion, amounting to 1.44 kcal/mole, Kamb /2/ found it difficult to account for the whole amount in terms of moderate strains. Without entering into all the pros and cons, let us assume tentatively that half of it should be attributed to the breaking of bonds, and see where that takes us. If we assume 4 kcal per bond, we get $\alpha = 0.18$, and if we assume further that the relative abundance of complementary defect pairs on the same molecule is just α^2, we end up with abundances of 45%, 40%, and 13% for 4-, 3-, and 2-coordinated molecules, respectively, at 273°K. If we assume only 3 kcal per break, the proportions become 33%, 42% and 20%.

The appropriate values of K exceed the Boltzmann factors after doubling on account of molecular symmetry, by factors of about 7 and 6, respectively. Such estimates seem compatible with the spectroscopic evidence and also with the observed entropy of fusion. The large gain per break becomes more plausible when we realize that a fully four-connected network must contain a lot of 5- and 6-rings if shorter ring connections are ruled out (in view of the bond strain). However, such a ring imposes severe constraints on the possibilities for placing topologically distant parts of the network close to it. Consider the rigid packing in the self-clathrates formed by ices VI and VII!

1.5 TRANSPORT PROCESSES AND RELAXATION

The dielectric constant of water at the freezing point is a little lower than that of the ice, despite the greater density. The difference is easily explained in terms of a somewhat disorganized structure, and possibly a slightly smaller dipole moment due to the slightly longer hydrogen bonds. The dielectric relaxation is more than a million times faster in the liquid. While this clearly imples a significant difference in the mechanism, it is tempting to seek a common explanation for the remarkable simplicity of both dispersion curves. Very well, the half Bjerrum defects in water are just as fit to carry a displacement current as are their bigger cousins in the solid: the half D is transferred when an uncoordinated proton swings in to displace a previous participant in another bond; the analogous transfer of a half L defect involves just a rotation of the intervening molecule. In both cases the topology of the network is changed as one link is tied while another is severed. Viscous relaxation also requires a rearrangement of the network topology, and if that is the rate-determining part of the process, then the identity of elementary steps accounts for the observed close correlation between viscous and dielectric relaxation in the liquid; the ratios of relaxation times in D_2O and H_2O is very nearly equal to the ratio of the viscosities /5/. This ratio decreases with increasing temperature over the range 0°-70°C, so as to suggest that both rotations and translations of the molecules are involved, and that the former are relatively more critical at the lower temperatures.

Possibly the most paradoxical kinetic property of water is the decrease of the viscosity with increasing pressure, observed over the range up to 1kb at the freezing point and over decreasing ranges of pressure and the temperature is raised to 40°C. At 0.6 kb the compression amounts to 2.3%, and the viscosity is reduced by more than 6%. About the only explanation in sight is this, that the additional crowding places a larger premium on added flexibility, hence more defects! The need for new defects on compression could not readily explain the abnormal absorption of sound; the associated reorganization of the network might.

REFERENCES

1 J. S. Rowlinson, Trans. Faraday Soc. 47 (1951) 150

2 B. Kamb, p. 507 Structural Chemistry and Molecular Biology, Alexander Rich and Norman Davidson, eds., W. H. Freeman and Company, San Francisco and London 1968

3 N. Bjerrum, Kgl. Druske Vid. Selskob Mat-fys. medd. 27 (1951) 1

4 C. Jaccard, Helv. Phys. Acta 32 (1959) 89

5 C. H. Collie, J. B. Hasted and D. M. Ritson, Proc. Roy. Soc. (London) 60 (1948) 145

I. 2. RESTRICTIONS FOR AN ACCEPTABLE MODEL FOR WATER STRUCTURE

Henry S. Frank
Department of Chemistry
University of Pittsburgh
Pittsburgh, Pa. 15260 U.S.A.

ABSTRACT

As an introduction to the record of the Symposium, it is explained that many things about the structure of water and of aqueous solutions are still only poorly understood, and that we require as much information as we can get, and from as many different sources as possible, in order to solve the problems which are still outstanding. Some kinds of information are at present more useful than others since they are able to establish logically rigorous conditions, or restrictions, or constraints, to which theoretical water models must conform if they are to be acceptable. Examples of such restrictions are found in some long-known peculiarities of water, such as the fact that it has a density-maximum, and these, as well as more recent examples, are discussed from the point of view of the controversy that still exists between those who consider that water should be described as a mixture and those who wish to think of it as a "continuum". Arguments on both sides of this controversy are analyzed, and stress is placed on the fact that the recent *ab initio* quantum-mechanical calculations of the hydrogen bond give reason for thinking that there is some cooperativity between the several H-bonds into which a given water molecule can enter. One of the most successful of the recent theoretical water models is discussed in some detail.

1. Introduction. Water is unique among chemical substances in its relationships to the human body and to the needs and activities of the human race. It is therefore not surprising that the scientific community should have subjected water and its properties to intense scrutiny in a continuing quest for better understanding, and thus control, in such diverse fields as sea-water desalination, the biochemical processes essential to life, and petroleum recovery (e.g. in the relation between the flows of oil and of water in rock formations).

Against this background, a person who had had no special concern with "the water problem" could be excused for finding it strange that although "in the last decade numerous scientific groups in many countries have been applying modern physical techniques to the determination of the structures of water and aqueous solutions",(1) it could still have been necessary, or even reasonable, for "150 scientists from 20 countries to (spend nine days to) discuss critically the different methods used in studying the structure of water, to ascertain the present position regarding the nature of water and its anomalous properties." It must therefore be pointed out that water, in addition to its enormous practical importance, is a very special substance in a scientific sense. Not only are many of its properties anomalous in the extreme, when compared with those of the vast majority of other liquids. In addition, there have grown up conflicting schools of thought regarding the point of view from which its structure should be considered. Moreover, many of the experiments from which decisive answers have been hoped for have given results which have been interpreted in conflicting ways by workers holding conflicting views. Indeed, it has happened more than once, as we shall see below, that, between the publication of a paper reporting some purported fact and the subsequent appearance of a disproof (or retraction or tacit abandonment), there has been a period during which most people believed things about water which turned out not to be true.

One reason for the persistent lack of consensus among experts is the high degree of sophistication of many of the newer experimental methods. Even

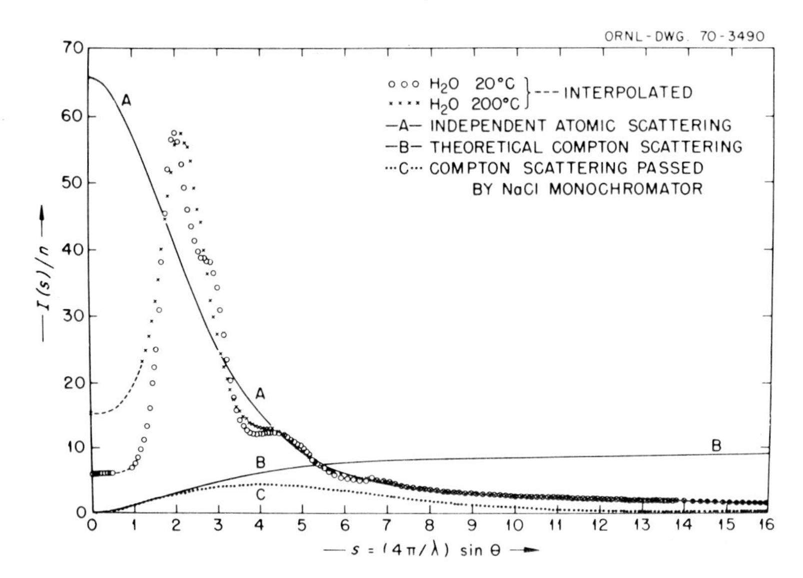

Figure 1 Total scattered intensities of X-rays for water.

Units: electron units per molecule.

From Ref. 2.

a cursory glance at the later chapters of this book will suffice to convey the (justified) impression that many of the experiments they describe give results that can safely be interpreted only be experts. For example, Fig. 1, which shows the results of an experimental study of the scattering of x-rays by liquid water,[2] also shows some "corrections" that had to be subtracted from the scattered intensity actually measured in order to get the remainder to which a structural interpretation could be assigned. The expertise to a) make the measurements, b) make the correct "corrections"[3], and c) arrive at a reliable interpretation of the corrected data is difficult to master, and the number of these fields is therefore limited in which any one individual can himself be authoritative. In consequence it is often difficult for one man to understand clearly something that another man knows. This was the primary reason for calling the Marburg Conference of 1973 - to enable experts in different fields to talk to each other in person, in the hope of laying a broader foundation for an agreed body of fact and opinion about water and its nature.

The search, then, is for agreed statements about water which will reduce the range still open for speculative conjecture - i.e., which place reliable constraints upon what can reasonably be believed. Every time a constraint of this sort can be established the number of acceptable alternatives is reduced. The eventual purpose is to "back water into a corner" in which no alternatives are left, so that we can see it for what it is.

2. Macroscopic Constraints. Some of the constraints on an acceptable model can be established by inspection and are, in a sense, trivial - e.g. that any model must represent the molecules as moving in such a way as to allow the liquid to flow. When this idea is followed up, however, it quickly leads to a constraint that is far from trivial, because of the unique way in which the viscosity changes with changes of temperature and of pressure. A correspondingly non-trivial set of constraints emerges from the study of phenomenologically simple thermodynamic properties such as C_p, the heat capacity at constant pressure, and PVT behavior (e.g. a table of volumes occupied by a gram of water at different pressures and temperatures).

These macroscopic peculiarities of water have, for the most part, been known for a long time - one of them, the existence of a density maximum, for over 300 years.[4] It is, nevertheless, worth-while to outline them briefly here, both as true constraints and also as essential background for the newer information reported in the later chapters.

The first anomaly is the thermodynamic one just referred to, namely, the fact that when water at 0°C and 1 atmosphere pressure is progressively warmed it shrinks slightly in volume before proceeding with the expected thermal expansion. That is, there is a temperature at which its density is a maximum. This temperature of maximum density (TMD) is near 4°C for ordinary H_2O, near 11.2°C for the heavy water D_2O,[5] and near 13.4°C for the recently studied isotopic species T_2O[5] (T is the symbol used here for tritium, the isotope of hydrogen which has a mass number of 3). This behavior is illustrated in Fig. 2, which also displays two additional features

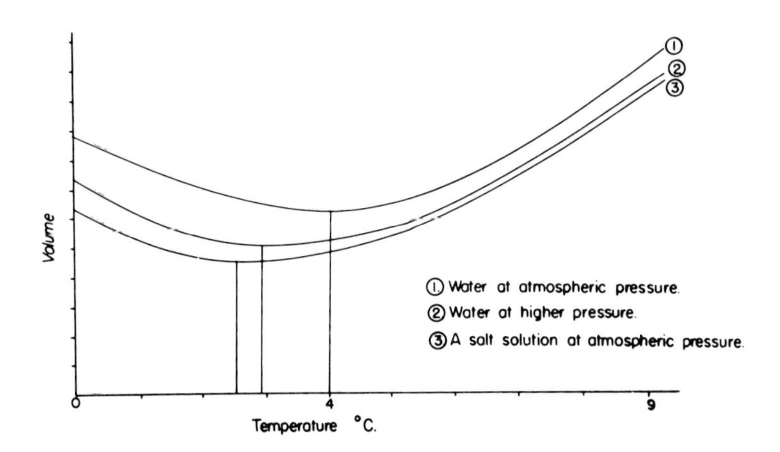

Figure 2 The phenomenon of maximum density. Effect of changing temperature upon the volume of a water sample. How this is affected by increase of pressure or by dissolving a salt.

of the phenomenon, namely, that the TMD is lowered when the water is under pressure (for H_2O, by about 0.020°C per atmosphere, so that the TMD of water in equilibrium - i.e., in the absence of undercooling - disappears below the freezing point at a pressure of some 400 atm)[(6)] and also when the water contains some salt in solution (different salts depress the TMD at different specific rates; for NaCl the rate is about 13°C per mole per liter.[(7)] Small amounts of some other solutes - e.g. tertiary butanol - raise the TMD).[(8)]

That the peculiarities just described have counterparts at temperatures other than 4°C is illustrated in Fig. 3, which shows how water compares with several other liquids in its coefficient of thermal expansion, $[(1/V)(\partial V/\partial T)_p]$ at 25°C, and over a range of pressures.[(9)] At 25°C the

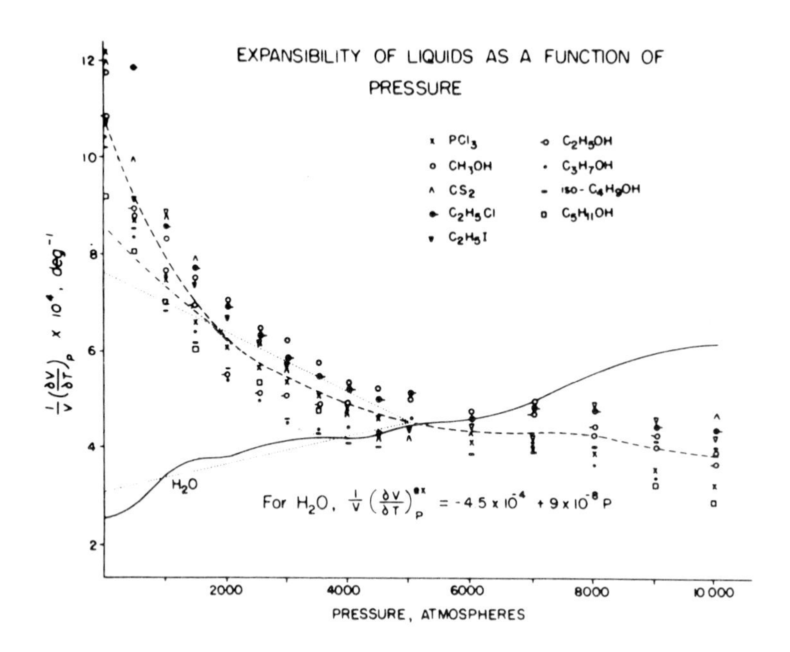

Figure 3 The coefficient of thermal expansion of water and of several other liquids as functions of pressure. Calculated from data of Ref. 9.

expansibility of water is not, indeed, negative, as it is at 2.5°C, but it is still only about 1/4 as large as it would be if water were like normal liquids. The positive slope of the expansibility - pressure plot, which is the 25°C counterpart of the lowering of the TMD by application of pressure, is here seen also to be the opposite of what normal liquids display. An implication of this slope for the molecular make-up of water will be remarked upon in a later paragraph.

Two additional macroscopic anomalies of water are the unusually large value, and near constancy between 0 and 100°C, of its heat capacity at constant pressure, C_p, and the peculiar way in which its viscosity, if measured below 30°C, decreases when a small pressure is applied. The latter is displayed in Fig. 4,[10] and is, in some sense, the strangest of all of

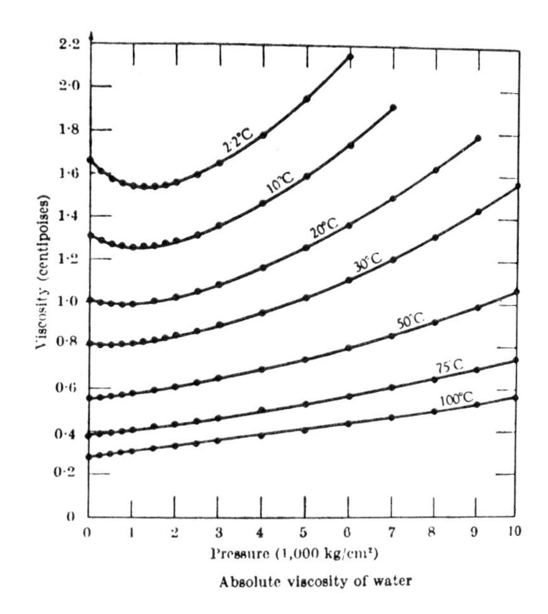

Figure 4

Effect of pressure upon viscosity of water.

Data of Ref. 10.

the properties of water. Research using more recently devised techniques shows that this peculiarity is likewise displayed by the motions of individual molecules - their rates of diffusion and reorientation also increase initially when cold water is subjected to pressure.(11) The inference suggested by the curves of Fig 4 that water behaves more nearly "normally" at higher temperatures and pressures seems generally to be correct, though rather high temperatures and pressures are needed before some of its anomalies disappear.(12)

Even this statement, however, needs qualification. When a liquid is described as behaving normally at any temperature and pressure there is the implied understanding that its properties are related in some expected way to those of some standard fluid, the test fluid and the standard one being taken in so-called corresponding states - i.e. at temperatures and pressures which are equal fractions of the respective critical temperatures and pressures.(13) For this purpose the critical temperatures and pressures are accepted as given, and the point now being made is that water can be said to approach "normality" at high temperatures and pressures only on these terms - i.e., if no attention is paid to the fact that it has a critical temperature and a critical pressure which are strikingly high (T_c = 374.15°C; P_c = 218.3 atm. for H_2O) when compared with, say, CH_3OH (T_c = 185°C; P_c = 66.2 atm.) NH_3 (T_c = 132.4°C; P_c = 113.1 atm) and HF (T_c = 188°C; P_c = 66.2 atm.). That water displays such high values of its critical temperature and pressure [and so interesting a change in going from H_2O to D_2O (T_c = 370.9°C; P_c = 215.7 atm whereas at room temperature D_2O has a lower vapor pressure - i.e., presumably stronger hydrogen bonds - than H_2O)][14] is therefore another constraint upon - i.e., something else that must be accounted for by - any model which may be put forward to explain the energetic interactions between its molecules.

3. Rival Points of View. In the days when structural theory had not yet been highly developed it was possible to be satisfied with explanations in terms of chemical and thermodynamic analogy, and from this point of view it was not difficult to find a qualitative interpretation for the above-mentioned macroscopic anomalies of cold water. A fully explicit discussion along these lines seems first to have been published by Röntgen in 1892,[15] though he remarks that it was, even then, "by no means new". His discussion starts out from the fact that ice shrinks upon fusion, and proceeds to assume that the melting that takes place to form liquid water at 0°C is in some sense incomplete, so that still further melting, and therefore additional shrinkage, occur as 0°C water is warmed. By the time the temperature has risen to 4°C the rate of this supplementary shrinkage is just balanced by the rate of the "normal" expansion which also takes place, so that the resultant $(\partial V/\partial T)_P$ is zero at that temperature, and becomes progressively more positive about it. In Röntgen's words cold water is a solution of ice in a more truly fluid material; in modern language this model represents it as an equilibrium mixture of a low-energy, bulky, "ice-like" constituent and a higher-energy, dense, "un-ice-like" one:

$$(H_2O)_b = (H_2O)_d \qquad (1)$$

It is the shifting of this equilibrium in conformity to the LeChatelier-Braun Principle, then, which causes the anomalous shrinkage on warming between 0 and 4°C and makes a negative contribution to the expansibility even at 25°C and above. Analogously, the equilibrium (1) is shifted to the right in response to an increase in pressure, and this both lowers the TMD (Fig 2) [reduces the negative contribution to the expansibility (Fig 3)] and makes the water flow more freely (Fig 4), since it now contains a larger proportion of the more fluid constituent. That the heat capacity should be anomalously high is also qualitatively accounted for by the need to supply heat as $(H_2O)_b$ "melts" with rising temperature.

The idea that cold water is an equilibrium mixture of distinguishable constituents has since been put forward in many different detailed forms[16] but in essence these can all be regarded as paraphrases of, or elaborations upon, the idea originally outlined by Röntgen.

With the development in the 20th century of the experimental methods and the theoretical viewpoints of chemical physics an opposing school of thought sprang up which directs attention primarily upon what individual water molecules are like and how they interact with each other. Adherents of this school are reluctant to concede that there can be objective criteria for classifying the molecules in a water sample into two categories, or populations, which coexist at any given instant of time, and satisfy Röntgen's definitions of "ice-like" and "un-ice-like" species. The modern movement may be regarded as having originated in a paper published by Bernal and Fowler in 1933,[17] and it is in every sense appropriate that that landmark work, drawing as it does upon X-ray structural studies of solids and of liquids, upon the theory of dipolar molecules and polar liquids, and upon the results of Raman spectroscopy, and employing the mathematical methods of quantum mechanics and of statistical mechanics, should have appeared as a contribution to Volume 1 of the Journal of Chemical Physics.

It is not surprising that the "uniformist" or "continuum" representations of water which have been put forward in this newer tradition should be, by and large, particularly useful in accounting piecemeal for the results of the newer physical measurements upon water which have become accessible in recent decades, but comparatively clumsy in treating of the macroscopic anomalies, discussed above, of which the mixture concept disposes, as it were, with a single stroke. As more and more is learned, however, about the details of molecular interaction in water, it is becoming less and less acceptable to postulate the existence of a mixture simply "because it is needed", in the absence of some additional support for this representation from chemical physics sources, in the form of data which either require it (e.g., by showing that when the molecules in cold water undergo re-orientation they do so in correlated groups),or at least permit of it (e.g. by showing that liquid water contains enough "broken hydrogen bonds" to allow for the existence of some molecules which are "unbonded".)

4. The Hydrogen Bond. It has long been believed, and has recently been confirmed by *ab initio* quantum mechanical calculations,[18] that the water molecule can, for many purposes, by represented by an oxygen atom at the center of a regular tetrahedron which has hydrogen nuclei (protons, deuterons or tritons) at two of its vertices and "lone pair" electrons in orbitals directed toward the other two (Fig. 5a). The elementary

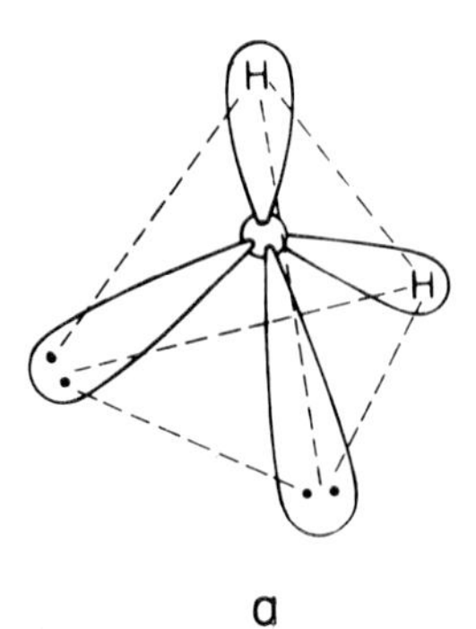

Figure 5 a. Approximate geometric relation of nuclei and electron charge distribution in a water molecule.

nature of the proton (deuteron or triton) and the exposed position of the lone-pair electrons make it possible, in addition, for two water molecules to be held together by a special interaction, in this case typically 4-6 kcal $mole^{-1}$ in strength, known as the hydrogen bond,[19] or H-bond (Fig 5b).

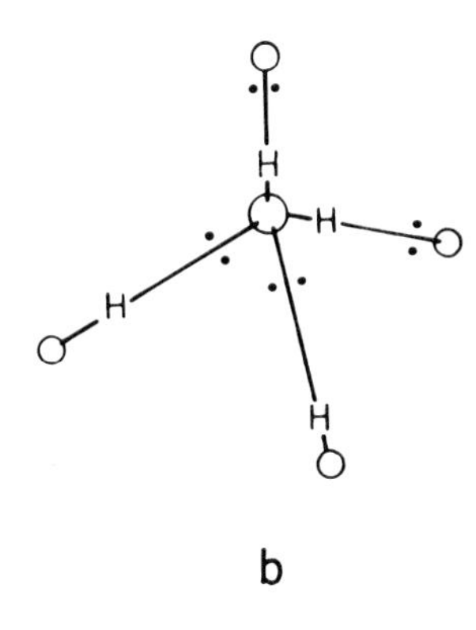

Figure 5 b. The hydrogen bond as the attraction of an H from one molecule to a lone pair of another.

The quantum-mechanical nature of the H-bond has also recently been established by <u>ab initio</u> calculations[20]. Ordinary ice, or Ice Ih, has structural features (Fig 6) determined by the properties of the hydrogen

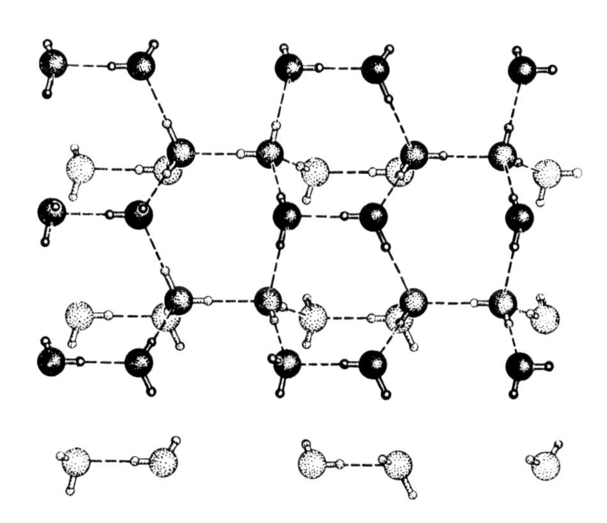

Figure 6

The structure of ice.

Refs. 31 and 17.

bond, in particular tetrahedral coordination and the tendency of a proton to lie exactly, or almost exactly, in the O-O line of centers, though in other crystals, where other packing forces are at work to push it out of the O-O line (or N-H---O, or O-H----N, or F-H----O etc. line ; the typical H bond can be written A-H----B where both A and B are electronegative atoms of elements of the right-hand side of the first period of the periodic table) "bent" H-bonds are also known.[21] Other interesting features of the H-bond revealed by the quantum-mechanical calculations[20b,c] include the extraordinary sensitivity of its binding energy to changes in geometry, and the transfer of a fraction of an electronic charge from the electron donor (basic) molecule to its electron-acceptor (acidic) partner. This charge transfer is presumably associated with the property of a sequence of H-bonds "in series" - e.g.,

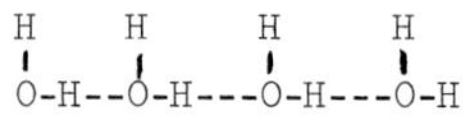

```
H     H      H      H
|     |      |      |
O-H--O-H---O-H---O-H
```

of reinforcing each other (when in favorable geometric configurations in relationship to each other; when in unfavorable configurations or when not "in series" - e.g.

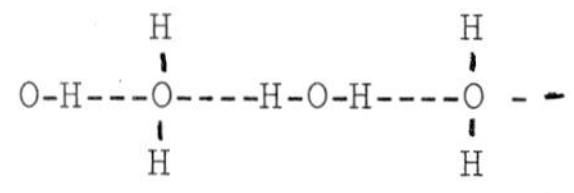

```
        H                H
        |                |
O-H---O----H-O-H----O  - -
        |                |
        H                H
```

they can mutually weaken each other) so that each is stronger than it would be if the other(s) did not exist, producing an effect which has been referred to as "non-addivity" or "cooperativity".

Although earlier quantum-mechanical calculations made by approximate methods[22,23] had led to many of these conclusions and the property of cooperativity in liquid water had been inferred on qualitative grounds[24], it continued for a long time to be the fashion in many quarters to represent the H-bond in water essentially as a simple electrostatic phenomenon. This was partly on grounds of amenability to quantitative manipulation, but also in good part of the then state of valence theory.[25] There were, moreover such successes as those of Bjerrum[26], who made coulombic interactions between + and - charges situated in tetrahedral positions in a model water molecule [he placed a tetrahedral skeleton similar to that of Fig 5a at the center of a sphere of radius 1.38 Å - half the O-O distance in ice; each leg of the tetrahedron was 0.99 Å long and carried a charge of absolute magnitude 0.171[26] electron units, two + and two -; these molecules were placed in the known geometry of the ice lattice (Fig 6)] the basis of a calculation of the lattice energy of ice, in excellent agreement with experiment; and of Pople,[27] who showed that an average degree of bending of such a coulombic "bond" could be calculated by the methods of statistical mechanics with results that seemed to account well both for the density and for the temperature dependence of the dielectric constant of real water. This latter work in particular encouraged the adherents of the uniformist school in the widely held view that there was no such thing as a "broken" H-bond in liquid water - a view in any case encouraged by the fact that such coulombic energy as is here in question decreases in a continuous manner when the rigid charge frameworks are rotated out of their equilibrium positions, so that "breaking" the "bond" must involve assigning some arbitrary degree of bending as the "breaking point". The question whether, or in what sense and to what extent, such hydrogen bonds as are present in ice are broken in liquid water is thus an important one from several points of view and, as will be seen, the investigation of this problem over the past decade has involved controversies which made it instructive in more than one respect.

The conventional experimental criteria of hydrogen bonding have been spectroscopic and crystallographic. When a hydrogen group -A-H enters into a hydrogen bond - AH---B- the spectrum -- infrared absorption (IR) or Raman scattering -- associated with the stretching vibration of the -A-H valence bond is characteristically altered.[28] Its frequency is lowered, the intensity of IR absorption is enhanced, and the width of the IR or Raman band is increased. Moreover, the degrees to which these effects are produced are in some sense proportional to the strength (energy required to dissociate) the H-bond, with the exception that the width of a band in a liquid material is, comparatively, greater than in the corresponding solid.

For H_2O the fundamental O-H stretch frequencies,[29] which are 3657 cm^{-1} (ν_1, symmetric) and 3756 cm^{-1} (ν_3, antisymmetric) in the vapor coalesce upon formation of ice into a single broad band with some structure, centered about 3220 cm^{-1}, and extending (at half-height) some 500 cm^{-1} in width. In the liquid at 25°C the band, still displaying some structure, has shifted upward from its position in ice to center near 3400 cm^{-1} and has a width which is again some 400-500 cm^{-1}. These data are interpreted as meaning that there is hydrogen bonding in both liquid and solid water, but that it is weaker in the liquid.

The same inference can also be drawn using the crystallographic criterion. Here the convention is that the existence of the AH---B hydrogen bond causes, or is revealed by, a shortening in the A-----B distance, to an extent, again, in some sense proportional to the H-bond strength.[30] Thus, the O-O distance in ice,[31] 2.76Å, is short enough in relation to the accepted van der Waals radius of the H_2O molecule,[32] say, 1.5 Å, to correspond to the rather strong ice H-bond. The fact then, that the nearest-neighbor O-O distance in liquid water can be inferred[33] from the x-ray scattering pattern (Fig 1) to be between 2.8 and 2.9 Å also corresponds to the "expected" weakening of the H-bond on melting. Neither the spectroscopic nor the x-ray data just quoted, however, throw any light on the question at issue, namely, whether, or how many, H-bonds are "broken" on melting. (see chapter III.3.3.)

In order to simplify this problem Wall and Hornig[34] employed the device of studying dilute solutions of H_2O in D_2O and of D_2O in H_2O.[35] In each of these cases the rapid exchange of isotopic atoms produces HOD molecules which (a) are free of the spectroscopic complications introduced by molecular symmetry, and (b) possess fundamental frequencies ($\nu_1 \approx 2450\text{-}2600$ cm^{-1} and $\nu_3 \approx 3400\text{-}3600$ cm^{-1})[29] which in each case is far enough removed from strong absorptions in the solvent to permit it to be observed, and also to make it unnecessary to worry about coupling effects. Wall and Hornig were rewarded by obtaining, as the traces of the Raman scattering of the O-D stretch in H_2O, and of the O-H stretch in D_2O, curves which in each case had the simple

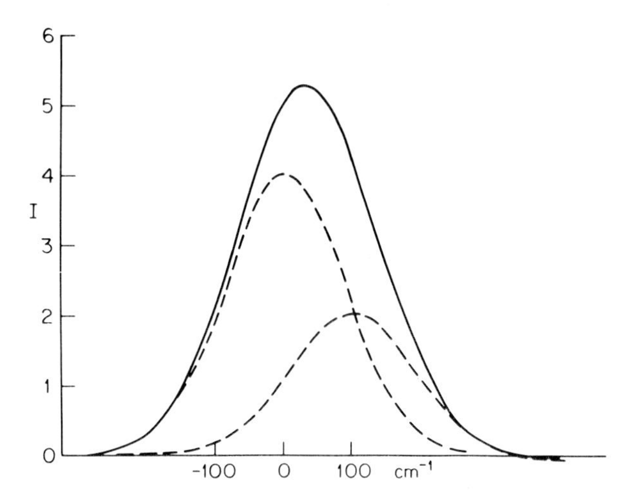

Figure 7 A simulated Raman or IR spectral band, showing how a simple appearance can result when two Gaussians are added together.

bell-shaped appearance of the fully-drawn line of Fig 7. This lent itself nicely to the inference that both the O-D group, vibrating outward into an H_2O medium, and the O-H group into a D_2O medium, were "looking" in each case at a single environment in which a wide, but continuous, variety existed among the instantaneous distances and orientations of the surrounding water molecules. This strongly suggested the essential correctness of the purely electrostatic interpretation of the hydrogen bond, with the corollary that however great the degree of bending of such a bond (the average would determine the position of the maximum of the symmetrical curve) there was nc point beyond which the bond "broke" and the vibration had to (or could) be considered as looking into a qualitatively different environment, or as belonging to a different population. The possibility that the band concealed two sub-bands, such as those dashed in in Fig 7, which would admit the interpretation that there were in fact two populations, a "bonded" and "unbonded" kind of O-H or of O-D, was explicitly rejected. To the contrary, these authors offered additional support for their uniformist interpretation by showing that their bell-shaped intensity curve could be combined with a relation between O-H vibration frequencies and O-O distances in hydrogen-bonded crystals[36][37] to obtain a reasonable-seeming continuous distribution of O-O nearest-neighbor distances in liquid water, suggesting that it was the existence of this structural feature which was responsible for the shape of their measured Raman band. Similar experimental results, with a similar interpretation, were subsequently reported by Falk and Ford[38] on the basis of an IR study of the same systems.

5. Wishful Thinking. At this point several groups of workers who had favored the mixture representation of water undertook new experiments to test whether the researches just reported really did necessitate abandonment of the mixture hypothesis. That experiments should be undertaken either to prove or to disprove a point violates one of the widely held popular views of what a sound scientific attitude ought to be. According to that view scientists are, in their scientific work, "without passion" - are, on the contrary, indifferent to the implications of any given experimental result, so long only as it has been properly arrived at. In particular, they are supposed to be ready to abandon any "preconceived" idea as soon as an experimental result is found to be inconsistent with it. That this notion of how science actually works is too simple was, of course, known to the scientists who took part in the Marburg Symposium, but it may not be familiar to all who read this book, so that a brief elaboration may be permissible. In point of fact, scientists as a group are strongly motivated men and women, and would not be able to put forth the effort involved in performing a careful experiment or in puzzling through the complexities of a line of theoretical reasoning if they were not able to visualize at least the possibility of a concrete outcome. They do not work at random. And between visualizing a particular outcome and desiring it, the boundary is very nebulous. Why, then, should such outcomes not be discredited as reflecting the "will to believe" of the worker rather than the objective "truth"? Because in addition to such a

will to believe the true scientist also possesses a penetrating critical faculty, along with a certain ruthlessness which enables him, so to speak, to slaughter his (intellectual) offspring without pity should they be found to be defective. It is only when he permits his critical faculties to be lulled into inattentiveness that the scientist is in danger from his wishful thinking. So long as the critical faculties are active and functioning without impairment, the capacity for wishful thinking is not only permissible - it is actually indispensable if work of the highest quality is to be done. Moreover, as will be illustrated below, the existence of competing (wishful) interpretations can, at its best, result in sounder and more rapid progress than might otherwise be possible.

In the case now under discussion, among the reactions to the Wall and Hornig paper were additional Raman effect experiments by Walrafen, and an infra-red study by Senior and Verrall, both looking for evidence that the possibility suggested by the dashed curves of Fig. 7 was in fact a reality. Walrafen concluded that it was, on the basis of curve analysis and the observation of an isosbestic point displayed by a family of curves representing a dilute solution of HOD in H_2O at a variety of temperatures.[39] When it became possible, moreover, to obtain spectrophotometric tracings of Raman scattering from this system using laser excitation he found that the higher-frequency sub-band even produced an overt shoulder (Fig. 8.[39][40] Meanwhile, Senior and Verrall[41] had subjected the corresponding IR absorption band to a numerical analysis which made it possible to plot the derivative of the contour curve. This plot with the parent contour, is displayed in Fig. 9, and shows that in this mode also evidence exists for two subbands. Senior and Verrall, like Walrafen, found the higher-frequency, nonhydrogen-bonded, component centered near 2650 cm^{-1} to grow in intensity at the expense of the lower-frequency, H-bonded, one near 2550 cm^{-1} as temperature is raised, the rates of change being those which a simple van't Hoff calculation whould give for a heat of bond-breaking of about 2.5 kcal $mole^{-1}$. The existence of sub-bands, their change in relative intensity with change in temperature, and the assignment to H-bonded and non-H-bonded stretching

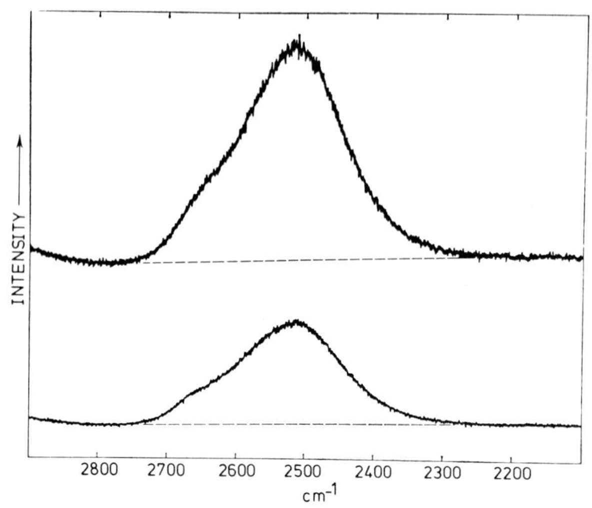

Figure 8 The O-D valence-stretching band of HOD in 1% solution in H_2O, as recorded in an experiment using laser excitation. The two traces are duplicates except that the amplification in the upper one is twice that in the lower. From Ref. 39.

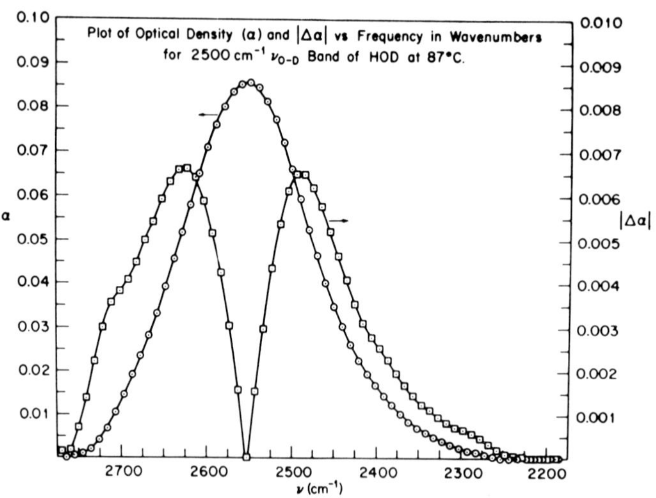

Figure 9 The IR absorption band corresponding to the O-D valence-stretching motion of HOD in H_2O solution. Circles, parent bond; squares, its derivative as obtained by numerical analysis. From Ref. 41.

motions of the O-D group of HOD in H_2O was placed beyond doubt by the IR studies of Franck and Roth[42] and the Raman experiments of Franck and Lindner,[43] in which the water was carried to (supercritical) temperatures up to 500°C and pressures to 4 kbar (Roth) and 400°C and 5 kbar (Lindner). Lindner's'results, shown in Fig. 10 and Fig. 11, and discussed by Tödheide,[12] make it clear that there is a new, and additional kind of interaction between H_2O molecules which is not H-bonding, but imitates H-bonding in its effect on the valence-stretching frequency. It was this, in a sense, unanticipated fact, together with the conventional identification of downward shift of O-H (or O-D) valence-stretching frequency as, so to speak, constituting the H-bond, which was presumably responsible for at least some of the difficulty that had been experienced in some quarters in accepting the existence in

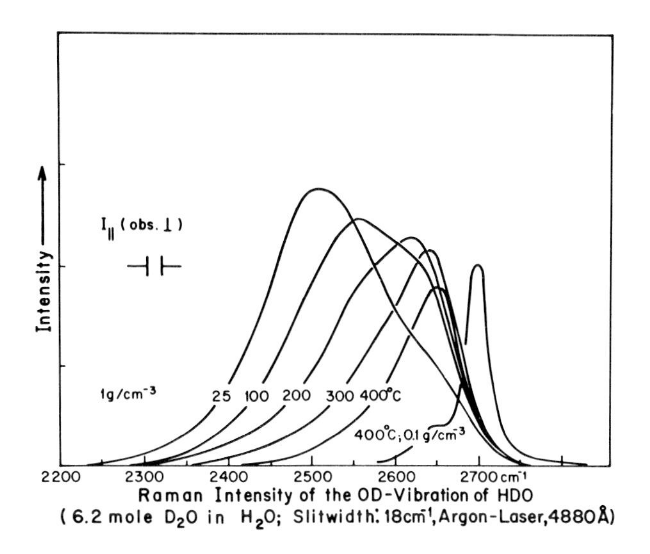

Figure 10 (legend on photo) From Ref. 43.

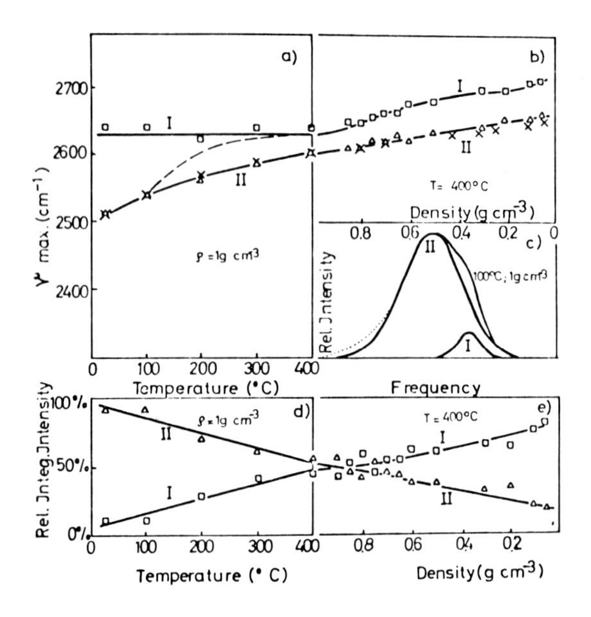

Figure 11 Splitting of the Raman O-D stretching vibration band of HOD into two Gaussian components. Plate (c) shows I as high-frequency component and II as low-frequency component (cf Figs 7 and 9) Plate (a) shows how the frequency of maximum intensity of each changes with temperature at constant density. Plate (b) shows changes in ν_{max} with density at constant temp. of 400°C. Plates (d) and (e) show the corresponding influences upon integrated intensities. From Ref. 43.

cold water of broken H-bonds. It also seems possible that the nature of this non-hydrogen-bonding attraction is responsible for the otherwise paradoxical fact, noted above, that D_2O, though a stronger H-bond former than H_2O at room temperature, has the lower critical temperature.

In any case, this spectroscopic controversy, and the new experiments which it stimulated, had the important result of establishing what seems now to have become a genuine constraint upon water models. Any acceptable model, from this time forward, must account for the existence in cold water both of intact and of "broken" H-bonds.[44]

In accepting this conclusion it must be emphasized again that it would not have been good science to reject experimental results which seemed to conflict with it simply because we did not like them, or even because we possessed contrary evidence from a different source. It is only after we have succeeded in identifying where or how any given piece of work is wrong that we are clearly entitled to call it so. It is for this reason in particular that the work of Walrafen, Senior and Verrall, Roth,and Lindner is so important. It had for some years seemed highly probable, from the work of Luck, using overtone and combination bands in the near IR,[45] that liquid water contains both intact and broken H-bonds at all temperatures, and that a rise in temperature causes bonds to be broken. That work, however, and the uniformist interpretation of the fundamental bands, did not meet on common grounds where, so to speak, they would have been able to come to grips. This had the result that the former could not, of itself, disprove the latter. The interpretation of the near IR spectra has, therefore, also been strengthened by the clearing up of the contradictions in the fundamental region. (see chapter fundamental and overtone region pages 221 and 247).

This does not, of course, mean that all of the interpretations that have been given of near IR spectra have been right. One very definite assignment in particular[46] was at one time widely accepted, which specified the fractions of H_2O molecules "bonded through two H's", "bonded through one H", and "unbonded", on the basis of a set of simultaneous equations which related the relative concentrations and the putative extinction coefficients of the presumed species to the total absorption at several specified frequencies. This set of equations was, however, subsequently shown[47] to have no unique solution, which robs the assignment of significance.

Another purported proof that liquid water could contain no "unbonded" molecules proceeds from a comparison of the electronic absorption band in the vacuum ultraviolet spectrum of H_2O vapor with the long-wavelength tail of the corresponding band for the liquid.[48] The finding that down to about 1800 Å the absorption of the liquid at any wave-length is less intense than that of the vapor by factors of about 0.0013 at 25°, 0.003 at 50° and 0.009 at 75°C was interpreted as meaning that these were the approximate fractions of "gas-like" molecules in the liquid at these temperatures. Again actuated by "wishful thinking" Verrall and Senior[49] performed new experiments in which they succeeded in obtaining the complete absorption band of liquid water, with the result shown in Fig. 12. The essential correctness of the shape and position of their

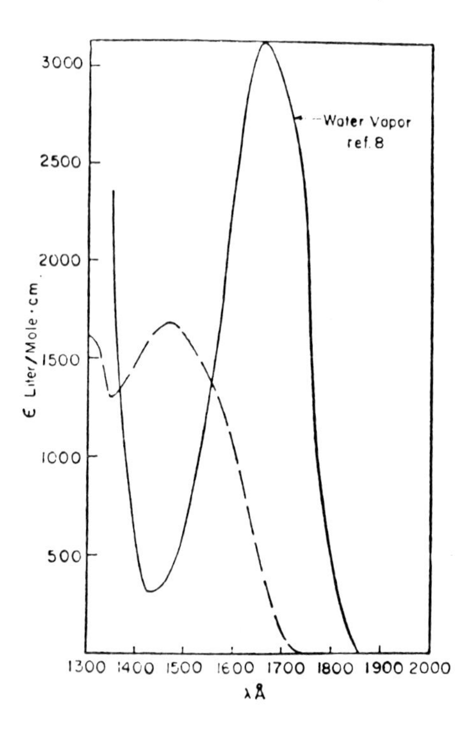

Figure 12 Legend on photograph. From Ref. 49.

liquid band was confirmed independently by the reflectance measurements of Painter, Birkhoff, and Arakawa.[50] The detailed implications of this shift of some 7000 cm^{-1} upon changing from vapor to liquid have been discussed in another place,[51] the conclusion, in brief, being that since these curves prove that condensation to the liquid must result in considerable changes in the energies both of the ground and of the excited electronic states they permit no inference, one way or the other, about the existence of "gas-like" molecules in the liquid.

6. <u>"Bulky and Dense"</u> Another feature of many models in the "mixture" tradition has recently come under experimental attack, this time successfully. That is, the freedom of the theorist to postulate the existence in liquid water of dense and bulky regions has been severely restricted by the results of low-angle X-ray scattering experiments. These are included in Fig. 1.[2] There $s = (4\pi/\lambda)\sin(\theta/2)$, where θ is the angle by which scattering has deflected the incident X-ray beam, and the limiting value of $I(s)/n$ as $s \rightarrow 0$ agrees well with the requirements of scattering theory and statistical mechanics in corresponding to the value of κ_T, the isothermal compressibility of water at the temperature of the experiment. The point of special interest for our present purpose is the absence of "features" in the intensity curve from $s = 0$ up to $s \approx 1 A^{-1}$. This was interpreted by Narten and Levy[33] as meaning that there are no bulky or dense "patches" in water - i.e., no volume elements in which the density is significantly larger or significantly smaller than the average - if by that term we wish to refer to regions more than a few molecular diameters across.

This result was both important and new, since Fisher and Adamovich had in 1963 published[52] a curve which indicated that there were significant density fluctuations in room-temperature water, extending over ranges of tens of Angstroms. Their method was to employ the equations[53]

$$\langle \delta N_G^2 \rangle = \bar{N}_G[1 + 4\pi \int_0^D [g(r)-1](1 - \frac{3r}{2D} + \frac{r^3}{2D3})r^2 dr] \quad (2)$$

and

$$\langle \delta N_G \delta N_G' \rangle = \bar{N}_G \; \pi\rho[\int_0^D [(g)r-1](1 - \frac{r}{D} + \frac{r^2}{5D^2}) \frac{r^2}{D^2} r^2 dr$$

$$+ 4\int_D^{2D}[g(r-1)-1](- \frac{2D}{5r} + 1 - \frac{r}{2D} - \frac{r^2}{4D^2} + \frac{r^3}{4D^3} - \frac{r^4}{20D^4})r^2 dr] \quad (3)$$

where $<\delta N_G^2>$ is the mean square fluctuation in the number of molecules in a spherical volume element of diameter D, $\bar{N}_G$ is the average number of molecules in such an element and $<\delta N_G \delta N_G'>$ is the average correlation between the fluctuations in two such spherical elements which are tangent to each other. g(r) is the radial distribution function, which describes an average over the instantaneous relative locations of all pairs of molecules which are separated by the distance r. It is obtained by Fourier transformation of the structural part of the I(s)/n of Figure 1 - i.e. from X-ray scattering data. Fisher and Adamovich had taken the earlier and less complete X-ray data of Danford and Levy,[54] supplemented by an asymptotic expression for the range $8.2Å < r < \infty$ ($r \to \infty$ corresponds to $s \to 0$). Their results can be represented by curves of Fig. 13, where K(D) is the ratio of

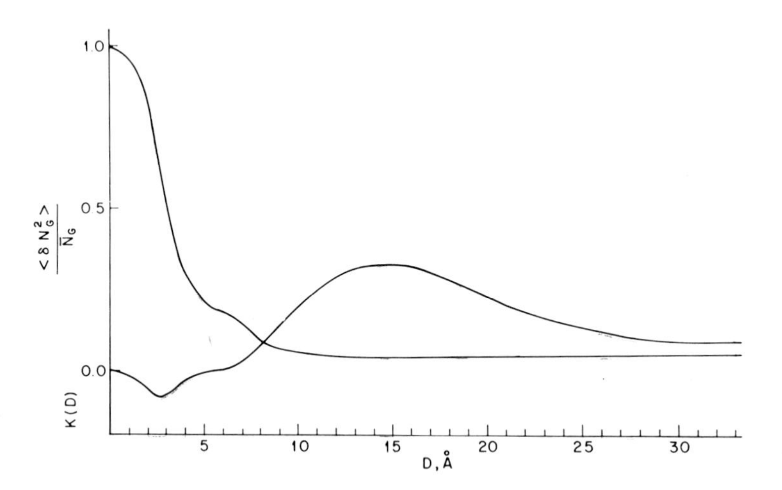

Figure 13 Mean square fluctuation in number density in water in a sphere of diameter D Å. K(D), the ratio to this of the correlation of fluctuation in two such spheres tangent to each other. After Ref. 52.

density correlation to mean square fluctuation in tangent spheres of diameter D- i.e., $K(D) = \langle \delta N_G \delta N_{G'} \rangle / \langle \delta^2 N_G \rangle$. It is in the K(D) curve that the property can be observed which was referred to above, a sizeable correlation between fluctuations in spheres each over 15Å in diameter, and therefore both parts of a single patch which is observably either denser or bulkier than average. In order to resolve the contradiction between this result and the later conclusion of Narten and Levy,[33] Chay[55] converted equations (2) and (3) into integrals in s, over the corrected scattering intensity itself, thus eliminating one step of numerical integration. She also used the data of the more recent Oak Ridge tables[2] of I(s) against s from which the curves of Fig. 1 had been constructed, thus employing the extrapolation to s = 0 that had already been performed, and circumventing the need for inserting an asymptotic form of g(r) as $r \to 0$. The new result is shown in Fig. 14, where the solid lines refer to H_2O at 25°C. From the K(D) line it may be

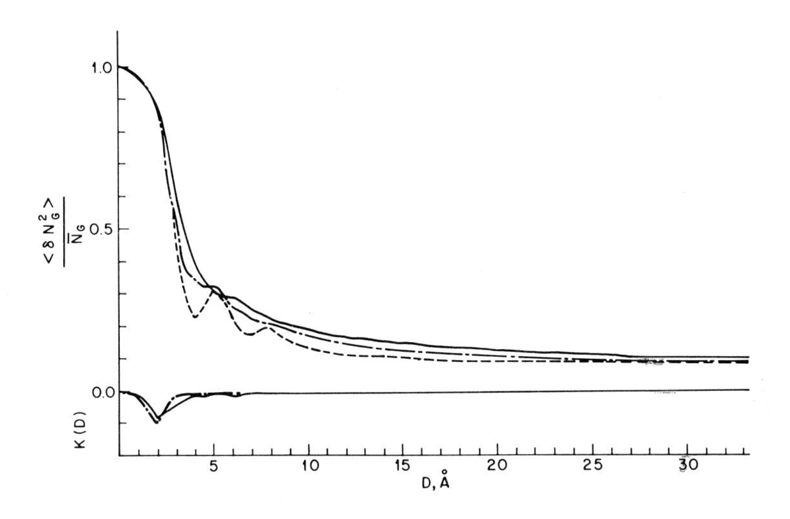

Figure 14 Solid lines have the same meaning as in Fig. 13, recalculated as in Ref. 55.

inferred that the correlation in density fluctuations between adjacent volume elements is never positive, and is zero, within the limits of experimental determination, for elements larger than 6 or 7 Å in diameter. This confirms quantitatively the qualitative interpretation of Narten and Levy[33] earlier alluded to, and, considering the high degree of reliability that can be assigned both to the low-angle data and to the reasoning based upon them, seems to entitle that conclusion to be accepted as another firm constraint upon water-model-making.

7. How Two Constraints May Be Better Than One. This new constraint, that water can contain no sizeable "dense" or "bulky" patches, if, indeed, any patches are present at all, has been explained in some detail because it impinges so sharply upon the macroscopic constraints, namely, the need to account for the existence of a maximum in density, and for the increase in $\frac{1}{V}(\frac{\partial V}{\partial T})_P$ with increasing pressure. To see how sharp the confrontation is, the latter condition may be recast in molecular terms by making use of the identity $\frac{\partial}{\partial P}[\frac{1}{V}(\frac{\partial V}{\partial T})_P] \equiv \frac{\partial}{\partial T}[\frac{1}{V}(\frac{\partial V}{\partial P})_T]$. The earlier observation therefore requires that the coefficient of isothermal compressibility, κ_T, defined as $-\frac{1}{V}(\frac{\partial V}{\partial P})_T$, become larger with falling temperature. That it does so for water has been known as an empirical fact from direct measurements, but its relation to the TMD phenomenon has not always been recognized. Also comparatively new is the observation[55] that below 20°C or so this effect is so strong that the product $\rho kT\kappa_T$ (ρ = number density) also grows, and with increasing rapidity, as the temperature falls, through 0°C and into the range of supercooling. But this product, as follows from statistical-mechanical reasoning[53] of the kind that led to Eq. (2) equals $\langle\delta^2 N_G\rangle/\bar{N}_G$ for bulk water ($D \rightarrow \infty$). That is, we have extended the group of phenomena associated with the existence of the density maximum to include the assertion that below about 20°C the fluctuations in density in liquid water become the more pronounced the lower the temperature of observation. The well-known tendency of fluctuations in most systems to increase with rising temperature (cf. the curves for other liquids in Fig. 3) lends additional weight to the need for a water model to provide some special explanation - and now

a molecular explanation - for its PVT anomalies (which include, by the way, the fact that even at room temperature its $\rho kT\kappa_T$ product is several-fold larger than those of normal liquids).[55]

The invocation for this purpose of the idea of bulky and dense patches is now seen to be not only sufficient, as illustrated by Röntgen's hypothesis, but also necessary, since this is exactly what fluctuation in density means, that the matter occupying a given element of fixed volume be found to be sometimes bulkier and sometimes denser than the average - or, alternatively, that among the volume elements into which the sample can be thought of as being divided up some will, at any instant, be found to contain more molecules and some fewer. Of course there are density fluctuations in all liquids, so that the mere fact that they also exist in water is not worth talking about. What is of interest is that in water the fluctuations are of a very special kind, being much larger than normal, and possessing the unique property of becoming still larger as the temperature goes down. It therefore now seems safe to assert that cold water contains patches which are denser and patches which are bulkier than average, and that this is true to an extent, and in a sense, quite different from anything that can be said of normal liquids.

The conclusion drawn from the low-angle X-ray scattering, therefore, is not only a new constraint, it is one which constrains an old constraint. As a result we can now say not only that the dense or bulky patches in cold water must exist, but that they must also be small - i.e., on a molecular scale, be highly localized. Between them, therefore, these two pieces of information seem to require that the geometrical structure of cold water possess many of the features which characterize the so-called interstitial class of models.[56] In such models (see chapter II.2 and reference 16) there are regions in which a certain fraction *f* of the water molecules are tied together in a more or less perfectly tetrahedrally H-bonded framework (ice-like, clathrate-like,[57] or less regular) in which the void spaces left by the inefficiency of space-filling in 4-coordinated structures (cf. Fig 6) enclose the remaining (1-f) of the molecules of that region as non-H-bonded, monomeric "interstitials". If the fraction of the interstitial sites which are occupied is allowed to vary with changes in temperature and pressure - i.e., if f = f(P,T) - it has been shown[58] that such a model can reproduce the phenomenon of maximum density. This means that it contains bulky and dense patches, but in this case these are of a special type which automatically satisfies the size constraint. That is, a bulky "patch" in such a model consists of the framework molecules which surround (and define) an unoccupied interstitial site, whereas a dense patch differs only in that the interstitial site is now occupied. In such a model, also, f (the proportion of the molecules which are used in building the framework) is expected to increase as temperature falls and, provided the numerical value of f satisfies certain conditions, (which involve the specific geometry of the framework) the compressibility, and thus the mean square fluctuation in number density, will also increase with falling temperature, satisfying another of the macroscopic constraints. Still another constraint which such a model will satisfy is the spectroscopic one that the fraction of broken H-bonds shall increase with rising temperature.

Whether or not an interstitial model can be made tolerable in valence-theoretical terms seems to be entirely a matter of how much cooperativity can be admitted in the hydrogen bond system and how severely H-bonds can be considered to be strained by bending. As has been set forth in another place [59] if cooperativity and bending-strain are great enough, the idea seems at least statable that, if too many molecules happen to be too close together in a small region, the bond system of lowest over-all energy may be one in which an occasional molecule is omitted from the pattern - i.e., unbonded - in order that the bonds which do form, giving rise to the surrounding framework, may enjoy the greater stability conferred by the more favorable (tetrahedral) angles between them. The group of molecules which at any instant consitute such an interstitial framework must not, of course, be too large, nor can the pattern have too long a lifetime, in order that other constraints may be satisfied (see chapter II.2 and reference 58).

8. Dynamical Constraints. The conclusion of the preceding paragraphs, that the structure of cold water must have important features in common with what would be expected in a model liquid of the interstitial class, seems to go about as far as will be possible until some further firm constraints can be established which are restrictive enough to help. Of particular importance will be constraints upon the nature of molecular motions in water, for the question of where the molecules are with respect to each other can not logically be separated from questions about how they get there and how long they stay there. Some qualitative information on these subjects is implicit in what has already been said - e.g. that some of the motions are solid-like oscillations which give rise to spectroscopic frequencies which bear appropriate similarities and dissimilarities to those found in ice. The sharing of other insights into various aspects of molecular motions was a good part of the occasion for holding this Conference, and information from experimental studies of, for example, dielectric relaxation,[60] nuclear resonance relaxation,[61] neutron scattering[62] and the depolarized Rayleigh scattering of laser light[63] will be found in later chapters of this volume. As will be seen there, some of this information is approaching the place where firm constraints can be derived from it. In much of it, however, there still remain uncertainties, encountered in other quarters as well, whether working up the data can be made clear-cut enough without requiring initial assumptions which have been too greatly oversimplified. For example, most of the equations which connect phenomenological relaxation times with molecular reorientation, or correlation, times treat the molecules as reorienting essentially one by one - i.e., independently - an assumption which has been called in question by indications[64][65] that the members of a group, or "cluster", of molecules relax simultaneously and cooperatively. The questions seem also not to have been fully settled whether water molecules typically change orientation in large or in small jumps, nor how much coupling there needs to be between trans-

lational and rotational motion.[66] The situation in these regards seems to be clearer for solid than for liquid water, and this lends particular interest to Onsager's suggestions (Chapter II.1) based on his insights into the situation in ice.

Before leaving the subject of molecular motions in water, notice must be taken of a recent theoretical treatment in which those motions are the feature of first-instance attention. This is the Molecular Dynamics (MD) calculation of Rahman and Stillinger[67] which should in any case be mentioned as the most ambitious and successful quantitative attack upon the water problem that has yet appeared. It exploits the resources of modern statistical-mechanical techniques and computer capabilities to simulate the translational and rotational motions of each molecule in a sample under the joint influence of the forces and torques exerted on it by (in principle) all of the other molecules. In these calculations the sample consists of 216 water molecules confined in a cube 1.862×10^{-7} cm on a side - i.e., at a density of 1 gm cm^{-3}. Wall effects are eliminated by use of periodic boundary conditions. The model molecules interact through coulombic forces between fractional + and - charges rigidly fixed in a tetrahedral pattern in each molecule (cf. Bjerrum's ice model)[26] as well as through a Lennard-Jones potential to take care of van der Waals attractions and repulsions. The details of the pair potential between two molecules are specified according to an ingenious criterion[68] for minimizing the errors arising from the fact that the model potential energy is pairwise additive whereas the quantum-mechanical H-bonding interaction in real water is known to contain some non-additivity (see above).

The successes of this calculation are remarkable, including the prediction of a density maximum near the expected temperature, quite good agreement between observed and calculated self-diffusion coefficients and dielectric relaxation times (all the more remarkable since the total real time represented by the 10^{4} jumps that constitute an "experiment" - jumps at intervals of 10^{-16} seconds - is only of the order of picoseconds, barely as long as the experimental correlation times themselves) and an empirical basis, which emerges naturally from the numerical results, for distinguishing between intact and broken "hydrogen bonds", which receive definitions in terms of pair-interaction energies and are found to vary in their proportions in the expected way with change in temperature. Since, moreover, an "intact" H-bond by this criterion corresponds to a + vertex of one molecule which is relatively near, in distance and orientation, to a - vertex of another molecule, and a "broken H-bond" to a + which does not point toward, nor come close to, any - vertex, it seems reasonable that they should correspond well enough to the spectroscopically "made" and "broken" H-bonds described above to justify a claim that this model satisfies that constraint.

A particularly interesting development arising from the MD calculation relates to the discussion given above of wishful thinking. At issue in this case was the fact that it had been widely accepted that the pattern of quasi-elastic scattering of cold neutrons from water could be explained only if it were assumed that water molecules diffuse by a "jump-and-wait" mechanism, in which the time required to get from one well-defined site to another, perhaps 1 Å or more away, is small compared to the time of "residence" in either place.[69] Since such a jump-and-wait motion would be very different from the motions which occur in the MD calculation, Rahman and Stillinger were motivated by wishful thinking to undertake a re-analysis of the reasoning that led from the actual neutron scattering to the jump-and-wait result. Since this subject is very recondite no attempt will be made here to follow their argument, but its conclusion was that the jump-and-wait hypothesis, while a conceivable way, was not the only conceivable way to account for the relevant features of the currently available neutron-scattering data. This means that for the present, at least, the "requirement" that water molecules diffuse by a jump-and-wait mechanism is not a valid constraint in the sense in which we are using the word. So far, therefore, as the present state of neutron-scattering experiments is concerned, the MD model is as acceptable as any other and we end up by knowing more about the inferences that can be drawn from neutron scattering than we did before the MD calculations were undertaken.[70] This is thus another example of the progress of science through suitably controlled wishful thinking.

It will be noted, however, that if the "experimental result" with which the calculated result disagreed had been one that would not go away when more closely examined, then the theory would have been in trouble, and, indeed, it seems likely that some such fate is in store for the MD model in its present form when enough more tests are made. For example, separate simulation runs could be made for some low temperature, say 0°C, and densities of 1.000 and 1.055 gm cm^{-3} (corresponding approximately to pressures of 1 and 1200 atm. respectively).[71] As shown in Fig. 4, real water near 0°C decreases in viscosity by some 7.5% when the density increases to this extent and, as previously noted, this behavior is reflected in the molecular reorientation time as studied by NMR methods. The calculations must therefore show a molecule reorienting itself faster at the higher density if they are to represent real water in this respect. Another test which earlier paragraphs of this chapter suggest is to examine whether, or how rapidly, the fluctuation in number density at a pressure of 1 atmosphere, as calculated by MD, <u>increases</u> as temperature <u>falls</u> from +20°C to -20°C. It will be noted that until these tests have been made it will not be possible to say, on the basis of MD results, that water can not have an interstitial structure.[72]

Should the MD model not pass these tests and others which, like them, involve genuine constraints, it would be necessary to conclude that, in spite of the wealth of new insights to which it has led, into the factors that govern the properties of water, it still does not tell us what the water molecules are actually doing. Even such an outcome could hardly be called a "failure", however, for it seems reasonable to hope that the MD treatment would, in the process, be able to render another, and very characteristic, service by displaying some sort of relationship between the <u>kind</u> of experimental result on which it broke down and the <u>kind</u> of over-simplification in the model which was responsible for the break-down. There might be many possibilities that needed to be examined, but one that can be mentioned here is that failure in either of the tests outlined above would seem to point to the existence in real water of more drastic consequences of cooperativity - or non-additivity - in H-bonding than can be provided for by adjusting the parameters of an additive pair potential.

9. Solutions. It would be a mistake to conclude this chapter without a brief mention of the fact that some properties of dilute aqueous solutions are regarded in many quarters as constituting pseudo- or semantic constraints. The Röntgen concept that water is a mixture of "structured" and "unstructured" regions has for a long time been used by chemists to account for some of the striking properties of salt solutions by ascribing "structure-making" or "structure-breaking" tendencies to the solutes. Thus, Lewis and Randall[73] suggested that the negative values of partial molal heat capacity typically displayed by, say, alkali halide salts resulted from a tendency of ionic solutes to act as structure-breakers - i.e., to shift the equilibrium of equation (1) toward the left. Bernal and Fowler,[17] also, offered a structural interpretation of the fact that dilute solutions of some alkali halide salts are more viscous (e.g. LiCl, NaF) while some are less viscous (e.g. CsBr, KI) than water itself, saying figuratively that the former solutes "lowered the structural temperature", whereas the latter "raised" it - i.e., made the solutions flow in the way that water would at temperatures lower and higher, respectively than that of the experiment.

Perhaps even more interesting is the fact that non-polar solutes, and non-polar groups in polyfunctional solutes, seem to surround themselves with "co-spheres"[74] which are on average more structured than the bulk water as a whole.[75] While the qualitative features of a water model suggested by Frank and Wen[24] to account for such behavior need at least such revision as is suggested by earlier sections of this chapter, it can hardly be gainsaid that whatever other constraints a water model must conform to it must also, in one way or another, provide for these reactions to the presence of solutes. Since such reactions seem to be of great significance both in electrolyte theory[76] and in solutions of many biochemically important solutes,[77] it is not surprising that, all of the successes of uniformist theorists not withstanding, there persists a feeling among many chemists that, even if water is not a mixture, it surely acts like one.

Relevant to this question is the statistical-mechanicl explanation of fluctuations as the basis of the phenomenologically continuous manner in which physical systems respond to changes in temperature, pressure, etc. Thus the ability to absorb heat smoothly as temperature is raised, no matter by how little, is accounted for by the fact that a sample in a thermostat is continually absorbing or giving out heat as its internal energy fluctuates. This relationship is expressed in the well-known relation[78] $C_V = (1/kT^2)[\langle E^2\rangle - \langle E\rangle^2]$. Correspondingly, the ability of a fluid to undergo an appropriate decrease in volume when the external pressure is increased, no matter how slightly, is expressed by the equation[79] $(\partial V/\partial P)_T = (-1/kT)[\langle V^2\rangle - \langle V\rangle^2]$ - i.e., the volume is changing all the time anyhow in fluctuations which are simply biased by the change in pressure. By analogy it may not be too far-fetched to compare these cases with the experimental fact that the standard entropies of solution at 25°C of He, Ne, Ar, Kr and Xe become progressively more negative and the partial molal heat capacities progressively more positive, as the size of the inert atom increases,[75] plus the fact that Xe forms a clathrate hydrate under suitably elevated pressure and lowered temperature.[80] The sequence CH_4, C_2H_6, C_3H_8 --- also displays progressively more negative entropy of solution,[75] (and progressively larger $\overline{C}_{P2}$) and all three of the molecules named are clathrate-hydrate formers.[81] The suggestion here is that, in dilute solution in water at 25°C, the molecules in each of these series are occupying progressively larger "voids", surrounded by progressively more extensive (on average) "bulky patches" of water. But we have seen above that the magnitude and temperature-dependence of the compressibility of water can be interpreted to mean that fluctuations are continually producing and suppressing bulky patches (void spaces) anyhow. Simply entering such a "prefabricated" void space, then, might be the way that a molecule with so weak a force field as Ar, or CH_4 can be associated with a "change" of water structure great enough to produce the very sizeable extra negative contribution to its standard entropy of solution, and positive contribution to $\overline{C}_{P2}^{\circ}$.[81]

It seems appropriate, then, to end with the prediction that, as additional constraints are discovered or elaborated on the nature of an acceptable water model, some will have their origins in the behavior of water as a solvent.[82]

NOTES AND REFERENCES

1. This quotation and the one immediately following are from the preliminary announcement of the present volume.

2. A. H. Narten, "X-Ray Diffraction Data on Liquid Water in the Temperature Range 4°C - 200°C". Oak Ridge National Laboratory ORNL 4578 (1970).

3. Some 16 years ago a study was published [Discuss. Faraday Soc. 24, 200 (1957)] of X-ray scattering of water in which the conclusion was reached that the typical co-ordination number in liquid water is close to 6, rather than to 4, as previously and at present universally accepted. The error in that work seems to have been the applying of incorrect corrections to the experimental scattering intensity.

4. "Saggi dell' Esperienze Naturali fatte nell' Accademia del Cimento", Accademia del Cimento, Firenze, 1667. An English translation of this work is to be found in Phil. Trans. Roy. Soc. (London) 5, 2020 (1670).

5. G. S. Kell in F. Franks, Ed. "Water. A Comprehensive Treatise" Vol.I. Plenum Press, New York and London 1972 p. 376.

6. Ref. 5. p. 380.

7. L. De Coppet, C.r. Acad. Sci. (Paris) 115, 625 (1892).

8. C. Wada and S. Umeda, Bull. Chem. Soc. Japan 35, 646 (1962).

9. These values of $(1/V(\partial V/\partial T)_P$ were calculated by the late Dr. M. S. Tsao from the PVT data of P. W. Bridgman, Proc. Am. Acad. Arts. Sci. 48, 307 (1913).

10. K. E. Bett and J. B. Cappi, Nature 207 620 (1965). These are recent and accurate data.

11. H. G. Hertz and C. Rädle, Z. physik. Chem. (Frankfurt) 68, 324 (1969).

12. K. Tödheide in Ref. 5, p. 480.

13. see, e.g. K. S. Pitzer, J. Chem. Phys. 7, 583 (1939).

14. Ref. 5, p. 393.

15. W. K. Röntgen, Ann. Phys. (Wied.) 45, 91 (1892).

16. See G. Nemethy in Chapter II.2 of this volume; also H. S. Frank, Fed. Proc. 24, No. 2 Part III P.S-1 (1965).

17. J. D. Bernal and R. H. Fowler, J. Chem. Phys. 1, 515 (1933).

18. e.g. C. W. Kern and M. Karplus, Ref. 5. Chap. 2.

19. See, e.g. G. C. Pimentel and A. L. McClellan, "The Hydrogen Bond" W. H. Freeman and Company, San Francisco and London, 1960.

20. e.g. (a) C. N. R. Rao in Ref. 5, Chap. 3; but also see
(b) J. Del Bene, and J. A. Pople, J. Chem. Phys. 52, 4858 (1970).
(c) D. Hankins, J. W. Moskowitz and F. H. Stillinger, ibid, 53, 4544 (1970).

21. J. Donohue, in A. Rich and N. Davidson Eds. "Structural Chemistry and Molecular Biology", W. H. Freeman and Company, San Francisco and London, 1968, p. 443.

22. C. A. Coulson and U. Danielsson, Arkiv. Fysik. 8, 239 (1954).

23. Ref. 19, p. 236.

24. H. S. Frank and W. Y. Wen, Ref. 3, p. 133; H. S. Frank, Proc. Roy. Soc. (London) A247, 481 (1958).

25. e.g. L. Pauling, "The Nature of the Chemical Bond", 3rd. Ed. Cornell University Press, Ithaca, New York, 1960. Chapter 12.

26. N. Bjerrum, Kgl. Dansk. Vid. Selskab Mat.-Fys. Medd. 27, 1 (1951); also in Science 115, 385 (1952).

27. J. A. Pople, Proc. Roy. Soc. (London) A205, 163 (1951).

28. Ref. 19, p. 75.

29. D. Eisenberg and W. Kauzmann, "The Structure and Properties of Water", Oxford University Press, Oxford and New York, 1969. p. 234.

30. Ref. 19, p. 258.

31. W. H. Barnes, Proc. Roy. Soc. (London) A125, 670 (1929).

32. B. Kamb, J. Chem Phys. 43, 3917 (1965).

33. A. H. Narten and H. A. Levy, Science 165, 447 (1969).

34. T. T. Wall and D. F. Hornig, J. Chem. Phys. 43, 2079 (1965).

35. G. L. Hiebert and D. H. Hornig, ibid, 20, 918 (1952).

36. K. Nakamoto, M. Margoshes, and R. E. Rundle, J. Am. Chem. Soc. 77, 6480 (1955).

37. Ref. 19, p. 88.

38. M. Falk and T. A. Ford, Can. J. Chem. 44, 1699 (1966).

39. G. E. Walrafen, J. Chem. Phys. 48, 244 (1968).

40. The same effect was found in a laser-Raman study of this band by H. J. Bernstein, Raman Newsletter No. 1 (November,1968)

41. W. A. Senior and R. E. Verrall, J. Phys. Chem. 73, 4242 (1969).

42. E. U. Franck and K. Roth Ref. 3,43, 108 (1967).

43. E. U. Franck and H. Lindner quoted by K. Tödheide in Ref. 5, pp. 509-511.

44. H. S. Frank, in Ref. 5, p. 527.

45. W. A. P. Luck e.g. in Ref. 3, 43, 115 (1967).

46. K. Buijs and G. R. Choppin, J. Chem. Phys. 39, 2035 (1963).

47. G. Boettger, H. Harders, and W.A.P. Luck, J. Phys. Chem. 71, 459 (1967).

48. D. P. Stevenson, ibid. 69, 2145 (1965).

49. R. E. Verrall and W. A. Senior, J. Chem. Phys. 50, 2746 (1969).

50. L. R. Painter, R. D. Birkhoff, and E. T. Arakawa, ibid 51, 243 (1969).

51. H. S. Frank, in Ref. 5, p. 524.

52. I. Z. Fisher and V. I. Adamovich, J. Struct. Chem. USSR Engl. Trans. 4, 759 (1963).

53. The basis for such relationships is set forth, for example, in T. L. Hill "Statistical Mechanics", McGraw-Hill Book Co., Inc. New York, 1956, e.g. p. 236.

54. M. D. Danford and H. A. Levy, J. Am. Chem. Soc. 84, 3965 (1962).

55. T. R. Chay and H. S. Frank, J. Chem. Phys. 57, 2910 (1972).

56. a. O. Ya. Samoilov, Zh. Fiz. Khim 20, 12 (1946)
b. L. Pauling, in D. Hadzi, Ed. "Hydrogen Bonding", Pergamon Press, New York, 1951 p. 1.

57. M. von Stakelberg and H. R. Müller, Z. Electrochem. 88, 25, 1954.

58. H. S. Frank and A. S. Quist, J. Chem. Phys. 34, 604 (1961).

59. H. S. Frank, Ref. 5, p. 521.

60. See Chap. VI.1. of this volume, or Ref. 5 p. 277.

61. See Chap.VII.1.of this volume, or Ref. 5, p. 239.

62. See Chap. V.1. of this volume or Ref. 5, Chap. 9

63. T. A. Litovitz, Chap.X.V.4. of this volume.

64. W. M. Slie, A. R. Donfor, Jr. and T. A. Litovitz, J. Chem. Phys. 44, 3712 (1966).
65. J. C. Hindman and A. Svirmickas, J. Phys. Chem. 77, 2487 (1973).
66. The temperature dependence of the ratio of the viscosity of D_2O to that of H_2O seems to argue for different flow mechanisms in cold and in hot water. This ratio is about 1.30 at 5°C and drops rapidly with rising temperature, passing 1.2 near 35°C. [Ref. 5, pp 406 and 408; I. Kirschenbaum, "Physical Properties and Analysis of Heavy Water", McGraw-Hill Book Co., Inc. New York, 1951 p. 33] The square root of the ratio of the masses of these species is $(20/18)^{1/2} = 1.055$, whereas the square root of the mean ratio of the moments of inertia [Ref. 5, p. 54] is 1.38. This makes it look as if one molecule passes another, near the ice point, by rolling, but in hot water by sliding.
67. (a) A. Rahman and F. H. Stillinger, J. Chem. Phys. 55, 3336 (1971); (b) F. H. Stillinger and A. Rahman, ibid, 57, 1281 (1972); (c) F. H. Stillinger and A. Rahman in press.
68. (a) F. H. Stillinger, J. Phys. Chem. 75, 3677 (1970).
    (b) A. Ben-Naim and F. H. Stillinger in R. A. Horne, Ed. "Water and Aqueous Solutions", Wiley-Interscience, New York, 1972. Chap. 8.
69. K. S. Singwi and A. Sjölander, Phys. Rev. 119, 863 (1960).
70. A whole new analysis of neutron scattering from water has since been worked out [F. H. Stillinger and A. Rahman, "Molecular Motions in Liquids", Proceedings of the 24th International Meeting of the Societé de Chimie Physique, Paris. June 3-6, 1973. In press] in which MD calculations also show that, to the order of approximation implied in the assumptions of the method, the common distinction between inelastic and quasielastic scattering of neutrons from water is meaningless from an operational standpoint.
71. Ref. 5 pp 384-5.
72. There is some question whether a sample of 216 molecules with periodic boundary conditions would in any case be able to display the fluctuations which this test rests upon (cf Fig. 14), and whether 10^4

jumps would suffice to bring them out completely. It is possible, therefore, that this fluctuation test can not be meaningfully applied at the present time. This supports the statement in the text that the fact that the molecular patterns generated by the present calculations do not include interstitial configurations does not mean that they are not present in real water.

This statement is additionally supported by the consideration that the MD calculations as they now stand are purely classical in nature whereas the difference between the vapor-pressure vs. temperature curves of H_2O and D_2O (see above) requires a quantal explanation. That is, differences in the zeropoint energies of the "external", as well as of the "internal" motions of the molecules must be invoked in the theory of the Vapor Pressure Isotope Effect [J.Bigeleisen, J.Chem.Phys. 34, 1485 (1961), and, e.g., H.Wolff in "Physics of Ice", N.Riehl, B.Bullemer and H.Englehardt, Eds., Plenum Press, New York, 1969, p. 305].

73. G. N. Lewis and M. Randall, "Thermodynamics", McGraw-Hill Book Co., Inc., New York, 1923, p. 85; 2nd Ed. Revised by K. S. Pitzer and L. Brewer, 1961, p. 378.

74. R. W. Gurney, "Ionic Processes in Solution" McGraw-Hill Book Co., Inc. New York, 1953, p. 251.

75. H. S. Frank and M. W. Evans, J. Chem. Phys. 13, 507 (1945).

76. e.g. H. L. Friedman in Ref. 5, Vol. 3, chap. 1.

77. e.g. W. Kauzmann, Adv. Protein Chem. 14, 1 (1949); C. Tanford, J. Amer. Chem. Soc. 84, 4240 (1962); Ref. 5, Vol. 4, in press.

78. Ref. 53, p. 101.

79. Ref. 53, p. 102.

80. Ref. 25, p. 471.

81. H. S. Frank and F. Franks, J. Chem. Phys. 48, 4746 (1968).

82. Prof. O. Ya. Samoilov was unable to attend the conference, but a communication from him was read.

I.3. WATER AND AQUEOUS MIXTURES AT ELEVATED AND SUPERCRITICAL TEMPERATURES AND PRESSURES

E.U. Franck

Institut für Physikalische Chemie und Elektrochemie
Universität Karlsruhe

Abstract

A survey is given of recent experimental results on properties of fluid water and aqueous mixtures over a range of pressures and temperatures extending to several kbar and several hundred degrees C. In certain cases comparisons are made with similar compounds in dense supercritical states.

For supercritical water, PVT-data, viscosity and thermal conductance are discussed. The static dielectric constant of water, hydrogen chloride and acetonitrile at corresponding conditions is presented and discussed in terms of correlation factors. Structural conclusions to be derived from infrared and Raman spectra to 400°C are reviewed. Critical curves of binary aqueous systems with nonpolar components and the supercritical miscibility are described. Examples of ionic conductance of dilute and very concentrated electrolyt solutions are shown in connection with fused salt properties. Infrared spectra of high temperature water-sodium hydroxide mixtures are presented. Finally the self ionization of water and ammonia at very high pressures and temperatures is discussed.

I.3.1.

INTRODUCTION. THERMODYNAMIC AND TRANSPORT PROPERTIES OF PURE WATER.

The study of the properties of water over a wide range of temperatures and pressures extending into the supercritical region obviously has great practical interest, particularly for power plant operation, chemical technology and the geosciences. In addition, however, the knowledge of high temperature, high pressure water properties can help considerably to understand structural features of water and aqueous sc. tions at ordinary conditions. Certain relatively weak interactions become unimportant at conditions of high thermal energy and only the strong types of intermolecular interactions prevail. Pressure variations permit the detection of volume dependent interaction phenomena.

As an introduction, Fig. 1 gives a temperature-density diagram of water which extends to 1000°C and 1.6 gcm^{-3} [1)2)]. Critical point, CP (374°C, 221 bar), and triple point, TP, are indicated on the gas-liquid coexistence curve. The points on the broken line extending from TP to the right denote the transitions between the different high pressure modifications of ice. To about 10 kbar the isobars are based on static experiments [2)3)4)]. At pressures above 25 kbar, water densities have been derived from shock wave measurements [5)]. At 500°C and 1000°C pressures of about 8 and 20 kbar are needed to produce the triple point density of liquid water.

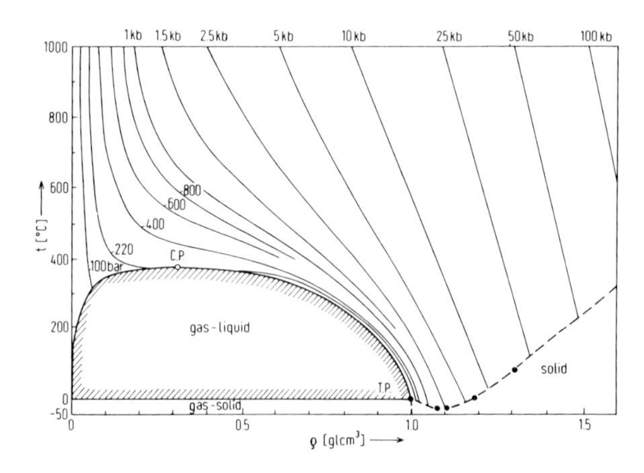

Fig. 1 Temperature-density diagram of water

The difference between densities in the liquid and gaseous states, $\varrho_L - \varrho_G$, at not too great distances from the critical temperature, T_c, can be expressed as

$$(\varrho_L - \varrho_G) = \text{const} \times (T_c - T)^{\beta}$$

For a classical or "van der Waals" fluid one would obtain $\beta = \frac{1}{2}$. An analysis of real fluids [6)7)], for example argon, carbon dioxide and methane, however, gives values for β close to 0.35. Even for highly polar fluids as NH_3 and HF one finds $\beta \approx 0.34$. Only very few examples have so far been found, which have $\beta \approx 0.50$, among these are fluid sodium and fluid ammonium chloride [8)]. An analysis of water date gives β close to 0.35 [9)], which shows that the peculiar interaction of water molecules does not greatly affect the difference of the coecisting densities at high temperatures.

Other thermodynamic functions, the heat capacity for example, do give considerable evidence for intermolecular association of water molecules even at high temperatures and have been discussed in this context Transport properties, which are related to some extent to the thermodynamic properties and which furnish structural information are also of importance for the kinetics of chemical reactions. Unfortunately very little experimental data about diffusion coefficients for water at high temperatures and pressures are available. Viscosity data, however, have been determined to 550°C and 3.5 kbar [10)]. It is interesting to observe the results plotted as supercritical isotherms from very low to liquid-like densities as shown in Fig. 2. Since the temperature dependence in the dilute gas is positive, and negative in the dense fluid, there is a region at about 0.8 g/cm^3 where the temperature dependence is

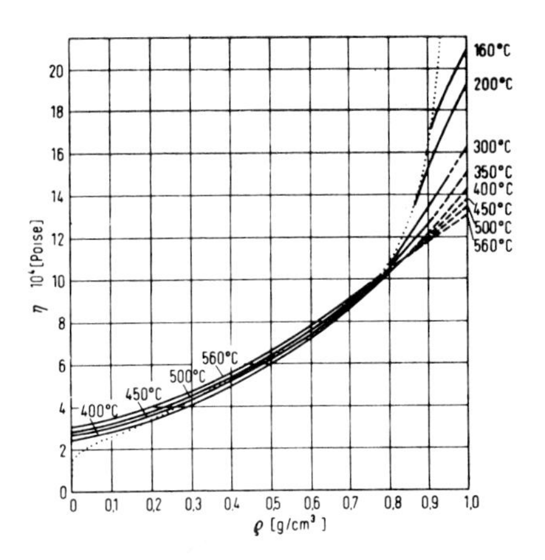

Fig. 2. The viscosity of water as a function of density and temperature.

very small. The total increase of viscosity from nearly zero to high density at 400°C and 500°C is only by a factor of about five. Thus approximate estimates of viscosities at supercritical conditions with empirical equations are quite possible. According to a detailed discussion it appears unlikely that at pressures around 100 kbar the viscosity of fluid water reaches values six orders of magnitude higher than at ordinary conditions as has been suggested from shock wave observations [2)11)]. Attemps to represent the experimental viscosity data for supercritical water quantitively on the basis of Chapman Enskog equations using a hard sphere model remained unsuccessful,however [10)].

It is an experimentally established fact, that diffusion coefficients and molar conductances of dissolved neutral and ionic species vary proportionally to the inverse viscosity of the water. Since this should be valid for the dense fluid at supercritical temperatures too, the viscosity gives at least a basis for an approximate evaluation of transport velocities of dissolved particles in fluid water at extreme conditions.

Anomalies of the shear viscosity in the critical region of several gases have been found, although they are very weak. Very pronounced anomalies, however, have been found for the specific heat and also for the thermal conductivity. This applies also to polar gases such as steam. Recently the thermal conductivity of steam between 370°C and 520°C to pressures of 500 bar could be measured [12)]. In Fig. 3 the "anomalous" thermal conductivity $\Delta\lambda$ of steam is plotted as a function of density at constant temperatures. $\Delta\lambda$ is the amount, to which the conductivity exceeds the "normal" values obtained by a prescribed extrapolation based on the data far away from the critical density into this region. Even at 440°C, that is 70 degrees above the critical temperature, a maximum at the critical density is clearly visible. The "scaling law" concept has been applied to anomalous transport properties. It could be shown, that the scaled thermal conductivity of steam in the range of Fig. 3 is indeed a single valued function of a variable x, which is a combination of reduced density and reduced temperature only [12)].

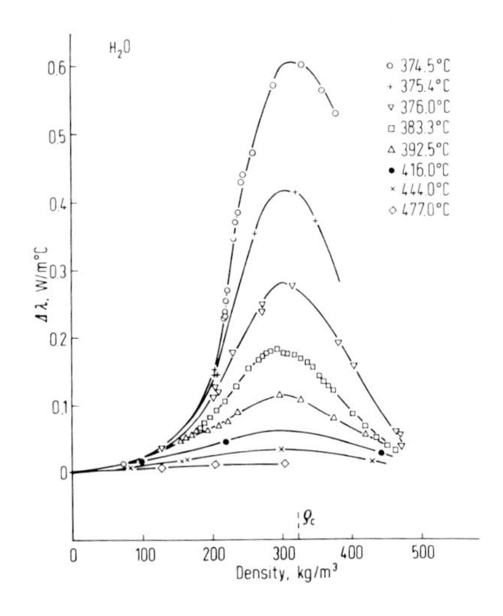

Fig. 3. Anomalous thermal conductivity $\Delta\lambda$ of steam as a function of density for several temperatures [12].

I.3.2.

DIELECTRIC PROPERTIES OF WATER AND RELATED COMPOUNDS

The static dielectric constant of polar fluids is to a certain extent determined by intermolecular interaction, that is by structural properties. This applies particularly to water, where hydrogen bond association in the supercritical dense phase should influence the dielectric constant. This quantity could be measured with frequencies to about 1 MHz to temperatures up to 550°C and to pressures extending in some cases to 5 kbar. Gold-palladium condensers of variable geometry within a corrosion resistent cylindrical autoclave were used. Water, [13][14] pure hydrogen chloride [15] and acetonitrile [16] have recently been investigated.

Earlier measurements of the dielectric constant of water at elevated temperatures were made up to 400°C and in part to 2 kbar, many of them at saturation pressure of the liquid (for a critical summary see [17][18]) Fig. 4 gives a compilation of earlier and recent results as curves of constant values for the dielectric constant, superimposed on the isobars of a temperature-density diagram of water. The very high values of the dielectric constant are restricted to the region of low temperatures and high densities. At supercritical temperatures and high pressures, values of the constant between about 5 and 25 can be obtained. This corresponds to the dielectric properties of the more polar organic liquids at ordinary conditions.

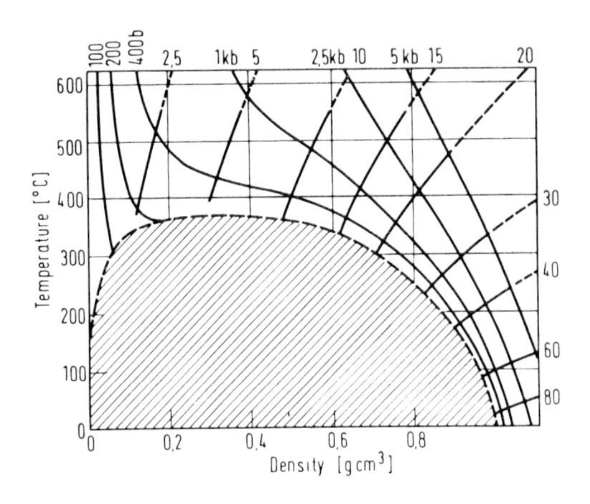

Fig. 4.
Curves of constant values for static dielectric constants of water as a function of temperature and density.

It should be possible to derive conclusions concerning the structure of supercritical water from the dielectric behaviour [2)19)]. The Onsager-Kirkwood [20)] equation for the dielectric constant of a liquid consisting of polarizable molecules with permanent dipoles contains a correlation factor g :

$$g = 1 + \sum_{i=1}^{\infty} N_i \langle \cos \gamma_i \rangle$$

which accounts for local ordering of the neighbouring molecules. N_i is the number of molecules in the i-th shell around a central molecule and $\langle \cos \gamma_i \rangle$ is the average cosine of angels between the dipole moments of molecules in the i-th shell and that of the central molecules. g should be equal to unity for random distribution of the molecular dipoles. For liquid water at room temperature g is about 2.8.

From the high temperature experimental results numerical values of correlation factors can be derived. It is interesting to compare such data with those for another fluid of small polar molecules which can not form hydrogen bonds. For this purpose the dielectric constant of hydrogen chloride at conditions corresponding to those used with water has been measured.

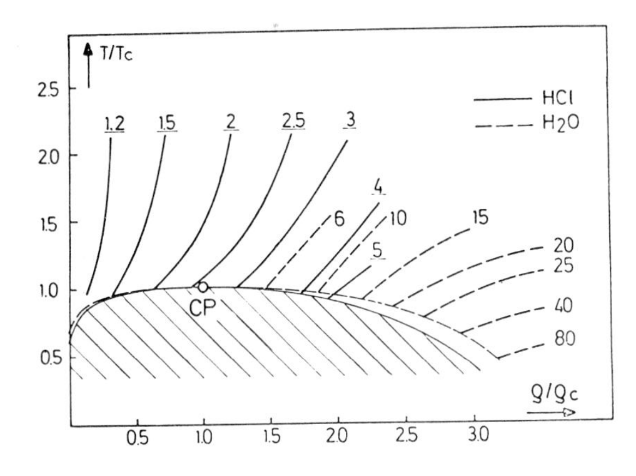

Fig. 5. Curves of constant values of the static dielectric constant of HCl and H_2O in a reduced temperature and reduced density diagram.

CP: Critical point

T_c: Critical temperature

ϱ_c: Critical density

Fig. 5 shows the results in comparison with those for water in a diagram of reduced temperatures and densities. Obviously hydrogen chloride has much lower dielectric constants than water at corresponding states. Fig. 6 shows g-factors for both substances as a function of density for supercritical temperatures. At the critical density and at a reduced temperature of 1.05, which is about 400°C for water and 60°C for hydrogen chloride, the correlation factors are about 1.5 and 1.05 respectively. This expresses the considerable higher degree of molecular orientation in water.

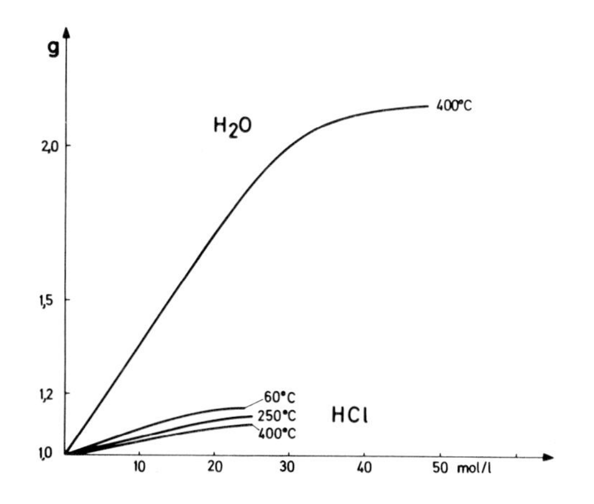

Fig. 6. Correlation factors g for H_2O and HCl as a function of density at supercritical temperatures.

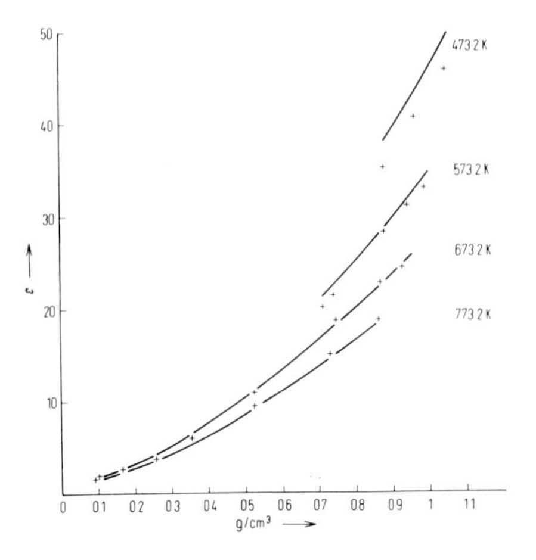

Fig. 7.
The dielectric constant of water, ε, as a function of density for several temperatures.
Experimental values: + + +
Calculated curves: ———

One should be cautious, however, with the derivation of quantitative conclusions about the extent of hydrogen bonding in dense supercritical water. General expressions for the static dielectric constant of a fluid of hard spheres with embedded point dipoles have been derived by Wertheim [21] on the basis of a "Mean Spherical Model" (MSM) treatment. Numerical results were derived for water from these expressions with the additional inclusion of polarizability effects, using only the dipole moment, polarizability coefficient and diameter of water molecules [14]. These results were compared with the experimental data for water extending to 800 K and 5 kbar. Fig. 7 gives a comparison of experimental and calculated values. The agreement between calculated and experimental data above 500 K is satisfactory and extends over the wide density range from 0.1 t 1.0 g/cm^3. At low temperatures the model overestimates the dielectric constant at higher densities. It should be stressed, that in no way the hydrogen bonding concept was explicitly introduced. It appears that hydrogen bonding effects are not necessary to describe the dielectric constant of water over a broad density range above 500 K. The simple dipole-dipole interaction already constitutes a considerable amount of short range structure (resulting in high g-factors). An analogous treatment of hydrogen chloride data gave also good results except for the room temperature region [22].

The isolated gaseous molecules of water and hydrogen chloride have dipole moments of 1.86 and 1.06 debyes. Very recently similar dielectric constant measurements were made with acetonitrile, CH_3CN, which has a dipole moment of 3.85 and is still a small, almost

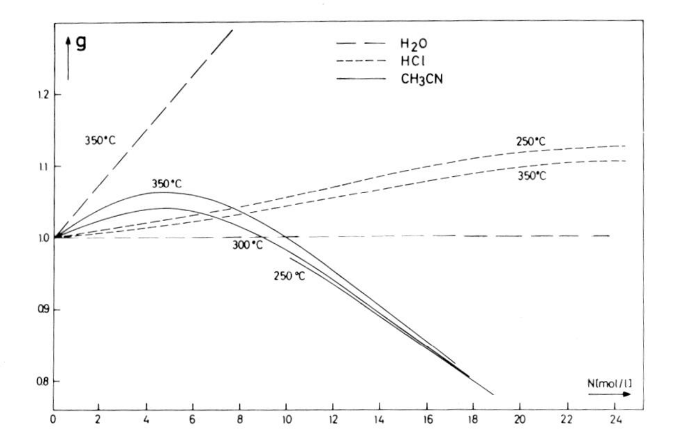

Fig. 8. Correlation factors g for acetonitrile, hydrogen chloride and water as a function of density at supercritical temperatures.

spherical molecule. The critical temperature is 275°C. The measurements could be extended to 350°C and 3 kbar [16]. The dielectric constant at 25°C and atmospheric pressure is 38. At 300°C and 2000 bar for example it is 18. In Fig. 8 isotherms of correlation factors g, derived from these experiments. are given in comparison with those of hydrogen chloride. The interesting feature of the acetonitrile g-factors is, that they proceed to values smaller than unity at high density. This should indicate the formation of associates with opposed dipole directions as observed in certain liquid alcohols. Apparently this does not occur in dense supercritical water.

I.3.3.

INFRARED AND RAMAN SPECTRA OF PURE WATER AT HIGH TEMPERATURES

More specific information on hydrogen bond formation in water can be obtained from absorption and Raman spectra than from the dielectric behaviour. This is clearly demonstrated in several other contributions of this volume. Only certain high temperature high pressure spectroscopic studies shall be discussed here.

A number of different infrared absorption cells have been designed and built which permit measurements with corrosive fluids in very thin layers to 500°C and to about 4 kbar of pressure [23]. Synthetic sapphire windows and one-window reflection type arrangements are particularly useful. A mirror behind the window is completely surrounded by the pressurized fluid. Thus the sample thickness is

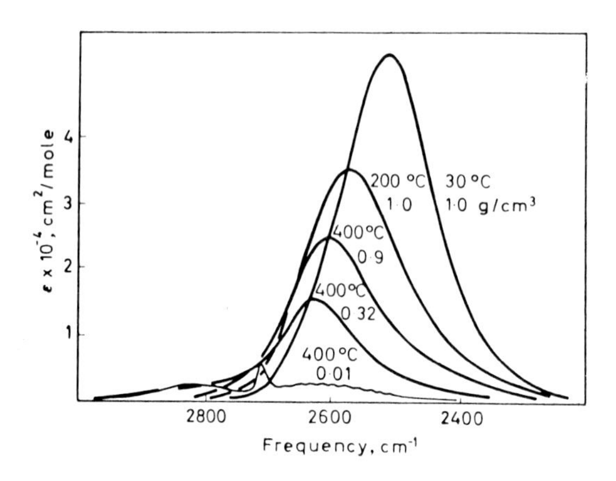

Fig. 9. Infrared OD-absorption bands of 9.5 mole per-cent HDO in H_2O at different temperatures and densities.

almost not affected by the applied pressure. Raman cells with two or more windows can also be built for the same temperature and pressure range.

The frequency of the oxygen-hydrogen stretching vibration is shifted towards lower frequencies if one passes from the dilute gas state to the liquid state. This shift is generally considered as at least partially caused by an increased influence of hydrogen bonding. The shift can be studied particularly well with the OD-stretching band of HDO diluted in normal water. In the dilute gas the peak of the Q-branch occurs at 2719 cm^{-1}, the band maximum in the liquid state at room temperature is at 2507 cm^{-1}. This change by more than 200 cm^{-1} has been followed by first heating up the sample at constant density of 1 g/cm^3 to 400^oC and then isothermally expanding to a low pressure gaseous state. Fig. 9 gives some of the results.

The high absorption intensity at 30^oC which is considered to be caused by the hydrogen bonded structure of the liquid, decreases with temperature but remains relatively high. The bond maximum gradually shifts towards lower frequencies and the band becomes more asymmetric. No new band or shoulder is observed. Rotational structure of the band does not appear, until the water density is reduced to less than 0.1 g/cm^3. Certain analogies can be observed between the behaviour of the OD-absorption of supercritical dense HDO and the vibrational absorption of HCl at high pressures [25].

Walrafen and other authors [26] have observed the OD-stretching band in dilute liquid mixtures of HDO in H_2O at moderate temperatures at normal and elevated pressure in the Raman spectrum. The Raman band

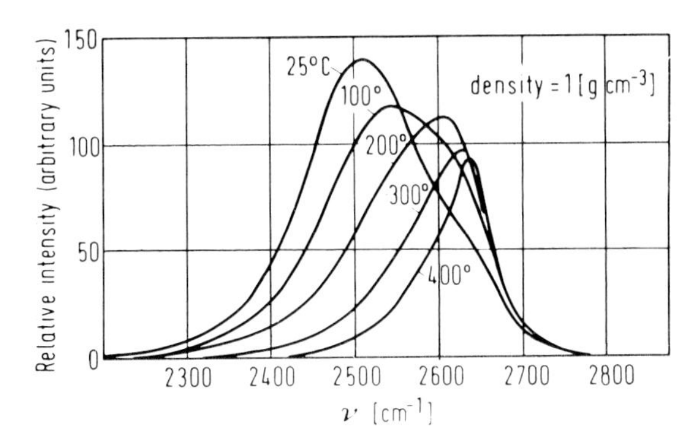

Fig. 10.
Raman intensity of the OD-vibration of HDO diluted in H_2O at constant water density of 1 g/cm^3 and at different temperatures.

with its maximum, as in the infrared absorption, at 2507 cm^{-1} has always a distinct shoulder at about 2650 cm^{-1}. This suggests to investigate the band over a wide range of temperature and density. An externally heated Raman cell with argon ion laser irradiation at 4880 Å was used for this purpose which could be operated to 400°C and 5 kbar or 500°C and 2 kbar [24]. In Fig. 10 five observed Raman bands for constant water density of 1 g/cm^3 are presented. These are intensities of the horizontally polarized component, scattered in the 90° direction. The Fig. 11 shows the variation of the same band at a constant temperature of 400°C with decreasing density. It appears,

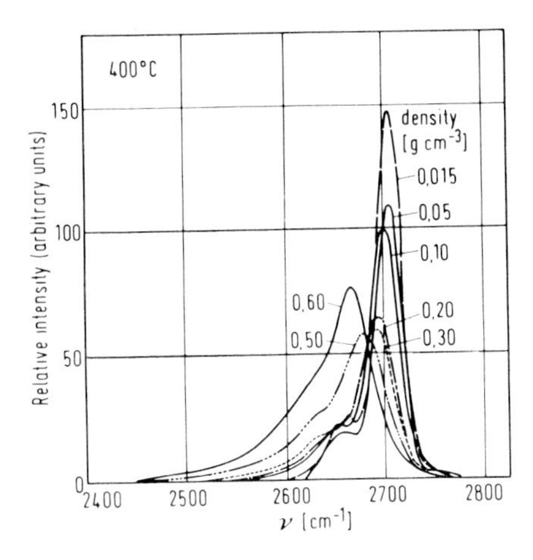

Fig. 11. Raman intensity of the OD-vibration of HDO diluted in H_2O at constant temperature of 400°C and at different water densities.

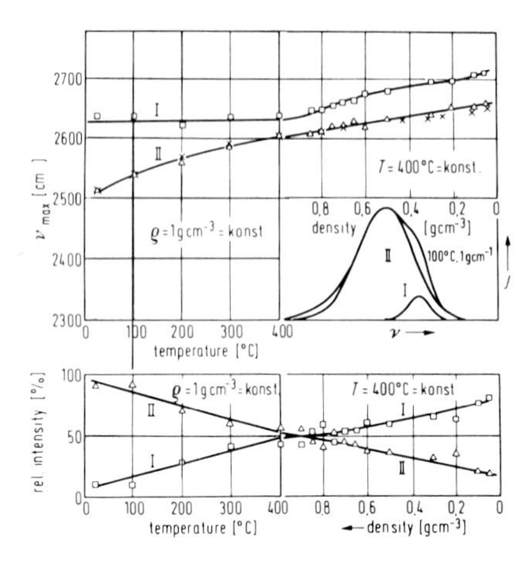

Fig. 12. Maximum frequencies and relative integrated intensities of the Raman band components I and II as functions of temperature at constant density of 1 g/cm^3 and as functions of density at constant temperature of 400°C.

that with increasing temperature a new band develops at the position of the shoulder at 2650 cm^{-1}. This band becomes dominant at 400°C and has acquired the shape of a very narrow sharp peak at the low density of 0.015 g/cm^3.

The shape of the bands of Figs. 10 and 11 suggests to attempt a decomposition into two components. This has been done by a curve analyzer assuming gaussian character for the components. At 25°C component II with a maximum at 2507 cm^{-1} dominates, while component I begins at the shoulder at 2650 cm^{-1}. Fig. 12 shows the maximum frequencies and relative integrated intensities of I and II as functions of temperature and density. The upper part of the figure demonstrates, that the maximum frequency of component I is virtually independent of temperature but is gradually shifted at 400°C with decreasing density towards 2720 cm^{-1}, the band center for single molecules. Component II is also shifted with temperature and with decreasing density, but the "dilute gas" value is not reached.

Consideration of Figs. 10 - 12 certainly suggests the assumption of two rather distinct states of the oxygen-hydrogen group: one which is associated with component II and is often considered as a "hydrogen-bonded" state and a second "non hydrogen bonded" state related to component I. Accepting this simplified description, one must admit, that even in the critical region a portion of the oxygen-hydrogen groups may remain hydrogen bonded. This follows from Figs. 10 and 11 for the critical density of 0.32 g/cm^3 and for temperatures from the critical temperature of 374°C to 400°C or more. Although quantitative statements may not be justified, the assumption of about 25% associated oxygen-hydrogen groups at 400°C and 0.3 g/cm^3 seems reasonable.

This is not in contradiction to infrared absorption. Apparently the extremely strong absorption of the "bonded" OD-groups masks the weaker absorption of the "non-bonded" groups at higher frequency to an extent, that only a growing asymmetry of the composite band at higher temperatures can be observed. Earlier measurements of the abnormal mobility of hydrogen ions in dense supercritical water seem to support this assumption. The success of the simple "Meam Spherical Model" approach in describing the dielectric constant in this region, as described in the previous section, should be a warning, however, that interactions in dense supercritical water may be more complex than the Raman spectrum suggests.

I.3.4.

AQUEOUS SOLUTIONS OF NONPOLAR COMPOUNDS

Since in dense supercritical water the intermolecular association is considerably reduced, one should expect a much higher miscibility with non-polar fluids than in ordinary water. This has indeed been observed. Three-dimensional pressure-temperature-composition diagrams are best suited to describe such phenomena. Binary systems have critical curves in such diagrams. They extend uninterrupted between the two critical points of the pure components if these are not too different, as for example ethane and hexane. If there are greater differences in size, polarity or otherwise, the critical curve is often divided in two branches. The upper branch is of particular interest. It begins at the critical point of the higher boiling pure component and proceeds to lower mole fractions and higher pressures. If this branch of the critical curve remains above the critical temperature of this component, it is often classified as of "Type I", while it would be of "Type II" if it passes through a temperature minimum. Many examples have been reviewed in a comprehensive survey [27]. A detailed analysis of the types of critical curves and of possibilities of predictions has recently be given [28]. Such predictions are as yet very restricted, however, if polar components are involved.

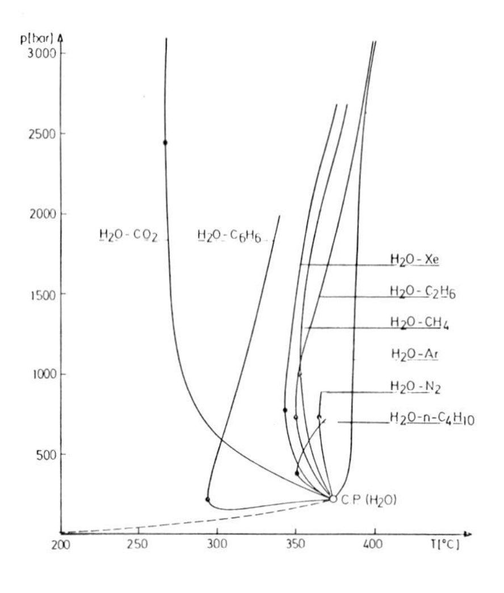

Fig. 13.
Critical curves of several binary aqueous systems (upper branches, beginning at the critical point of pure water).

Fig. 13 shows a number of critical curves of binary aqueous systems projected on the pressure-temperature plane. These critical curves limit the region of two-phase behaviour. The region of complete miscibility is on the high temperature side. One can have homogeneous mixtures of liquid-like densities at all concentrations, if the pressure can be raised to about one or two kbar in this temperature range. For all these systems a portion of the boundary surface of the two-phase region has been determined, which extends to the left of these curves. All the curves of Fig. 13 except the one for water-argon are of Type II. Two of the systems of Fig. 13 deviate significantly from the remaining group: water-benzene and water-carbon dioxide. The

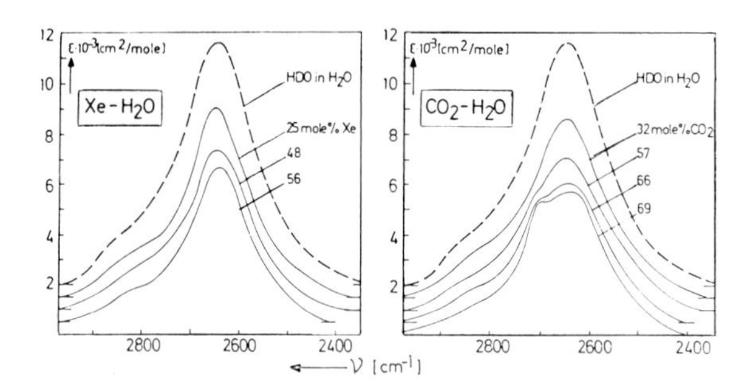

Fig. 14.
Infrared absorption bands of the OD-valence vibration of HDO diluted in H_2O with added Xe and CO_2 at 400°C. Water density constant at 0.17 g/cm^3 for both diagrams.

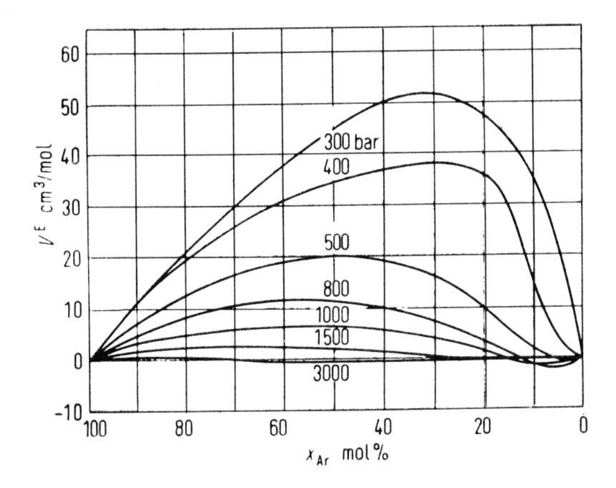

Fig. 15.
Isobars of the excess volume of the water-argon system at 400°C.

minimum temperature for water benzene is particularly low and occurs at 200 bar. Complete miscibility can be obtained easily. Comparable behaviour has been observed for other aromatic compounds. For water-carbon dioxide the critical temperatures remain relatively low even at high pressures, indicating a specific interaction. This is apparently verified by infrared absorption as shown in Fig. 14. Here the absorption caused by the OD-valence vibration of HDO in H_2O with added xenon and carbon-dioxide is shown [29]. The water density, 0.17 g/cm^3, is constant for all curves and the concentration of the second component is increased. Xenon changes the absorption band only slightly, while carbon dioxide in high concentrations produces a second band around 2700 cm^{-1}. Since Xe and CO_2 are comparable in size, the spectrum indicates a specific H_2O - CO_2 interaction. The nature of it is not yet sufficiently understood, however. Raman spectra at these conditions should be useful.

For several of the above mentioned systems the density in the supercritical homogeneous state has been measured and the deviation from the behaviour of an ideal mixture has been discussed. One way to do this, is by means of the excess volume, V^E, of the mixtures. Fig. 15 gives a diagram of the excess volume of water-argon mixtures at 400°C between 300 and 3000 bar [30]. V^E is defined by the relation:
$V^E = V - V_{H_2O} - x\,(V_{Ar} - V_{H_2O})$. V is the molar volume of the mixture. V_{H_2O} and V_{Ar} are the molar volumes of pure water and argon at the same conditions. x is the mole fraction of argon. The high V^E values for the lower pressures must be seen in relation of the relatively high absolute molar volumes at these conditions. Above 1500 bar the excess volume does not exceed a few percent of the molar volume of the

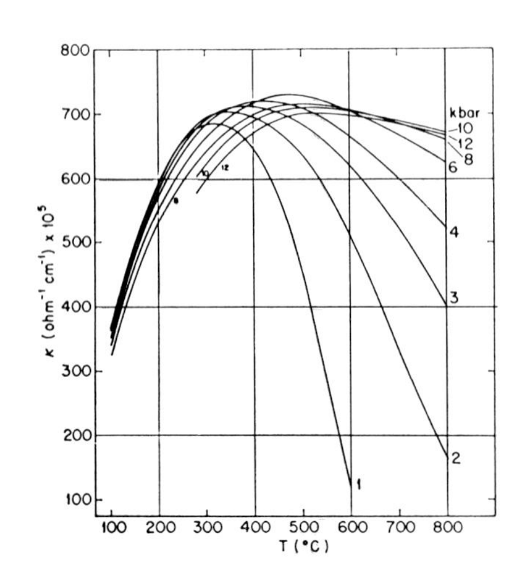

Fig. 16.
Specific conductance of 0.01 molal aqueous KCl solutions as a function of temperature for constant pressures.

mixture and may be even slightly negative. Such a small deviation from ideal mixture behaviour at high pressures has since been found for additional aqueous systems (H_2O - Xe and H_2O - CH_4) [31] and may well be typical for others, for example nitrogen-water.

I.3.5.

AQUEOUS SOLUTIONS OF ELECTROLYTES

Although the dielectric constant of dense supercritical water is much lower than that of ordinary water, it is still high enough to permit the electrolytic dissociation of dissolved salts. Ion hydration is still possible. Numerous electrolytic conductance determinations have been made with acids, bases and salts in supercritical water [32][33]. As an example only a summary of results for dilute aqueous potassium chloride solutions [34] obtained by several authors and laboratories is given in Fig. 16. Isobars of the specific conductance for pressures from one to twelve kbar are shown, which extend to 800°C. The initial increase of the isobars is to be expected as a consequence of the decreasing viscosity and increasing ion mobility. Above about 400°C the isobars decrease strongly at lower pressures and slightly at the highest pressures. Qualitatively it can be assumed, that ion pair association becomes important above 400°C and that the product of dielectric constant and temperature at constant density varies rather little at these conditions. Furthermore the viscosity does not change

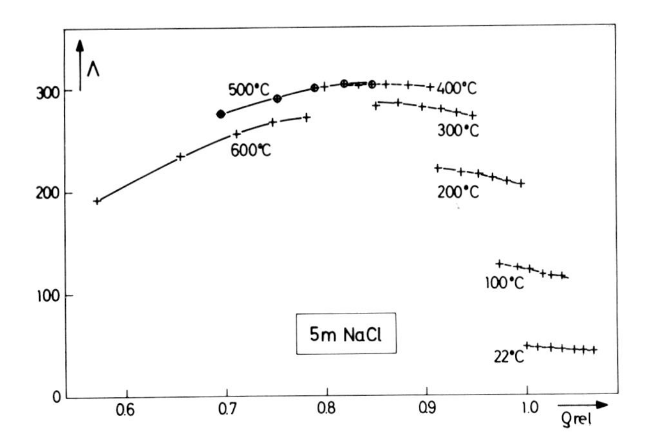

Fig. 17.
Isotherms of the molar conductance of aqueous NaCl-solutions as a function of the relative solution density.
Molality of the NaCl: 5.0.

very much at densities higher than about 0.5 g/cm^3 and at supercritical temperatures.

Recently conductance measurements could be performed also with very concentrated alkali halide solutions to 600°C and pressures of several kbar [35]. Concentrations were extended in some cases to 10 molal. In Fig. 17 a number of isotherms of molar conductances of about 5 molal sodium chloride solutions are presented as a function of the solution density [36]. As could be expected, the molar conductances in the whole region are considerably lower than those of dilute solutions. Beyond 300°C the temperature dependence above 0.5 g/cm^3 is not very pronounced at high temperatures. Energies and volumes of activation could be derived from these conductances.

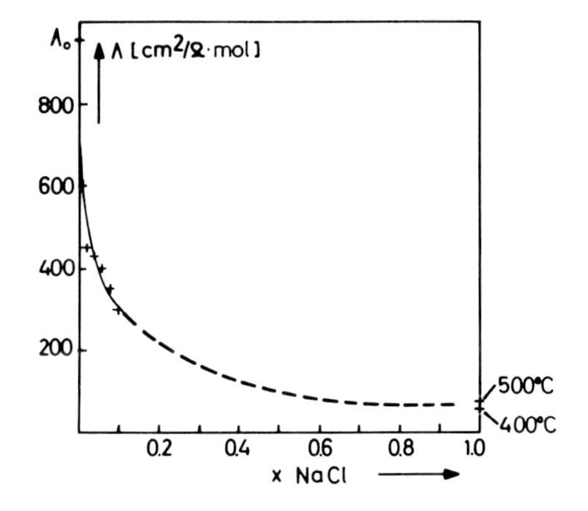

Fig. 18.
Molar conductance of aqueous NaCl-solutions at concentrations from zero to 5.0 molal, plotted as a function of the mol fraction x of the NaCl. +++ : Measurements taken at 400 and 500°C and at 2000 bar. Values at x= 1.0, see text.

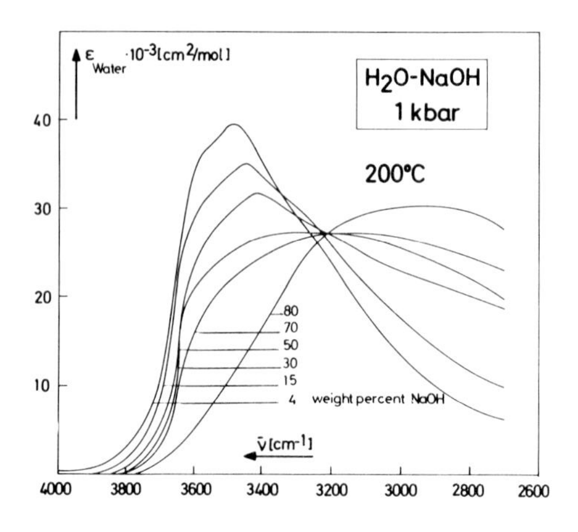

Fig. 19.
Molar extinction coefficient of water in the oxygen -hydrogen stretching region at 200°C for H_2O-NaOH mixtures of different concentrations.

It is interesting to observe, that at constant temperature the molar conductance with increasing temperature approaches fairly soon the corresponding values for the fused salt. This is demonstrated in Fig. 18. Here the molar conductance for NaCl at 400°C and 2 kbar has been plotted as a function of mole fraction in a diagram reaching from pure water to pure salt. The melting point of sodium chloride is 801°C. Since the temperature dependence of the molar conductance in the fused salt has been measured over a considerable range, however, these data have been extrapolated to lower temperatures to obtain the molar conductance for a ficticious supercooled melt of sodium chloride at 400°C and 500°C. This diagram suggests, that the conductance of solutions with salt mole fractions above 0.1 or 0.2 is not very different from that of the fused salt. Considerable amounts of water admixed to the fused salt, where this is possible, might affect the conductance of the ionic fluid only to a comparatively small degree.

It would be desirable to determine the variation of some physical quantity over the whole range from pure water to the pure fused salt in a system where there is complete miscibility between these two. This is difficult when the melting points are as high as those of the alkali halides. Sodium hydroxide, however, melts already at 318°C and is completely miscible with water at this temperature. This situation was made use of to measure the infrared absorption up to 350°C and at pressures up to 2000 bar of NaOH-H_2O mixtures over the whole range of possible concentrations [37)]. A modified reflection technique with a sapphire window cell was used.

pK_{NH_3} (p=0,68; -33°C) = 32,8	pK_{H_2O} (p=0,95; 100°C) = 12,3
pK_{NH_3} (p=0,68; 430°C) = 17,3	pK_{H_2O} (p=1,0; 1000°C) = 6,0
pK_{NH_3} (p=1,12; 430°C) = 3,6	pK_{H_2O} (p=1,5; 1000°C) = 2,2

Fig. 20.

pK-values (negative decadic logarithms of the ion product) of dense fluid ammonia and water at low and at high supercritical temperatures

Fig. 19 shows one set of absorption curves in the OH-stretching region. Vertically plotted is the molar extinction coefficient of the water in the mixture. It is obvious that a broad band around 2900 cm^{-1} develops with increasing sodium hydroxide concentration. It is preliminarily taken as indicating a distinct association of the hydroxide with water.

I.3.6.

IONIZATION AT VERY HIGH DENSITY-POLAR FLUIDS IN GENERAL

In conclusion, a short discussion of the ion product of water at very high temperatures and pressures, that is for conditions in the upper right corner of the temperature-density diagram of Fig.1 shall be given. The enthalpy of the ionic dissociation of water is +57 $kJ.mol^{-1}$ at standard conditions. The volume decrease connected with the dissociation at moderate pressures is -21 cm^3mol^{-1}. Although these quantities certainly do not remain constant, it is obvious, that a combination of temperature and pressure increases will produce very high values of the ion product of water. Analogous behaviour could be expected for similar small, stable and very polar molecules. One example is ammonia.

Electrical conductivities of water were measured in shock waves [38] to 130 kbar and 800°C and by a static method to 100 kbar and 1000°C [39]. Similar static measurements were made with ammonia to 40 kbar and 500°C [40][41]. The highest conductances reached in these experiments

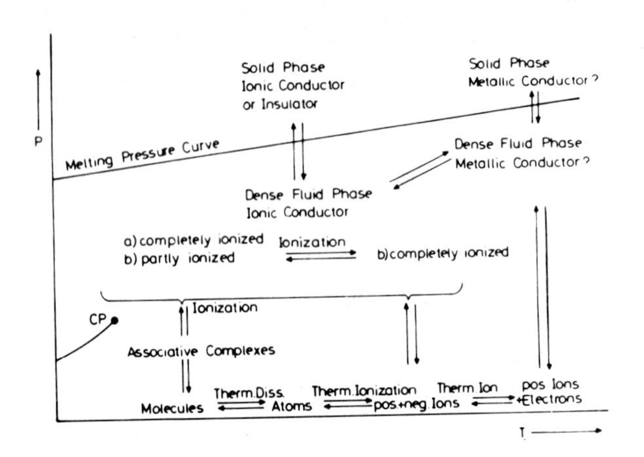

Fig. 21. Schematic presentation of phenomena to be expected in polar fluids at supercritical temperatures and pressures. Vapor pressure curve and critical point at lower left [42].

were almost 10^{0} $ohm^{-1}cm^{-1}$ for water and 10^{-2} $ohm^{-1}cm^{-1}$ for ammonia. For water the density at extreme conditions was about 1.6 g/cm^3, for ammonia about 1.1 g/cm^3. The results of the static measurements with water agree well with the shock wave data. Reasonable assumptions and estimates for the transport coefficients permit to derive at least approximate numbers for the ion products of water and ammonia at high temperatures and pressures. A selection of data is given in the table of Fig. 20. It contains pK-values, that is the negative decadic logarithms of the ion products. The numbers listed are not inconsistent with the existent relevant thermodynamic quantities ΔH, ΔS and ΔV of ionization in the low temperature liquids. Apparently in both fluids an increase of ion products by many orders of magnitude can be produced. In the case of water the conjecture has been made, [38] that complete ionization may occur at densities above 1.8 g/cm^3 and temperatures above 800°C. At the critical point of water, however, that is at 374°C and 0.32 g/cm^3, the quantity pK (H_2O) is probably not lower than 15.

The last Fig. 21, taken from Tödheide [42] gives a schematic survey of phenomena which can be expected with fluids of small, stable, polar molecules with increasing temperature and pressure.
At low pressure, increase of temperature leads to thermal dissociation, ionization and finally to plasma states. At higher pressures, molecular association and subsequently ion formation and ion solvation can be expected. Completely ionized dense fluids can occur in the

upper right region of the diagram. At extreme conditions even transitions to metallic states in certain cases have been predicted.

Water is a typical example for this group of polar fluids and the knowledge of its behaviour in such a wide range of states will improve our understanding of the intricate properties of liquid water within the limited conditions of our natural environment.

References :

1) E.U. Franck, Pure and Applied Chemistry, 24, 13 (1970).

2) K. Tödheide, "Water at High Temperatures and Pressures" in "Water a Comprehensive Treatise", Vol. I, F. Franks, ed., Plenum Press, N.Y., 1970

3) H. Köster, E.U. Franck, Ber.Bunsenges.Physikal.Chemie, 73, 716 (1969)

4) C.W. Burnham, J.R. Holloway, N.F. Davies, Amer.J.Sci. 267 A, 70 (1969)

5) M.H. Rice, J.M. Walsh, J.Chem.Physics 26, 824 (1957)

6) H.E. Stanley, "Introduction to Phase Transition and Critical Phenomena", Clarendon Press, Oxford, 1971

7) J.S. Rowlinson, Ber.Bunsenges.Physikal.Chemie 76, 281 (1972)

8) M. Buback, E.U. Franck, Ber.Bunsenges.Physikal.Chemie, 76 (1972) 350

9) G.S. Kell, "Thermodynamic and Transport Properties of Fluid Water" in "Water a Comprehensive Treatise" Vol. I. F. Franks, ed., Plenum Press, N.Y. 1970

10) K.H. Dudziak, E.U. Franck, Ber.Bunsenges.Physikal.Chemie, 70 (1966) 1120

11) V.N. Mineev, R.M. Zaidel, J.Exp. and Theor.Physics, USSR, 54 (1968) 1633

12) B. Le Neindre, R. Tufeu, P. Bury, J.V. Sengers, Ber. Bunsenges.Physikal.Chemie 77 (1973) 262

13) K. Heger, "Die statische Dielektrizitätskonstante von Wasser und Methanol im überkritischen Temperatur- und Druckbereich", Thesis, Karlsruhe, 1969.

14) V.M. Jansoone, E.U. Franck, Ber.Bunsenges.Physikal.Chemie, 76 (1972) 944

15) W. Harder, "Statische Dielektrizitätskonstante von Chlorwasserstoff im überkritischen Temperatur- und Dichtebereich bis 400°C und 2000 bar"., Thesis, Karlsruhe, 1972

16) W. Hill, "Die statische Dielektrizitätskonstante von Acetonitril bis 350°C und 3 kbar", Thesis, Karlsruhe, 1974.

17) E.U. Franck, Z.Phys.Chemie (Frankfurt) 8 (1956) 107

18) A.S. Quist, W.L. Marshall, J.Phys.Chem. 69 (1965) 3165

19) D. Eisenberg, W. Kauzmann, "The Structure and the Properties of Water", Oxford, University Press,(1969)

20) J.G. Kirkwood, J.Chem. Physics,7 (1939) 911

21) M.S. Wertheim, J.Chem.Physics, 55 (1971) 4291

22) V.M. Jansoone, Acta Physica Austriaca 37 (1973) 326

23) E.U. Franck, K. Roth, Disc.Far.Soc. 43 (1967) 108

24) H. Lindner, "Ramanspektroskopische Untersuchungen an HDO, gelöst in H_2O, an HDO in wässrigen KJ-Lösungen und an reinem H_2O bis 400°C und 5000 bar". Thesis, Karlsruhe,1970

25) M. Buback, E.U. Franck, Ber.Bunsenges.Physikal.Chemie 75 (1971)33

26) See the respective contributions in this volume.

27) G.M. Schneider: "Gas-Gas-Gleichgewichte" in "Topics in Current Chemistry" Vol. 13, P. 559, Springer, Heidelberg, New York (1970)

28) R.L. Scott, Ber.Bunsenges.Physikal.Chemie 76 (1972) 296
R.L. Scott, P.H. van Konynenburg, Discuss.Far.Soc. 49 (1970) 87

29) W.v.Osten, "Die Infrarotabsorption von HDO in H_2O in homogenen Mischungen mit CO_2 und Xe bei hohen Drucken und Temperaturen", Thesis, Karlsruhe, 1971.

30) H. Lentz, E.U. Franck, Ber.Bunsenges.Physikal.Chemie 73 (1969) 28.

31) H. Welsch, "Die Systeme Xenon-Wasser und Methan-Wasser bei hohen Drücken und Temperaturen", Thesis, Karlsruhe, 1973

32) E.U. Franck, Angew.Chemie 73 (1961) 309

33) E.U. Franck, Pure and Applied Chemistry, 24 (1970) 13

34) A.S. Quist, W.L. Marshall, E.U. Franck, W.v.Osten, J.Phys.Chem. 74 (1970) 2241

35) J.U. Hwang, H.D. Lüdemann, D. Hartmann, High Temperatures-High Pressures, 2 (1970) 651

36) W. Klostermeier, "Elektrolytische Leitfähigkeit konzentrierter wässriger NaCl-Lösungen bis 600°C und 3 kbar", Thesis, Karlsruhe, 1974

37) M. Charuel, "Infrared absorption of H_2O-NaOH-mixtures to 350°C and 2000 bar", Thesis, Karlsruhe, 1974.

38) H.G. David, S.D. Hamann, Trans.Far.Soc. 55 (1959) 72
S.D. Hamann, M. Linton, Trans.Far.Soc. 65 (1969) 2186

39) W. Holzapfel, E.U. Franck, Ber.Bunsenges.Physikal.Chemie, 70 (1966) 1105

40) D. Severin "Elektrische Leitfähigkeit und Ionendissoziation von wasserfreiem Ammoniak bis 500°C und 40 kbar"., Thesis, Karlsruhe, 1971

41) E.U. Franck, Ber.Bunsenges.Physikal.Chemie 76 (1972) 341

42) K. Tödheide, Naturwissenschaften, 57 (1970) 72

II. THEORY

II.1. RECENT STRUCTURAL MODELS FOR LIQUID WATER

George Némethy

Laboratoire d'Enzymologie Physico-Chimique et Moléculaire,
Université de Paris-Sud, 91405 Orsay, France

ABSTRACT

Some general concepts of importance for structural model studies are discussed the diverse meanings of the term "mixture" when applied to liquid water, the cooperativity of hydrogen bonding, the concept of hydrogen-bond breaking and its relationship with the reality of distinct molecular species in the liquid. Recent approaches to the development of structural models are reviewed. Models are classified into three groups, according to the basic approach taken. Among studies based on simple physical models, earlier two-state and cluster models, as well as their recent extensions, are reviewed. Some recent calculations based on simplified forms of liquid theory (such as cell theory) and on lattice statistics are discussed as a second type of approach. Third, results and potentials of molecular dynamics studies are analyzed.

II.1.1. INTRODUCTION

No unified general theory of liquid water exists. Many structural models have been proposed, especially since 1960. Their theoretical basis (including the simplifying assumptions), characteristic features, and the correlations with experimental data have been compared in several critical reviews /1-8/. Only recent reviews are cited here. They contain references to earlier ones. The intent of this paper is to give a comparative survey of some significant recent advances. New approaches to model studies are emphasized. Some older models will be mentioned briefly, for the sake of comparisons. Criteria for the evaluation of model proposals are discussed in more detail elsewhere in this volume /9/. (See also Sec. 4.8 of ref. /1/ and ref. /6/.) A few general concepts will be presented and clarified before particular models are discussed.

II.1.2. GENERAL CONSIDERATIONS

1. The Nature of Structural Models

All structural models proposed so far have been approximative. In order to be tractable, they are based necessarily on simplifications, both in the physical and the mathematical sense. Even so, each one can reproduce only a limited number of observed properties. In terms of basic approach, models can be classified into three groups.

(A) Physical model approach. Most older models, as well as several recent ones fall into this category. Their starting point is a more or less intuitive physical concept of what the dominant structural features in water ought to be like, in order to explain certain experimental observations. The model is evaluated quantitatively by an approximate statistical mechanical or thermodynamic treatment of the equilibrium relations between the structures proposed. This leads to the calculation of some physical properties. The structural entities assumed may be large or small clusters, including monomeric molecules /10-13/+, small oligomers, such as hexameric rings in different packings /14/, a hydrogen-bonded framework and interstitial molecules /15-17/, distorted or broken hydrogen-bonded network structures /18-20/. Physical properties were analyzed also in terms of local interactions (hydrogen-bond equilibria) between molecules, without assumptions about long-range structures /21/.

In general, these models can explain, at least qualitatively, the characteristic anomalous properties of liquid water ; every one gives a quantitative interpretation of a limited number of physical properties, thought to be the most important ones by its authors. On the other hand, every one has one or several shortcomings : (a) there are always some properties to which the model is not applicable, or which are calculated with large errors; (b) some basic assumptions of the model may not be realistic;

\+ without monomers !

(c) the quantitative treatment may require mathematically not admissible simplifications. The models elucidate characteristic features of water but do not give a complete picture of the liquid.

(B) Liquid theoretical and lattice approaches. In some recent studies, the simplified version of one of the general theories of liquids was taken as the starting point. No structural entities were proposed a priori. Only local ordering is assumed, arising from the tetrahedral orientation of hydrogen-bonded molecules. Sometimes an added hypothesis regarding molecular packing is used. The analysis is carried out in a formal manner. In some models, the cell theory of liquids, with local ordering, was used /22, 23/. The use of distribution functions was proposed without a specific application /24/. An alternative approach was taken in lattice statistical calculations. They treat only the breaking of hydrogen bonds on a lattice /25-28/.

Very few actual numerical parameters were calculated and tested against experiment in these models. Some important features of water may be obscured by the limitations of the approach chosen. (For example, in lattice models, Sec. II.5.) While these models lead to some useful insights, their results seem to describe in some cases what is happening in an idealized theoretical framework, rather than in the actual liquid.

(C) Molecular dynamics approach. In this approach, no structure is assumed a priori. The behavior of a certain number of water molecules is simulated, based on a simple physical picture of molecular structure and of nearest neighbor intermolecular interactions /29, 30/. Thus the model is essentially an "analog calculation". In principle, it is very powerful. With the proper initial assumptions, the behavior of an ensemble of molecules is reproduced quite well. In actual practice, approximations and simplifications have to be introduced.

2. What is a "Mixture" and a Continuum" ?

A continuing controversy in this field centers on the question of whether water is better represented by a "mixture" or by a "continuum" model. Usually, the former is meant to describe water in terms of distinct species of molecules (usually according to their hydrogen bonding arrangements), while the latter is taken to mean that no such distinctions are possible but that properties of the molecules change in a continuous manner over a certain range. Kell gave concise definitions of the two types of models in this sense : "A mixture model is understood to describe liquid water as an equilibrium mixture (or solution) of species that are distinguishable in an instantaneous picture," and "a continuum theory describes water as having essentially complete hydrogen bonding, at least at low temperatures, but as having a distribution of angles, distances, and bond energies." (Pp. 332-333 of ref. /4/.) A controversy arose because the necessarily simplified models adopt one or the other concept

Table I

Comparison of the Various Uses of the Term "Mixture" in Models for Liquid Water

(1) Structural inhomogeneities in space
(2) Distinct molecular species
(3) Hydrogen bond breaking equilibrium

	Mixture according to definition			Nature of model	Examples /references/
	(1)	(2)	(3)		
(a)	+	+	+	Most two-state models Cluster models	See list in /1/ /10-13, 31/
(b)	+	+	-	Not feasible	-
(c)	+	-	+	Special two-state models	/14/
(d)	+	-	-	Possible two-state models	?
(e)	-	+	+	Cell models Interstitial models Lattice statistics	/22, 23/ /15-17/ /20, 25-27/
(f)	-	+	-	Not feasible	-
(g)	-	-	+	Gel network model	/28/
(h)	-	-	-	Distorted network (continuum) model	/18/

exclusively and in its extreme form, i.e. species are taken with well-defined unique properties (in "mixture models"), or the existence of qualitatively different kinds of molecules in negated completely(in "continuum models"). Actually, both features must be present in the true liquid, which is more complex than any of the simple models proposed. In this sense, the controversy is reflects more the limitation of our understaning of the structure of water and of the ways in which to describe water. It should disappear as better models are devised.

Nevertheless, a further point must be clarified in the use of these two terms. The concept of a mixture has been used to describe several different things. These are not always equivalent (although sometimes they are considered as if they were), nor do they always imply one another.

(1) Since most models presuppose that water molecules exist in more or less well-defined structural entities, the concept of a "mixture" often is taken to imply the existence of differently structured regions in adjacent regions, i.e. a structural inhomogeneity on space. This conclusion is obvious in many models, but it is not always true. Sense (1) is not necessarily linked in a one-to-one manner with the two

following definitions of mixtures.

(2) Often the concept of a "mixture" indicates the postulation of different molecular species. These are generally differentiated in terms of their hydrogen-bonding status. The minimal assumption is that of two species : a mixture of fully hydrogen-bonded molecules which have tetrahedrally arranged neighbors and of non-hydrogen-bonded molecules which interact less specifically with their neighbors. In some models further species with intermediate numbers of hydrogen bonds are included.

(3) Sometimes the term "mixture" as applied to a model merely implies that a hydrogen-bond forming and breaking equilibrium is considered, and the existence of different molecular species is not assumed explicitly. For this use of the term, one needs merely a clear distinction of an existing and a broken hydrogen bond (see Sec. 2.4). Obviously, the occurence of a hydrogen-bonding equilibrium is a necessary prerequisite of the classification in the sense of (2), at least if the species are defined in terms of hydrogen bonding. (No other definition would seem to be reasonably useful.) However, the converse is not necessarily true.

Table I summarizes the ways in which these definitions can be combined, with their application to various models. It is clear that most "mixture models" represent mixtures in all three senses. However, structures may exist which form mixtures in senses (1) and perhaps (3) , even though they are composed of only one molecular species : for example, if they correspond to different ways of packing hydrogen-bonded oligomeric structures (lines c and d of the Table). On the other hand, the dependence of definition (2) on (3) precludes the possibilities listed in lines (b) and (f). Several alternatives are possible in principle when uniformity in space is presupposed [absence of a mixture in sense (1)]. The "continuum model", as introduced originally /18/ and as usually discussed /4/, implies the absence of a mixture in all three senses (line h). On the other hand, a more or less continuous spatial distribution can be thought of which contains different species (line e).

Some analyses of experimental data, especially spectroscopic studies /21, 32/ are concerned with the local environment of water molecules or with the status of individual hydrogen bonds only, and do not necessarily imply any conclusions about long-range structure. Thus they do not deal with definition (1). In principle, any of the two possibilities of the latter might be consistent with them.

The classification of Table I is a crude one. It is even somewhat arbitrary : some models might be placed into other groups as well. Nevertheless, the table may be helpful as a reminder that the various definitions of a "mixture" are not equivalent. Some confusion may have been caused by the neglect of this fact, exacerbating some discussions in an unnecessary way.

3. Cooperativity of Hydrogen Bonding

The concept that the strength of successive hydrogen bonds by a water molecule is not independent, was proposed in a qualitative way by Frank and Wen /33/. Recent quantum mechanical calculations confirmed the existence of cooperativity, even though they disagree about its numerical magnitude /34-37/. All kinds of models are consistent with such an effect. In fact, it was taken as a fundamental feature in many mixture models. However, it has not been included explicitly in the quantitative formulation of most models.

4. Hydrogen-Bond Distortion and Distinct Molecular Species

The concept of the breaking of hydrogen bonds is basic to many mixture models, according to definition (3) above. It is not easy to define when a hydrogen bond is broken, since the bond is easily deformable either by stretching or by bending. The potential energy is a continuous function of bond length and angles. For this reason, the applicability of the bond breaking concept was questioned by Eisenberg and Kauzmann (p. 246 of ref. /1/). As they pointed out, the concept is useful only if the potential energy is not a smoothly increasing function of the distortion coordinate (curve a of Fig. 1) but there is a secondary minimum corresponding to the nonbonded state (curve b) or at least there is a region in which the potential energy rises rather abruptly (curves c and d of Fig. 1). The actual form of this potential curve is not known. Form a is preferred by some workers /1/. However, spectroscopic evidence, cited elsewhere in this volume /38,39/, supports the idea that distinctly different states of the hydrogen bond exist in the liquid (curves b, c, or d).

The occurrence of different molecular species (definition 2 of a mixture) seems to require that hydrogen-bond breaking be a clear cut all-or-none process. However, this is not necessary. The presence of fairly well defined molecular species can also be due to packing differences, related to the tetrahedral four-coordination of hydrogen-bonded molecules and to the higher coordination of the non-hydrogen-bonded ones. It does not depend on the exact shape of the hydrogen-bond energy potential. No matter which of the curves in Fig. 1 is followed during the bending of an individual hydrogen bond, ultimately a geometry is reached which no longer assures strict tetracoordination of the molecule. When one or several of the original four neighbors are strongly displaced from their ideal position, the coordination may increase through the introduction of further neighbors which then interact with the central molecule /2, 10/. This lowers the overall energy. Thus the total interaction energy of a given molecule with its surrounding would display several minima (Fig. 2). This could occur with any of the curves of Fig. 1. The existence of more than one minimum defines various molecular species. Of course, the minima are rather broad, because the interaction energies with various neighbors change in

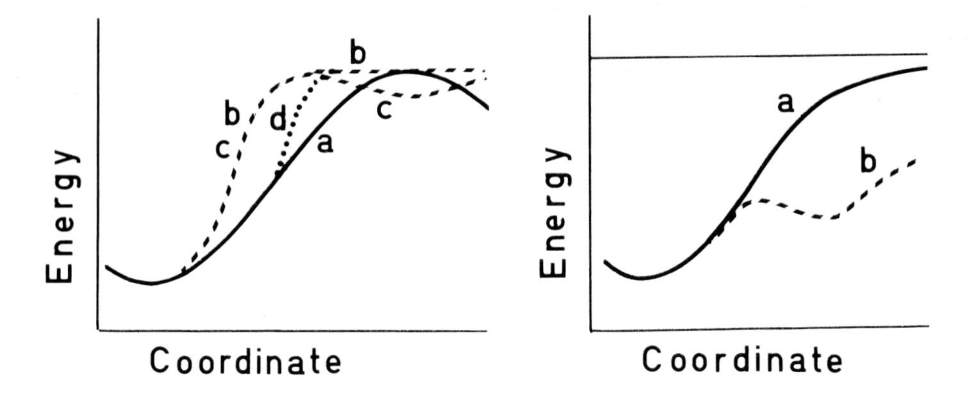

Figure 1. Possible forms of the potential energy curve of an O-H...O hydrogen bond in water in function of the relative rotation of two water molecules. It cannot be defined unequivocally when the "breaking of the hydrogen bond" occurs if curve a is the correct one. With curves b, c, or d, a clear definition could be given. (Adapted from Fig. 4.27 of Ref. /1/.)

Figure 2. Comparison of the potential energy for the distortion of an individual hydrogen bond (curve a) and of the total potential energy of a molecule when its coordination number increases, following a distortion or breaking of its hydrogen bonds (curve b).

---

a continuous fashion. Nevertheless, this phenomenon would allow the practical distinction of diverse species, in a manner related to hydrogen-bond breaking, yet not relying solely on the exact nature of the latter. Thus it may justify the use of mixture models in sense (2).

II.1.3. SURVEY OF RECENT MODELS

Table II summarizes some of the important features of several recent and older structural models /7/. A concise summary of this type is necessarily oversimplified and does not provide full comparisons. For example, the goodness of fitting of diverse properties varies within any one model. The number of properties fitted is indicated in the table. However, this in itself is not a comparative measure of the merits of different models, because the errors of fitting were not indicated, and because the range of variables considered (temperature and in some cases pressure) differs from one model to another. The table is merely a qualitative comparison, because no single parameter of the "goodness of the model" can be given. The number of parameters, given in the table, is usually that of the explicitly varied adjustable parameters of a model. Generally, there are more hidden parameters. They can be of two types : some parameters do not appear explicitly but are implied by the basic

Table I

Comparison of Models for Liquid Water

Model Type	Authors	Year	Ref.	Description	Method used	Properties calculated	Number of parameters	Main criticism
A	Némethy, Scheraga	1962	/10/	Cluster-monomer equilibrium, T-dependent single cluster size	SM	G, H, S, c_p, V, rad.distrib.fct.	3	Combinatorial error, single cluster size
	Hagler, Scheraga, and Némethy	1972	/11/	T-dependent distribution of cluster sizes, in equilibrium	SM	G, H, S	3	Mathem.restrictions, errors in small clust.
	Lentz, Hagler and Scheraga	1974	/12/	T-dependent distribution of cluster sizes, in equilibrium	SM	G, H, S, c_p	3	Very small cluster sizes
	Luck	1972	/13/	2-state H-bond breaking equilibrium, orientational lattice defects	TD	surface energy, V, c_p, ΔH_{vap}	3	Bulk properties assigned to structures
	Davis, Litovitz	1965	/14/	2-state equilibrium, hexagonal H-bonded rings in different packings	TD	V, relax.prop., rad.distrib.fct.	6	2-state implies infinite structures
	Frank, Quist	1961	/15/	2-state equilibrium, ice-like framework, interstitial monomers	TD	P-V-T prop.	3	Framework strain, too solid-like
	Narten, Danford, and Levy	1967	/17/	2-state equilibrium, ice-like framework, interstitial monomers	TD	rad.distrib.fct. V	7	Framework strain, too solid-like
	Walrafen	1968	/21/	2-state H-bond breaking equilibrium, only nearest neighbors considered	TD	T-dep of Raman, c_p, $\partial V/\partial P$	4	Limited to local interactions
	Pople	1951	/18/	Complete H-bonding, T-dependent bending of H-bonds	SM	rad.distrib.fct. diel.const.	4	Lack of fluidity, no broken bond state
B	Weres, Rice	1972	/23/	Cell theory. Local lattice ordering in different configurations	SM	G, H, S, c_p, V	4	Solid-like, no long range correlations
	Levine, Perram	1968	/25/	H-bond breaking statistics on a fixed lattice	SM	-	4	Only H-bonds in a solid considered
	Bell	1972	/27/	Distribution of H-bonded molecular configurations on a lattice	SM	critical prop., state surfaces	3	Lattice-gas model, dense structure wrong
	Gibbs, Cohen, Fleming, Porosoff	1973	/28/	Cooperativity of H-bond breaking in a gel-like network	Qu.	Qualit. discuss. several propert.	-	Solid-like, not a detailed model
C	Rahman, Stillinger	1971	/29/	Local interaction of molecules, with a tetrahedral potential	MD	Self-diffusion diel.relax., g	4	No cooperativity, H-bond too directed

SM = statistical mechanics, TD = thermodynamic equilibrium, Qu = qualitative discussion, MD = molecular dynamics

assumptions and restrictions of the model. Others seem to be fixed quantities. However, if they are not known perfectly, the values used are reasonable selections, but are not determined unequivocally (for example, vibrational frequencies occuring in many models). The criticisms listed, too, carry unequal weights: some refer to serious basic faults, some to quantitative details.

1. Simple Two-State and Cluster Models

The assumption is made often that the properties of liquid can be described in terms of additive contributions from two components whose relative concentration is a function of external conditions (usually only the temperature is considered) : a more open-packed, generally more ordered state (with higher extent of hydrogen bonding) and a denser, usually less ordered state. /e.g. 13,14/. Many of the early models were of this type, together with some more recent ones. Many were developed for the interpretation of a particular physical measurement. These models can be described with simple thermodynamic equations. They usually reproduce qualitatively the anomalous features of water; they can account quantitatively for a limited number of properties, using a few parameters only. However, an inherent contradiction is contained in most of them : in effect, it is assumed that the structural domains corresponding to the two states are infinite. This is implicit in the frequent use of the physical properties of known bulk phases to characterize the two states. However, scattering measurements /40/ and related experiments suggest that no large-scale inhomogeneities exist in the liquid. Thus any structural domain of one kind must be small. This means that most molecules are on or near the surface of a domain, and their properties differ from those of molecules in the bulk phases. This effect can be taken into account in two ways : either by introducing at least one more state in the model, but thereby impairing its simplicity and increasing the number of necessary parameters,or by using modified values for the characteristic physical properties which enter the model equations. In the latter case, these values are not obtained directly from any measured bulk phase properties. They are therefore somewhat arbitrary. In the absence of either of these corrections, the adjustable parameters obtained in the models become effective parameters : they are useful for the reproduction of the observed properties, but their physical meaning may be obscured. The value of the parameters no longer represent the quantity used in their original definition.

Cluster models were introduced to avoid these difficulties /10,41/. By assuming small hydrogen-bonded clusters, the liquid is described in terms of more than two species of molecules, so that the problem of structural domain interfaces is circumvented. On the other hand, an uncertainty regarding the properties of the diverse species remains. Even if reasonable properties are assigned, some arbitrariness is inherent to this treatment.

Various levels of approximation are possible in these models. This can be illustrated with the development of one particular cluster model approach. In the model introduced by Némethy and Scheraga in 1962 /10/, a single-size cluster was used, in equilibrium with non-hydrogen-bonded molecules ("monomer"), and forming an ideal mixture. The hypothesis of a single cluster size neglects an entropy contribution arising from the distribution of cluster sizes. The width of this distribution is not negligible if the mean cluster size is small (as was suggested by the scattering experiments cited above) because the "maximum term approximation" of statistical mechanics no longer holds. Additional criticisms of the model were pointed out recently. The most important weaknesses are : the neglect of cooperativity in hydrogen bond formation, in common with most other models ; the use of an incorrect simple form of the term which describes the arrangement of molecules : this leads to an overcounting /25/; disregard of changes of zero-point energies. These and other criticisms were discussed in detail elsewhere /2,7/. In a revision by Hagler, Scheraga, and Némethy /11/, some of these problems could be eliminated. A restricted, one-peaked distribution of cluster sizes was introduced, with a maximum which varies as a function of temperature. Zero-point energies were considered, and a better set of vibrational partition functions was used. The interaction of clusters was considered by deriving a translational partition function based on the Percus-Yevick theory. However, cooperativity of hydrogen bonding was still neglected. The model gives a good representation of the main thermodynamic parameters between 0 and 60° but the heat capacity is fitted poorly. There are errors in the properties of small clusters. This is a serious weakness because the size of clusters was computed to be rather small : the median size ranges from 11 to 6 molecules. This value might be questioned in the light of relaxation experiments. It also results in a very large value for broken hydrogen bonds. A further development of this model, by Lentz, Hagler, and Scheraga /12/, eliminates some approximations and errors of the previous version : it introduces cooperativity and uses a more realistic cluster size distribution, free of artificial restrictions of the earlier model. The agreement of the thermodynamic parameters is improved, but the favored aggregate sizes become even smaller : dimers and cyclic pentamers predominate. Small-aggregate models conflict with some experimental data (see p. 255 of ref /1/). On the other hand, the model implies strong restrictions of the rotation of clusters. It is suggested /12/ that these might be interpreted as cluster interactions involving distorted hydrogen bonds. This idea was not elaborated in detail. However, if it can be substantiated, it would represent a significant conceptual and quantitative advance towards the reconciliation of conflicts between simple models. It also would mark a major departure from the usual concepts /2, 10, 11, 33/ of what a "cluster" is like.

A nonspecific cluster model was presented recently by Luck /13/, based on an analysis of orientational defects. It can reproduce in a simple manner, without requiring specific assumption about clusters, some experimental properties, such as density, heat capacity, heat of evaporation. Since it is based chiefly on considerations of hydrogen-bond breaking, it might also be considered as a simplified and less restrictive version of lattice models (Sec. II.5). The general criticisms of two-state models are applicable to this model, too.

The advantages of cluster models are that they focus attention on the behavior of hydrogen bonds, that they may provide sufficient local irregularity to describe a fluid and that they can be used easily to account for properties of aqueous solutions. Their disadvantages are, besides the inherent errors of their mathematical approximations, that they usually consider structural entities which are too rigidly defined and which form an ideal mixture. They overemphasize hydrogen-bond breaking, to the exclusion of bending and of other interactions.

Walrafen's analysis of Raman intensities /21/ represents a different, special form of two-state models. His data provide information on the hydrogen-bonding status of molecules. Thus the properties calculated from the model are restricted to those which depend only on local interactions and hydrogen-bond breaking. Obviously, the model in its present form does not make any predictions about longer-range interactions and about structures beyond nearest neighbors. On the other hand, the experimental data on which it is based offer a criterion which must be taken into account by more specific structural models.

2. Interstitial models

They assume /15, 17/ that hydrogen-bonded molecules form a more or less regular framework, and that non-hydrogen-bonded molecules are located in the interstices of the latter. They present some appealing features: above all, they allow the presence of two kinds of water molecules, with different hydrogen bonding and other properties, without being incompatible with the lack of long-range structural inhomogeneities. Nor do they have to introduce small oligomers as clusters, an objectionable feature of some other models. They are able to reproduce pressure-volume behavior and the thermodynamic properties of some simple aqueous solutions.

On the other hand, they must face several critical problems : (i) there is the question why the interstitial structure would be stable enough to be the dominant feature of the liquid : the hydrogen bonds of an ice-like framework must be stretched if the interstitial water molecules are to be accomodated. Furthermore, it is not clear why the interstitial molecules do not interact strongly with their neighbors in the framework. It is true that structures in the liquid are considered as short-lived, i.e. "flickering" /33/. Still, a structure must be energetically favored if it is present in large proportions. (ii) An unbroken framework ought to show long-range order and correlations. (iii) If the existence of only two states, i.e. of interstitial and of fully hydrogen-bonded framework molecules, is taken literally, the model is too solid-like, both in a dynamic and in a static sense. The framework would have to be interrupted if the model were to account for fluidity and for lack of long range order. The model is more like a solid in its equilibrium thermodynamic properties, too. It was calculated /7/ that a structure corresponding to the model of Narten *et al.* /17/ would give values of the heat and entropy of fusion as well as of the heat capacity of the liquid which are much too low. The model lacks the liquid structural contribution. These faults could be eliminated by introducing a "state III" which would occur in regions in which the framework is disrupted, as suggested already in the original calculation of Frank and Quist /15/. However, in this way a combined cluster-interstitial model is obtained, in which the appealing features are lost.

3. Continuum (Distorted Hydrogen Bond) Model

Although many arguments have been brought forward recently in favor of a continuum model /1,4/, the only quantitative derivation is that by Pople /18/. Deviations from ice were described entirely in terms of a hydrogen-bond bending force constant. This is too simple to give a good representation of water /1/. Even a continuum model ought to be treated in a more detailed manner. As mentioned already, there is good evidence that broken bonds exist in the liquid /5, 38, 39/. Bond distortion should not be neglected, but it must be considered together with bond breaking.

4. Model Based on Cell Theory

Weres and Rice /23/ attempt to describe water in terms of the local interactions of a molecule with a cell formed by its neighbors. A given molecule is partially hydrogen-bonded to its neighbors. Diverse local structures are assumed, corresponding to molecules with various numbers of hydrogen bonds, and in analogy with an amorphous solid. The formalism accounts mainly for the interaction of a molecule with its nearest neighbors. The model is not supposed to result in long-range continuity of the structure, i.e. in correlations over several bond lengths. It seems probable, though, that this is implied actually in the model. A configurational energy for molecular arrangements is calculated by using the device of the local distribution of molecules over a lattice. In addition, the translational motion of molecules, with its potential energy contributions, is taken into account by using the cell theory of liquids : the molecules are assumed to move in a field defined by their neighbors in fixed average positions. A pair potential porposed by Ben-Naim and Stillinger /42/ was used. It had to be rescaled in order to account well for the properties of ice. Various packing arrangements of two- and three-bonded molecules are the most frequent local structures in the model. Over the 0 to 100°C temperature range, the free energy, enthalpy, and entropy of the liquid are calculated with about 10 % error. The heat capacity and its temperature dependence have much higher errors. The volume is obtained fairly well at high temperatures, but there is no tendency to give the minimum near 4°. This may result from the fact that the consideration of only local hydrogen-bonding interactions does not allow for the formation of a "bulky" structure. The low value and the small temperature coefficient of the entropy seem to indicate that the model is too solid-like, in spite of the translational contribution calculated from cell theory. This is also shown by the result that the extent of hydrogen bonding does not vary with temperature. All this may arise from limitations in the construction of local molecular arrangements which define the cells. As pointed out by the authors, the correlation of motions of neighboring molecules is taken care of incompletely. This prevents comparison of the calculated distributions with X-ray radial distribution curves. Some of the discrepancies of the model were ascribed by the authors to the known defects of the cell theory.

On the one hand, the main weakness of the model seems to be the neglect of long-range correlations which are required in water models. On the other hand, its important merit is the more adequate treatment of local non-hydrogen-bonded interactions than that provided by the simple models discussed in previous sections.

5. Lattice Model Calculations

Several such studies were published recently /25-27/. The study of Levine and Perram /25/ was initiated with the purpose of finding a consistent way to calculate the number of broken hydrogen bonds in a network structure. This was done to avoid the combinatorial error which occurs in the Némethy-Scheraga theory /10/ and in related simple cluster models. A network of hydrogen-bonded molecules on a lattice is assumed. A configurational entropy is calculated from the distribution of broken hydrogen bonds on the lattice. Cooperativity of hydrogen bonding was neglected in most lattice calculations. A recent computation by Perram and Levine /26/ is an exception : nearest neighbor interaction is considered in the quasi-chemical approximation of an Ising model.

In the study of Bell /27/, the lattice is introduced differently, in analogy with the general lattice theories for fluids. It is used to describe local ordering around molecules. Very few physical properties could be calculated with the model. Critical parameters are computed. However, the model essentially is that of a lattice gas, which is particularly unsuitable to represent the liquid near the critical point /23/. Furthermore, the structure proposed contains a large amount of ice-VII-like arrangements. This is inconsistent with X-ray diffraction data.

The most important results of lattice statistical calculations are the general insights provided. They demonstrate the important role of hydrogen-bond breaking in accounting for many properties, such as the configurational (=excess) heat capacity /20,25/. On the other hand, they represent rigid structures and do not account for the fluidity of a liquid. Above all, they are limited to one aspect of the behavior of water, and thus do not provide a complete model. For example, it was stated by Perram and Levine /26/ that their calculation gives the opposite of a cluster model, namely, an "extensive bond network, interrupted by regions containing relatively few bonds". This is natural for a system of bonds on a fixed lattice. However, two features of the liquid are omitted from consideration : (i) cooperativity of hydrogen bonding, as opposed to bond breaking as a random process, and (ii) the gain of translational degrees of freedom when individual molecules or aggregates are no longer fixed to lattice sites. Both features oppose the maintenance of a connected lattice /11/.

Except for Bell's results cited, no quantitative comparisons with experiment were attempted. It follows from the preceding discussion, no valid comparisons could be expected. Angell's model /20/ of random breaking of hydrogen bonds on a lattice is appropriate for a supercooled glassy state.

Interesting considerations were introduced by Gibbs et al. /28/ in a qualitative discussion which was aimed at the interpretation of the solid-liquid and liquid-vapor phase transitions in terms of hydrogen bonding. They drew analogies with other cooperative systems. (i) The helix-coil transition in polypeptides is compared with the disappearance of interconnected small ring structures in the melting of ice. (ii) Gelation in polymerization, i.e. formation of a tree-like, completely interconnected structure throughout the phase is compared with the condensation of water vapor. They conceive the liquid as a single hydrogen bonded structure, with chains and rings, but without three-dimensional networks. The analogy with helix-coil transition seems somewhat far-fetched, because the topological interconnectedness of rings in the two systems is very different. The existence of essentially infinite hydrogen-bonded chains at high temperatures might be questioned, too.

The model correspond to a rather rigid, amorphous solid-like structure. Accordingly, no quantitative calculation of any property was attempted. It was pointed out by the authors /28/ that the model is qualitatively consistent with several features of liquid water. It is stated that it could account for the radial distribution function, for the existence of two states of OH groups, with most of them engaged in hydrogen bonding (consistency with infrared spectra), for the minimum in molar volume, for local orientational correlations which explain the static dielectric constant and the dielectric relaxation time. However, none of the arguments is unique to this model. These properties are explained by any model which contains many water molecules which are interconnected by hydrogen bonds, in any geometry whatsoever. The interest of this study lies (i) pointing out the desirability of considering phase transitions in structural models and (ii) in the contribution of a novel concept : that of geometric cooperativity, due to the formation of interconnected three-dimensional hydrogen-bonded structures (in contrast to chains or single rings). This cooperativity is different from the quantum-mechanically determined cooperativity of successive hydrogen-bond formation /33/. (Sec. I.3.)

6. A Molecular Dynamics Study

The behavior of an assembly of water molecules was analyzed by Rahman and Stillinger /29,30/ by means of molecular dynamics simulation. This approach is fundamentally different from the models discussed above. As the starting point, only a pair potential, acting between water molecules, is assumed. The Ben-Naim-Stillinger potential /42/ was used. This is the combination of a nondirectional Lennard-Jones potential and of an electrostatic component, representing hydrogen bonding and arising from two positive and two negative partial charges, arranged tetrahedrally in the molecule. This component gives strong directional preference. However, no cooperativity of hydrogen bonding was included. Optimal orientation of two molecules corresponds to a hydrogen bond energy of -6.50 kcal/mol. An ensemble of 6^3=216 molecules was treated, in a fixed volume corresponding to the density of 1 g/cm^3. The motion of the molecules is calculated using classical dynamics. Computations were carried out at three temperatures, viz. 265, 307.5, and 588 K.

The static structure is described in terms of various radial and angular pair correlation functions. They provide a description of the system. A strong tendency towards tetrahedral coordination is seen. The correlations become weaker and the number of nearest neighbors increases with temperature, as expected. Second neighbors are also strongly correlated. This confirms the persistence of a hydrogen-bonded lattice in the liquid, albeit with strong distortions, as indicated by large angular deviations from the ideal hydrogen-bonding directions. However, is addition, there are also contacts between pair of molecules which do not form a hydrogen bond. From the analysis of the distribution of pair interaction energies and of a reasonable hypothesis on the energy of molecules of various species, it appears that 3-bonded molecules are most frequent. Some molecules have more than four neighbors. The picture obtained from the calculations is described /29/ as "a random, defective, and highly strained network of hydrogen bonds that fills space rather uniformly". The analysis of the numerical data and of stereoscopic pictures of the molecular configurations was summarized /29/ as follows : no long-range correlations, strong local tetrahedral coordination, the presence of some "dangling", i.e. non-bonded OH groups. No evidence was seen for a partitioning of molecules into two classes, viz. bonded and unbonded, as postulated in two-state theories, nor for the presence of definite clusters with anomalous densities or with ice-I-like structure, nor for interstitial molecules and clathrate-like structures, nor for interpenetrating hydrogen-bonded networks. Thus the calculations seem to disagree with some features of about every specific structural proposal in models of type (A).

On the other hand, the analysis of the distribution of effective pair interaction energies at the three temperatures reveals a quite clear-cut division into two intervals. The separation occurs at a pair interaction energy of -3.5 kcal/mol. (Fig. 11 of ref. /30/). With a rise in temperature, the number of pair interactions in the low energy interval decreases, with a corresponding increase of the numbers in the higher energy interval. It was suggested that this is due to the presence of a single basic hydrogen bond excitation mechanism, in which a strained hydrogen bond is broken, resulting in a more weakly interacting pair of molecules. The mean energy difference between the two states is about 2.5 kcal/mol. This number agrees well with the ΔH values obtained for hydrogen-bond breaking in many different experiments (see other chapters in this volume and the list of data in ref. /30/). This finding supports the arguments regarding the existence of distinct hydrogen-bonded and and nonbonded states, as discussed in Section II.4.

So far, only few properties could be calculated which permit comparisons with experimental data. There is only moderate agreement with X-ray scattering intensities. The dipole direction correlations give a reasonable value of the Kirkwood g factor ; it has a strong temperature dependence. This indicates the breakdown of order. Kinetic properties can be obtained from the model. The calculated self-diffusion coefficient and the dielectric relaxation time agree reasonably well with experimental results.

The method is quite powerful in principle. Since it depends only on molecular structure and the choice of molecular interaction potentials, it does not require *a priori* assumptions about intermolecular structure. With the proper choice of interactions, it should describe the behavior of an ensemble of molecules very well. It can deal with both static and kinetic properties ; the presence of simple solutes can also be simulated. The latter has been attempted in a recent continuation of the studies /43/. On the other hand, the model is only as good as the potentials used. Its results reflect strongly the errors and approximations of the latter. Rahman and Stillinger pointed out /29/ that the potential is too strongly tetrahedral, *i.e.* the hydrogen bond is too directional. This may bias the results towards too much local order. This defect could be remedied by simple modifications. On the other hand, the partial covalent contribution in the hydrogen bond, resulting in cooperativity (Sec. II.3), and leading to higher than pairwise interactions, was not taken into account at all. Correction for this would be difficult in practice, because it requires the introduction of three-body and higher interactions in the dynamic computation. At the present stage, therefore, it is an open question whether some features seen in the results actually correspond to true details of the structure of the liquid or whether they are merely consequences of the particular choice of potentials and parameters. Further studies are needed before it can be stated definitely whether some calculated correlations are a precise representation of those in the real liquid. On the other hand, the method can serve as a good testing ground for intermolecular potentials. It is also a critical check on simple models since it can point out the possible structures which are consistent with certain given potentials. In practice, though, the lengthiness of the computations may prevent the practical testing of many different hypotheses regarding interaction potentials.

II.1.4. CONCLUSIONS

Most models are limited in scope because they treat only some of the observed features of water. They give partial descriptions of what water may be like /6/. The simple models of type (A) contain inherent approximations. Some of their errors may be compensated in part by the existence of adjustable numerical parameters. Good agreement with certain experimental data may mean sometimes that approximations and simplifications in the setting up of the model were hidden in the numerical values of the parameters. Thus the parameters provide good fitting of data, but their values themselves may not correspond to the physical quantity which they are supposed to represent. Thus they become effective parameters instead of structural ones. In such cases, the reality of some concepts underlying the models may be in doubt. Models of type (B) are thought to be fundamentally more accurate. However, they, too, contain approximations and simplifications. Thus they may suffer from the same defects. They are set up sometimes to evaluate exactly certain aspects of the behavior only (e.g. the hydrogen bond distribution in lattice statistical calculations) ; therefore, they cannot always be tested against experimental data.

Obviously, the molecules in the real liquid exist in many different states. The construction of a model usually implies that these states are lumped into a few representative groups which are characterized by average properties, or that only some properties of the molecules in a distribution of states are considered. Both are done in order to make a calculation feasible. On the other hand, many experimental methods do not supply detailed information on the various states of the molecules and on their distribution in these states either. The results of such measurements correspond to a different lumping or averaging of the properties of the molecules into limited classes. The way in which molecular states are lumped also differs from one model to the next. A simple model may correspond to the lumping in one particular technique (e.g. a spectroscopic one) but not to the lumping when a different physical property is measured. (Cf. the discussion of various kinds of structures and their detection in Sec. 4.1 of ref. /1/.) Thus one cannot expect that a simple model would describe successfully a very large variety of experimental measurements. If some models appear realistic, they are at best different partial reflections of the nature of a more complex model which corresponds to the real liquid. Of course, simple models which can be handled with relatively simple mathematical techniques are necessary and useful for providing limited correlations between various properties of water and of solutions.

REFERENCES

1 D.Eisenberg and W. Kauzmann, The Structure and Properties of Water, Oxford University Press, New York - Oxford, 1969.

2 G. Némethy, Ann. Ist. Super. Sanità (Rome), 6, 489 (1970).

3 H.S. Frank, Science, 169, 635 (1970).

4 G. Kell, in : Water and Aqueous Solutions (R.A. Horne, Ed.), p.331, Wiley-Interscience, New York - London, 1972.

5 C.M. Davis and J. Jarzynski, in : Water and Aqueous Solutions (R.A. Horne, Ed.), p. 377, Wiley-Interscience, New York - London, 1972.

6 H.S. Frank, in : Water - A Comprehensive Treatise (F. Franks, Ed.), Vol. 1, p.515, Plenum Press, New York - London, 1972.

7 A.T. Hagler, H.A. Scheraga, and G. Némethy, Ann. N.Y. Acad. Sci., 204, 51 (1973).

8 W. Kauzmann, Advan. Phys. Chem., 24, (1973).

9 H.S. Frank, this volume.

10 G. Némethy and H.A. Scheraga, J. Chem. Phys., 36, 3382 (1962).

11 A.T. Hagler, H.A. Scheraga, and G. Némethy, J. Phys. Chem., 76, 3229 (1972).

12 B. Lentz, A.T. Hagler, and H.A. Scheraga, to be published.

13 W. Luck, Advan. Molec. Relaxation Processes 3, 321 (1972).

14 C.M. Davis, Jr. and T.A. Litovitz, J. Chem. Phys., 42, 2563 (1965).

15 H.S. Frank and A.S. Quist, J. Chem. Phys., 34, 604 (1961).

16 O. Ya. Samoilov, Zh. Fiz. Khim., 20, 12 (1946).

17 A.H. Narten, M.D. Danford, and H.A. Levy, Discussions Faraday Soc., 43, 97 (1967).

18 J.A. Pople, Proc. Roy. Soc. (London), A205, 163 (1951).

19 J.D. Bernal, Proc. Roy. Soc. (London), A280, 299 (1964).

20 C.A. Angell, J. Phys. Chem., 75, 3698 (1971).

21 G.E. Walrafen, in : Hydrogen-Bonded Solvent Systems (A. K. Covington and P. Jones, Eds.), p. 9, Taylor and Francis Ltd., London, 1968.

22 M. Weissman and L. Blum, Trans. Faraday Soc., 64, 2605 (1968).

23 O. Weres and S.A. Rice, J. Amer. Chem. Soc., 94, 8983 (1972).

24 A. Ben-Naim, J. Chem. Phys., 57, 3605 (1972).

25 S. Levine and J.W. Perram, in : Hydrogen-Bonded Solvent Systems (A. K. Covington and P. Jones, Eds.), p.115, Taylor and Francis Ltd., London, 1968.

26 J.W. Perram and S. Levine, Mol. Phys., 21, 701 (1971).

27 G.M. Bell, Proc. Phys. Soc. London (Solid State Phys.), 5, 889 (1972).

28 J.H. Gibbs, C. Cohen, P.D. Fleming, III, and H. Porosoff, J. Solution Chem., 2 , 277 (1973).

29 A. Rahman and F.H. Stillinger, J. Chem. Phys., 55, 3336 (1971).

30 F.H. Stillinger and A. Rahman, J. Chem. Phys., 57, 1281 (1972).

31 M.S. Jhon, J. Grosh, T. Ree,and H. Eyring, J. Chem. Phys., 44, 1465 (1966).

32 M. Falk and T.A. Ford, Canad. J. Chem., 44, 1699 (1966).

33 H.S. Frank and W.-Y. Wen, Discussions Faraday Soc., 24, 133 (1957).

34 K. Morokuma and L. Pederson, J. Chem. Phys., 48, 3275 (1968).

35 J. Del Bene and J.A. Pople, J. Chem. Phys., 52, 4858 (1972).

36 D. Hankins, J.W. Moskowitz, and F. Stillinger, J. Chem. Phys., 53, 4544 (1970).

37 F.A. Momany, R.F. McGuire, J. F. Yan, and H.A. Scheraga, J. Phys. Chem., 74, 2424 (1970).

38 G.E. Walrafen, this volume.

39 W. Luck, this volume.

40 A.H. Narten and H.A. Levy, Science 167, 1521 (1970).

41 V. Vand and W.A. Senior, J. Chem. Phys., 43, 1869 (1965).

42 A. Ben-Naim and F.H. Stillinger, in : Water and Aqueous Solutions (R.A. Horne, Ed.), p. 295, Wiley-Interscience, New York - London, 1972.

43 F.H. Stillinger, personal communication.

II. 2. RECENT DEVELOPMENTS IN THE MOLECULAR THEORY OF LIQUID WATER

A. Ben-Naim*

Department of Inorganic and Analytical Chemistry
The Hebrew University of Jerusalem
Jerusalem, Israel

ABSTRACT

The molecular theory of liquid water has been developed along essentially two routes: the first, and the older one, is referred to as the modelistic approach. The second, more recent one, is often referred to as the *ab-initio* approach.

At present both approaches assume many simplifying approximations, without which little progress could have been achieved.

Some recent development along the two routes are surveyed and discussed. The main conclusion is that both methods provide complementary information on the molecular origin of the outstanding properties of liquid water as well as aqueous solutions.

* Present address: Laboratory of Molecular Biology, NIH, Bethesda, Maryland 20014 U.S.A.

II. 2.1 INTRODUCTION

The theoretical study of liquid water and aqueous solutions has been pursued along two main routes. The first, and the older one, views water as a mixture of several components, and then applies a simplified partition function (PF) for the system in order to compute various thermodynamic quantities. This approach has been referred to as the Mixture-Model (MM) approach, and we shall return to discuss its merits later.

The second, and more recent one, is often referred to as ab-initio approach /1/ and it is considered, by some authors to be a more fundamental approach, since it begins with an ensemble of water molecules as the fundamental entities of the system. The two routes are schematically shown in fig. 1.

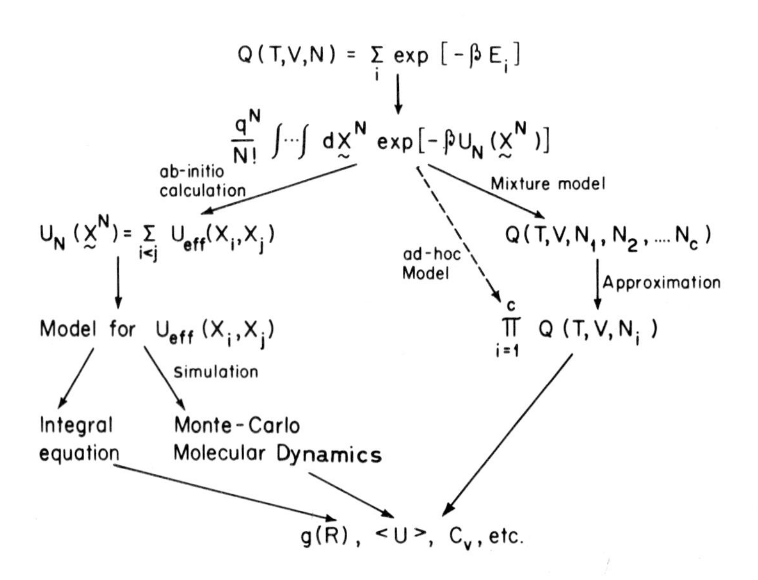

Fig. 1 Schematic illustration of the two main theoretical routes along which liquid water has been studied.

Q is the canonical partition function, $\beta = (kT)^{-1}$ with k the Boltzmann constant and T the absolute temperature X^N stands for the configuration of the whole system i.e. $X^N = X_1 \ldots X_N$, with X_i the configuration of the i-th molecule. $U_N(X^N)$ is the total potential energy of interaction of the system at the configuration X^N. $N_1 \ldots N_c$ are the numbers of molecules of each species in the mixture model approach. The left hand route of fig. 1 is often referred to as the ab-initio method. The right hand route is known as the modelistic approach, or the mixture-model approach. Both methods aim at computation of average properties such as the energy, heat capacity or pair distribution function. The purpose of this paper is to point out both the difficulties and the virtues of the two methods of studying liquid water. One of the main conclusions is that, at present, it is impossible to critically assess the various approximations introduced by each of the methods. Therefore one cannot establish an absolute preference for one approach over the other.

II. 2.2 AB INITIO CALCULATION

We consider first the *ab initio* route depicted on the left hand side of fig. 1. The most fundamental starting point should be the quantum mechanical partition function PF of the system. However, little can be done in the direction of progress with such a PF, and one immediately resorts to the so called classical version of the PF. This step invokes already an important assumption, namely the factorization of the internal PF of the single molecules from the configuration PF. Clearly, because of the strong intermolecular forces operating among the water molecules, one might expect that the internal degrees of freedom of a single molecule be dependent on the type of environment in which it is temporarily found.

The next important approximation, involves the assumption of pairwise additivity of the total energy of the system. Today, direct quantum mechanical calculations have established that the nonadditivity effects for a group of water molecules are important. Unfortunately neither the pair potential nor higher order potentials for water are known in any details and one must resort to a so called "effective pair potential". By this choice we already commit ourselves to working with a *model system*, obeying the pairwise additivity assumption, with this effective pair potential.

The next question is how to chose the pair potential itself. Although we know that we are going to study water molecules, we are still quite far from knowing the form of the pair potential operating between a pair of water molecules.

Furthermore, even if we knew the *exact* pair potential $U(\underset{\sim}{X}_1, \underset{\sim}{X}_2)$ for a *pair* of water molecules at the configuration $\underset{\sim}{X}_1$, $\underset{\sim}{X}_2$, this information is likely to be irrelevant to the study of *liquid water*. The reason is that in the liquid higher order potentials play an important role therefore if all these are to be accounted for by constructing an effective pair potential, then it is likely that the latter would not bear a great similarity to the bare pair potential.

The choice of an effective pair potential is essentially the same as adoption of a model for the water molecule. The extent of likeliness to real water molecules is not yet clear, and therefore all the results that one obtains, strictly pertain to this particular model, though some features of these may also be of relevance to real liquid water.

Finally it must be pointed out that even after we have agreed upon an effective pair potential, the execution of any computation of a thermodynamic quantity is usually a very time consuming project. This fact imposes a severe limitation on the accuracy of the results obtained by any of the presently available methods of statistical mechanics.

$$\rho^{(2)}(\underset{\sim}{R}_1, \underset{\sim}{R}_2) = \rho^2 g^{(2)}(\underset{\sim}{R}_1, \underset{\sim}{R}_2) \tag{1}$$

$$\rho^{(2)}(\underset{\sim}{R}_1, \underset{\sim}{\Omega}_1, \underset{\sim}{R}_2, \underset{\sim}{\Omega}_2) = \left(\frac{\rho}{8\pi^2}\right)^2 g^{(2)}(\underset{\sim}{R}_1, \underset{\sim}{\Omega}_1, \underset{\sim}{R}_2, \underset{\sim}{\Omega}_2) \tag{2}$$

$$g^{(2)}(\underset{\sim}{R}_1, \underset{\sim}{R}_2) = \frac{1}{(8\pi^2)^2}\int g^{(2)}(\underset{\sim}{R}_1, \underset{\sim}{\Omega}_1, \underset{\sim}{R}_2, \underset{\sim}{\Omega}_2)\, d\underset{\sim}{\Omega}_1\, d\underset{\sim}{\Omega}_2 \tag{3}$$

$$g(R) = 0 \qquad \text{for} \quad R < \sigma \tag{4}$$

$$g(R) \longrightarrow 1 \qquad \text{for} \quad R \longrightarrow \infty \qquad (R \approx 5\sigma) \tag{5}$$

Eq. (1) - (3) are basic relations between the pair distribution function $\rho^{(2)}$ and the pair correlation function $g^{(2)}$. The orientational averaged pair correlation function has the two fundamental properties eq. (4) and eq. (5).

Most of the present paper will be devoted to the various computations of the radial distribution function (RDF) for water. The relevant definitions are given in eq. (1) - (3). We note that, in principle, theory could provide the full angle dependent pair distribution function, however from experimental sources we can only get the angle-average <u>spatial</u> pair correlation function or the RDF.

Figure 2 depicts the RDF for water (4°C and 1 bar) and for argon (84.25°K and 0.71 bar) as a function of the reduced distance R/σ, where σ is the effective diameter of the molecule (σ = 2.82 Å for water and σ = 3.4 Å for argon). There are two important differences between water and argon. 1. The location of the second peak of RDF in argon is at about 2σ whereas in water it shifts leftward to $R \sim 1.6\sigma$. 2. The coordination number computed to, say the location of the first minimum, is about 10 in argon and 4.4 in water.

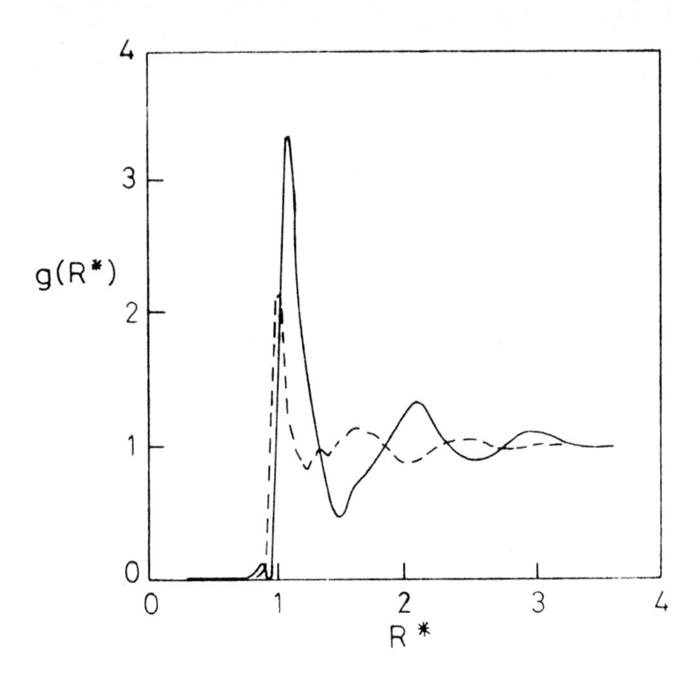

Fig. 2 The pair correlation function of argon (full curve) and water (dashed curve) as a function of the reduced distance R = R/σ. σ being the effective diameter of the molecules.

These two features manifest the strong directional forces operating between water molecules which dictates a particular mode of packing in the liquid. The situation is schematically described in figure 3.

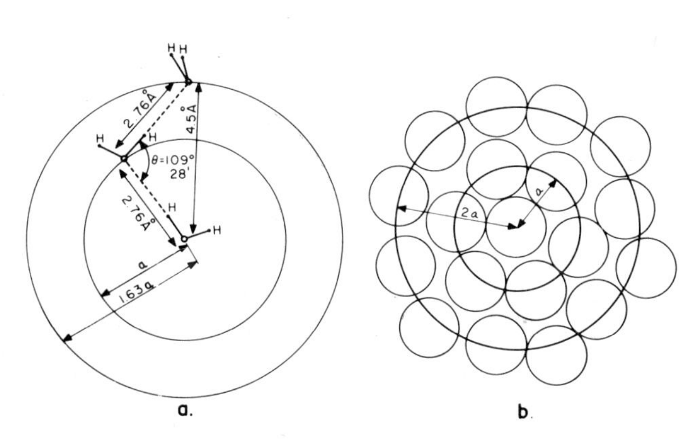

Fig. 3

A schematic comparison between the mode of packing of molecules in water (a) and in liquid consisting of simple spherical particles (b)

The most important conclusion that may be drawn from these findings is the following. If we were sitting at the center of a water molecule, and observing the "structure" of the immediate surroundings (say up to a distance of $R \sim 5.0$ Å from the center) then most of the time we would observe a situation similar to that of ice I. i.e. there are about four nearest neighbors and these are located at the vertices of a regular tetrahedron. It must be emphasized however that this information pertains only to the very local surroundings of a water molecule, one cannot reach any sound conclusions regarding the structure of the liquid beyond the first coordination shell from this data.

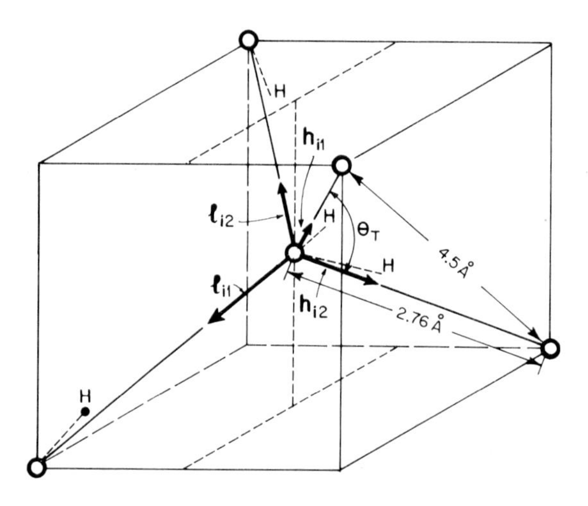

Fig. 4 Local geometry around a water molecule in ice. Hydrogen bonding occurs along four preferential directions as indicated by the vectors $\underset{\sim}{h}_{i1}$, $\underset{\sim}{h}_{i2}$, $\underset{\sim}{\ell}_{i1}$ and $\underset{\sim}{\ell}_{i2}$.

Figure 4 shows the basic tetrahedral geometry around a water molecule. One may define four unit vectors emanating from the center of the molecule and pointing toward the four vertices of a regular tetrahedron. Along these directions we know that there is a large probability of finding the center of another molecule at a distance of about $R \sim 2.8$ Å.

Accordingly we now proceed to construct a pair potential based on the above information. The simplest picture of a water molecule is the so called Bjerrum four-point-charge model, depicted in figure 5. The details of the construction of the potential function operating between a pair of such waterlike particles has been described by Ben-Naim and Stillinger /2/. The analytical form of this function for a pair of water molecules is shown in eq. (6) - (8) the various ingredients are schematically depicted in figures 6 and 7.

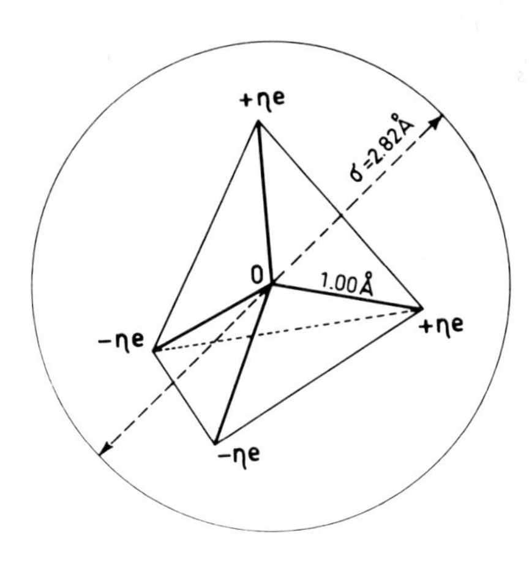

Fig. 5

Four point charges model used as a simple and a crude representative of a water molecule. The van-der-Waals diameter of the molecule is $\sigma = 2.82$ Å.

Pair Potential Function

$$U(1,2) = V_{LJ}(R_{12}) + S(R_{12})\, V_{HB}(1,2) \tag{6}$$

$$V_{LJ} = 4\varepsilon\left[\left(\frac{\sigma}{R_{12}}\right)^{12} - \left(\frac{\sigma}{R_{12}}\right)^{6}\right] \tag{7}$$

$\varepsilon = 7.21 \cdot 10^{-2}$ Kcal/mole

$\sigma = 2.82$ Å

$S(R_{12})$ is a switching function to consider the repulsion at close distance.

$$V_{HB}(1,2) = (\eta e)^2 \sum_{\alpha_1\alpha_2=1}^{4} \frac{(-1)^{\alpha_1+\alpha_2}}{d_{\alpha_1\alpha_2}(1,2)} \tag{8}$$

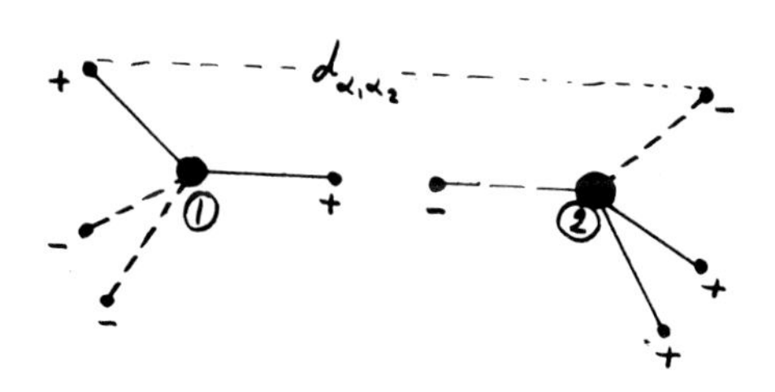

Fig. 6

The distances $d_{\alpha_2\alpha_1}$ are the distances between the point charge, α_1, on particle 1 and the point charge, α_2, on particle 2. The total electrostatic interaction between the sixteen pairs of point charges is denoted by V_{HB}.

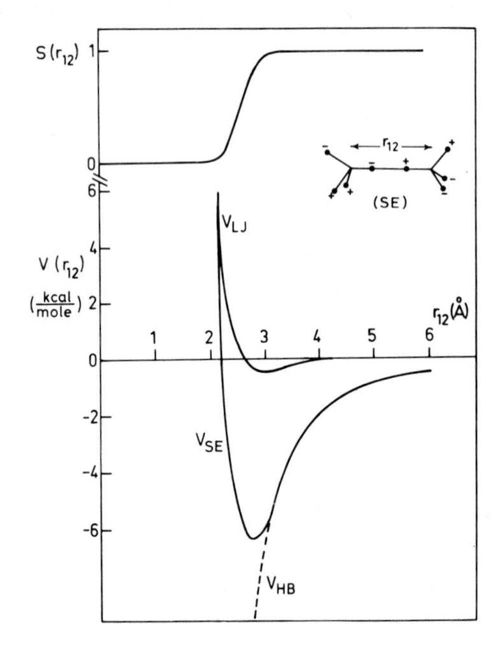

Fig. 7

The full pair potential V_{SE} and its three components V_{LJ}, V_{HB} and S as a function of the intermolecular separation for the symmetrical eclipsed configuration (SE) depicted at the upper right side of the figure.

Note that the strength of the "hydrogen bond" is of the order of 6 Kcal/mole compared with about 1/2 Kcal/mole for the interaction between nonpolar molecules.

In order to fix the various parameters of the model potential we have endeavored to obtain the experimental values of the second virial coefficient and its temperature dependence, this is shown in figure 8. It must be stressed that such an agreement may be obtained by a great variety of potential functions and this in itself does not lend any support to the validity of the effective pair potential. The only reason for doing that is to be sure that the parameters of the model were chosen within some reasonable limits.

A different form of the pair potential may constructed by a combination of Gaussian functions /3/ eq. (9) - (13). We shall return to a simpler version of this function later on.

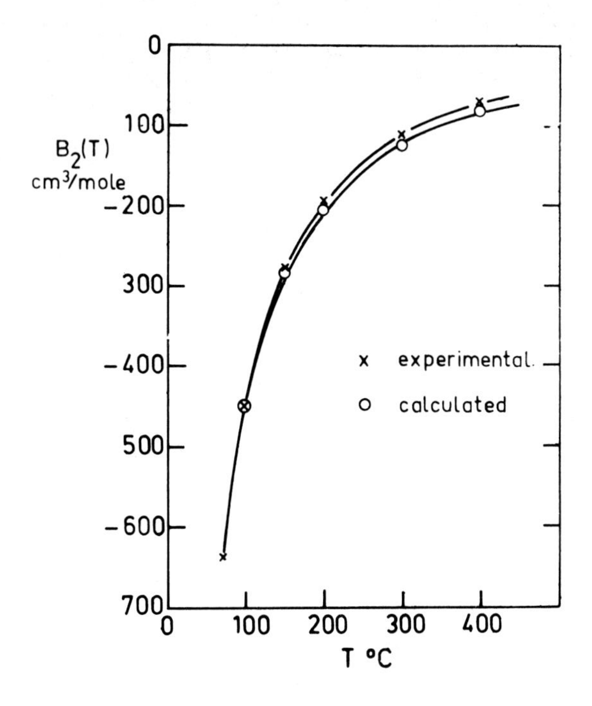

Fig. 8

The second virial coefficient for water as a function of temperature.

POTENTIAL FUNCTION

$$\underset{\sim}{X}_i = R_{ix}, R_{iy}, \alpha_i \tag{9}$$

$$U_{ij}(\underset{\sim}{X}_i, \underset{\sim}{X}_j) = U_{LJ}(R_{ij}) + U_{HB}(\underset{\sim}{X}_i, \underset{\sim}{X}_j) \tag{10}$$

$$U_{HB} = \varepsilon_H G(R-R_H) \sum_{k,l} G(\underset{\sim}{i}_k \cdot \underset{\sim}{n} - 1)\, G(\underset{\sim}{j}_i \cdot \underset{\sim}{n} + 1) \tag{11}$$

$$\underset{\sim}{n} = \underset{\sim}{R}_{ij} / |R_{ij}| \tag{12}$$

$$G(x) = \exp\left[-x^2/2\sigma^2\right] \tag{13}$$

Eq. (11) gives an alternative pair potential for a pair of water molecules (fig. 9). The hydrogen-bond part of the potential is built up of a product of three Gaussian functions governing the direction and orientation of the two molecules. (Fig. 9)
i_k and j_i are the unit vectors in the directions between the charges and the molecule center in the point charge model, $\underset{\sim}{n}$ direction of bond.

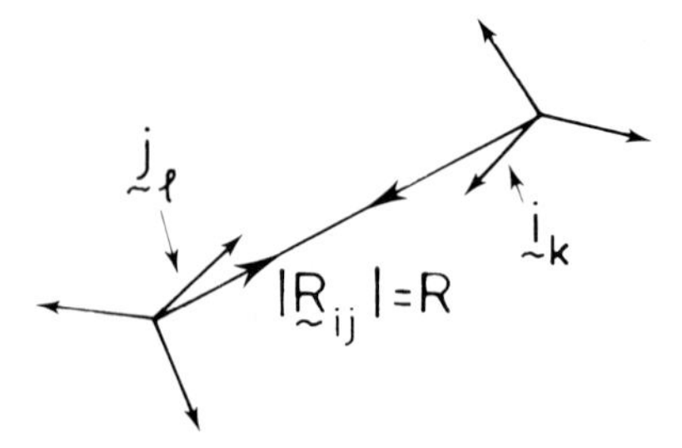

Fig.9
Relevant geometry for the "Gaussian" pair potential described in eqs. 9 - 13.

Let us summarize some of the results obtained by various methods of statistical mechanics.

Barker and Watts /4/ have computed g(R) for a sample of waterlike particles using the Monte Carlo technique. Figure 10 exhibits their results. We note that the characteristic peak at R = 4.6 Å does not appear in this function. In fact the location of the second peak at R ~ 6 Å is closer to the location of this peak for spherical molecules.

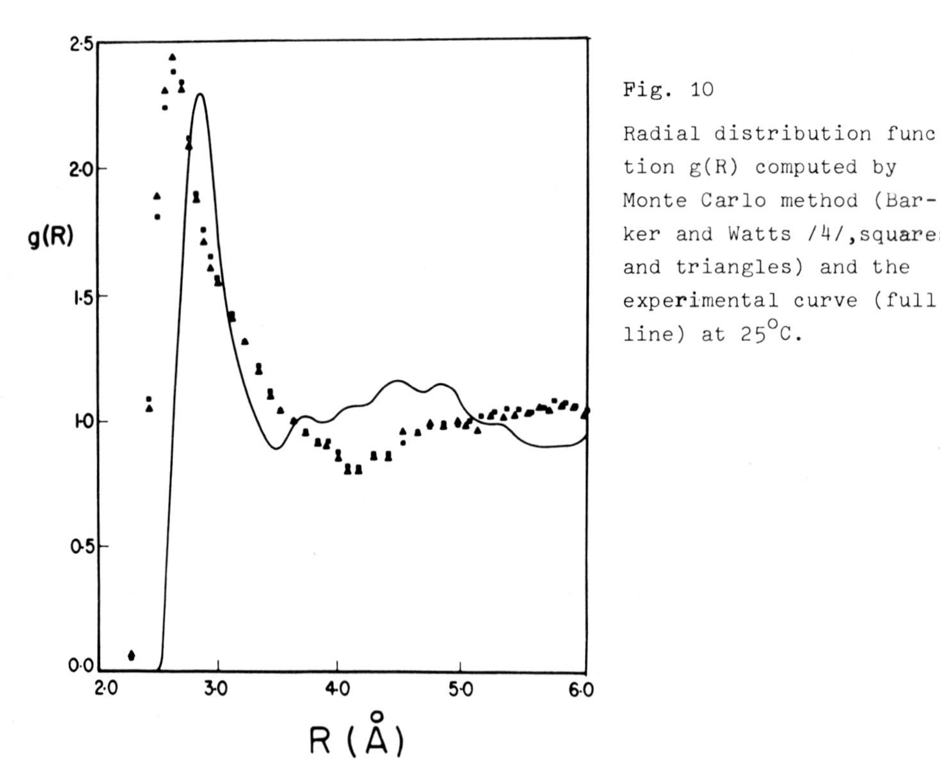

Fig. 10
Radial distribution function g(R) computed by Monte Carlo method (Barker and Watts /4/, squares and triangles) and the experimental curve (full line) at 25°C.

An attempt to apply the Percus-Yevick integral equation for water has been carried out by Ben-Naim /5/. The relevant equations are described in eq. (14) - (17) and the corresponding results in figure 11.

ORNSTEIN-ZERNICKE RELATION

$$h(1,2) = c(1,2) + \frac{\rho}{8\pi^2} \int h(1,3)\ c(2,3)\ d(3) \tag{14}$$

$$h(1,2) = g(1,2)-1 \tag{15}$$

h = total correlation function
c = direct correlation function

$$y(1,2) = g(1,2)\ e^{\beta U(1,2)} \tag{16}$$

$$\beta = \frac{1}{RT} \qquad g \longrightarrow e^{-\beta U} \quad \text{as} \quad \rho \longrightarrow 0$$

The Ornstein-Zernicke relation eq. (14) - (16), is a relation between the total and the direct correlation function. The Ornstein-Zernicke relation is a relation between the total correlation function, h, defined as h = g-1 and the direct correlation function c which is defined through this relation. The numbers 1, 2 and 3 stand for the configuration of the corresponding particles.

PERCUS-YEVICK APPROXIMATION

$$c(1,2) = y(1,2)\ f(1,2) \tag{17}$$

$$f(1,2) = e^{-\beta U(1,2)}-1 \tag{18}$$

$$y(1,2) = 1 + \frac{\rho}{8\pi^2}\int y(1,3)\ f(1,3)\ y(2,3)\ f(2,3)\ d(3) + \frac{\rho}{8\pi^2}\int y(1,3)\ f(1,3)\left[y(2,3)-1\right] d(3) \tag{19}$$

The Percus-Yevick equation is obtained by employing the approximation c = y f. The last line presents the Percus-Yevick integral equation for the function y, the pair correlation function.

The application of the socalled Percus-Yevick approximation eq.(17) - (19) $[c = y \cdot f]$ leads to the Percus-Yevick integral equation.

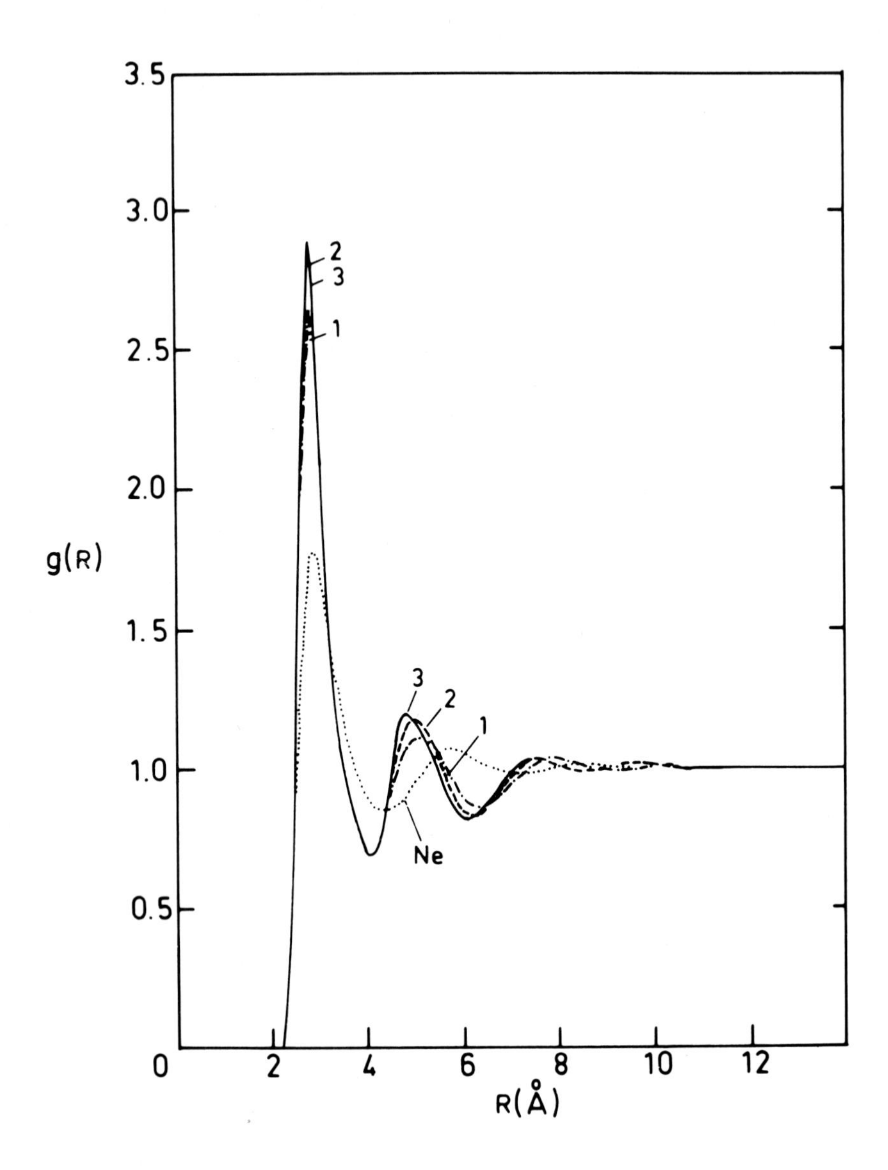

Fig. 11 The radial distribution function g(R) computed by the Percus-Yevick equation for neon (Ne) and for system of particles interacting via a pair potential defined in eq. (6)-(8).The numbers 1, 2, 3 indicate successively larger coupling of the angle-dependent part of the pair potential /5/ (V_{HB}).
The dotted curve is the solution of the Percus-Yevick equation for Neon, i.e. when the angular part of the pair potential is switched off.

The various curves in the figure 11 correspond to different coupling parameters λ which serve to couple the angel-dependent part of the potential function. The most important feature of this coupling process is the gradual shift of the second peak leftward For $\lambda \approx 0.3$ one gets a second peak at about R = 4.8 Å.

Because of the strong attractive potential, the Percus-Yevick equation seems to fail for $\lambda \geq 0.3$, and indeed convergent solutions are very difficult to obtain for larger values of λ.

Another technique that has been applied by Rahman and Stillinger /6/, /7/ is the Molecular Dynamics, Figure 12 reports their results for g(R) and its temperature dependence. One clearly observes the characteristic peak at R $\approx$ 4.6 Å = 1.63 σ and the coordination number at room temperature is about 5.

The monotonously increasing function of R is the average coordination number as a function of the upper limit of the integral

$$n_{CN}(R) = \int_0^R \rho g(r)\, 4\pi r^2\, dr \tag{20}$$

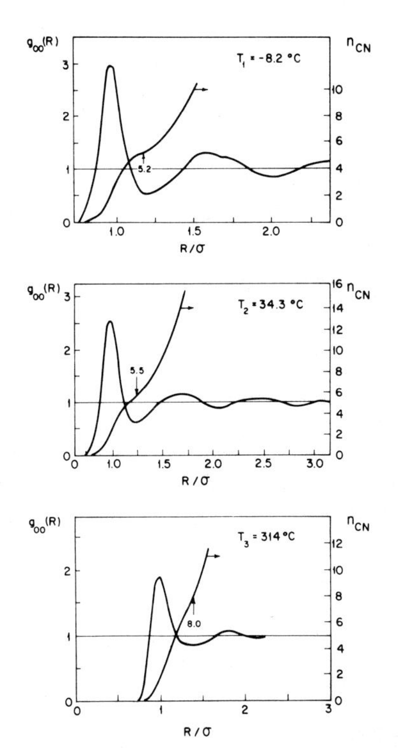

Fig. 12

Radial distribution function for water, computed by Rahman and Stillinger /6/, /7/ by the Molecular Dynamic method at three different temperatures.

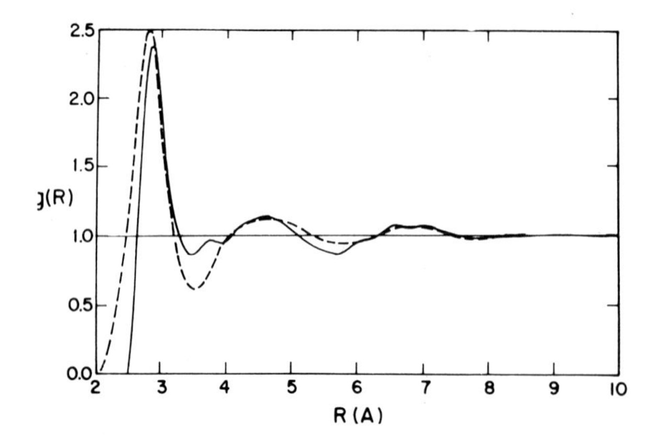

Fig. 13

Comparison of the computed radial distribution function (by Rahman and Stillinger /6/, /7/, fig. 12) with the experimental curve.

Figure 13 shows the computed and the experimental g(R). One sees that the agreement is quite satisfactory.

The comparison is made here between the experimental curve at 25°C and the computed curve corresponding to 34.3°C.

Recently a new method has been devised by Andersen to compute g(R) for a potential function similar to that described in eq. 11.(Instead of Gaussians one may use square well potential to simplify the computations). A preliminary result is shown in figure 14.

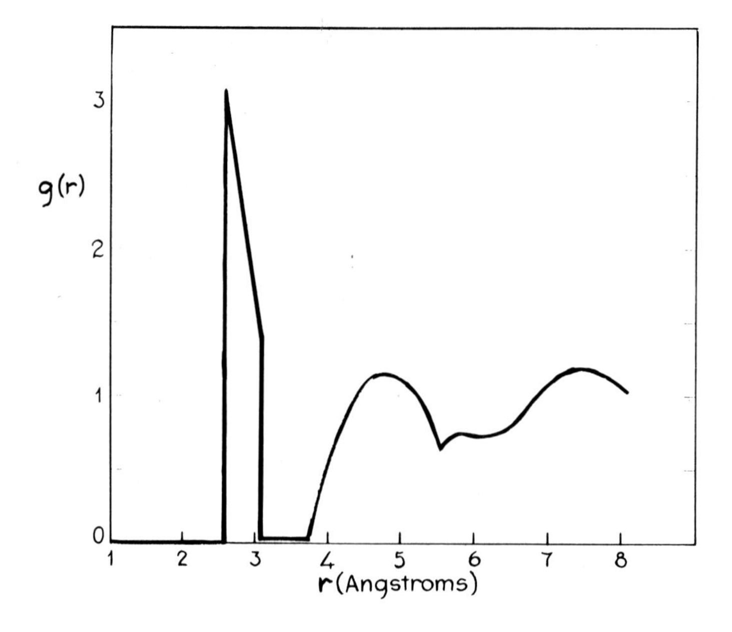

Fig. 14 Computed radial distribution function for water (Andersen/8/).

II. 2.3 MODELISTIC APPROACH

We now turn back to the second route, the so-called modelistic approach to the study of liquid water. In recent years a vigorous debate has been taking place regarding the validity of this approach. This approach raises a few questions which are of fundamental importance: 1) Can one view water as a mixture of various species? 2) is such an approach useful, and if so, in what sense? 3) Can one justify the factorization of the partition function into a product of terms each of which pertain to one of the species?

We shall try to answer, very briefly the first two questions (for more details the reader is referred to reference /9/). To do that we define the distribution function $x(\nu)$ in eq. 21, where $x(\nu)d\nu$ is the probability of finding a specific molecule having binding energy (BE) between ν and $\nu + d\nu$. By BE we mean the total interaction energy of a given molecule with the rest of the system (eq. 23)

$$x(\nu) = \int \ldots \int d\underset{\sim}{X}^N \, P(\underset{\sim}{X}^N)\, \delta\left[B_1(\underset{\sim}{X}^N) - \nu\right] \tag{21}$$

$$\int_{-\infty}^{\infty} x(\nu)d\nu = 1 \tag{22}$$

$P(X^N)$ is the fundamental distribution function for observing the configuration X^N of a system, say, in the canonical ensemble. δ is the Dirac delta function.

A definition of the distribution of particles is given according to their binding energy (BE), defined by B_1.

$$B_1(\underset{\sim}{X}^N) = \sum_{j=2}^{N} U(\underset{\sim}{X}_1, \underset{\sim}{X}_j) \tag{23}$$

where U is the pair potential.

$x(\nu)d$ is the mole fraction of particles with binding energy between ν and $\nu + d\nu$. A two structure model may be constructed from $x(\nu)$ to obtain the two mole-fractions x_1 and x_2

$$x_1 = \int_{-\infty}^{\nu^*} x(\nu)d\nu, \qquad x_2 = \int_{\nu^*}^{\infty} x(\nu)d\nu \tag{24}$$

We have computed the form of this function for a two dimensional system of water-like particles, interacting via a pair potential similar to the one shown in eq.(9)-(13). A sample of such particles computed by the Monto-Carlo method is depicted in figure 15. Clearly the

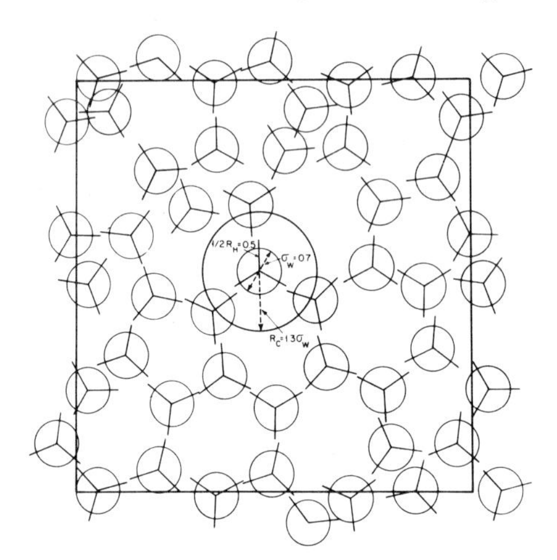

Fig. 15 A sample of "water-like" particles in two dimensions /9/. Each molecule is essentially a spherical disk interacting through a Lennard-Johns pair potential. In addition there are three selected directions along which "hydrogen bond" may be formed. This situation resembles the one for real water as indicated in fig. 9. The precise definition of the pair potential is given in eq. (9) - (13) (for more details see ref. /9/).

tendency to form "hydrogen-bonds" is the feature which makes the particles similar to water. Moreover we note here that regions of low local density are also characterized by strong binding energy. We stress in passing that this feature of the mode of packing of water molecules may be the most important one in the determination of the unique properties of both pure water and aqueous solutions /9/.

Figures 16 and 17 shows the distribution function $x(\nu)$ for spherical particles and its dependence of the total density-ρ, and on the energy parameter ε/kT (where ε is the energy parameter in the Lennard-Jones potential). We note that in all of the cases examined the function $x(\nu)$ for simple particles has been found to have only one peak. For the waterlike particles the situation is drastically different. The

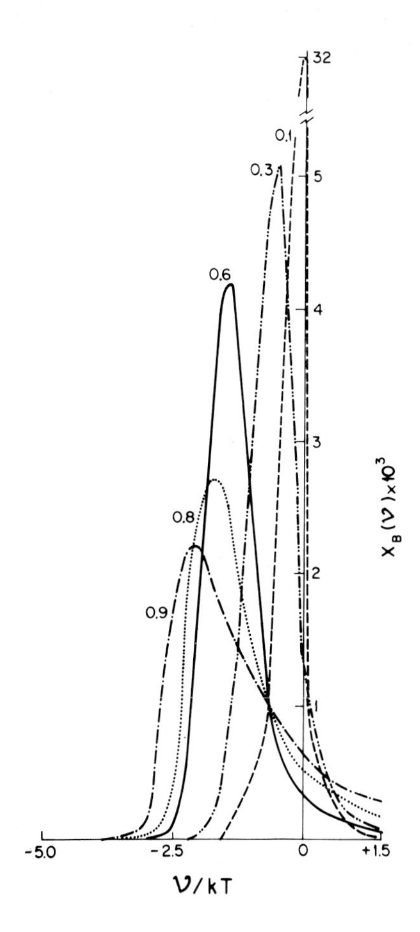

Fig. 16 The distribution function $X(\nu)$ for a system of spherical particles in two dimensions, interacting via Lennard-Jones pair potential. The total number densities ($\rho\sigma^2$) are indicated next to each curve.

function $x(\nu)$ shows various peaks corresponding to waterlike particles with zero, one, two and three bonds. The relevant results and the dependence on the hydrogen-bond energy is shown in figure 18. Consider for example the curve with $\varepsilon_H/kT = -5.0$. The locations of the minima ν_i^* may be used to define an exact mixture model (MM) for this system-i.e.

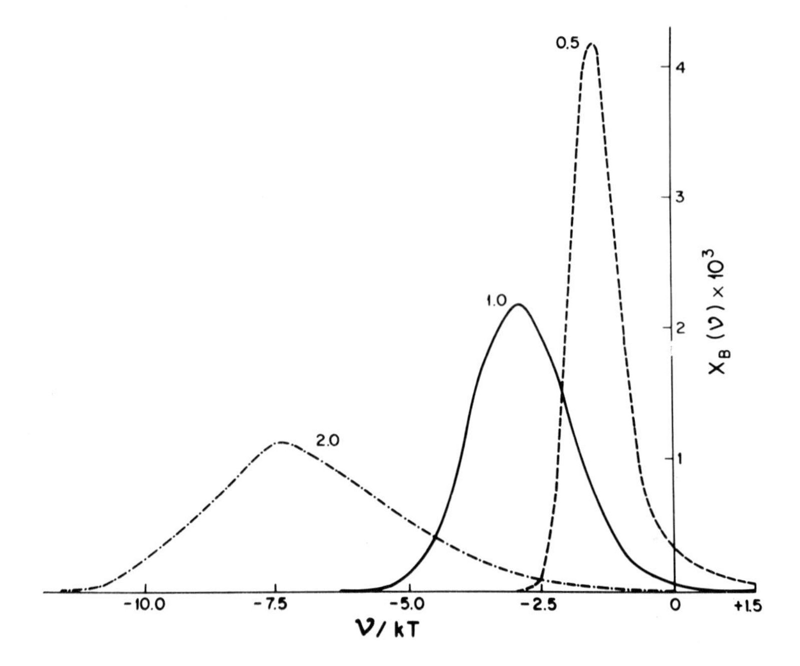

Fig. 17 The distribution function $X(\nu)$ for a system of spherical particles in two dimensions, interacting via Lennard-Jones pair potential. The energy parameter ε/kT is indicated next to each curve.

$$x_1 = \int_{-\infty}^{\nu_1^*} x(\nu)d\nu, \quad x_2 = \int_{\nu_1^*}^{\nu_2^*} x(\nu)d\nu, \quad x_3 = \int_{\nu_2^*}^{\nu_3^*} x(\nu)d\nu, \quad x_4 = \int_{\nu_3^*}^{\infty} x(\nu)d\nu \tag{25}$$

Other possibilities of constructing an exact MM for such a system exist /9/. Though we have demonstrated the point in two dimensions it is clear that the same function may be defined for a three dimensional system as well. Such a computation may reveal to what extent there exists a clear cut resolution between the various "components" defined in terms of their binding energies.

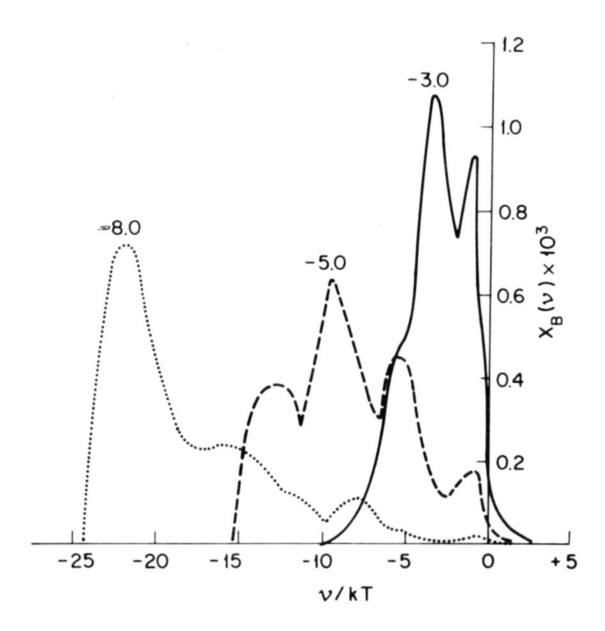

Fig. 18 The distribution function $x(\nu)$ for a system of water-like particles in two dimensions (fig. 15). The "hydrogen bond" energy ε_H/kT is indicated next to each curve. ε_H is the maximum strength of the interaction as defined in eq.(9)-(13). When a particle is fully bonded its interaction with its environment would be $3\varepsilon_H$ (excluding the Lennard-Jones interaction).

As a final illustration we turn to the "coupling" between binding energy and local density (the distribution function for coordination number $x_c(k)$ serves as a measure of the distribution of local densities). Figure 19 shows the (singlet) distribution function for the first coordination number $x_c(k)$. Figure 20 show the conditional distribution function of the binding energy BE for each coordination number. For spherical particles the behavior of this function is "normal", in the sense that a large coordination number correspond to stronger binding energy. The reverse is true for the waterlike particles as is demonstrated in figure 21. We believe that this feature is relevant to real water as well, namely molecules with a strong binding energy are likely to be found in regions of low local density.

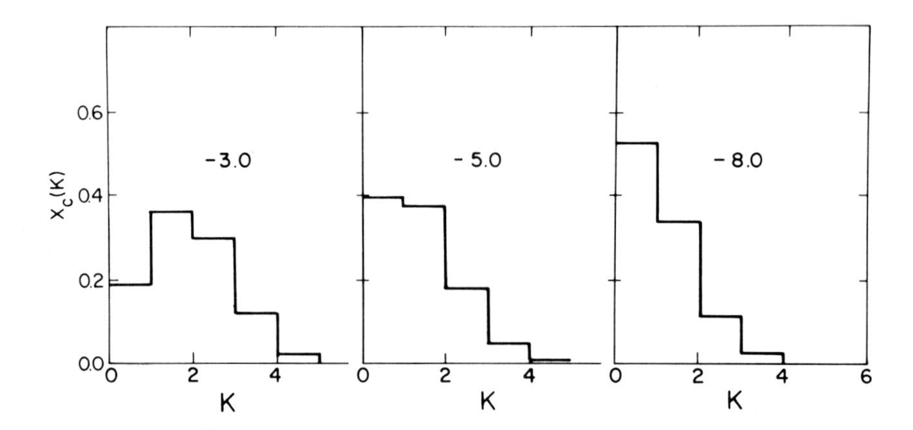

Fig. 19 Distribution of coordination number $x_c(K)$ for a system of water-like particles as in figure 18. Note that as the strength of the hydrogen bond energies increases from $\varepsilon_H/kT = -3$ to $\varepsilon_H/kT = -80$ the coordination number distribution function shifts to the left. This is an important feature of the pair potential, which resembles liquid water. Namely as temperature decreases (i.e. $|\varepsilon/kT|$ increases) the local density around the particles becomes smaller. Which means an open mode of packing of water molecules.

Regarding the question of the usefulness of the mixture model approach we should like to present one conclusion that has emerged from a study of the thermodynamics of a system viewed as a mixture of several components. For simplicity we consider here only a two-component system.

One finds that in order that such a model be useful, one must have two well separated peaks in $x(\nu)$ in such a way that the area under the two peaks are of comparable magnitude (a more detailed analysis is given in reference /9/).

As a conclusion to the whole talk we may say that at present each method is highly modelistic in character and involves quite drastic approximations. The two routes seem to furnish complementary information on water on both the molecular and on the macroscopic.

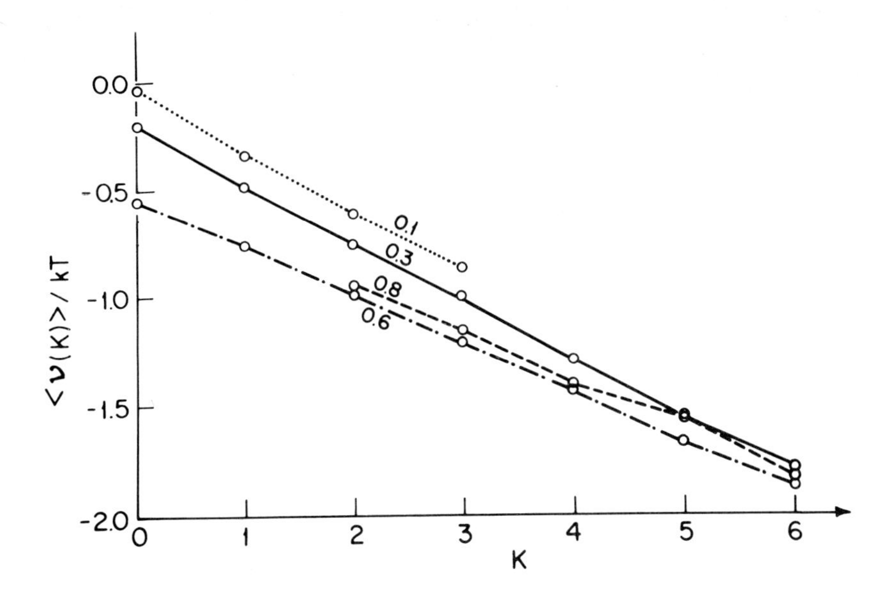

Fig. 20 The conditional average binding energy of particles having a fixed coordination number. The curves pertain to spherical particles in two dinensions. The total density ($\rho\sigma^2$) is indicated next to each curve.

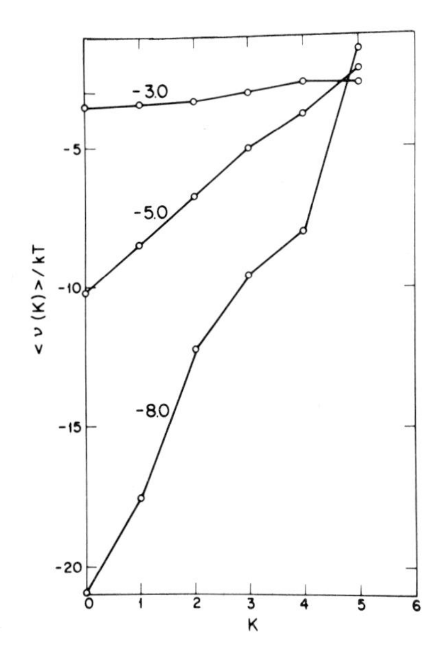

Fig. 21

The conditional average binding energy of water-like particles in two dimensions, having a fixed coordination number. The "hydrogen bond" energy (ε_H of eq. 11) is indicate next to each curve.

REFERENCES

1 N. H. Fletcher, Reports on Progress in Physics 34, 913 (1972).

2 A. Ben-Naim and F. H. Stillinger: Water and Aqueous Solutions, ed. by R. A. Horne, Wiley, New York 1972.

3 A. Ben-Naim: Water, a Comprehensive Treatise, vol. I, ed. by F. Franks, Plenum Publ. 1972.

4 J. A. Barker and R. O. Watts, Chem. Phys. Letters 3, 144 (1969).

5 A. Ben-Naim, J. Chem. Phys. 52, 5531 (1970).

6 A. Rahman and F. H. Stillinger, J. Chem. Phys. 55, 3336 (1971).

7 F. H. Stillinger and A. Rahman, J. Chem. Phys. 57, 1281 (1972).

8 H. C. Andersen, Preprint.

9 A. Ben-Naim: An Introduction to a Molecular Theory of Water and Aqueous Solutions, Plenum Publ. (in press 1974).

II.3. STRUCTURE AND PROPERTIES OF THE WATER MOLECULE

Peter Schuster

Institut für Theoretische Chemie, Universität Wien
Währingerstraße 17, A-1090 Wien, Austria

ABSTRACT

The structure, the energies and the other most important properties of the isolated water molecule are obtained by a straightforward application of quantum mechanics. The theoretical results are compared with the available experimental data and in this way a critical summary of the most reliable values is presented. The success of more or less extended quantum mechanical methods in the calculation of various observable quantities is discussed.

II.3.1. INTRODUCTION

Water plays a predominant role in chemistry, biochemistry and biology. The unique and interesting properties of this molecule and its aggregates in the gaseous, liquid and solid state have fascinated theorists from the very beginning of theoretical chemistry. Hence all kinds of quantum mechanical methods were applied to describe the properties of the water molecule. During the last ten years an enormous progress in the facilities for numerical calculations was made. Together with the computers the numerical methods were improved and nowadays it became possible to calculate almost all the important quantities of small molecules by purely theoretical "ab initio" methods. In many cases the calculated values are of comparable or even higher accuracy than the experimental data. In the last four years two extensive books on water were published. They both contain a chapter on the properties of the isolated water molecule (1,2). Quite recently a number of very extended calculations on the water molecule were performed (3-8). In this contribution only a short outline of the material summarized before (1,2) is presented. The main emphasis is laid on the most recent theoretical papers.

The time dependent Schrödinger equation is the starting-point for any quantum mechanical ab initio calculation of molecular properties. The basic assumption of molecular spectroscopy uses the fact that the mass of nuclei is more than three powers of magnitude larger than the mass of the electron. Hence nuclear motion is much slower than electronic motion. Using the approximation of infinitely slow nuclear motion, which is well founded in the most cases, the Schrödinger equation can be separated into two differential equations, one for the electronic motion and the other describing the motion of the nuclei (table 1). As long as the individual electronic states have sufficiently different energies the off-diagonal coupling terms between electronic and nuclear motion are not important. Depending on the consideration of the diagonal coupling terms the whole procedure is called either adiabatic or Born-Oppenheimer approximation. In the first case the diagonal terms are taken into account, in the latter case they are neglected[+)].

Since we are primarily interested in stationary states of molecules and hence our Hamiltonian is no explicit function of time, the space and time dependent factors of the wave functions can be separated. Now we are left with the problem to solve the stationary

[+)] For a more detailed discussion see (9,10).

Table 1

The Schrödinger equation for stationary states of the water molecule

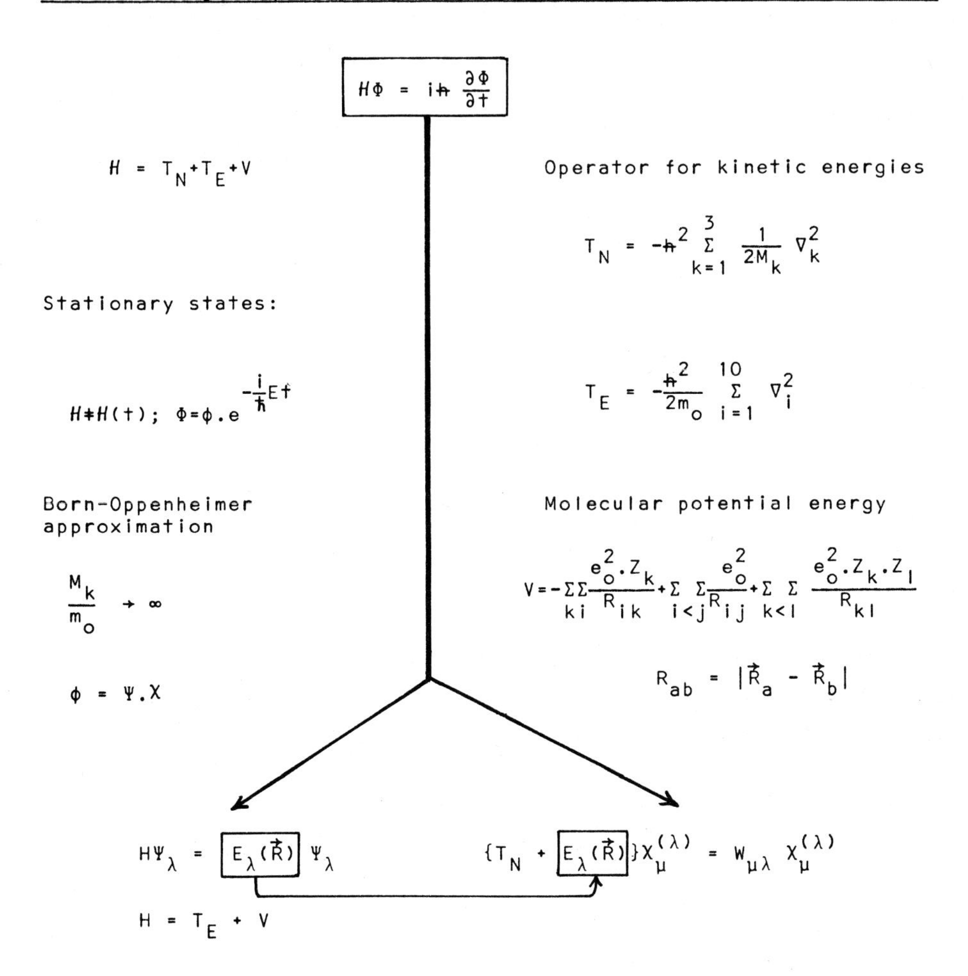

$E_\lambda(\vec{R})$: Energy hyper surface; $W_{\mu\lambda}$: Molecular energy levels
M_k, Z_k, $\vec{R}_k$: Mass, charge and position of nucleus "k"
$\vec{R} = (\vec{R}_1, \vec{R}_2 \; . \; . \; . \; \vec{R}_N)$: Molecular geometry
$m_o, e_o, \vec{R}_i$: Mass of the electron, elementary charge and position of the electron "i"

Schrödinger equations for electronic and nuclear motion. The electronic eigenvalue equation has to be solved first, because the energy hypersurface $E(\vec{R})$ enters as potential energy into the Schrödinger equation for nuclear motion. In principle the eigenvalues and the eigenfunctions of both differential equations contain all information for a full and detailed description of the isolated molecules.

II.3.2. THE ELECTRONIC EQUATION

For all molecules besides the hydrogen molecular ion (H_2^+) no analytical solution of the electronic Schrödinger equation has been found. Various approximative methods were developed, but unfortunately all these methods involve extensive numerical calculations. Hence, the energy surface $E(\vec{R})$ has to be calculated point by point by a tedious and time consuming numerical procedure. Among all the methods, which have been proposed during the development of quantum chemistry, the LCAO-MO formalism[+)] turned out to be most efficient for numerical calculations on small molecules consisting of light atoms (atoms approximately up to Cl or K).

Nowadays two levels of approximation are commonly used for accurate LCAO-MO calculations. At the lower level of accuracy, well known as Hartree-Fock (HF), self consistent field (SCF) or independent particle approximation, the molecular wave function is represented by a single determinant (Slater determinant), which is built up from one electron MO's (table 2). These one electron orbitals describe the independent motions of single electrons. The correlation between the motions of the individual electrons is neglected at this stage of sophistication. Many molecular properties can be approximated directly from the knowledge of the one electron orbitals. According to Koopmans' theorem (11) the HF eigenvalues are approximately equal to the ionization potentials of the molecule. All expectation values of one electron operators can be split into a sum of individual orbital contributions within the HF scheme.

The one electron MO's ψ_k obtained by the iterative HF procedure are commonly called canonical MO's (fig.1). These orbitals diagonalize the HF matrix, they are delocalized and each of them belongs to a particular irreducible representation of the symmetry group of the

[+)] Linear Combination of Atomic Orbitals to Molecular Orbitals. Most frequently Gaussian Orbitals (GTO's) or Slater Orbitals (STO's) are used as basis sets.

Table 2

Solution of electronic Schrödinger equation:

LCAO-MO method

One electron molecular orbitals: $\psi_k = \sum_{i=1}^{m} C_{ik}\phi_i$

Basis set: $(\phi_1, \phi_2 \ldots \phi_m)$

Minimal basis set: Inner shell and valence orbitals; schematic notation (as,bp,cd,df/es,fp,gd) means a s-type, b p-type c d-type and d f-type functions at the oxygen atom and e s-type, f p-type and g d-type at the hydrogen atom.

Variational coefficients: C_{ik}

Spin orbitals: $\psi_k = \psi_k\alpha$, $\overline{\psi}_k = \psi_k\beta$

α,β are spin functions for the m_s values $\pm\frac{1}{2}$ rsp.

HF many electron wave function (for 2n electrons)

$$\Psi = \frac{1}{\sqrt{2n!}} \begin{vmatrix} \psi_1(1)\psi_1(2)\psi_1(3) & \ldots & \psi_1(2n) \\ \overline{\psi}_1(1)\overline{\psi}_1(2)\overline{\psi}_1(3) & \ldots & \overline{\psi}_1(2n) \\ \vdots & & \vdots \\ \overline{\psi}_n(1)\overline{\psi}_n(2)\overline{\psi}_n(3) & & \overline{\psi}_n(2n) \end{vmatrix} \quad ; \quad m>n$$

water molecule: n=5

HF equations: $F\,\psi_k = \varepsilon_k\,\psi_k$; ε_k ... HF eigenvalue

$$F(1) = -\frac{1}{2m_o}\cdot\nabla_1^2 - \sum_{k=1}^{3}\frac{e_o^2 Z_k}{R_{1k}} + \sum_{j=1}^{10}(J_j(1)-K_j(1))$$

$$J_j(1)\psi_k(1) = \{e_o^2\int\frac{\psi_j^+(2)\psi_j(2)}{R_{12}}d\tau_2\}\,\psi_k(1)$$

$$K_j(1)\psi_k(1) = \{e_o^2\int\frac{\psi_j^+(2)\psi_k(2)}{R_{12}}d\tau_2\}\,\psi_j(1)$$

Configuration interaction:

$\underline{\underline{\Psi}}_l = \sum_{k=1}^{n}\gamma_{kl}\Psi_k$; Ψ_k . . . Slater determinants

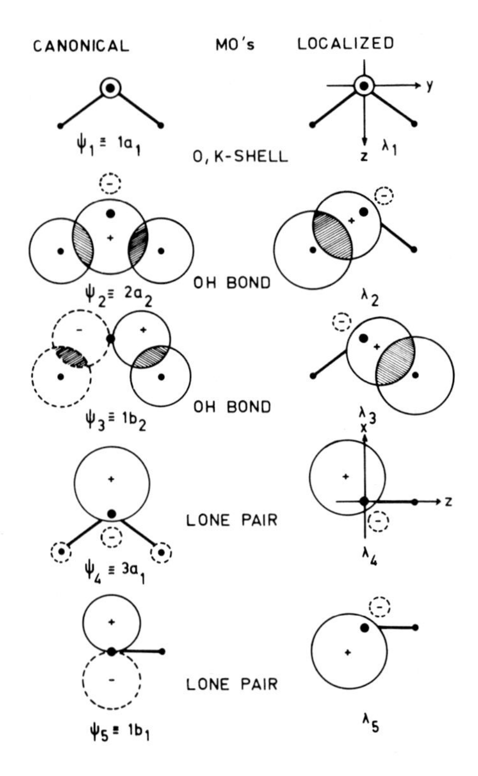

Fig. 1: Canonical and localized one electron MO's of the water molecule (schematicly).

water molecule (C_{2v}). The canonical orbitals can be used directly in various approximate methods to calculate electronic excitation or ionization spectra. The chemists, in contrary, are accustomed to think in terms of bonds and lone pairs. The canonical orbitals can be converted into a new set of localized orbitals λ_k by a unitary transformation (30). This transformation, however, does not change the many electron determinantal wave function. Accordingly all observable quantities remain the same. These localized orbitals correspond very well to the idea of molecules built up from inner shell, bond and lone pair orbitals (fig.1) and can be used for comparison of analogous parts in different molecules. The localized orbitals are also useful for the construction of simple electrostatic models simulating the molecular potential. Usually the individual localized orbitals have local symmetry, which does not coincide with the symmetry of the whole molecule. For a more detailed discussion of the shape of one electron MO's see ref.(2).

The difference between the lowest energy value, that can be obtained within the independent particle approximation using a sufficiently large basis set (HF-limit), and the exact non relativistic energy is called correlation energy. In principle the exact solution of the non relativistic Schrödinger equation can be constructed as a linear combination of an infinite number of determinants provided the atomic basis set was large enough (table 2). However, this expansion of the wave function commonly called configuration interaction (CI) shows extremely slow convergence and in general is far away from being an economic procedure. Many different methods for the calculation of correlation energies have been proposed. At the present stage two methods starting with pairwise correlation of electron motions seem to be most economic for very accurate calculations on many center, many electron molecules. Both methods use transformed MO's the so called pseudo natural orbitals (PNO's;12,13). The first method stops at the level of pairwise electron correlation and hence is named characteristicly Independent Electron Pair Approximation (IEPA;14,15). In the second method, the Coupled Electron Pair Approximation (CEPA;16) configuration interaction between different PNO's is included.

In the many determinantal approach, the one elctron MO's have lost their physical meaning. In order to find out, if the HF orbitals have still a reasonable degree of physical reliability, the HF one electron energies are compared with the vertical ionization energies calculated by more sophisticated methods or determined experimentally by photoelectron spectroscopy and ESCA (tabie 3). Especially the HF eigenvalues

Table 3

Calculated and experimental vertical ionization potential of the water molecule (4)

Ionized state			Vertical ionization potential (eV)			
Symmetry	half occupied+) orbital "ν"	$-\varepsilon_\nu^{SCF}$ (eV)	I_ν^{SCF}	I_ν^{IEPA}	I_ν^{CEPA}	I_ν^{EXP}
B_1	$1b_1$	13.86	11.10	12.92	12.48	12.62
A_1	$3a_1$	15.87	13.32	15.09	14.68	14.73
B_2	$1b_2$	19.50	17.59	19.19	18.85	18.55
A_1	$2a_1$	36.77	34.22	32.48	32.35	32.19
A_1	$1a_1$	559.50	539.13	539.82	539.62	539.73

+)The shape of the individual MO's is shown schematicly in fig.1.

of the valence orbitals agree very well with the more exact values and experimental data. However, regarding the most important implicit errors of Koopmans' theorem more closely, we find that this agreement is brought about by error compensation. At first Koopmans' theorem neglects electron relaxation which occurs, when one electron is removed during the ionization process. This error can be corrected within the HF theory by an independent calculation of the ionized state. The ionization energy is then calculated as the difference in energy between the ion and the ground state of the water molecule. For the removal of an electron from the orbital ψ_ν we obtain:

$$I_\nu(H_2O) = E_{(5-\nu)}(H_2O^+) - E_o(H_2O) \sim -\varepsilon_\nu$$

The second error of Koopmans' theorem originates from the neglect of the difference in correlation energy between the ion and the molecule. In the case of valence orbitals the contributions of electron relaxation and electron correlation to the ionization energy are of similar size but opposite sign. The contribution of electron relaxation to the ionization from lower lying orbitals, however, exceeds the contribution of electron correlation and hence the agreement between the HF eigenvalue and the ionization energy is worse. Nevertheless, the one electron energies represent the ionization energies fairly well within certain error limits. The use of one electron MO's is therefore

Table 4

Molecular geometries of H_2O^+ in various states and adiabatic ionization energies (4)

	Symmetry+)	Geometry (SCF) R_{OH} (Å)	α (°)	Ionization energies (eV) I_{vert}^{CEPA}	I_{adiab}^{CEPA}	I_{adiab}^{exp}
H_2O	A_1	0.941	106.4	-	-	-
H_2O^+	$B_1(1b_1)$	0.978	112.5	12.48	12.31	12.62
	$A_1(3a_1)$	0.976	180	14.68	13.40	13.69
	$B_2(1b_2)$	1.101	57.5	18.85	16.49	17.22

+) The half occupied MO is given in paranthesis. The shape of the individual MO's is shown schematicly in fig.1.

well justified at least for a more qualitative discussion of molecular properties. A comparison of the HF, IEPA and CEPA or experimental ionization energies in table 3 shows, that the error of the HF method is overcompensated by the IEPA approach. The CEPA results on the other hand agree very well with the data from photoionization spectroscopy and ESCA.

The equilibrium geometry of the ionized state, however, does not coincide with the geometry of the ground state. Depending on the extent of nuclear relaxation, the vertical ionization energies differ from the adiabatic ionization energies (table 4). As expected, a removal of an electron from the π lone pair orbital ($1b_1$) causes the least change in the geometry of H_2O and hence the vertical and the adiabatic ionization energies are almost identical. On the other hand ionization of σ-electrons leads to appreciable nuclear relaxation.

The excited electronic states of the water molecule were discussed in a previous review article (2) and the spectroscopic data were summarized by Herzberg (26). The interested reader is refered to these two books.

Besides LCAO-MO methods multiple scattering theory has been applied to the water molecule too. Recently ground state properties (37), ionization energies (37) and electron excitation spectra (38) were calculated. At the present stage of development the ground state properties of water are reproduced much better by extended LCAO-MO calculations than by the MSXα-method (for comparison the authors in ref.(37) performed an LCAO-MO calculation using a very poor minimal basis set - GTO: (7s,3p/3s) contracted to (2s,1p/1s) - and hence came to a different conclusion). On the other hand the Rydberg states of the water molecule (38) are reproduced quite well by the MSXα-calculation.

Most excitation energies agree with the experimental values within 0.5 eV. Only for one band a difference of 1.4 eV was found. In the case of the water molecule the MSXα-method, however, has lost its ab initio character, since scaled "empty" spheres have to be introduced into the "muffin-tin" potential in order to describe the lone pairs. Actually, the critical problem of the present multiple scattering calculations seems to be the construction of molecular potentials and electron densities by the "muffin-tin" approximation. Very recent calculations try to avoid this crucial point and we can expect an interesting development in this field in the near future.

II.3.3. ENERGIES AND DERIVED QUANTITIES

In principle the energy hypersurface of the water molecule is of nine dimensions, since there are nine degrees of freedom for nuclear motion. Six degrees of freedom can be separated from the remaining three if they are chosen in such a way, that they describe the translation and rotation of the water molecule as a rigid entity, The coupling between molecular rotation and the residual three internal degrees of freedom is rather complicated. In the case of the water molecule the coupling terms are not negligible if a fairly correct description of the rotational and vibrational spectra is desired. For a detailed discussion the reader is refered to (2) and the references given therein.

The energy surface of the water molecule is thus reduced to the three internal coordinates, which are appropriately chosen as R_1, R_2 and α (fig.2). For short we represent the three independent variables by vectorial notation: $\vec{R} = (R_1,R_2,\alpha)$. The first result obtained directly from the energy surface of a particular electronic state $E_k(\vec{R})$ is the equilibrium geometry of the molecule ($\vec{R}_e$). The ground state geometry of the water molecule is thus determined by the three equations:

$$\left(\frac{\partial E_o}{\partial R_1} = 0,\ \frac{\partial E_o}{\partial R_2} = 0,\ \frac{\partial E_o}{\partial \alpha} = 0\right) \longrightarrow \vec{R}_e = (R_1^e, R_2^e, \alpha_e)$$

From the equilibrium geometry the moments of inertia can be calculated without further difficulty. Hence the most important quantity for the rotational spectrum is obtained directly from the energy surface. In table 5 calculated and experimental equilibrium geometries of H_2O, H_3O^+ and OH^- are compared. In general the equilibrium geometries calculated by the SCF method agree well with the experimental values. However, there are some distinct differences, which can be removed when correlation effects are taken into account. As in the case of ionization energies the IEPA method trends to overcompensate the errors of the independent particle approximation.

The equilibrium geometries of molecules built up from different isotopes are expected to be the same as long as the Born-Oppenheimer approximation is valid. Hence, a comparison of the experimental $\vec{R}_e$ values in H_2O, HDO and D_2O obtained from microwave spectra provides a proof of the validity of the Born-Oppenheimer approximation for the water molecule in its ground state. As table 10 shows, the equilibrium geometries of the three molecules are practically identical.

Table 5

Calculated and experimental equilibrium geometries of H_2O, H_3O^+ and OH^-

Molecule	Symmetry	SCF R_e(Å)	SCF α(°)	IEPA R_e(Å)	IEPA α(°)	EXP. R_e(Å)	EXP. α(°)
OH^-	$C_{\infty v}$	.942[a]	-	.974[b]	-	.970[a]	-
H_2O	C_{2v}	.941[c]	106.6[c]	.973[b]	103.3[b]	.958[d]	104.5[d]
		.944[e]	105.3[e]	.960[e]	103.8[e]	-	-
H_3O^+	C_{3v}	.959[e]	113.8[e]	.972[e]	112.3[e]	-	-
		.960[f]	115.1[f]	.983[f]	111.6[f]	-	-

a) See ref.(19) b) See ref.(7) c) See ref.(6) d) See ref.(12)

e) SCF and SCF-CI calculations by Roos and Diercksen using 3304 configurations for H_2O and 10010 configurations for H_3O^+, ref.(18)

f) See ref. (17)

The total energy of the water molecule in its ground state, $E_o(\vec{R}_e)$, represents the energy for complete separation of all electrons and nuclei. Therefore $E_o(\vec{R}_e)$ is a very large quantity. A comparison of the various contribution to the experimental total energy with the results of LCAO-MO calculations is given in table 6. The correlation energy is only about 0.5% of the total energy and hence the independent particle approximation seems to be well justified. However, regarding the absolute value ($\Delta E_{corr}(H_2O)$ = 229.4 kcal/mole) more closely, we find, that the largest energy changes in chemical reactions, e.g. the binding energy of H_2O, are of the same order of magnitude as correlation effects. Hence, electron correlation has to be taken into account if accurate values for energy differences are desired.

For a discussion of the properties of water a detailed knowledge of dissociation and proton transfer energies is inevitable. In table 7 the most important calculated and experimental values are compared. A general shortcoming of the HF method is, that it fails completely to describe the energy curve for the dissociation of a covalent bond. The one determinantal wave function is not split correctly into the wave

Table 6

Partitioning of the total energy of the water molecule at the experimental equilibrium geometry (R_e=0.9572 Å, α_e=104.52°)[a]

Exp. energy of separated atoms (2H+O)	-76.1101 a.e.u.[b]
Binding energy	- 0.3496
Zeropoint energy	- 0.0211
Exp. total energy	-76.4808
Relativistic energy correlation	+ 0.0494
Exp.non relativistic total energy	-76.4314
HF energy limit	-76.0659[c]
LCAO-MO-CI energy	-76.283 [d]
	-76.3683[e]
CEPA energy	-76.3829[b]
IEPA energy	-76.4208[f]

[a] Geometry taken from (23)

[b] all values are taken from (4) unless stated otherwise
1 a.e.u. = 627.6 kcal/mole

[c] Lowest SCF energy, see ref.(5)

[d] LCAO-MO-CI calculation including 3304 configurations (18)

[e] Lowest variational energy obtained by CI based on PNO's (4)

[f] Because of additional model assumptions the IEPA and CEPA energies are not obtained by a straight-forward application of the variational principle and hence are no upper limits to the exact value.

functions of the two radicals at infinite separation. Usually the binding energies are not reproduced correctly by HF calculations too. In case of the water molecule only 70% of the binding energy are obtained by SCF calculations at the HF limit (table 7). Correlation effects do not contribute very much to proton affinities (ΔE_{PA}) or

$$\Delta E_{PA}(H_2O) = E_o(H_3O^+) - E_o(H_2O)$$

to energies for proton transfer. The HF values are of reasonable accuracy. Previously it was found, however, that for a calculation of reliable proton affinities large basis sets leading to energies near the HF limit are necessary (21,22). Small basis sets or less rigorous LCAO-MO methods produce large errors in this case. On the other hand proton affinities are very difficult to obtain experimentally.

Table 7

Dissociation and proton transfer energies of the water molecule

Reaction	ΔE_{calc} (kcal/mole) SCF	IEPA (12,13)	ΔE_{exp} [+)] (kcal/mole)
$H_2O \rightarrow H + OH$			125.8
$H_2O \rightarrow 2H + O$	159.7 (6)		232.6
$OH^- + H^+ \rightarrow H_2O$	-408.0 (7)	-396.4 (7)	-(370-392)
$H_2O + H^+ \rightarrow H_3O^+$	-176.6 (7)	-174.0 (7)	-(151-181) -168 [++)]
$H_3O^+ + OH^- \rightarrow 2H_2O$	-231.4	-222.4	-219

+) The experimental data are taken from (1), (21) and (26) and corrected for zeropoint energy contributions ($E_{oo}(H_2O)$ = 13.25 kcal/mole, $E_{oo}(OH)$ = 5.28 kcal/mole)

++) More recent determination (20)

Table 8

Force constants for the water molecule

Constants	Calculated values (10^5 dyn/cm) SCF [a)]	SCF [b)]	IEPA [c)]	Exp. values (10^5 dyn/cm)
f_{RR}	9.88	9.31	8.37	8.45
$f_{RR'}$	-0.079	0.395	-0.55	-0.101
$f_{R\alpha}/R_e$	0.258	0.232	0.222	0.227
$f_{\alpha\alpha}/R_e^2$	0.881	0.821	0.722	0.761
f_{RRR}	-10.51			-9.55
$f_{RRR'}$	-0.0004			-0.32
$f_{RR\alpha}/R_e$	-0.033			0.16
$f_{RR'\alpha}/R_e$	-0.526			-0.66
$f_{R\alpha\alpha}/R_e^2$	-0.149			0.15
$f_{\alpha\alpha\alpha}/R_e^3$	-0.145			-0.14

a) Basis set: GTO (9,5,2/4,1);(4,3,2/2,1) (24)

b) Basis set: GTO (11,7,1/6,1);(7,4,1/4,1) (7)

c) Independent electron pair approach (13-15) applied to H_2O (7)

d) Experimental values were taken from (25)

the data measured by various techniques or extrapolated from thermodynamic cycles differ largely. Values obtained recently by mass spectroscopic methods seem to be more reliable (20). As expected, the energy for proton transfer from one water molecule to the other is very large in the vapour phase ($\Delta E_{2H_2O \rightarrow H_3O^+ + OH^-} \sim 220$ kcal/mole) the major contribution to this value is the high coulomb energy, which originates from the separation of oppositely charged ions.

The total energy at a particular nuclear configuration can be expanded in a Taylor series at the equilibrium geometry:

$$E_o(\Delta\vec{R}) = E_o(\vec{R}_e) + \frac{1}{2}f_{RR}(\Delta R_1^2 + \Delta R_2^2) + f_{RR'}\Delta R_1 \Delta R_2 + f_{R\alpha}(\Delta R_1 + \Delta R_2)\alpha + \frac{1}{2}f_{\alpha\alpha}\Delta\alpha^2$$

$$+ \frac{1}{R_e}\{f_{RRR}(\Delta R_1^3 + \Delta R_2^3) + f_{RRR'}(\Delta R_1 + \Delta R_2)\Delta R_1 \Delta R_2 + f_{RR\alpha}(\Delta R^2 + \Delta R_2^2)\Delta\alpha$$

$$+ f_{RR'\alpha}\Delta R_1 \Delta R_2 \Delta\alpha + f_{R\alpha\alpha}(\Delta R_1 + \Delta R_2)\Delta\alpha^2 + f_{\alpha\alpha\alpha}\Delta\alpha^3\} + \ldots$$

$$\Delta\vec{R} = (\Delta R_1, \Delta R_2, \Delta\alpha) = (R_1 - R_1^e, R_2 - R_2^e, \alpha - \alpha_e)$$

Since the first derivatives of the energy vanish at the equilibrium geometry, the first non zero contributions originate from the quadratic terms. The expansion coefficients are called force constants and contain all information on the energy hypersurface. Hence it is a straight-forward test for any theoretical model to compare the calculated force constants with the experimental values (table 8). The SCF values of force constants for the second order terms show a general over all agreement with the experimental data, which is improved by the consideration of correlation energy. Correct constants for the third order terms are much more difficult to obtain. Large errors have to be expected for both theoretical and experimental values. In the harmonic approximation, the terms of higher than second order are neglected. Then the four harmonic force constants are represented by the second derivatives of the energy surface:

$$\left(\frac{\partial^2 E_o}{\partial R_1^2}\right)_{R_2,\alpha} = \left(\frac{\partial^2 E_o}{\partial R_2^2}\right)_{R_1,\alpha} = f_{RR} \; ; \qquad \left(\frac{\partial^2 E_o}{\partial R_1 R_2}\right)_{\alpha} = f_{RR'}$$

$$\left(\frac{\partial^2 E_o}{\partial R_1 \partial\alpha}\right)_{R_2} = \left(\frac{\partial^2 E_o}{\partial R_2 \partial\alpha}\right)_{R_1} = f_{R\alpha} \; ; \qquad \left(\frac{\partial^2 E_o}{\partial\alpha^2}\right)_{R_1,R_2} = f_{\alpha\alpha}$$

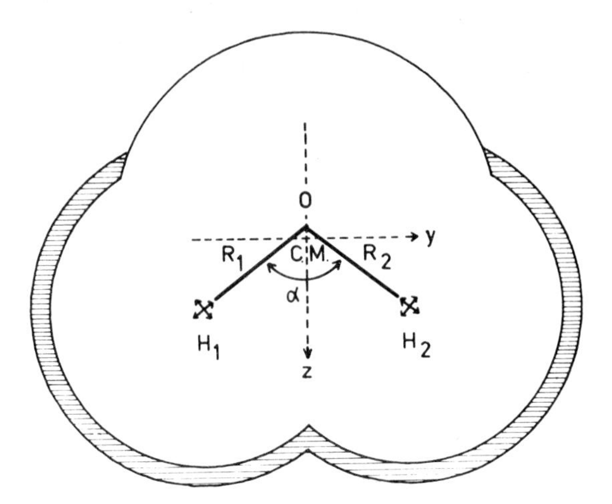

Fig. 2: Equilibrium geometry of H_2O (The spheres around the atoms show the van der Waals radii, $R_O^o = 1.4$ Å and $R_H^o = 1.2$ Å. Arrows at the hydrogen and hatched areas demonstrate the mean square root amplitudes of zero point vibration (24,29). The oxygen atom was fixed for the sake of simplicity. C.M. indicates the center of mass, which was chosen as the origin of the coordinate system.

Using the equilibrium geometry ($\vec{R}_e$) and the harmonic (second order) part of the energy surface (f_{RR}, $f_{RR'}$, $f_{R\alpha}$ and $f_{\alpha\alpha}$) the normal coordinates and the frequencies of the normal vibrations can be obtained by standard techniques, e.g. the well known GF-matrix method (27). In the case of H_2O and D_2O this procedure is largely simplified by molecular symmetry. The three modes correspond closely to symmetric and antisymmetric combinations of OH rsp. OD bond stretching and the bending of the ∢ HOH angle:

$$Q_1(A_1) \propto \Delta R_1 + \Delta R_2; \qquad Q_2(A_1) \propto \Delta\alpha; \qquad Q_3(B_2) \propto \Delta R_1 - \Delta R_2$$

For lack of symmetry the normal modes of HDO are represented quite well by the individual changes in bond lengths and the bending of the ∢ HOH angle (36)

$$Q_1(A') \propto \Delta R_1 ; \qquad Q_2(A') \propto \Delta\alpha ; \quad Q_3(A') \propto \Delta R_2$$

Normal modes and frequencies for H_2O are summarized in fig.3. Again we find good agreement between theory and experiment provided electron correlation has been taken into account. Corrections

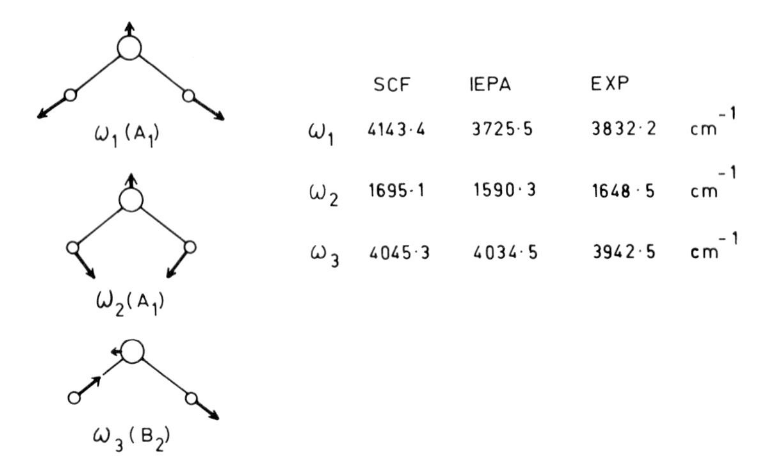

	SCF	IEPA	EXP	
ω_1	4143·4	3725·5	3832·2	cm^{-1}
ω_2	1695·1	1590·3	1648·5	cm^{-1}
ω_3	4045·3	4034·5	3942·5	cm^{-1}

Fig. 3: Normal modes and harmonic frequencies of the water molecule (the calculated frequencies were taken from (7) and the experimental frequencies from (1).

according to anharmonicity of the energy surface can be introduced also into the formula of vibrational energy by the use of anharmonic constants x_{ij}. These constants were determined from the observed vibrational frequencies (28) and are shown in table 9.

$$E_o(v_1,v_2,v_3) - E_o(\vec{R}_e) = \sum_{i=1}^{3} \omega_i(v_i+\frac{1}{2}) + \sum_{i=1}^{3} \sum_{j\geq i}^{3} x_{ij}(v_i+\frac{1}{2})(v_j+\frac{1}{2})$$

Using this set of constants all frequencies of transitions to the lower lying vibrational states can be calculated with high accuracy. The zero point energy of the water molecule is obtained easily by putting $v_1=v_2=v_3=0$. For H_2O one obtains:

$$E_o(0,0,0)-E_o(\vec{R}_e) = 4634.32\ cm^{-1} = 13.25\ kcal/mole$$

Finally, we regard the influence of zero point vibration on the geometry of the water molecule more closely. The center of gravity of nuclear motion and the mean square root of the deviation from the equilibrium values were calculated from theoretical (24) and experimental data (29). The excellent agreement is shown in table 10.

$$R_1^g = \langle\Delta R_1\rangle = \int \chi_{oo}^{*}(R_1-R_1^e)\ \chi_{oo}d\tau; \quad \langle\Delta R_1^2\rangle = \int \chi_{oo}^{*}(R_1-R_1^e)^2\ \chi_{oo}d\tau$$

For R_2 analogous definitions are used.

$$\alpha_g = \langle\Delta\alpha\rangle = \int \chi_{oo}^{+}(\alpha-\alpha_e)\ \chi_{oo}\, d\tau; \quad \langle\Delta\alpha^2\rangle = \int \chi_{oo}^{+}(\alpha-\alpha_e)^2\ \chi_{oo}\, d\tau$$

The extend of the mean square root deviations due to zero point vibration (~10%) is somewhat surprising and was included in fig. 2 therefore.

Table 9

Anharmonic constants for the vibrational spectrum of H_2O (28)

X_{ij} in cm^{-1}	j = 1	2	3
i = 1	-42.6	-15.9	-165.8
2		-16.8	-20.3
3			-47.6

Table 10

Molecular geometries in the electronic and vibrational groundstates for H_2O, HDO and D_2O (1,24,29)

	H_2O		D_2O		HDO	
	calc.	exp.	calc.	exp.	calc.	exp.
$\overline{R}_e$ (Å)	.9413	.9572	.9413	.9575	.9413	.9571
α_e (°)	106.11	104.523	106.11	104.474	106.11	104.529
$\langle\Delta R_1\rangle$ (Å)	.0147	.0140	.0101	.0102	.0148	
$\langle\Delta R_2\rangle$ (Å)	.0147	.0140	.0107	.0102	.0108	
$\langle\Delta\alpha\rangle$ (°)	.12	.18	.09	.12	.11	
$\langle\Delta R_1^2\rangle^{1/2}$ (Å)	.0651	.0677	.0556	.0578	.0651	.0677
$\langle\Delta R_2^2\rangle^{1/2}$ (Å)	.0651	.0677	.0556	.0578	.0556	.0578
$\langle\Delta\alpha^2\rangle^{1/2}$ (°)	8.68	8.72	7.43	7.49	8.13	

II.3.4. QUANTITIES DERIVED FROM THE WAVE FUNCTION

A number of important properties of the water molecule can be calculated from the wave function. It seems reasonable to devide these properties into three groups. The first group includes all ground state properties of the isolated molecule, mainly the electrostatic multipole moments. The second group is represented by the quantities describing the behaviour of the molecule in presence of an external field. The most important quantities of this type are electric polarizabilities and magnetic suszeptibilities. In the third class fall all transition moments which are important for calculations of spectral band intensities.

The most important quantity derived from the wave function is the electron density distribution $\rho(\vec{R})$, which represents the probability to find an electron at the position $\vec{R}$:

$$\rho(\vec{R}) = \int \Psi^{+}\Psi d\sigma_1 d\tau_2 \ . \ . \ . d\tau_n$$

σ_1 represents the spin coordinate of electron 1, τ_i contains both, spin and space coordinates of electron i. The density function can be observed experimentally e.g. by X-ray diffraction. For the water molecule, the density function shows three maxima. The by far highest lies at the oxygen nucleus ($\rho(\vec{R}_O)$~300), two smaller local minima are found at the positions of the hydrogen atoms ($\rho(\vec{R}_H)$~0.4).

For the calculation of intermolecular energies of interaction the electrostatic potential of the water molecule is of some importance. A direct calculation from the density function yields:

$$\phi(\vec{R}) = e_o \left\{ \sum_{k=1}^{3} \frac{Z_k}{|\vec{R}-\vec{R}_k|} - 2 \sum_{i=1}^{5} \int \phi_i^{+}(2) \frac{1}{|\vec{R}-\vec{R}_2|} \phi_i(2)\, d\tau_2 \right\}$$

The electrostatic potential of H_2O was calculated recently (31). It was used for a calculation of the intermolecular electrostatic energy in the water dimer. In most practical examples this calculation is rather time consuming and hence several approximations were introduced.

The electrostatic potential of a molecule can be described by multipole expansion (32). The individual coefficients of that expansion are the well known multipole moments, which can be defined by the equations summarized in table 11.

Table 11

Definitions and formulas for the calculation of electrostatic moments

Dipole moment: $\mu_\alpha = \sum_{k=1}^{3} Z_k R_{k\alpha} - 2 \sum_{i=1}^{5} <\phi_i | R_\alpha | \phi_i>$

Second moments:

$$Q_{\alpha\beta}(C) = \sum_{k=1}^{3} Z_k (R_{k\alpha}-C_\alpha)(R_{k\beta}-C_\beta) - 2 \sum_{i=1}^{5} <\phi_i | (R_\alpha - C_\alpha)(R_\beta - C_\beta) | \phi_i>$$

Quadrupole moment:

$$\Theta_{\alpha\beta} = \frac{1}{2}(3Q_{\alpha\beta} - \delta_{\alpha\beta} \sum_\gamma Q_{\gamma\gamma})$$

Third moments:

$$R_{\alpha\beta\gamma}(C) = \sum_{k=1}^{3} Z_k (R_{k\alpha}-C_\alpha)(R_{k\beta}-C_\beta)(R_{k\gamma}-C_\gamma) - $$
$$- 2 \sum_{i=1}^{5} <\phi_i | (R_\alpha - C_\alpha)(R_\beta - C_\beta)(R_\gamma - C_\gamma) | \phi_i>$$

Octupole moments:

$$\Omega_{\alpha\alpha z}(C) = \frac{1}{2}\left(5R_{\alpha\alpha z}(C) - (1+2\delta_{\alpha z}) R_z(C)\right)$$

α = x,y or z; β = x,y or z; γ = x,y or z

$R_{k\alpha} = X_k, Y_k$ or Z_k; R_α = X, Y or Z; $C_\alpha = X_c, Y_c$ or Z_c

Suffix β or γ leads to analogous expressions; C represents the center of mass

Calculated and experimental multipole moments of the water molecule are compared in table 12. In the case of an uncharged dipole molecule like water the quadrupole and higher moments depend on the choice of the origin. The center of mass (fig.2) was used as origin in all calculations reported here. In general good agreement is found. At the HF limit the calculated dipole moment is somewhat too large. Unfortunately no calculations of multipole moments beyond the HF level have been reported in the literature until now.

The electrostatic potential of the water molecule can be simulated also in a different way by a number of point charges. Two models

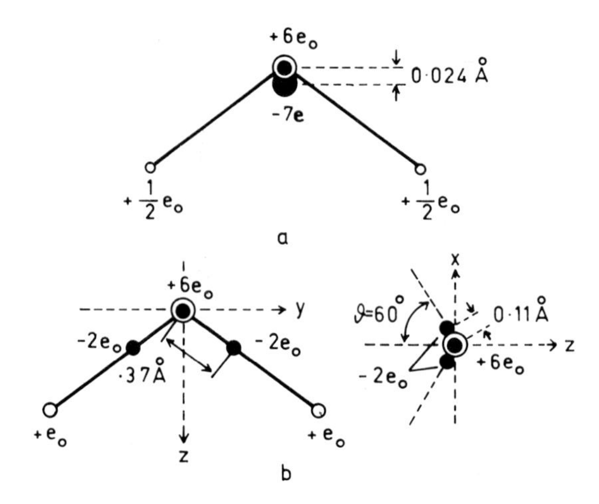

Fig.4: Point charge model for the electrostatic potential of the water molecule (a: four point charges (33); b: seven point charges (34)).

one with four (33) and the other with seven (34) point charges are shown in fig.4. Recently two point charge models for the water molecule were constructed by fitting the calculated energy of interaction with cations and anions to the LCAO-MO energy surface (35, see also chapter).

A combination of both expansions was suggested recently (39,40). The potential of the water molecule is represented by fife multipole expansions up to the second moments. The origins of these expansions are chosen at the nuclei and in the middle of the OH bonds. The expansion coefficients are determined from the density function $\rho(\vec{R})$ or from population analysis (39). Comparison of exact and approximated potential show excellent agreement at distances sufficiently large. In the sourrounding of the expansion centers, however, large errors might occur. Hydration studies on a number of large molecules were performed using this potential for the water molecule (41,42).

Calculations of electric polarizabilities, paramagnetic suszeptibilities and chemical shifts are somewhat more involved. Usually second order perturbation theory is applied (43-45,8). In order to determine electric polarizabilities MO calculations simulating the presence of an electric field in the Hamiltonian were performed too (47).

Table 12

Multipole moments of the water molecule

Moment	Unit	Calculated Value Ref(48)[a]	Ref(45)[b]	Ref(6)[c]	Experiment Value	Ref.
$\mu=\mu_z$	D	1.995	1.926	2.03	1.85±0.02	49
Q_{xx} [d]		-7.482				
Q_{yy}	10^{-26} esu.cm^2	-4.180				
Q_{zz}		-5.939				
Θ_{xx} [d]		-2.422	-2.741	-2.47	-2.50±0.02	
Θ_{yy}	10^{-26} esu.cm^2	2.531		2.66	2.63±0.02	50,51
Θ_{zz}		-0.108	-0.039	-0.20	-0.13±0.03	
R_{xxz} [d]		-0.112				
R_{yyz}	10^{-34} esu.cm^3	1.191				
R_{zzz}		0.282				
Ω_{xxz} [d]		-0.960	-1.170	-0.84		
Ω_{yyz}	10^{-34} esu.cm^3	2.297		2.19		
Ω_{zzz}		-1.337	-1.523	-1.35		

a) Basis set: GTO (10s,6p,2d/4s,2p) b) Basis set: STO (5s,2p,1d/2s1p)

c) Basis set: GTO (11s,7p,2d/5s,1p) contracted to (6s,5p,2d/3s,1p)

d) Calculated relative to the center of mass of $H_2{}^{16}O$.

The results are summarized in table 13. The theoretical values agree well with the available experimental data. In some cases additional information is provided by MO calculations, e.g. no reliable experimental data are available for the anisotropy of polarizability and diamagnetic suszeptibility of the water molecule.

Other interesting quantities obtainable from the wave functions are the intensities of spectral transitions. These intensities are proportional to the square of the transition moments, which can be calculated for electric dipole transitions by the following formula:

$$\vec{\mu} = \int \Psi_e^* e_o \vec{R} \Psi_k d\tau; \quad \text{transition: } k \rightarrow e$$

Hence for a reliable calculation of the intensities of electronic transitions accurate wave functions for the ground state (k=0) and the excited states are necessary. Usually wave functions for excited states are not known well enough. On the other hand the experimental determination of oscillator strengths is rather inaccurate too in cases where the bands for the individual electronic transition overlap as they usually do.

Table 13

Electric and magnetic properties of the water molecule

Quantity		Unit	Calculation Value	Calculation Ref.	Experiment Value	Experiment Ref.
Polarizability:	α_{xx}		1.069			
	α_{yy}		1.279			
	α_{zz}	$\mathring{A}^3$	1.162	45		
	$\bar{\alpha}$		1.170		1.456	52
Suszeptibility[a]:	$\bar{\chi}$		-14.373		-13.0 ± 1.9	
	χ_{xx}		-14.350			
	χ_{yy}		-14.222			
	χ_{zz}	10^{-6} cm^3 /mole	-14.547	45		
paramagnetic	χ^p_{xx}		2.061		2.3794	
contribution	χ^p_{yy}		0.540		0.7286	51,54
	χ^p_{zz}		1.034		1.4033	
Chemical shift[b]:	$\bar{\sigma}$		28.30		30.20	48
paramagnetic	σ^p_{xx}		0.65			(55,56)
contribution	σ^p_{yy}		6.26			
	σ^p_{zz}	p.p.m	3.66	8		
diamagnetic	σ^d_{xx}		37.08			
contribution	σ^d_{yy}		14.04			
	σ^d_{zz}		23.21			

[a] Gauge origin at the center of mass

[b] Chemical shift of H in the water molecule relative to the free proton. Gauge origin at the center of negative charge.

For transitions between vibrational levels of the same electronic state (e.g. the ground state) the transition moment becomes proportional to the change of the dipole moment along the particular normal mode. Starting from the zero vibrational level we obtain:

$$|\mu| \propto \left| \frac{\partial \mu(\vec{R})}{\partial Q_i} \right|_{R=R_e} \qquad i=1,2,3$$

This formula has been used to calculate changes in IR-intensities due to hydrogen bonding (ref.(3), see also chapter). For a detailed analysis of complete ab initio calculations of IR spectra of molecules and molecular aggregates the reader is refered to (46).

In any case it seems that a good deal of further work is necessary before reliable absolute values for band intensities can be given.

II.3.5. CONCLUSION

At the present stage of quantum chemistry all important ground state properties of small molecules including rotational and vibrational frequencies can be calculated with reasonable accuracy. In most cases it is necessary to consider electron correlation in order to obtain values of an accuracy comparable with the data of experimental measurements. Especially the anisotropies of certain quantities are easily obtained by theoretical calculations. In this case the MO method presents information which is not accessible by the present experimental techniques. The excited electronic states are much less well known. More extensive calculations are necessary before an accurate description similar to that of the ground state is possible.

ACKNOWLEDGEMENTS

The author is indebted to Dr.E.Clementi, Dr.R.Janoschek, Dr.H.Lischka, Dr.A.Pullman and Dr.B.Roos for sending their most recent results prior to publication. The assistence of our technical staff, Mrs.J.Dura and Mr.J.Schuster with the preparation of the manuscript is gratefully acknowledged.

REFERENCES

1 D.Eisenberg, W.Kauzman, "The structure and properties of water" Clarendon Press, Oxford 1969, p.1-35

2 C.W.Kern, M.Karplus, "The water molecule", in "Water", ed.F.Franks Vol.1, Plenum Press, New York 1972, p.21-91

3 G.H.F.Diercksen, Theoret,chim,Acta (Berl.) 21, 335 (1971)

4 W.Meyer, Internat.J.QuantumChem.5S, 341 (1971)

5 E.Clementi, H.Popkie, J.Chem.Phys.57, 1077 (1972)

6 Th.H.Dunning,jr., R.M.Pitzer, S.Aung, J.Chem.Phys.57, 5044 (1972)

7 H.Lischka, Theoret.chim.Acta(Berl.) 31, 39 (1973)

8 M.Jaszunski, A.J.Sadlej, Theoret.chim.Acta(Berl.) 27, 135 (1972)

9 A.Messiah, "Quantum Mechanics", Vol.2, North Holland Publ.Co. Amsterdam 1970, p.781-793

10 J.O.Hirschfelder, W.J.Meath, "The nature of the intermolecular Forces" in "Advances in Chemical Physics", Vol.12, Interscience New York 1967, p.8-17

11 T.Koopmans, Physica 1, 107 (1933)

12 W.Kutzelnigg, Theoret.chim.Acta(Berl.) 1, 327 (1963)

13 M.Jungen, R.Ahlrichs, Theoret.chim.Acta(Berl.) 17, 339 (1970)

14 R.K.Nesbet, Phys.Rev.155, 51 (1967)

15 O.Sinanoglu, J.Chem.Phys.36, 706 and 3198 (1962)

16 W.Meyer, J.Chem.Phys.58, 1017 (1973)

17 H.Lischka, Chem.Phys.Letters, in press

18 B.Roos, G.H.F.Diercksen, to be published

19 P.E.Cade, J.Chem.Phys.47, 2390 (1967)

20 S.L.Chong, R.A.Myers,jr., J.L.Franklin, J.Chem.Phys.56 2427 (1972)

21 A.C.Hopkinson, N,K.Holbrook, K.Yates, I.G.Csizmadia, J.Chem-Phys. 49, 3596 (1968)

22 P.Schuster, Theoret.chim.Acta(Berl.) 19, 212 (1970)

23 L.E.Sutton,ed. "Tables of Interatomic Distances and Configurations in Molecules and Ions", The Chemical Society, Burlington House, London 1958

24 W.V.Ermler, C.W.Kern, J.Chem.Phys.55, 4851 (1971)

25 K.Kuchitsu, V.Morino, Bull.Chem.Soc.Japan 38, 814 (1965)

26 G.Herzberg, "Molecular Spectra and Molecular Structure III. Electronic Spectra and Electronic Structure of Polyatomic Molecules", Van Nostrand, Princeton, New Jersey 1966, p. 489

27 E.B.Wilson, J.C.Decins, P.C.Cross, "Molecular Vibrations", McGraw Hill Book Co., New York 1955

28 W.S.Benedict, N.Gailar, E.K.Plyler, J.Chem.Phys.24, 1136 (1956)

29 K.Kuchitsu, L.S.Bartell, J.Chem.Phys.36,2460 (1962)

30 C.Edmiston, K.Ruedenberg, Rev.Mod.Phys. 35, 457 (1963)

31 R.Bonaccorsi, C.Petrongolo, E.Scrocco, J.Tomasi, Theoret.chim.Acta (Berl.) 20, 331 (1971)

32 J.O.Hirschfelder, C.F.Curtiss, R.B.Bird, "Molecular Theory of Gases and liquids", J.Wiloy, New York 1954, p.836-851

33 E.J.W.Verwey, Recl.Trav.Chim.Pays-Bas Belg.60, 887 (1941)

34 J.A.Pople, Proc.Roy.Soc.A 205, 163 (1951)

35 H.Kistenmacher, H.Popkie, E.Clementi, J.Chem.Phys., in press

36 C.W.Kern, R.L.Matcha, J.Chem.Phys.49, 2081 (1968)

37 J.W.D.Conolly, J.R.Sabin, J.Chem.Phys.56, 5529 (1972)

38 M.Boring, J.H.Wood, J.W.Moskowitz, J.W.D.Conolly, J.Chem.Phys. 58, 5163 (1973)

39 M.Dreyfus, Thèse 3è Cycle, University of Paris, 1970

40 G.Alagona, R.Cimiraglia, E.Scrocco, J.Tomasi, Theoret.chim.Acta (Berl.) 25, 103 (1972)

41 G.Alagona, A.Pullman, E.Scrocco and J.Tomasi, Internat.J.of Peptide and Protein Chemistry, in press

42 G.N.J.Port, A.Pullman, to be published

43 G.P.Arrighini, M.Maestro, R.Moccia, Chem.Phys.Letters 1,242 (1967)

44 G.P.Arrighini, M.Maestro, R.Moccia, J.Chem.Phys.49, 882 (1968)

45 G.P.Arrighini, C.Guidotti, O.Salvetti, J.Chem.Phys.52, 1037 (1970)

46 R.Janoschek, "Calculated vibrational spectra of hydrogen bonded systems", in "Recent Developments in Hydrogen Bonding", Ed. P.Schuster, G.Zundel, C.Sandorfy, North Holland Publ.Co. in press

47 P.Drossbach, P.Schmittinger, Z.f.Naturforschung 25a, 834 (1970)

48 D.Neumann, J.Moskowitz, J.Chem.Phys.49, 2056 (1968)

49 W.H.Kirchhoff, D.R.Lide,jr., Natl.Std.Ref.Data Ser.Natl.Bur.Std. 10 (1967)

50 J.Verhoeven, A.Dymanus, H.Bluyssen, J.Chem.Phys.50, 3330 (1969); Phys.Letters 26 A, 424 (1968)

51 H.Taft, B.P.Dailey, J.Chem.Phys.51, 1002 (1969)

52 H.H.Landolt, R.Börnstein, "Zahlenwerte und Funktionen", Springer Verlag, Berlin 1951, Vol1, pp.510

53 D.Eisenberg, J.M.Pochan, W.H.F.Flygare, J.Chem.Phys.43 , 4531 (1965)

54 S.G.Kukolich, J.Chem.Phys.50, 3751 (1969)

55 J.A.Pople, W.G.Schneider, H.J.Bernstein, "High Resolution NMR", McGraw Hill Book Co., New York 1959

56 N.F.Ramsey, Phys.Rev.78, 699 (1950)

II.4. THEORY OF HYDROGEN BONDING IN WATER AND ION HYDRATION

Peter Schuster

Institut für Theoretische Chemie, Universität Wien
Währingerstraße 17, A-1090 Wien

ABSTRACT

Calculated energies, geometries and spectral properties of water dimer, higher aggregates of water molecules and solvated hydroxonium ions are summarized and compared with the available experimental material. Ab initio LCAO-MO-SCF results obtained with extended basis sets agree well with the measured values. Since experimental investigations on defined oligomers of water are extremely difficult, the theoretical data very often present the only reliable information available now. In one-dimensional chains of water molecules the mean hydrogen bond energy increases with chain length. In three-dimensional tetrahedral clusters, however, the deviations from pairwise additivity are drasticly diminished. Ion hydration has been investigated too by MO methods. Again the theoretical results present an additional insight into the molecular details of solvation. The calculated energies of interaction agree well with the enthalpies of ion water complex formation measured by mass spectroscopy. The electron density function of $(LiOH_2)^+$ was analyzed in detail. Predominant electrostatic interaction and strong polarization was found. Charge transfer is extremely small. The structure of anion water complexes is determined by a competition between hydrogen bonding and classical ion dipole interaction. The deviation from the linear X-H-O arrangement increases with increasing radius of the anion.

II.4.1. INTRODUCTION

Hydrogen bonds determine the properties of water in all three states of aggregation. Actually, hydrogen bonds are responsible for all the unusual properties of water and hence, theorists and experimentalists became interested in hydrogen bonding very long ago. Many theories have been developed and were applied to hydrogen bonded systems. During the last few years LCAO-MO calculations were performed on structures with hydrogen bonds. Some progress was achieved recently because very accurate numerical calculations have become routine now. For almost all interesting properties of hydrogen bonds purely theoretical "ab initio" results are available.

Several review articles were published recently (1-4). In this paper only the most accurate older calculations are summarized and more emphasis is put on the more recent literature (5-32). For a more extended review on hydrogen bonding or ion solvation the reader is refered to two papers in preparation (33,34).

Any theoretical treatment of hydrogen bonding or ion hydration leads to the more general problem of the intermolecular interactions. Among all quantum mechanical procedures perturbation theory and MO methods turned out to be most successful in calculations of intermolecular energies.

II.4.2. PARTITIONING OF INTERMOLECULAR ENERGIES

In perturbation theory the wave functions of the individual molecules at infinite distance, Ψ_A^o and Ψ_B^o represent the starting point of the calculation. The total energy of interaction is obtained as a sum of different contributions, which can be classified as first, second or higher order terms. For a detailed discussion the reader is refered to the references (35-37):

$$\Delta E = E_{AB}-(E_A+E_B) = \underbrace{\Delta E_{COU}+\Delta E_{EX}^{(1)}}_{\text{1st order}} + \underbrace{\Delta E_{POL}+\Delta E_{CHT}+\Delta E_{DIS}}_{\text{2nd order}} + \ldots$$

contributions

The first order contribution to the energy of interaction can be split into two parts, the coulomb energy, ΔE_{COU}, which is identical with the classical electrostatic energy of interaction and the exchange energy $\Delta E_{EX}^{(1)}$, which represents the non classical exchange repulsion between the two closed shell molecules. The superscript indicates that in the framework of usual intermolecular perturbation theory only exchange

TABLE 1

Partitioning of hydrogen bond energy in perturbation theory (37)

$R_{AB}^{a)}$ (Å)	$\Delta E_{COU}^{b)}$	$\Delta E_{EX}^{b)}$	$\Delta E_{POL}^{b)}$	$\Delta E_{CHT}^{b)}$	$\Delta E_{DIS}^{b)}$
2.65	-12.7	10.7	-2.6	-6.4	-2.9
2.91	- 5.3	4.3	-1.7	-3.3	-1.7
3.18	- 2.2	1.7	-1.0	-0.3	-1.0
3.44	- 1.0	0.6	-0.6	-0.1	-0.6
3.70	- 0.5	0.2	-0.4	-0.0	-0.4

[a] For the calculation general model orbitals were assumed for a hydrogen bond A-H ... B.

[b] All energies in kcal/mole

of single electrons between the molecules is taken into account. Both quantities, ΔE_{COU} and $\Delta E_{EX}^{(1)}$, are obtained directly from the wave functions of the isolated molecules Ψ_A^o and Ψ_B^o. The second order contribution represents the energetic effects resulting from first order changes in the wave functions. It consists of three major parts: The polarization energy ΔE_{POL}, the charge transfer energy ΔE_{CHT} and the dispersion energy ΔE_{DIS}. The polarization energy, ΔE_{POL}, describes an additional stabilization brought about by displacement of the electron distribution in the field of the second molecule. In classical electrostatics this term is described by polarizability tensors and electric fields of molecules. On the other hand, charge transfer energy, ΔE_{CHT}, has no analogue in classical physics. It can be interpreted as partial electron transfer from one molecule to the other as a consequence of mixing of molecular orbitals from both molecules. In hydrogen bonded structures electron transfer always occurs from the molecule with the lone pair to the molecule with the HX bond. At short distances charge transfer energy becomes very important and often determines the energeticly favoured structures of molecular aggregates. Dispersion energy, ΔE_{DIS}, again has no classical analogue. It represents the universal stabilization resulting from the attractive London or dispersion forces.

Numerical values of the individual contributions in a typical hydrogen bonded complex are shown in table 1 (37). The most striking result is the different dependence on intermolecular distance. We find that the first order exchange energy and the charge transfer energy fall of much more strongly with increasing distance than the other three contributions. Both are proportional to the square of intermolecular overlap, S_{AB}, which on its side is proportional to exp(-ξR). At larger distances ΔE_{COU}, ΔE_{POL} and ΔE_{DIS} are proportional to some inverse power of intermolecular distance, R^{-n}. In the case of usual hydrogen bonds all contributions mentioned above with the exception of exchange energy are negative quantities. Around the energy minimum the first order energies, ΔE_{COU} and ΔE_{EX}, represent the most important contributions. They are of similar size but opposite sign and hence almost cancel each other. At the most stable arrangement the electrostatic contribution, ΔE_{COU}, is comparable to the total ΔE. This is the reason for the success of simple electrostatic theories in the calculation of hydrogen bond energies.

A very recent application of SCF perturbation theory to the water dimer (9) led to the opposite conclusion that only covalent contributions and steric hindrance are important for the stability of the complex and coulomb contributions are negligible. This result strikingly contradicts all the other calculation based on perturbation and MO theory and seems to be an artifact of the unusual method of partitioning applied by the authors. The curves for the total energy ΔE agree well with results of full SCF calculations.

Besides its heuristic value the partitioning of intermolecular energies is also very useful for a qualitative discussion of the most stable geometries of associated molecules by simple MO theory (52). An illustrative example is presented by the equilibrium geometry of $(HF)_2$. The angular dependence of the energy of interaction is essentially determined by the sum of coulomb and charge transfer energy since the other contributions are more or less independent of the intermolecular angle ω (cf.fig.4) as long as the molecules considered are approximately spherical. In fig.1 ΔE_{COU} and ΔE_{CHT} are obtained roughly by regarding the angular dependence of the largest term. Depending on the distance between the HF molecules the coulomb energy is more or less determined by the dipole/dipole interaction. In the case of HF the molecular dipole of course coincides with the local dipole of the HF bond. This fact makes a consideration of $(HF)_2$ much simpler than an estimate of the prefered geometry of $(H_2O)_2$. The dipole/dipole term predicts that the linear arrangement ($\omega=0^o$) is the most stable one. At $\omega=90^o$ the coulomb energy is very small ($\Delta E_{COU}(90^o)\sim 0$). The charge transfer energy as mentioned before is proportional to the square of the overlap between the two corresponding MO's. In the case of $(HF)_2$ the most important contribution is brought about by charge transfer from the π lone pair orbital of one HF molecule to the antibonding σ^* orbital of the second HF molecule. As fig.1 shows this charge transfer

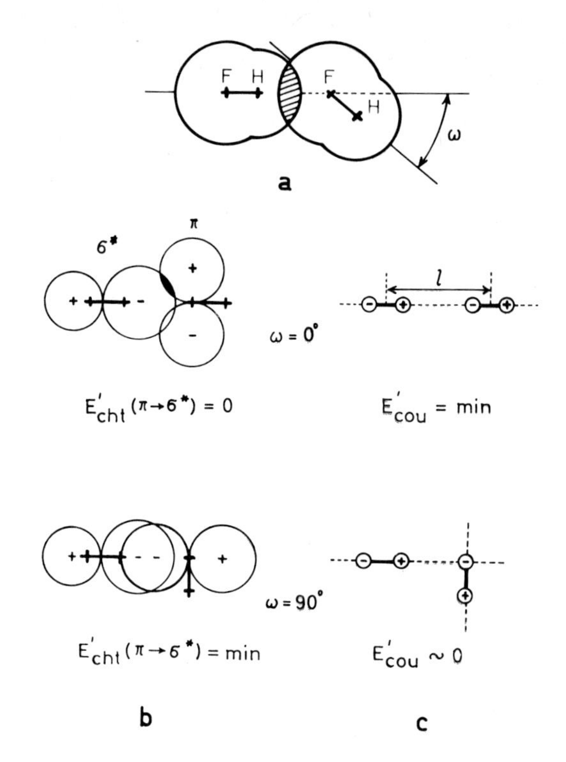

Fig.1: Geometry of hydrogen fluoride dimer (a: experimental data (56) Van der Waals radii are used to illustrate the shape of the molecules (R_F=1.4 Å, R_H=1.2 Å); b: most important contribution E'_{CHT} to the charge transfer energy ΔE_{CHT}; c: dipole/dipole interaction E'_{COU} as the most important contribution to the coulomb energy ΔE_{COU}).

cannot occur at $\omega=0^\circ$, because the overlap integral is zero. At $\omega=90^\circ$ it reaches a maximum. Hence charge transfer energy favours an arrangement with an angle $\omega=90^\circ$. Finally we have to expect a compromise between the angles favoured by ΔE_{COU} and ΔE_{CHT} leading to an intermediate value, $0<\omega<90^\circ$. Spectroscopic measurements gave a value of $\omega=40\pm5^\circ$(56) in good agreement with the results of accurate calculations $\omega=40^\circ$, ref.57). Similar arguments were used to explain the favoured geometries of many other hydrogen bonded molecular aggregates., e.g. the dimer of water (52). In this particular example, however, the estimate is more complicated, because the direction of the local dipole of the OH bond does not coincide with the direction of the dipole moment of the whole water molecule. Therefore the angular dependence of ΔE_{COU} is not as simple as in the previous example.

The perturbational treatment of interactions between molecules is characterized by one serious deficiency. Since only exchange of single electrons between the molecules is taken into account, perturbation theory can describe intermolecular interaction accurately only at sufficiently large distances between the molecules. Several calculations on structures with weak hydrogen bonds were performed, e.g on water dimer (8-10,58). The results agree fairly well with the results of MO calculations and with the available experimental data. On the other hand the overlap integrals between the molecules are much larger in structures with strong hydrogen bonds and hence these systems seem to be beyond the limits of reliability for an application of intermolecular perturbation theory (S_{AB}<0,15, see ref.35,37).

LCAO-MO-calculations can be applied directly to all kinds of intermolecular interactions. The energy of interaction is obtained directly as the difference in energy between the whole "supermolecule" AB and the isolated molecules A and B:

$$\Delta E = E_{AB} - (E_A + E_B)$$

this formula explains already the difficulties we have to expect in the application of MO methods to calculations of intermolecular energies. Since E_{AB}, E_A and E_B are very large quantities (e.g. $E((H_2O)_2)$~95 500 kcal/mole) and ΔE is usually of the order of a few kcal/mole ($\Delta E((H_2O)_2)$~5kcal/mole), extremely high numerical accuracy is an unescapable condition for reliable energies of interaction. Therefore no MO calculations on structures with hydrogen bonds were possible before large electronic computers could be used. Noway calculations with high numerical accuracy have become routine and many calculations on intermolecular energies were performed during the last five years.

In the case of MO calculations a partitioning of energies of interaction is possible too (59,60). From the wave functions of the isolated molecules (Ψ_A^o and Ψ_B^o) or the corresponding density functions (ρ_A^o and ρ_B^o) the coulomb energy of interaction ΔE_{COU} is obtained directly (for a definition of the symbols used see chapterII.3.table 1):

$$\Delta E_{COU} = \sum_{k \in A} \sum_{l \in B} \frac{Z_k Z_l e_o^2}{R_{kl}} + \iint \rho_A^o \rho_B^o \frac{1}{R_{12}} d\tau_1 d\tau_2 - $$

$$- \{ \int \rho_A^o \sum_{l \in B} \frac{Z_l e_o}{R_{1l}} d\tau_1 + \int \rho_B^o \sum_{k \in A} \frac{Z_k e_o}{R_{1k}} d\tau_1 \}$$

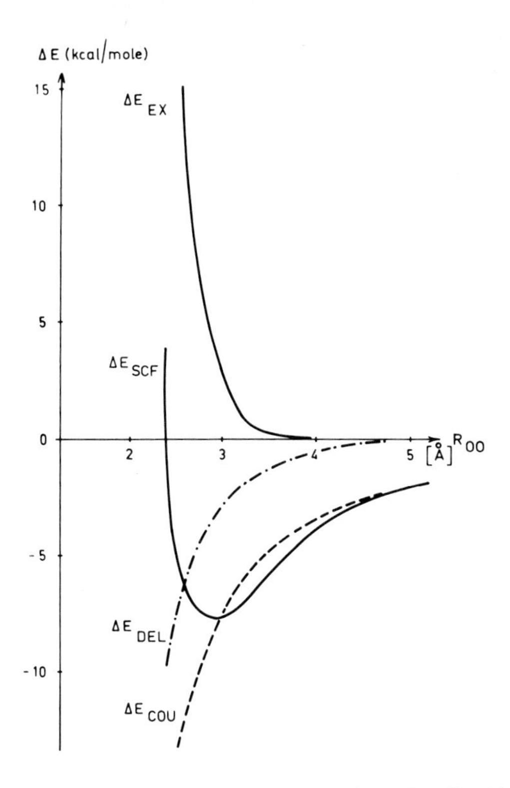

Fig.2: Partitioning of the energy of interaction in $(H_2O)_2$ calculated by LCAO-MO-methods (schematicly).

The total first order energy can be calculated with an antisymmetrized wave function Ψ_o of the isolated molecules and the complete Hamiltonian H of the supermolecule AB.

$$\Delta E_1 = \Delta E_{COU} + \Delta E_{EX} = \int \Psi_o^{*} H \Psi_o d\tau; \qquad \Delta E_{EX} = \Delta E_1 - \Delta E_{COU}$$

$$H = H_A + H_B + V_{AB}(\vec{R}); \qquad \vec{R} = (R,\Theta,\Phi); \qquad \Psi_o = \mathcal{A}\{\Psi_A^o \Psi_B^o\}$$

H consists of the Hamiltonians of the isolated molecules (H_A, H_B) and the interaction potential V_{AB}, which depends on the intermolecular distance (R) and the relative orientation (Θ,Φ). A complete antisymmetrized wave function Ψ_o is used in the MO ansatz and in contrary to the perturbational treatment the exchange contribution ΔE_{EX} obtained here is not restricted to single electron exchange. Therefore MO theory can be used equally well to calculate weak or strong intermolecular interactions.

The difference between the energy of interaction at the Hartree-

Fock (HF)level, ΔE_{SCF} and ΔE_1 is brought about by electron delocalization and might be called accordingly delocalization energy ΔE_{DEL}:

$$\Delta E_{SCF} = E_{AB}^{SCF} - (E_A^{SCF}+E_B^{SCF}) \ ; \ \Delta E_{DEL} = \Delta E_{SCF}-\Delta E_1$$

In fig.2 the dependence of the individual contributions to ΔE_{SCF} on the distance between the molecules is shown. Again we find that the exchange energy depends much more strongly on the intermolecular distance than the coulomb energy. The curve of the delocalization energy is more complicated since it corresponds to the sum of both polarization and charge transfer energy of the perturbational treatment:

$$\Delta E_{DEL} \sim \Delta E_{POL} + \Delta E_{CHT}$$

In MO calculations beyond the HF level we obtain an additional contribution of correlation energy to the energy of interaction:

$$\Delta E = \Delta E_{SCF}-\Delta E_{COR}; \quad \Delta E_{COR} = \Delta E_{AB}^{COR} - (\Delta E_A^{COR}+\Delta E_B^{COR})$$

Kollman and Allen concluded that in interactions between closed shell molecules the major part of ΔE_{COR} results from dispersion forces and hence ΔE_{COR} should correspond to ΔE_{DIS} of perturbation theory (59). More recent calculations, however, show, that the situation is more complicated. Unfortunately no calculations including electron correlation have been performed on water dimer. Without regarding calculations on H_2/H_2 and He/He very accurate energy surfaces were obtained recently with the IEPA-method$^{+)}$ only for the systems He/HF and He/H_2O (61), H_2/HF and H_2/H_2O (62) as well as HF/HF (63). In the case of hydrogen bonding in $(HF)_2$ an analysis of the influence of electron correlation shows that two contributions of opposite sign occur in the linear configuration HF..HF. One contribution results from a reduction of the electrostatic energy, ΔE_{COU}, since the dipole moment is obtained too large in SCF calculations and a consideration of electron correlation reduces it to the correct value. On the other hand, dispersion energy leads to the expected additional stabilization. From these very recent results we have to conclude that the approximate correspondence of correlation and dispersion energy suggested above (59) has to be examined with great care in every particular case.

In table 2 the partitioning of the energy of interaction in water dimer is shown. Unfortunately the LCAO-MO-SCF calculation (65) was performed with a rather small basis set and hence the electrostatic contribution seems to be overemphasized. Another striking difference is found among the two perturbational calculations. In the first

$^{+)}$For an explanation of "IEPA" see chapter II.3.

TABLE 2

Comparison of the partitioning of hydrogen bond energies in $(H_2O)_2$ calculated by perturbation theory and LCAO-MO-SCF method

R_{OO} (Å)	Quantity[a)]	Perturbation theory Ref.(58)	Perturbation theory Ref.(10)	Electrostatic Potential(64)	LCAO-MO-SCF calc.(59,65)
2.9	ΔE_{COU}	-12.0	-5.80	-5.6	
	ΔE_{EX}	+13.1	+5.55		
	ΔE_1	+1.1	-0.25		-5.15
	ΔE_{POL}	-1.5	-0.37		} -2.25
	ΔE_{CHT}	-1.4	-2.73		
	ΔE_{DIS}	-1.7			
	ΔE	-3.4	-4.4		-7.40
3.2	ΔE_{COU}	-7.2	-3.65	-3.7	
	ΔE_{EX}	+5.0	+1.55		
	ΔE_1	-2.2	-2.1		-5.3
	ΔE_{POL}	-0.8	-0.2		} -1.1
	ΔE_{CHT}	-0.3	-1.0		
	ΔE_{DIS}	-1.0			
	ΔE	-4.3	-3.6		-6.4

a) All energies are given in kcal/mole

example (58) the calculation is restricted to the OH..O fragment and both coulomb energy and exchange repulsion are largely exaggerated. In the full perturbational treatment (10) the whole molecules are included and as expected the coulomb energy agrees very well with the purely electrostatic calculations based on a wave function of comparable quality. Interesting enough there is an excellent agreement between the second order contributions of all three methods applied (10,58,59).

There are two needs for partitioning of intermolecular energies. First of all a comparison between simple electrostatic theories and the more elaborate treatments becomes possible. The most important advantage, however, results from the fact, that the individual contributions, ΔE_{COU}, ΔE_{EX}, ΔE_{POL}, ΔE_{CHT} and ΔE_{DIS}, can be approximated much easier by semiclassical or empirical formulas than the total energy, ΔE. Especially ΔE_{COU}, ΔE_{EX} and ΔE_{DIS} can be calculated very fast from approximate analytical expressions. In clusters of many molecules or in complexes between large systems direct MO calculations are extremely time consuming and cannot be performed with reasonable accuracy even on the largest available computers. For these examples reliable empirical potentials, which have been examined on the interaction between two small molecules are extremely important.

II.4.3. THE WATER DIMER AND HIGHER AGGREGATES OF WATER MOLECULES

Before accurate ab initio LCAO-MO calculations became possible, a number of semiempirical calculations mainly of the CNDO type (66) were performed. Concerning energies of interaction and calculated geometries of molecular aggregates, the CNDO/2 (67) results agree very well with the values of more sophisticated methods. Nowadays calculations on this level seem to be appropriate for theoretical investigations on large molecular clusters and on the interaction of large molecules (3,4,33).

During the last three years very accurate ab initio calculations (near the HF limit) on water dimer were published (5-7). A comparison of various calculated properties with the available experimental data is presented in table 4. The main problem of this comparison arises from the selection of appropriate experimental results, since there is no evidence for the predominant existence of independent water dimers under certain conditions. The only possible exceptions are investigations of highly diluted water vapour included in N_2 or Ar matrices at very low temperatures (20^oK, ref.68). Hence the most accurate theoretical calculations (5-7) represent also the most reliable information on the properties of isolated water dimers. At present calculations on $(H_2O)_2$ including electron correlation are in progress (63) and we can expect results of similar accuracy as previously described for the water molecule (chapter II.3) in the near future. Most calculated properties, however, will not be effected very much by correlation effects and hence are also reliable at the HF limit..

The energy of interaction in water dimer obtained by near HF calculations is about $-3.6>\Delta E_{SCF}>-4.2$ kcal/mole (table 3). Careful estimates on the contribution of correlation energy to the energy of interaction at the minimum lead to a value $+1.5<\Delta E_{COR}<+2.0$ kcal/mole (59,68). Therefore we would expect a value of $\Delta E\approx-5$ to -6 kcal/mole. The geometry of the water dimer is shown in fig.3. Hydrogen bonding leads to an appreciable mutual penetration of the Van der Waals hard spheres. In the most stable aggregate the two oxygen atoms and the hydrogen atom in the hydrogen bond lie on a straight line. Deviations from linearity lead to a pronounced increase in energy. With the

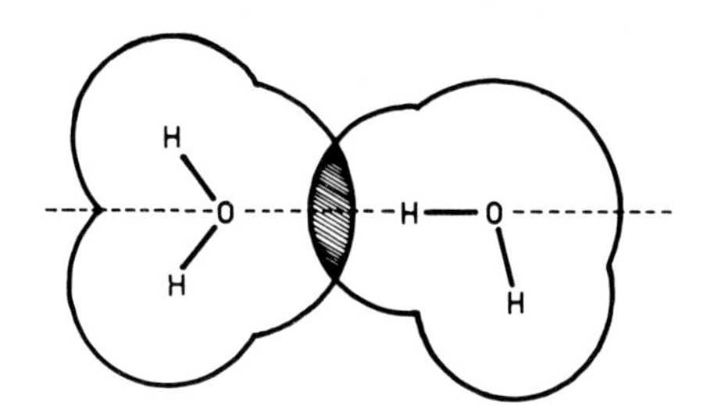

Fig.3: Geometry of the water dimer (Van der Waals radii are used to illustrate the shape of the molecules; R_O^0=1.4 Å, R_H^0=1.2 Å).

TABLE 3

Structure and energies of water dimer, hydrated hydroxonium cation hydroxide anion

Aggregate	Basis set[a)]	Ref.	ΔE_{SCF} (kcal/mole)	R_{OO} (Å)	$R_{OH}^{b)}$ (Å)	$\Delta R_{OH}^{b,c)}$ (Å)	ΔE_{EXP} (kcal/mole)
$H_2O..HOH$	(531/21)	6	-4.72	3.00	0.945	0	-5[d)]
	(541/31)	5	-4.84	3.00	0.948	0.004	
	(8521/421)	7	-3.90				
$H_2O..HOH_2^+$	4-31G	16	-43.8	2.36	1.18	0.22	-36[e)]
	(42/21)	17	-36.9	2.38	1.19		-31.6[f)]
	(541/31)	18	-32.2	2.39	1.15	0.19	
$HO^-..HOH$	4-31G	16	-40.7	2.45	1.11	0.16	-22.5[g)]
	(541/31)	18	-24.3	2.51	1.04	0.10	

a) The contracted basis sets are symbolized as follows: (ABCD/EFG) means A s-type, B p-type, C d-type, D f-type contracted GTO's at the oxygen and E s-type, F p-type and G d-type contracted GTO's at the hydrogen atom. 4-31G is an abbreviation for Slater orbitals approximated by a series of GTO's (38,39).

b) Hydrogen atom in the hydrogen bond.

c) ΔR_{OH} is difference in OH bond lengths between the isolated molecule and the hydrogen bonded aggregate.

d) See ref.(40); e) See ref.(41); f) See ref.(42); g) See ref.(45).

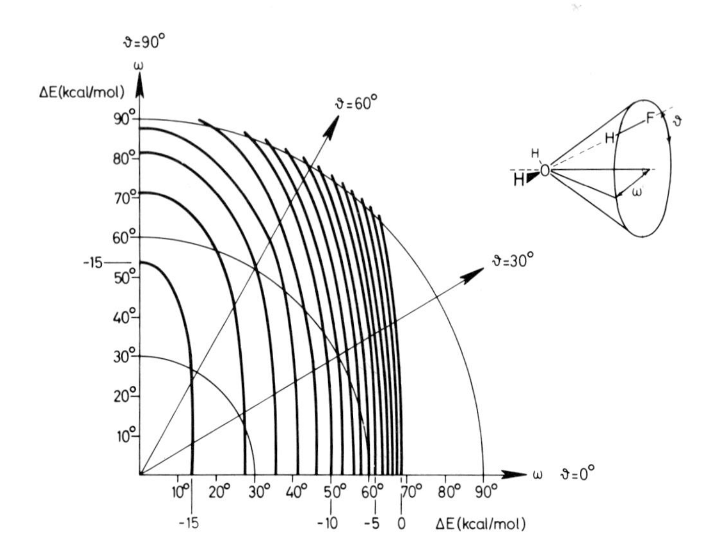

Fig.4: Stereochemistry of hydrogen bonding to the water molecule (The energy surface E=E(ω,δ) was calculated for fixed OF and HF distances).

assumption of a linoar hydrogen bond The additional angular dependence can be characterized by the two angles ω and δ (fig.4). A detailed discussion of the stereochemical properties of hydrogen bonds based on semiempirical MO calculations was presented previously (4,69). Here we concentrate exclusively on the directing properties of the water lone pairs, which can be derived best from the energy surface E=E(ω,δ) in fig.4. In order to avoid further complication by an additional degree of freedom the water molecule was replaced by hydrogen fluoride In the plan of the H_2O molecule ($\delta=0^\circ$) the energy increases strongly with increasing values of ω. Perpendicular to this plane ($\delta=90^\circ$), however, the energy surface is very flat. The energy of the interaction does not change appreciably with certain limits ($\omega<80^\circ$). The total energy minimum calculated with various basis sets is always found in this plane ($\delta=90^\circ$). The most accurate calculations gave the values $\omega=40^\circ$ (6) and $\omega=30^\circ$ (7). In the latter case a difference in energy of only 0.23 kcal/mole between the two structures $\omega=0^\circ$ and $\omega=30^\circ$ was obtained. For comparison the lone pairs of a tetrahedral model for the water molecule point out in the directions ($\delta=\pm90^\circ$, $\omega=54.7^\circ$). The directive properties of the H_2O lone pairs in hydrogen bonding can be summarized best by high flexibility for hydrogen bonds in the plane perpendicular to the OH bonds and fairly strong restrictions to $\omega=0^\circ$

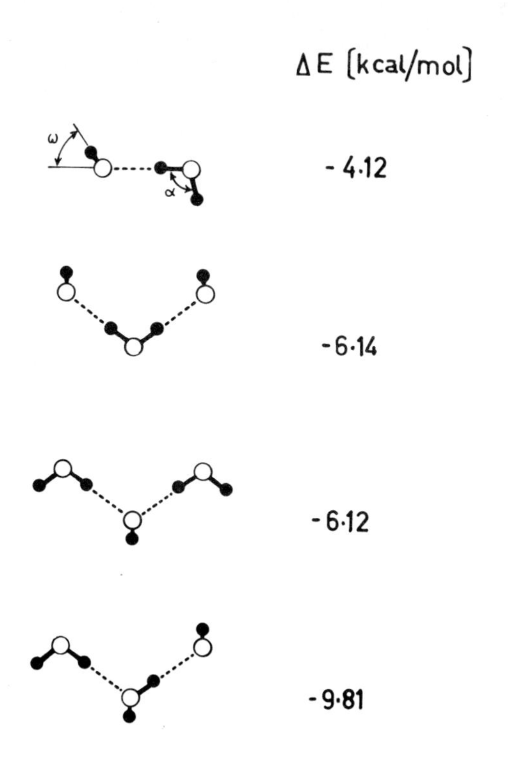

Fig.5: Energy of interaction in water trimer for different intermolecular orientations (6)(from the top of the bottom: dimer, double acceptor, double donor and sequential trimer).

in the plane of the molecules.

Dipole moments and spectral properties of the water dimer are shown in table 4. Most values were calculated from the less accurate energy surface and wave functions of ref.(65). In general all the essential criteria for hydrogen bonding are reproduced very well by the calculations. For quantitatively correct dipole moments and force constants, however, correlation effects have to be taken into account as we have seen already in the case of the isolated water molecule (chapter II.3.)

Hydrogen bonded aggregates built up from 3 to 17 water molecules have been investigated by MO methods. The results are summarized together with the values obtained from crystal orbital calculations (75) on infinitely long H_2O chains in fig.5 and table 4. One of the most crucial difficulties of hydrogen bond statistics in water concerns the question of pairwise additivity of molecular interaction. The trimer of water has been investigated in detail by extensive ab initio

TABLE 4

Calculated and experimental data of water dimer

Quantity	LCAO-MO-SCF-Calc.[a]			Experimental data (71)			
	Monomer	Dimer	Ref.	Monomer	Dimer[f)]	Liquid water (1atm, 20°C)	Ice I
R_{OH} (Å)	.944 .966	.948 .976	5 65	.957			1.01
∢HOH (°)	105.3		5	104.5			
R_{OO} (Å)		3.0	5,6,7		-5.0	2.82	2.74
ΔE(kcal/mole)		-4.83 -3.90	5 7		-5.0	-(1.3-4.5)	-(5.7- -6.7)
ω(°)[b)]		40 30	6 7				
μ(D)	2.21	4.29	5	1.83			
Δμ(D)[g)]		0.32	5				
f_{OH}(10^5dyn/cm)	12.45	9.96	65	8.45			
ν_{OH}(cm^{-1})	4020	3806	c,70	3707	3619	3400	3277
$\frac{d\mu}{dR_{OH}}$ (D/Å)	0,7	2.0	65				
$\left(\frac{d\mu}{dR_{OH}}\right)^2$ dimer / $\left(\frac{d\mu}{dR_{OH}}\right)^2$ monomer [d)]		8.2 5.3	65 5		12		
f_{OO} (10^5dyn/cm)		0.18	65				0.17- -0.19
ν_{OO} (cm^{-1})		152	c,70				
σ_H[e)] (p.p.m.)	28.3	27.6	11	30.20		25.62	

a) The following basis sets were applied in ref.65: GTO (10,5,5) contracted to (31/1); in ref.6: GTO (10,5,1/4,1) contracted to (531/21); in ref.5: GTO (11,7,1/6,1) contracted to (541/31) and in ref.7: GTO (13,8,2,1/6,2,1) contracted to (8521/41). The short notation for basis used here is defined in chapter II.3, table 2 (see also table 3).

b) The intermolecular angle ω is defined in fig.4.

c) The frequencies for the decoupled stretching vibration of the OH bond were calculated from the data of ref.5.

d) This quantity is proportional to the increase in band intensity of the OH stretching vibration due to molecular association.

e) Chemical shift relative to the free proton.

f) Diluted water vapour in solid N_2 at 20°K (72).

g) Increase in dipole moment relative to two isolated H_2O molecules in the geometrical arrangement of the dimer.

TABLE 5

Hydrogen bond energies in different aggregates of water molecules

Structure	n[a)]	h[a)]	$\Delta E_{n-1,n}$ (kcal/mole)[a)] ab initio calc. Ref.(6)	Ref.(15)	CNDO/2 calc. (73) min.g.[b)]	exp.g.[b)]
$(H_2O)_n$	2	1	-4.72	-6.10	-8.68	-4.8[d)]
Sequential chains	3	2	-5.09	-8.56	-10.59	
..OH..OH..OH.. (H H H)	4	3		-9 91	-11.11	
	5	4			-11.34	
	6	5			-11.44	
	7	6			-11.50	
	8	7			-11.54	-6.71
	∞	∞			-15.27[c)]	-7.58[c)]
$(H_2O)_n$	5	4		-7.28	- 9.00	
Tetrahedral clusters	8	7			- 9.15	
	11	10			- 9.25	
$(H_2O..H-)_2O$ (..H–OH, ..H–OH)	14	13			- 9.30	
	17	16			- 9.30	

a) $\Delta E_{n-1,n}$ is the energy of the last hydrogen bond:

$$\Delta E_{n-1,n} = E\big((H_2O)_n\big) - \big(E\big((H_2O)_{n-1}\big) + E(H_2O)\big)$$

n represents the number of water molecules, h the number of hydrogen bonds in the cluster. For tetrahedral clusters the average hydrogen bond energy is given.

b) "Min.g." means that the geometry at the CNDO/2 energy minimum of H_2O, "exp.g." that the experimental geometry of H_2O was used in the calculations (73)

c) CNDO/2-Crystal orbital calculations up to 5th neighbour interactions (73)

d) See also ref. (74)

calculations (6). The ΔE values obtained differ largely from the energy for two hydrogen bonds in independent dimers (fig.5). Additionally, a striking difference in energy is found for the three possible arrangements of hydrogen bonds to the central water molecule. Characteristicly these structures are called double acceptor, double donor and sequential trimer. In both double acceptor and double donor the hydrogen bond energy is appreciable smaller than in the sequential trimer. This finding can be explained easily by charge transfer arguments. The first hydrogen bond in the double acceptor increases the electron density at the central water molecule. Hence the second proton is less acidic and consequently forms a weaker hydrogen bond. An analogous argument can be used for the double donor, where the basicity of the lone pair is reduced by charge transfer from the first hydrogen bond. Only in the sequential trimer the first hydrogen bond increases the capability of the central water molecule to form a second hydrogen bond. Accordingly, the mean hydrogen bond energy is appreciable larger in this structure than in the dimer.

In sequential chains of water molecules, $H^{OH}\cdot\cdot(_H{}^{OH})_k\cdot\cdot{}_H{}^{OH}$, the hydrogen bond energy increases with increasing chain length until it reaches a certain limit. Around k~4 it becomes more or less constant. The mean hydrogen bond energy obtained by crystal orbital (CNDO/2) calculations is always somewhat larger than the values extrapolated from MO calculations on chains with different length to infinity. Tetrahedral arrangements of water molecules similar to the geometry found in ice I were investigated too. Interestingly, in these three dimensional clusters the deviations from pairwise additivity of hydrogen bond energies are much smaller. Only a small residual extra stabilization energy of the oligomers with respect to the dimer remains. This result can be interpreted withour further difficulty. In the tetrahedral structure we do not find exclusively the favourable sequential arrangements of water molecules but also a high number of double donor and double acceptor subsystems.

II.4.4. HYDRATED HYDROXONIUM AND HYDROXIDE IONS

During the last few years many theoretical and experimental investigations on hydroxonium ions were performed. The most important contributions to the knowledge of these structures came from X-ray and neutron scattering (76,77), IR (77) and mass spectroscopy (41-43,45,46) and from ab initio calculations (16,18,21). The most accurate calculations were performed on the system $H_5O_2^+$ (18) and $H_3O_2^-$ (21). In both cases strong hydrogen bonds are formed (table 3).

The most stable structure of $H_5O_2^+$ shows an usually short OO-distance R_{OO}=2.39 Å and an energy of interaction of ΔE=-32.2 kcal/mole. At this OO-distance a symmetric double minimum potential with an extremely small potential barrier for proton motion is obtained (ΔE < 0.02 kcal/mole). Calculations with less extended basis sets led to a

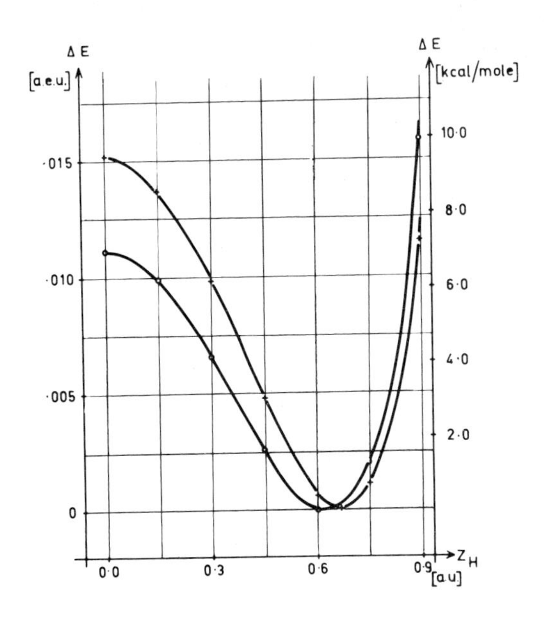

Fig.6: Electron correlation effects of the energy curve for proton transfer in $H_5O_2^+$ (R_{OO}=2.74 Å; -+-+-+-+-: SCF results, -o-o-o-o- CEPA-results (20). For an explanation of "CEPA" see chapter II. 3.

single minimum potential (16). In any case the zero point level lies above the energy maximum and we have to expect a symmetric structure with the proton in the middle between the two oxygen atoms for $H_5O_2^+$. Model studies on $H_5O_2^+$ at larger distances were performed in order to simulate an average environmental effect of the liquid or solid state (19,20). The vibrational spectra and the polarizabilities of $H_5O_2^+$ were investigated in detail with a medium size basis set (19). The influence of electron correlation on the potential curve for proton transfer in $H_5O_2^+$ was studied at fixed OO distances by the "CEPA" method (20). Actually a remarkable decrease of the potential barrier was found (fig.6). At R_{OO}=2.74 Å the barrier height is reduced from the HF value of ΔE^+=9.5 kcal/mole to about 7.0 kcal/mole. Hence correlation effects have to be taken into account if accurate energy surfaces for proton transfer are desired (78).

The hydrogen bond in $H_3O_2^-$ is not as strong as in $H_5O_2^+$ (ΔE=-24.3 kcal/mole). Accordingly the OO distance at the energy minimum is somewhat larger in $H_3O_2^-$ (R_{OO}=2.51 Å) and an asymmetric structure with the proton much nearer to one oxygen is obtained. The

TABLE 6

Hydration of hydroxonium and hydroxide ions (16,18,21)

n	Structure	Calculated energy minima					
		Sym.	E_{SCF} (a.e.u.)	R_{OO} (Å)	$\Delta E^{SCF\,c)}_{n-1,n}$ (kcal/mole)	$\Delta E^{EXP\,c)}_{n-1,n}$ (kcal/mole)	Ref.
0	H_3O^+	D_{3h}	-76.32798 a)				
		D_{3h}	-76.2006 b)				
1	$H_5O_2^+$	C_{2v} (D_{2d})	-152.42848 a)	2.39	32.24 d)	32.6	42
						36	41
		D_2	-152.1791 b)	2.36	43.87	32.3	43
2	$H_7O_3^+$	C_{2v}	-228.1370 b)	2.46	30.94	19.5	42
						22.3	41
3	$H_9O_4^+$	C_{3v}	-304.0872 b)	2.54	26.11	17.5	42
						17	41
4	$H_{11}O_5^+$	C_s	-380.0240 b)	2.54 2.70	17.70	15.3	41
0	HO^-	$C_{\infty v}$	-75.40671 a) -75.2298 b)				
1	$H_3O_2^-$	C_s	-151.49744 a)	2.51	24,31	22.5	45
		C_s	-151.2033 b)	2.45	40.73	34.6	46
2	$H_5O_3^-$	C_{2v}	-227.1599 b)	2.53	30.50	16.4	45
3	$H_7O_4^-$	C_{3v}	-303.1053 h)	2.61	23.10	15.1	45
4	$H_9O_5^-$	C_s	-379.0469 b)	2.70	20.71	14.2	45

a) Basis set: GTO (11,7,1/6,1) contracted to 541/31 , (18,21), see also table 3

b) Basis set: STO, 4-31G, (16,38), see also table 3.

c) The energy difference $\Delta E_{n-1,n}$ is defined by the following equation (see also table 4).

$$\Delta E_{n-1,n} = E\big(X(H_2O)_n\big) - \{E\big(X(H_2O)_{n-1}\big) + E(H_2O)\}$$

$$X = H_3O^+ \text{ or } HO^-$$

d) Using the numbers given in ref.(18) as slightly different value is obtained for the energy difference: $\Delta E_{0,1}(H_3O^+) = 30.44$ kcal/mole)

OH bond in $H_3O_2^-$ is about 0.1 Å larger than in the isolated water molecule.

Less accurate calculations with smaller basis sets give too large energies of interaction and too small OO distances in both ions $H_5O_2^+$ and $H_3O_2^-$. As in the case of the water dimer extended basis sets are necessary for a reliable description of these strong hydrogen bonds.

TABLE 7

Proton transfer energies in higher aggregates of water

Reaction	ΔE_{SCF} (kcal/mole)	Ref.	ΔE_{EXP}	Ref.
$2H_2O \rightarrow H_3O^+ + OH^-$	231.4[a]	53	219	22,54
	231.8[a]	5,18,21		
	242.8[c]	16		
$2(H_2O)_2 \rightarrow H_5O_2^+ + H_3O_2^-$	186.7[b]	5,18,21		
	174.6[c]	16		
$2(H_2O)_3 \rightarrow H_7O_3^+ + H_5O_3^-$	145.4[c]	16		
$H_2O(l) \rightarrow H_3O^+.aq + OH^-.aq$			23.0	55

[a] Basis set: GTO (11,7,2/6,2) contracted to (742/42)

[b] Basis set: GTO (11,7,1/6,1) contracted to (541/31) see also table 3

[c] Basis set: STO, 4-31G (16,38)

Aggregates of H_3O^- and OH^- with up to four water molecules were studied by ab initio methods with a medium size basis set (table 6.) As expected the hydrogen bond energy, $\Delta E_{n-1,n}$, decreases gradually with increasing n. Accordingly, the OO distance at the energy minimum becomes longer and longer when more water molecules are bound to the central ion. The calculated ΔE values were compared with experimental enthalpies obtained from the temperature dependence of equilibrium constants measured by mass spectroscopy (41-43,45,46). The agreement of the most accurately calculated ΔE values (18,21) with the experimental data is very good. The results of medium size basis set calculations do not agree as well but reproduce the trends correctly. An interesting question concerns the relative stability of the highly symmetric structures $H_9O_4^+$ and $H_7O_4^-$ which have been postulated as the predominant ionic species in liquid water. In the vapour phase no additional stabilization of these ions was found. The hydration energies decrease gradually and no break in the series is obtained.

In table 7 energies for proton transfer between water molecules in different stages of hydration are compared. The largest systems calculated describe proton transfer between water trimers. As expected the energy differences between the ionic and neutral species decrease with increasing number of solvating molecules. The aggregates considered, however, are much too small for a simulation of the situation in liquid water, as a comparison of calculated and experimental ΔE values show.

II.4.5. HYDRATION OF MONOATOMIC IONS

Recent progress in the experimental studies of ion/molecule complexes in the vapour phase (47-51) and in the liquid state (e,g.79, 80) encouraged theorists to start extended ab initio calculations in this field. Now quite a number of very accurate calculations on complexes of monoatomic ions with water molecules have been published (22-32,81).

Calculated structures of ion/water compexes are shown in fig.7. In the cationic complexes the structure predicted by purely electrostatic ion/dipole interaction is found to be most stable. As expected the distance between the metal and the oxygen atoms (R_{OX}) increases in the series Li, Na and K. In order to obtain more information on the metal ligand bond in these complexes the wave function of the $Li(OH_2)^+$ complex was analyzed in detail. The integral density difference function, $\Delta\bar{\rho}(z)$ was found to be most appropriate for this purpose (23):

$$\Delta\bar{\rho}(z) = \int\int\Delta\rho(\vec{R})dxdy;\ \Delta\rho(\vec{R}) = \rho_{Li(OH)_2^+} - \{\rho_{Li^+}\rho_{H_2O}\}^{+)}$$

As fig.8 shows the water molecule is polarized remarkably in the electrostatic field of the Li^+ cation. The K shell electrons of Li are displaced only slightly. There is no hint for an appreciable charge transfer in this complex (23). The structure of the complexes of anions with water is more complicated. In small ions a structure corresponding to an X-H-O hydrogen bond is prefered. The hydrogen bond in the $F(H_2O)^-$ ion is rather strong and the water molecule is more or less fixed with one OH bond. Deviations from linearity cause an appreciable increase in energy. NMR relaxation studies (79,80) confirmed that this hydrogen bonded structure with two not equivalent hydrogen atoms in the water molecule actually occurs in the hydrated F^- ion. In $Cl(H_2O)^-$ this hydrogen bond is much weaker and the arrangement shown in fig.7 is only slightly more stable than the electrostaticly favoured symmetric structure, $Cl^- {\cdot}{\cdot}{\cdot} \overset{H}{\underset{H}{>}}O$. Hence great mobility has to be expected for the water molecule in the hydrated Cl^- ion (30,31).

+) The isolated subsystems are arranged in exactly the same geometry as in the complex. For a definition of the density function $\rho(\vec{R})$ see chapter II.3.

TABLE 8

Structure and energies of ion water complexes (32)

Quantity	Complex				
	$Li(H_2O)^+$	$Na(H_2O)^+$	$K(H_2O)^+$	$F(H_2O)^-$	$Cl(H_2O)^-$
$\Delta E_{HF}^{a)}$ (kcal/mole)	-35.20	-23.95	-16.64	-23.70	-11.85
$\Delta E_{COR}^{b)}$ (kcal/mole)	+0.26	+0.26	+0.33	+0.75	0.57
$\Delta E_{00}^{c)}$ (kcal/mole)	2.24	1.74	1.51	3.40	1.80
ΔH_{298}^{calc} (kcal/mole)	-34.13	-23.32	-16.20	-22.18	-11.37
ΔH_{298}^{EXP} (kcal/mole)	-34.0	-24.0	-17.9	-23.3	-13.1
R_{OX} (Å)	1.89	2.25	2.69	2.51	3.31
∢ $\alpha^{d)}$	180°	180°	180°	47.9°	37.8°

a) The energies were obtained with the following contracted GTO basis sets for water (431/21) and for the ions: Li^+ (31) , Na^+ (741), K^+ (961), F^- (741) and Cl^- (961), see also table 3.

b) Correlation energy obtained by an estimate according to Wigner's method (82).

c) Zero point energy.

d) α is the angle between the twofold symmetry axis of the water molecule and the OX connection line (the four atoms H,H,O and X lie in a plane).

The energies of interaction calculated with basis sets of near HF quality agree very well with vapour phase enthalpies (24,26,27,28, 31). This good agreement encouraged a more careful estimate of the other much smaller contributions from vibrational and correlation energy (32). The contribution from electron correlation is very small (ΔE_{COR}<0.3 for cation/water complexes, ΔE_{COR}<0.8 kcal/mole for anion/ water complexes). As table 8 shows excellent agreement between calculated and experimental data is found for the small ions, Li^+, Na^+ and F^- slightly too small ΔH values were computed for K^+ and Cl^-.

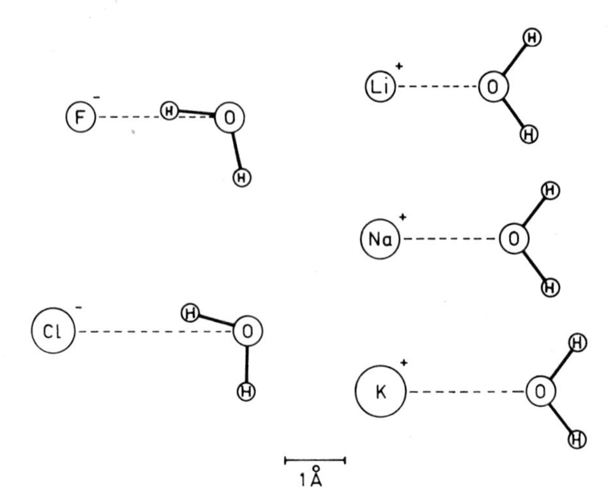

Fig.7: Geometries of ion water complexes

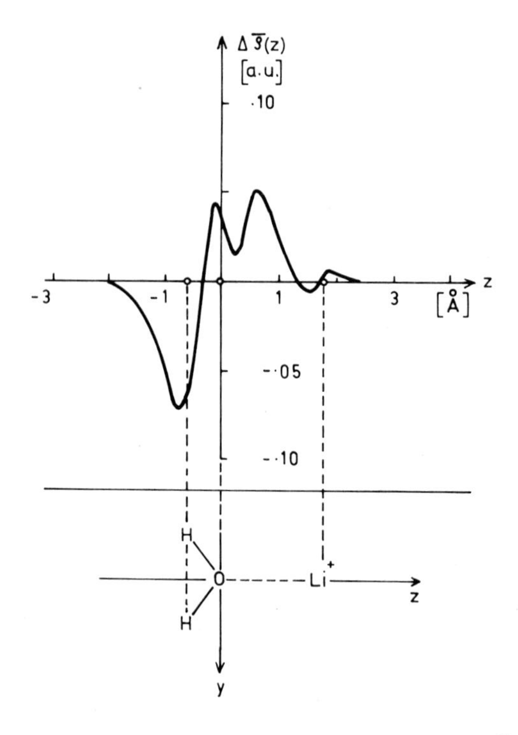

Fig.8: Integral electron density difference function $\Delta\bar{\rho}(z)$ in $Li(H_2O)^+$ (23).

TABLE 9

Complexes of ions with two and more water molecules

n	Quantity[a)]	$Li^+(OH_2)_n$ Calc.	Ref.	Exp.	Ref.	$F^-(H_2O)_n$ Calc.	Ref.	Exp.	Ref.
1	R_{OX}	1.89	24	1.92[c)]	47	2.52[b)]	28	2.46[c)]	49
	$\Delta E_{0,1}$	36.02		34.0		24.1		23.3	
2	R_{OX}	1.92	25	1.94	47	2.55	29		49
	$\Delta E_{1,2}$	31.52		25.8		20.8		16.6	
	$\overline{\Delta E}/2$	33.77		29.9		22.5		20.0	
3	R_{OX}			1.96	47				49
	$\Delta E_{2,3}$			20.7				13.7	
	$\overline{\Delta E}/3$			26.8				17.9	
4	R_{OX}			2.02	47				49
	$\Delta E_{3,4}$			16.4				13.5	
	$\overline{\Delta E}/4$			24.2				16.8	
5	R_{OX}			2.10	47				49
	$\Delta E_{4,5}$			13.9				13.2	
	$\overline{\Delta E}/5$			22.2				16.1	
6	R_{OX}			2.16	47				
	$\Delta E_{5,6}$			12.1					
	$\overline{\Delta E}/6$			10.5					

a) All distances are given in Å, all energies in kcal/mole. For the definition of $\Delta E_{n-1,n}$ see tables 4 and 5. $\overline{\Delta E}/n$ represents the mean energy of hydration per water molecule:

$$\overline{\Delta E} = \sum_{k=1}^{n} \Delta E_{k-1,k}$$

b) The value R_{OF}=2.41 Å in the abstract of ref.(28) seems to be a typographical error, since it does not fit to the other data presented in the paper. See also ref.(31)

c) The "experimental" distances were calculated by fitting an electrostatic model to the experimental energies (47,49). In the case of F^-H_2O the value is not reliable, since a structure with C_{2v}-symmetry was applied (49).

A very recent calculation on $Li(OH_2)^+$ including electron correlation (81)[+)] gave a somewhat larger contribution of correlation energy which additionally shows opposite sign: ΔE_{SCF}=-36.1, ΔE_{COR}=-1.0 and ΔE = -35.1 kcal/mole.

Complexes of ions with two or more water molecules have been calculated by semiempirical methods (83-86). These CNDO/2 calculations on ionic complexes, however, turned out to be rather unreliable, because charge transfer is extremely overemphasized by the semiempirical procedure. The only ab initio results available now concern the system $Li(H_2O)_2^+$ (25) and $F(H_2O)_2^-$ (29). In table 9 calculated and experimental results are compared. It seems that calculations overemphasize the stability of higher substituted complexes a little. As expected the mean energy of interaction decreases gradually with an increasing number of ligands. Together with the decrease in energy an increase in the ion oxygen distance R_{OX} is found.

II.4.6. CONCLUSION

Theoretical calculations on molecular interactions present a very important contribution to an understanding of this very complex field. LCAO-MO calculations, nowadays, can be performed with an accuracy comparable to experimental measurements. Provided extended basis sets are used the results are reliable and no serious discrepancies with the experimental data were found. In general calculations of near HF quality will be sufficient for a discussion of intermolecular geometries and energies of interaction. For many other purposes, e.g. the calculation of vibrational spectra, or for very accurate values correlation effects have to be taken into account.

ACKNOWLEDGEMENTS

The author is indebted to Dr.E.Clementi, Dr.J.P.Daudey, Dr.R. Janoschek, D.H.Lischka and Dr.B.Roos for sending their most recent results prior to publication. The assistence of out technical staff, Mrs.J.Dura and Mr.J.Schuster with the preparation of the manuscript is gratefully acknowledged.

[+)] SCF-CI calculation including 5613 configurations for the $Li(H_2O)^+$ complex.

REFERENCES

1 S.Bratoz, "Electronic Theories of Hydrogen Bonding" in Advances of Quantum Chemistry 3, 209 (1967), ed. P.O.Löwdin, Academic Press, New York

2 P.A.Kollman, L.C.Allen, Chem.Rev. 72, 283 (1972)

3 C.N.R.Rao, "Theory of Hydrogen Bonding in Water", in "Water", Ed.F.Franks, Vol.1, Plenum Press New York 1972, p.93-114

4 P.Schuster, Z.f.Chemie 13, 41 (1973)

5 G.H.F.Diercksen, Theoret.Chim.Acta (Berl.) 21, 335 (1971)

6 D.Hankins, J.W.Moskowitz, F.H.Stillinger, J.Chem.Phys. 53,4544 (1970)

7 H.Popkie, H,Kistenmacher, E.Clementi, J.Chem.Phys, 59, 1325 (1973)

8 J.G.C.M. van Duijneveldt-van de Rijdt, F.B. van Diijneveldt, J.Am.Chem.Soc. 93, 5644 (1971)

9 R.Sustmann, F.Vahrenholt, Theoret.Chim.Acta (Berl.) 29, 305 (1973)

10 J.P.Daudey, to be published

11 M.Jaszunski, A.J.Sadlej, Chem.Phys.Letters 15, 41 (1972)

12 L.R.Hoyland, L.B.Kier, Theoret.Chim.Acta (Berl.) 15, 1 (1969)

13 A.Karpfen, J.Ladik, P.Russegger, P.Schuster, S.Suhai, Theoret. Chim.Acta (Berl.) submitted for publication

14 C.Guidotti, U.Lamanna, M.Maestri, Theoret.Chim.Acta (Berl.) 26, 147 (1972)

15 J.DelBene, J.A.Pople, J.Chem.Phys. 52, 4858 (1970)

16 M.D.Newton, S.Ehrenson, J.Am.Chem.Soc. 93, 4971 (1971)

17 P.A.Kollman, L.C.Allen, J.Am.Chem.Soc. 92, 6101 (1970)

18 W.P.Kraemer, G.H.F.Diercksen, Chem.Phys.Letters 5, 463 (1970)

19 R.Janoschek, E.G.Weidemann, H.Pfeiffer, G.Zundel, J.Am.Chem.Soc. 94, 2387 (1972)

20 W.Meyer, W.Jakubetz, P.Schuster, Chem.Phys.Letters 21, 97 (197)

21 W.P.Kraemer, G.H.F.Diercksen, Theoret.Chim.Acta (Berl.) 23, 398 (1972)

22 P.Schuster, H.W.Preuss, Chem.Phys.Letters 11, 35 (1971)

23 P.Russegger, P.Schuster, Chem.Phys.Letters 19, 254 (1973)

24 G.H.F.Diercksen, W.P.Kraemer, Theoret.Chim.Acta (Berl.) 23, 387 (1972)

25 W.P.Kraemer, G.H.F.Diercksen, Theoret.Chim.Acta (Berl.) 23, 393 (1972)

26 E.Clementi, H,Popkie, J.Chem.Phys. 57, 1077 (1972)

27 H.Kistenmacher, H.Popkie, E.Clementi, J.Chem.Phys. 58, 1689 (1973)

28 G.H.F.Diercksen, W.P.Kraemer, Chem.Phys.Letters 5, 570 (1970)

29 W.P.Kraemer, G.H.F.Diercksen, Theoret.Chim.Acta (Berl.) 27 265 (1972)

30 L.Piela, Chem.Phys.Letters 19, 134 (1973)

31 H.Kistenmacher, H.Popkie, E.Clementi, J.Chem.Phys. 58,5627 (1973)

32 H.Kistenmacher, H.Popkie, E.Clementi, J.Chem.Phys., in press

33 P.Schuster, "Energy surface for hydrogen bonded systems", in "Recent Developments in Hydrogen Bonding" ,Ed.P.Schuster, G.Zundel, C.Sandorfy, North Holland Publ.Co., Amsterdam in press

34 P.Schuster, W.Jakubetz and W.Marius, "Model Studies on Ion Solvation", Progress in Chemistry, Springer Verlag, Heidelberg in press

35 J.N.Murrell, M.Randic, P.R.Williams, Proc.Roy.Soc.(London) A 284 566 (1965)

36 J.N.Murrell, G.Shaw, J.Chem.Phys. 46, 1768 (1967)

37 F.B.van Duijneveldt, J.N.Murrell, J.Chem.Phys. 46, 1759 (1967)

38 R.Ditchfield, W.J.Hehre, J.A.Pople, J.Chem.Phys. 54, 724 (1971)

39 W.J.Hehre, R.Ditchfield, J.A.Pople, J.Chem.Phys. 56, 2257 (1972)

40 G.Pimentel, A.D.McClellan, "The Hydrogen Bond", W.A.Freeman and Co., San Francisco 1960

41 P.Kebarle, S.K.Searles, A.Zolla, J.Scarborough, M.Arshadi, J.Am.Chem.Soc. 89, 6393 (1967)

42 A.J.Cunningham, J.D.Payzant, P.Kebarle, J.Am.Chem.Soc. 94, 7627 (1972)

43 M.DePaz, J.J.Leventhal, L.Friedman, J.Chem.Phys. 51, 3748 (1969)

44 P.Kebarle, J.Chem.Phys. 53, 2119 (1970)

45 M.Arshadi, P.Kebarle, J.Phys.Chem. 74, 1483 (1970)

46 M.DePaz, A.G.Giardini, L.Friedman, J.Chem.Phys. 52, 687 (1970)

47 I.Dzidic, P.Kebarle, J.Phys.Chem. 74, 1466 (1970)

48 J.D.Payzant, A.J.Cunningham, P.Kebarle, Can.J.Chem. 51, 3242 (1973)

49 M.Arshadi, R.Yamdagni, P.Kebarle, J.Phys.Chem. 74, 1475 (1970)

50 R.Yamdagni, P.Kebarle, J.Am.Chem.Soc. 94, 2940 (1972)

51 R.Yamdagni, J.D.Payzant, P.Kebarle, Can.J.Chem. 51, 2507 (1973)

52 P.A.Kollman, J.Am.Chem.Soc. 94, 1837 (1972)

53 H.Lischka, Theoret.Chim.Acta (Berl.) 31, 39 (1973)

54 A.C.Hopkinson, N.K.Holbrook, I.G.Csizmadia, J.Chem.Phys. 49, 3596 (1968)

55 G.L.Hofacker, private communication

56 D.F.Smith, J.Mol.Spectroscopy 3, 473 (1959)

57 G.F.F.Diercksen, W.P.Kraemer, Chem.Phys.Letters 6, 419, (197)

58 J.G.C.M.van Duijneveldt-van de Rijdt, F.B.van Duijneveldt, Theoret.Chim.Acta (Berl.) 19, 83 (1970)

59 P.A.Kollman, L.C.Allen, Theoret.Chim.Acta (Berl.) 18, 399 (1970)

60 M.Dreyfus, A.Pullman, Theoret.Chim.Acta (Berl.) 19, 20 (1970)and M.Dreyfus, These 30 Cycle, University of Paris, 1970

61 H.Lischka, Chem.Phys.Letters 20, 448 (1963)

62 H.Lischka, Chem.Phys., in press

63 H.Lischka, personal communication

64 R.Bonaccorsi, C.Petrongolo, E.Scrocco, J.Tomasi, Theoret.Chim.Acta (Berl.) 20, 331 (1971)

65 P.A.Kollman, L.C.Allen, J.Chem.Phys. 51, 3286 (1969)

66 J.A.Pople, D.L.Beveridge, "Approximate Molecular Orbital Theory" McGraw Hill, New York 1970

67 J.A.Pople, G.A.Segal, J.Chem.Phys. 44, 3289 (1966)

68 E.Clementi, personal communication

69 P.Schuster, Theoret.Chim.Acta (Berl.) 19, 212 (1970)

70 R.Janoschek, "Calculated vibrational spectra of hydrogen bonded systems" in "Recent Developments in Hydrogen Bonding", Ed. P.Schuster, G.Zundel, C.Sandorfy, North Holland Publ.Co. Amsterdam, in press

71 D.Eisenberg, W.Kauzmann, "The Structure and Properties of Water", Clarendon Press, Oxford 1969

72 M.van Thiel, E.D.Becker, G.Pimentel, J.Chem.Phys. 27, 486 (1957)

73 A.Karpfen, J.Ladik, P.Russegger, P.Schuster, S.Suhai, Theoret. Chim.Acta (Berl.), submitted for publication

74 A.S.N.Murthy, C.N.R.Rao, Chem.Phys.Letters 2, 123 (1968)

75 G.Del Re, J.Ladik, G.Biczo, Phys.Rev. 155, 997 (1967)

76 J.O.Lundgren, I.Olovsson, "The hydrated proton in solids", in "Recent Developments in Hydrogen Bonding", Ed. P.Schuster, G.Zundel, C.Sandorfy, North Holland Publ. Co., Amsterdam in press

77 J.M.Williams, "Spectroscopic studies of the hydrated proton species. $H^+(H_2O)_n$ in crystalline compounds", in "Recent Developments in hydrogen bonding". Ed. P.Schuster, G.Zundel, C. Sandorfy, North Holland Publ.Co., Amsterdam, in press

78 P.Schuster, W.Jakubetz, G.Beier, W.Meier and B.M.Rode, "Potential curves for proton transfer along hydrogen bonds; in "Chemical and Biochemical Reactivity, the VIth Jerusalem Symposium on Quantum Chemistry and Biochemistry", Ed.E.Bergmann, B.Pullman, Academic Press, New York, in press

79 H.G.Hertz, M.D.Zeidler, "Nuclear Magnetic Relaxation in Hydrogen Bonded Liquids", in "Recent Developments in Hydrogen Bonding", Ed.P.Schuster, G.Zundel, C.Sandorfy, North Holland Publ.Co., Amsterdam, in press

80 H.G.Hertz, C.Rädle, Ber.Bunsenges.Physik.Chemie 77, 521 (1973)

81 B.Roos, G.H.F.Diercksen, to be published

82 E.P.Wigner, Phys.Rev. 46, 1002 (1934)

83 H.Lischka, Th.Plesser, P.Schuster, Chem.Phys.Letters 6, 263 (1970)

84 P.Russggger, H.Lischka, P.Schuster, Theoret.Chim.Acta (Berl.) 24, 191 (1972)

85 R.E.Burton, J.Daly, Trans.Faraday Soc. 66, 1281 (1970)

86 R.E.Burton, J.Daly, Trans.Faraday Soc. 67, 1219 (1971)

II.5. SOLUTE-SOLUTE INTERACTIONS AND THE EXCESS PROPERTIES OF AQUEOUS SOLUTIONS*

Harold L. Friedman, C. V. Krishnan and Lian-Pin Hwang
Department of Chemistry
State University of New York, Stony Brook, New York 11790

ABSTRACT

Beginning with models which specify the forces between pairs of solute molecules in a solvent, one can calculate the thermodynamic excess functions (activity coefficients, heats of dilution, etc.) of the model systems at concentrations up to about 1M about as accurately as the same coefficients of real systems can be measured. This is true for 1-1 and 1-2 electrolytes and their mixtures, for non-electrolytes, and for mixtures of electrolytes and non-electrolytes.

The potential $u_{ab}(r)$ of the force between solute particles a and b is well known at small r and, for ions, at large r as well. However, there are poorly known contributions to u_{ab} from the molecular structure of the solvent in the range of r from "contact" of the solute particles to where one to three solvent molecules will fit between them. Such effects have in the past been recognized and given various names such as solvation, dielectric saturation, and structure-breaking.

By adjusting the models until their thermodynamic excess functions agree with those of various real aqueous solutions we learn about the factors which determine these properties in real solutions. The conclusion is that the data are accounted for by quite weak solvation effects compared to what has been believed until now. Moreover, solvation mostly contributes an attractive term to u_{ab}.

Very recent calculations of a non-thermodynamic property of models of this kind, namely the NMR relaxation of Li^+ by paramagnetic ions in aqueous solution, have also been compared with experiment and tend to support the conclusion from the thermodynamic evidence.

---

*Grateful acknowledgement is made of the support of the work reported here by the National Science Foundation.

II.5.1. INTRODUCTION

The various phenomena which characterize solutions stem from the forces between molecules rather than from the chemical bonds that hold the molecules together. Because the chemical bond is the central concept in the part of chemistry that has developed at a spectacular rate, we are all very sophisticated in interpreting phenomena in terms of chemical bonds and interpreting the bonds themselves in terms of electronic theory. By comparison, our knowledge of intermolecular forces is in its infancy.

Intermolecular interactions are not so sharply sensitive as chemical bonds to distance and orientation as shown by the fact that, within limits which are well known, the internal structure of a molecule is the same in the gas, liquid and solid phases. Also the concept of valence in chemical bonding is scarcely applicable to intermolecular forces. Of course, there are interactions which are intermediate in many respects between chemical bonds and intermolecular interactions, for example, hydrogen bonds.

It is convenient to classify solution phenomena as solvation phenomena, which are determined by the interaction of the solute molecules with a mass of solvent, or excess phenomena, which are determined by solute-solute interactions in the solvent. Thus, in terms of the potential U_{ij} of the interaction of a molecule of species i with one of species j , if we call the solvent species w and the solute species x , then solvation phenomena depend upon U_{xw} and U_{ww} while excess phenomena depend upon U_{xx} (or $U_{xx'}$, for two solute species) as well. /1/ This classification is quite natural for unsymmetrical solutions in which the solute and solvent play distinctive roles, as in ionic and most aqueous solutions, rather than for symmetrical solutions, which are typified by mixtures of isotopic molecules.

Studies of solvation, especially in water, have been reviewed by us elsewhere /1,2/ so the present review is devoted to some studies of the use of data for thermodynamic excess functions and certain magnetic relaxation processes to investigate solute-solute interactions in aqueous solutions.

The thermodynamic excess functions, such as activity and osmotic coefficients, heats of dilution, and volume changes on dilution, can be calculated quite accurately for hypothetical model systems in which the solute-solute forces as functions of solute coordinates and solvent temperature and pressure are specified. By comparing the excess functions of the models with those of real systems we may hope to learn about the solute-solute forces in the real systems.

In the models studied here, the force between a solute particle of species i and one of species j , both immersed in a mass of solvent, is assumed to be derived from a pair-potential function of the form illustrated in Fig. 1. Since the two solute

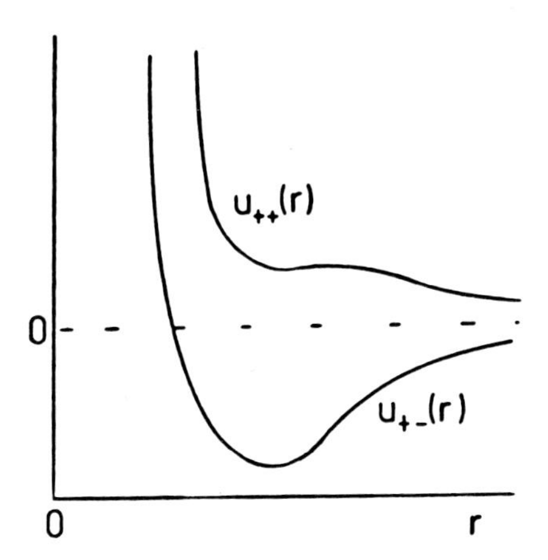

Fig. 1. Pair potentials for ionic solutes. Schematic.

particles cannot occupy the same space, the potential u_{ij} rises to very large positive values for $r < r^*_i + r^*_j$ where r^*_j is the radius of particle j estimated from crystallographic data. This important contribution to u_{ij} is called COR_{ij} ; it is described in more detail in § 2. We also know that if i and j are both charged (charges $z_i e_p$ and $z_j e_p$ where e_p is the charge on a proton) then at large enough r the dominant term in the pair potential is

$$COUL_{ij} = z_i z_j e_p^2/\varepsilon r \tag{1.1}$$

where ε is the dielectric constant of the pure solvent. Thus we know the behavior of u_{ij} at small r and at large r and our ignorance is limited to the intermediate region. From a practical point of view this means that our freedom to adjust the function $u_{ij}(r)$ to make a model which agrees with experiment for a particular solution is limited to the range

$$r^*_i + r^*_j < r < r^*_i + r^*_j + nr^*_w \tag{1.2}$$

where r^*_w is the mean radius of a solvent molecule and n is a small integer; whether 2 or 20 needs to be established.

In the work reviewed here a particular form is assumed for the part of the potential which accounts for the intermediate region, mainly the so-called Gurney (GUR) term, as well as about the magnitude of n . To some extent the assumptions are justified a-posteriori by the quality of the agreement of the computed model properties with the measured properties of real systems. Although one has a function to adjust in the model, the pair-potential u_{ij} is, by definition, independent of the solute concentrations, so getting agreement for the composition dependence of solution properties is a non-trivial test of the model. However, attempting to prove that a set of thermodynamic data can be uniquely interpreted in terms of model pair-potential functions is a losing game and it is preferable to focus attention on finding what can be learned about the solute-solute interactions that is independent of the uncertain features of the models studied.

The principal conclusion of the comparison of model calculations with thermodynamic properties is that for most pairs ij of solute species in aqueous solution, the contribution to $u_{ij}(r)$ in the intermediate region of r (cf. Eq.(2)) is most often negative and in any case rather small in magnitude in models that have the same thermodynamic excess functions as real systems. The study of a variety of aqueous solutions led to this conclusion as described in §4.

As already noted, it is not expected that one can determine uniquely the real pair potential by adjusting models to fit the experimental data. However, the actual studies show that very different models can be adjusted to fit the thermodynamic excess function data almost equally well. For example, as described in more detail in §4, one can fit the data for magnesium halide solutions either with models in which the hydration shell of Mg^{++} is readily penetrated by other ions or with models in which a layer of water on the magnesium is not penetrated when Mg^{++} collides with other ions in solution. Rigid solvation shells must contribute to the COR term of the pair potentials which then extend out much further than the COR term if the ions have penetrable hydration shells. However, it is found that for thermodynamic properties, this effect can be compensated by moderate changes in the parameter of the GUR term of the potential, the assumed intermediate range term mentioned above and defined in §2. When the two models are adjusted to fit the thermodynamic excess function data the calculated pair correlation functions* $g_{ij}(r)$, which are not directly measurable but which characterize the solution structure, are quite different. /3,4/

---

*The pair correlation function $g_{ij}(r)$ multiplied by the bulk concentration c_i of one of the species gives the local concentration of i particles at distance r from a j particle. Thus $g_{ij}(r)$ approaches zero as $r \to 0$ and $g_{ij}(r)$ approaches unity as $r \to \infty$. The average force on i due to its interaction with j is $-(\partial/\partial r)kT \ln g_{ij}(r)$. The pair correlation function depends upon the pressure and temperature and the composition of the solution as well as upon r although only the latter is shown in the conventional notation. For further information see references /3/ or /4/ or a current statistical mechanics text.

Ionic solutions with different structures can have closely similar thermodynamic excess functions up to moderate concentrations, say 1M, because any such function can be expressed in the form

$$f(c_1,\ldots,c_i,\ldots) = \Sigma_i \Sigma_j \; c_i c_j \int F_{ij}(r) \; g_{ij}(r) \; d^3r \qquad (1.3)$$

where i and j are solute species indices, c_i is the concentration of i, and $F_{ij}(r)$ is a weight function. It depends upon the thermodynamic excess property being calculated in Eq.(1.3) but for any such property it has the same form at large r, namely it varies as $1/r$. Clearly, this makes the convergence of the integral depend upon the large-r behavior of $g_{ij}(r)$ which in turn is determined mainly by the Coulomb interactions. Thus the COUL term of the potential tends to make a large contribution to the integral in Eq.(1.3) and thereby to obscure the contributions of the shorter-range interactions of the solvation shells of the ions.

We also note that the sum over pairs of species in Eq.(1.3) tends to obscure the effect of changes in any one pair potential function $u_{ij}(r)$. Under these circumstances it requires great accuracy, both in the experimental coefficients and in the model calculations, to get information about $u_{ij}(r)$ in the intermediate range of r. These observations prompt us to examine the theory of various non-thermodynamic solution properties which might be more suitable for learning about the solute-solute pair potential functions.

If we were to formulate the rate constant k_{ij} for the activation-controlled reaction of solute species i with solute species j in a manner parallel to Eq.(1.3) we would write

$$k_{ij}(c_1,\ldots,c_i,\ldots) = \int F_{ij}(r) \; g_{ij}(r) \; d^3r \qquad (1.4)$$

where F_{ij} depends upon the reaction but g_{ij} is the equilibrium pair correlation function as in Eq.(1.3), at least in a certain limiting case in which the rate is not too great. If the reaction occurs only when the separation of i and j has a certain value R then we have

$$F_{ij} = \overline{F}_{ij}\delta(R-r) \qquad (1.5)$$

where $\overline{F}_{ij}$ does not depend on r and $\delta(r)$ is the Dirac delta function. More generally, we probably have

$$F_{ij}(r) \simeq \overline{F}_{ij}e^{-\alpha r}/r^n \qquad (1.6)$$

where n is a small integer; this form has the typical distance dependence of overlap integrals. In either case, the weight function in Eq.(1.4) is much shorter-range than the weight function in Eq.(1.3). Thus, as a basis for comparing model calculations of a coefficient with experimental values, Eq.(1.4) suffers from neither of the disadvantages emphasized for Eq.(1.3): the behavior of the integrand at large r and the sum over all pairs of species. Unfortunately, the theory of chemical kinetics presently does not provide a basis for sufficiently accurate calculation of the weight function $F_{ij}(r)$. Indeed the promising application of Eq.(1.4) is more the other way around: When we have realistic models for the calculation of accurate $g_{ij}(r)$, we shall be able to advance the theory of the concentration-dependence of rate constants in solution by using Eq.(1.4) together with reasonable estimates for $F_{ij}(r)$.

Finally, we note that in the case that k_{ij} is the rate constant of a certain magnetic relaxation process rather than a chemical reaction, then it becomes possible to calculate the weight function quite accurately. In such cases, the weight function tends to have the form

$$F_{ij}(r) = \overline{F}_{ij}/r^6 \qquad (1.7)$$

modified by various complicating effects. Now it does indeed become possible to use the magnetic relaxation data to discriminate among models which fit the thermodynamic data equally well. This work, which is the first step in extending the calculations for fairly realistic models from thermodynamic coefficients to dynamic coefficients, is reviewed in §5.

II.5.2. MODELS

The solute-solute pair potential in the models studied here is /5/

$$u_{ij}(r) = COUL_{ij} + COR_{ij} + CAV_{ij} + GUR_{ij} \qquad (2.1)$$

The first term has been defined in §1. The second term has been assumed to have either the inverse power form /5/

$$COR_{ij} = B_{ij}[(r^*_i + r^*_j)/r]^n \qquad (2.2)$$

or the exponential form /6,7/

$$COR_{ij} = kT^*\exp[(r^*_i + r^*_j - r)/R] \qquad (2.3)$$

In either case, the parameters of the potential function are determined largely from crystal data. In the case of the exponential function, molecular beam scattering data establish the range of R values that may be relevant. The much studied hard-sphere repulsive potential is the limiting case of either of these forms, as $n \to \infty$ for the first and as $R \to 0$ for the second. The reason one may determine the parameters of COR_{ij} from non-solution data is that the repulsive force it represents develops from the operation of the Pauli exclusion principle when i and j approach closely enough so their electron clouds tend to overlap; this effect is approximately independent of the medium.

The term CAV_{ij} represents a particular known dielectric effect: A cavity in a dielectric medium is polarized by an electric field, in a sense opposite to the familiar picture of the polarization of a dielectric sphere in an electric field. That is, the polarization results in a force pushing the cavity toward a region of lower field. Now if one considers a model in which each ion is a cavity in the dielectric medium of radius r^* (the crystal radius of the ion) and dielectric constant ε_c, with the ionic charge at its center, then the potential of the ion-cavity interaction of two ions is given by electrostatics as:

$$CAV_{ij} = (e_p^2/2\varepsilon r^4)[(\varepsilon-\varepsilon_c)/(2\varepsilon+\varepsilon_c)](z_i^2 r^*_j{}^3 + z_j^2 r^*_i{}^3) + \cdots \qquad (2.4)$$

where ε is the dielectric constant of the solvent and the omitted terms are of order r^{-5}.

While it is of interest to find the effect of this term on the thermodynamic properties, there are some respects in which it is unrealistic, even for the alkali halides. /5/

In the model considered here, the missing parts of the cavity term, together with other poorly understood effects related to the molecular structure of the solvent, are considered to contribute to a catch-all term with an adjustable parameter. In the hope that this parameter would show enough regularity to be useful for predictive and interpretive purposes, the term is formulated according to a concept described by Gurney /8/ and Frank /9/ according to which one supposes that an ion which is far from the others is surrounded by a cosphere of solvent in which the solvent has different properties than the solvent further away. Then we can picture in the following simple way the mechanism by which these cospheres contribute to the pair potential.

When two ions come close enough together the sum of their cosphere volumes is reduced by overlap, if the cospheres are otherwise fixed, as in Fig. 2. The mutual volume of the cospheres represents the volume of solvent which must return to its normal state. The free energy change accompanying this process gives rise to the term GUR_{ij} in Eq.(2.1), the Gurney potential. We assume: /5/

$$GUR_{ij} = A_{ij} V_w^{-1} V_{mu}(r_i^* + w,\ r_j^* + w, r), \qquad (2.5)$$

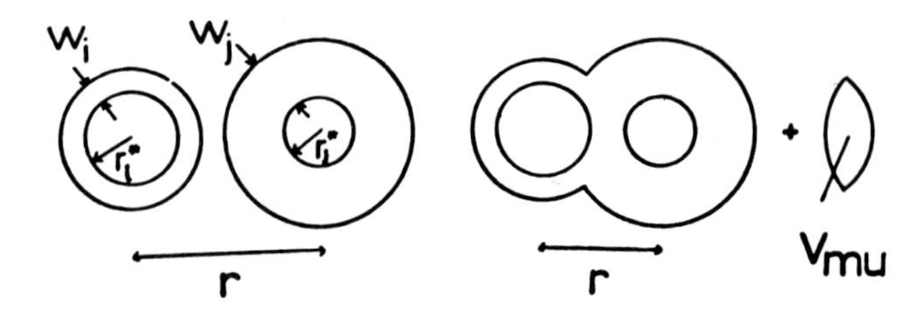

Fig. 2. Explanation of the Gurney term in the pair potential. In the configuration on the left the cospheres do not overlap so both V_{mu} and GUR are zero. The configuration on the right shows the overlapping cospheres and the volume of the displaced cosphere material.

where A_{ij} is the molar free energy change of the cosphere solvent returning to its normal state, V_w is the molar volume of the normal solvent (cubic angstroms/mole if r is in angstroms), and V_{mu} is the mutual volume function which is deduced by geometric considerations. The cosphere thickness w is taken to be the diameter of a water molecule, 2.76A in the results reported here, except as otherwise specified.

In this model A_{ij} is the only parameter in Eq.(2.1) which is adjusted on the basis of comparison with experimental data for electrolyte solutions. The general shape of each contribution to u_{+-} in this model has been represented for typical cases. /5-7/ A model of this kind does not completely neglect the molecular nature of the solvent which enters not only in parameters ε and w, but also in the adjustable parameter A_{ij}. However, it does neglect the effects which make the potential of interaction of several solute particles in a given configuration differ from the sum of the pair potentials. /5/

II.5.3. COMPARISON OF MODEL PROPERTIES

The basic technique /3,10/ used here is to apply the hypernetted-chain integral equation approximation to calculate the pair correlation functions $g_{ij}(r)$ at various solution compositions from the pair potentials u_{ij} which specify the model. This requires extensive numerical computations. /3,10,11/ In those cases in which the concentrations are low enough or in which models for non-ionic solutions or for ion,non-electrolyte interactions are studied, the HNC approximation can be replaced by a simpler one, namely the calculation of the second virial coefficient of the osmotic pressure, the so-called Poirier or DHLL+B_2 approximation. /3,6,12/

In comparing model properties with experiment, we encounter the problem that the model properties are calculated in the first instance in the McMillan-Mayer (MM) system of variables; $\underset{\sim}{c}$,T,P_0 while the experimental data are found (preferably) in what we now choose to call the Lewis and Randall (LR) system of variables, $\underset{\sim}{m}$,T,P. Here $\underset{\sim}{c}=c_1,c_2,\ldots$ is the set of solute molarities in the solution in osmotic equilibrium with pure solvent at pressure P_0 while $\underset{\sim}{m}=m_1,m_2\ldots$ is the set of solute molalities and the solution is at a pressure P near 1 atm which is independent of composition. It seems most convenient to convert experimental data in the LR system to the MM system, using the required thermodynamic experimental data for the conversion. This has been done with the help of the thermodynamic theory of the MM - LR conversion. /13,14/

II.5.4. SUMMARY OF RESULTS FOR THERMODYNAMIC EXCESS FUNCTIONS OF AQUEOUS SOLUTIONS

When the model in Eq.(2.1) is fitted to the data for the aqueous alkali halides using either the r^{-9} core potential or the exponential core potential it is found that the model can be fitted to the data quite satisfactorily up to an electrolyte concentration of 1M . /5/ A typical set of results is shown in Fig. 3 while a set

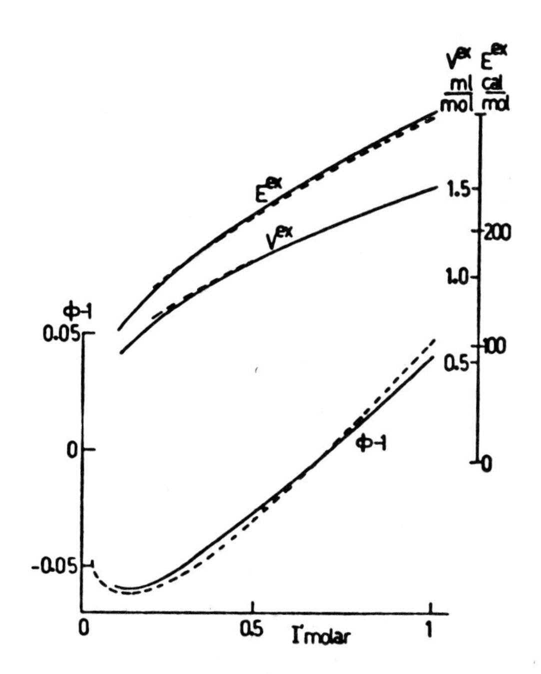

Fig. 3. Comparison of model and experimentally observed properties for LiCl(aq) at 25°C. --- Computed from a model. —— Experimental, converted to MM variables. I' is the molar ionic strength. The results are from /5/.

of Gurney A_{ij} parameters which is adequate for the alkali halides at 25° is given in Fig. 4. To fit the derivatives of excess free energy functions one must

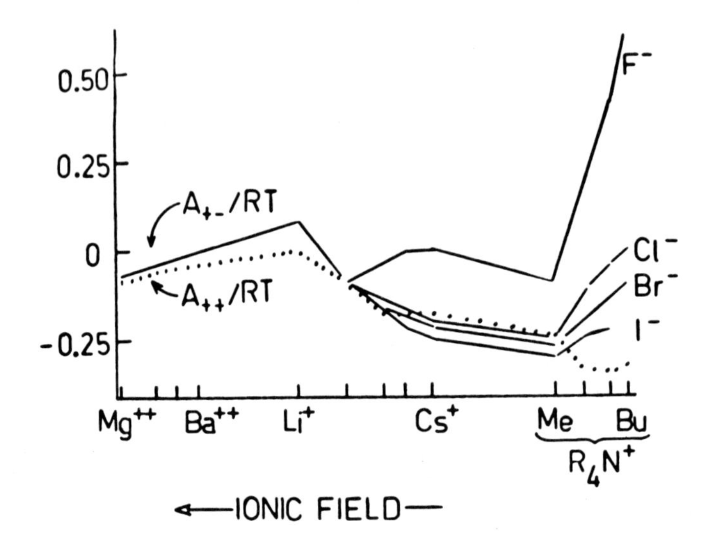

Fig. 4. Gurney free energy parameters for aqueous solutions at 25C. The "ionic field" variable is $\log(z_+\mu_w(r_+^* + 1.38)^{-1})$ increasing to the left. It gives a convenient scale for the cationic species. In the series Mg^{++} to Ba^{++} the cospheres on the cations are "2 soft", while for the rest the cospheres have a thickness of one water molecule. It is assumed throughout that $A_{--}=0$ for all the anions. /5/ There are no F^- data for the cations to the left of Na^+ while for these cations A_{+-} is the same for Cl^-, Br^-, and I^- . /6/

specify the corresponding derivatives of the model's A_{ij} . These derivatives for models which fit the data also have been reported. /5/

Because of the way they are defined (cf. Fig. 2) the Gurney parameters are readily discussed in chemical terms. The aspect chosen for emphasis here is the uniformly small magnitude and mostly negative sign found for the A_{ij} coefficients. This shows that the hydration shells of ions have little effect upon the ions' equilibrium distribution functions $g_{ij}(r)$ in solution. Such effect as there is tends mostly to make the ions attract each other more than would otherwise be expected or, in the case of pairs such as Cs-Cs (or Li-Cs, /5/, A_{ij} ~-200 cal/mol) perhaps it is better to say that they repel each other less than we might have expected. This is evidence for the hypothesis which we now proceed to test extensively,namely that solvation of solutes in aqueous solution most often tends to make the solute species attract each other more (or repel each other less) than would have been expected if the medium were an ideal structureless dielectric.

Earlier it had been found that for COR terms of the hard sphere form, adjusted to fit crystal radii, and for GUR terms of the square well type, positive values of A_{ij} were required in most cases to fit alkali halide data, /16,17/ but they were so small that the calculated correlation functions g_{ij} showed that in the models the solvation spheres did not keep i and j apart. When the COR potential is made more realistic, its contribution to the excess free energy increases /5/ and then negative values of A_{ij} are required to fit the models to the data for most of the alkali halides.

The same models have been fit to the properties of alkaline-earth halides /6/ and tetraalkyl ammonium halides. /7/ In each case there are some new problems as discussed in the original reports, but it is interesting to note that if these are disregarded for the moment then the Gurney A_{ij} parameters of all of these electrolytes may be compared as shown in Fig. 4, where the abscissa is proportional to the logarithm of the cation's contribution to the electric displacement (Maxwell's D) at the center of an adjacent water molecule. This simple correlation shows that the excess function data, reduced to A_{ij} values, show little evidence of specific structural contributions of any kind, except that it has been assumed that the cospheres on the alkaline earth ions have a thickness of two water molecules rather than one. /6/

Similar conclusions are reached from fitting models to the data for Setchenow coefficients (salting-out coefficients) of aqueous solutions /18/ except that there seems to be evidence for a contribution from dispersion or charge-polarizability interactions which are not represented in the present model and not clearly discernible in the data discussed above. Moreover the Setchenow results indicate that the solvation layer on Li^+ is qualitatively different from that on the other alkali metal ions; perhaps the thickness of the Li^+ cosphere is two water molecules rather than one.

Both the Setchenow coefficient studies and the studies for aqueous solutions of the alcohols /12/ show that A_{ij} near -100 cal/mole of water displaced characterizes the cosphere interaction of hydrophobic solute particles. This figure is largely in agreement with the results shown in Fig. 4 for A_{++} for R_4N^+ ions and with more recent calculations for the same systems by Wen and Streng. /19/

Since the A_{ij} parameters are mostly negative it follows that the effect on hydration of solute species is mostly to make pairs of solute particles attract each other more when they are fairly close together.

For some solutes, other models in which part of the hydration shell is represented as a rigid layer, not displaced in collisions with other solute particles, also have been studied. /6/ Several such models for aqueous $MgCl_2$ solutions are specified in Table 1, using a more descriptive nomenclature than in the original study. With the Gurney parameters given, these three models fit the osmotic coefficient data for aqueous $MgCl_2$ up to ionic strength I=1M about equally well. However, as

Table 1. Models for $MgCl_2$(aq) at 25°C.[a]

Model name	r_+^*	w_+	w_-	A_{++}	A_{+-}	A_{--}	$g_{+-}(3A)$ at I=1.2M
1 soft	r_+^P	d	d	0	310	-110	1.5
1 rigid, 1 soft	r_+^P+d	d	d	-145	-82	0	0.0
2 soft	r_+^P	2d	d	-52	38	0	0.5

a. r_i^* is the size parameter in Eq. (2.3) while r_i^P is the Pauling crystal radius. In all these models $r_-^* = r_-^P$. The diameter of a water molecule is d = 2.76A. The cosphere thickness parameter in Eq. (2.5) is w_i. The A_{ij} coefficients are given in units of calories per mole of water displaced.

expressed in terms of the pair correlation functions, the solution structures of the three models are quite different. For example, the Mg^{++},Cl^- correlation function at r=3.0A and at ionic strength I=1.2M has the values given in Table 1. Thus, if Mg^{++} and Cl^- reacted with each other chemically at a separation of 3.0A (cf. discussion of equations (1.4) and (1.5)) we would have significantly different expectations for the reaction rates on the basis of the three models, all of which fit the osmotic coefficient data.

One way to resolve this unsatisfactory situation is by theoretical calculations of the potentials u_{ij} on the basis of ion-solvent interactions, as described in §6.1. Another is to compare experimental and calculated values of NMR relaxation rates, as mentioned in §1 and described in the following section.

II.5.5. NMR RELAXATION COEFFICIENTS

Models of the kind described above do not provide sufficient information for the calculation of most dynamical properties of solutions, whether they be transport coefficients, rate constants, or relaxation coefficients. The situation can be discussed in a simple way in terms of the information required to construct the weight function $F_{ij}(r)$ in equations (1.3) to (1.7). For a thermodynamic excess function the weight function is a functional of the pair potentials $u_{ij}(r)$ and their dependence upon the pressure of the solvent and the temperature. /5/ The weight function for the conductivity of an ionic solution depends upon the $u_{ij}(r)$ and, in addition, upon the fluctuating motion (brownian motion) of the individual ions in the solvent and upon the hydrodynamic interactions among the ions. /20,21/ More generally the weight function for the calculation of a dynamic quantity depends upon all of these features and also on other aspects of the i,j interaction; this is obvious in the case of the weight function for the rate constant of a chemical reaction.*

Because the actual solute brownian motion and hydrodynamic interaction are not very well known for small solute molecules (because of the molecular structure of real solvents) it is highly desirable to be able to calculate the weight function for a dynamical coefficient such that the actual F_{ij} depends only very weakly upon these motional quantities. The search for suitable coefficients for solutions led to the study of the theory of the NMR relaxation of $^7Li^+$ (nuclear spin I) by Mn^{++} or Ni^{++} (electron spin S) in aqueous solution. /22,23/

In the calculation of the I-nucleus relaxation times T_{1I} and T_{2I} it is assumed that the IS interaction is dipolar and that it is modulated by the change in the distance r_{IS} due to diffusion and by the electron relaxation processes measured by T_{1S} and T_{2S}. The electron relaxation processes are attributed to the zero-field splitting modulated by fluctuations in the solvation of M^{++} with the correlation time τ_v. /24/ For Ni^{++} we have $T_{nS}/\tau_v \simeq 1$ so the process must be treated as non-Markoffian. /23/ This leads to a so-called dynamic shift in the S and I resonances even when the IS interaction is dipolar.

The various parameters for the electron spin relaxation may be adjusted to fit data for the EPR, proton, and ^{17}O relaxations† in the aqueous M^{++} solution. /22,23/

---

*In the present formulation the weight function for the rate constant becomes insensitive to the brownian motion of the particles and to their hydrodynamic interaction if the rate is activation-controlled with a barrier much greater than kT and in some other circumstances as well, as exemplified by one of the systems considered in detail in this section.

†In this case the scalar term in the IS interaction is included.

The measurable coefficients of interest are defined as

$$k_{IS}^{(n)} = (\partial T_{nI}^{-1}/\partial c_S), \; n = 1 \text{ or } 2 \tag{5.1}$$

where c_S is the concentration of S spins and the relevant variation corresponds to replacing the paramagnetic ion M^{++} by a similar diamagnetic ion, say Mg^{++} .* These rate constants are given by the equation (cf. Eq. (1.4))

$$k_{IS}^{(n)} = \int F_{IS}^{(n)}(r) \, g_{IS}(r) \, d^3r \tag{5.2}$$

where g_{IS} is the pair correlation function for the particle carrying the I spin, here Li^+ , and the particle carrying the S spin, here Mn^{++} or Ni^{++} and the weight function is

$$F_{IS}^{(n)} = (\gamma_I \gamma_S \hbar)^2 S(S+1) \frac{4}{15\pi d r^3 n} [3f_0(\omega_I) + 7f_1(\omega_I) + (n-1)(4f_0(0) + 6f_1(0))] \tag{5.3}$$

where the spectral density function is

$$f_m(\omega) = \int_0^\infty [2k^2 D + 1/T_{1+m,S} - i(\omega - Z_m - m\omega_S)]^{-1} j_1(kd) \, j_2(kr) \, k \, dk \tag{5.4}$$

These equations are obtained from those previously given /22/ by rearranging the order of operations. Here ω_I is the nuclear Larmor frequency and ω_S is the electronic Larmor frequency, both in the experimental magnetic field, and Z_m is the dynamical shift of the EPR spectrum (also a function of the magnetic field) which is a novel and important feature of this theory. The parameters d and D are the distance of closest approach and the mean diffusion coefficient, just as in Abragam's theory of the translational diffusion mechanism for the dipolar relaxation of an I spin by an S spin. /25/ If the relaxation rates $1/T_{1S}$ and $1/T_{2S}$ and the EPR dynamical shift are all neglected, then the present theory reduces to his.

In the bracketed factor in Eq.(5.4), k^2D gives the contribution to the spectral

---

*Thus if $1/T_{1I}$ is plotted as a function of c_S for a series of experiments in which $c_S + c_{Mg^{++}}$ is constant, the slope of the plot at any c_S is $k_{IS}^{(1)}$. The calculation described here applies when the c_S is small enough so that the slope doesn't depend on c_S . Similarly, $k_{IS}^{(2)}$ is determined by data for T_{2S} . In terms of the classification in §1, the slope is a dynamic excess function while the intercept at $c_S=0$, extrapolated also to $c_{Mg^{++}} = 0$ is a dynamic solvation coefficient. /2/

density due to* $\dot{r}_{IS}$ while $1/T_{1+m,S}$ gives the contribution due to electron relaxation. With suitable definitions of T_{1S} and T_{2S} the theory is valid even if the electron spin correlation function does not decay exponentially (i.e. the Bloch equations do not apply), a relevant consideration for Ni^{++}(aq) . /23,24/ The last term in the bracket shifts the frequency at which spectral density $f_m(\omega)$ is effectively evaluated from ω to $\omega - Z_m - m\omega_S$.

To get the results described below, the EPR parameters T_{1S}, T_{2S}, and Z_m are evaluated /22/ on the basis of the observed proton relaxation rates in the aqueous solutions of the paramagnetic ions and the integrals giving $F_{IS}^{(n)}$ are evaluated analytically.

The remaining integral is evaluated numerically using $g_{IS}(r)$ calculated for various models which (1) are consistent with the thermodynamic excess functions for the separate LiCl and MCl_2 solutions and (2) give values for the mixing coefficients w_0 which are in agreement with experiment /26/ for LiCl, $MnCl_2$ and LiCl,$NiCl_2$ mixtures.

Some results are shown in Figs. 5 and 6 for models for Mn^{++} and Ni^{++} , respectively, which are chosen in analogy with the models for Mg^{++} in Table 1. The Li^+ and Cl^- each have a single soft cosphere in all these models. The Gurney A_{ij} parameters have been adjusted to fit the thermodynamic data for the single electrolyte solutions and for the MCl_2,LiCl mixtures. While the $k_{IS}^{(1)}$ and $k_{IS}^{(2)}$ coefficients were actually calculated as described above, it is the nuclear relaxation rates $1/T_{nI}$ which are shown on the graph.

The importance of the assumed independent diffusion mechanism for $\dot{r}_{IS}$, which certainly is not very realistic, may be tested by varying the diffusion coefficient parameter D in the theory. Thus it is found that increasing D by a factor of about 1.6 for the 2 soft model lowers the model's relaxation curves in Fig. 5 until they agree with those shown for the 1 rigid, 1 soft model, while in Fig. 6 the corresponding effect is negligible. The reason is that the electron relaxation times T_{1S} and T_{2S} of Mn^{++}(aq) are so long that the spectral density is controlled mainly by $\dot{r}_{IS}$ while for Ni^{++}(aq) the electron relaxation times are so short that r_{IS} is nearly unchanged during the time it takes for the electron spin correlation to decay. At any rate, the unrealistic approximation for $\dot{r}_{IS}$ is much less important

---

*Here $\dot{r}_{IS} = dr_{IS}/dt$ refers to the fluctuating rate of change of r_{IS} due to the thermal motions of all of the molecules in the solution. While this fluctuating motion is most simply represented by assuming that the spins I and S are in particles which independently undergo Fick's Law diffusive motion /25/, this assumption about the motion is far from accurate when r_{IS} is relatively small. /20,21/

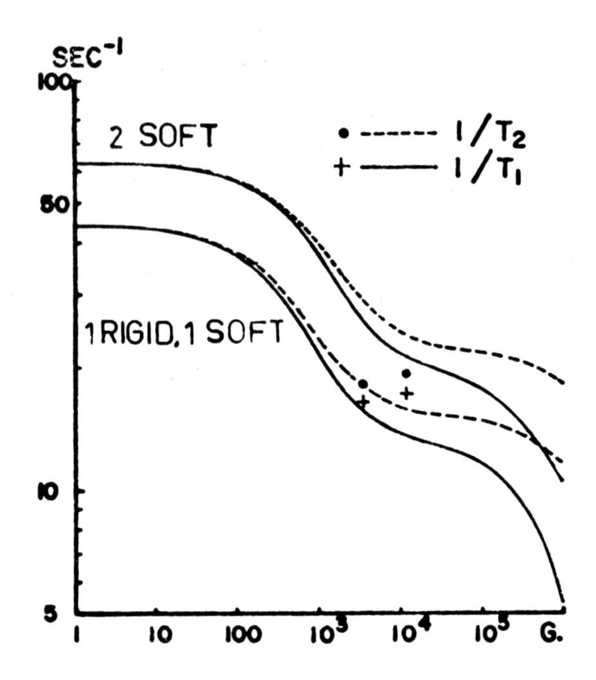

Fig. 5. Relaxation rates of $^7Li^+$ as a function of magnetic field for aqueous 0.1M $MnCl_2$, 1M LiCl at 25C. The curves are calculated. See /22/ for the source of the experimental data shown by discrete points.

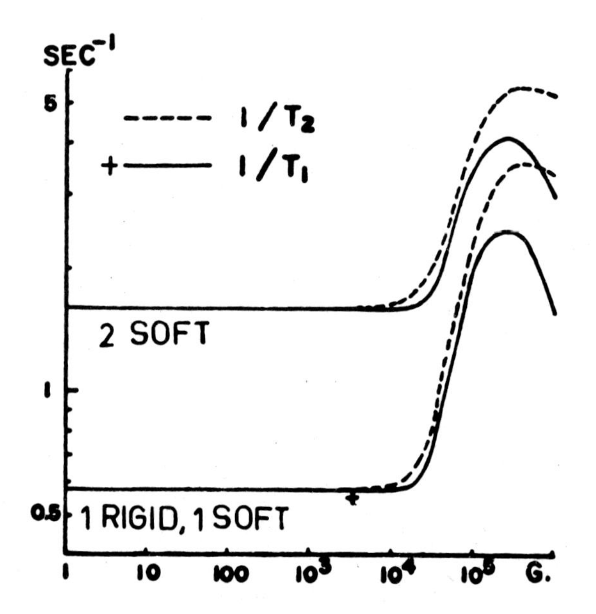

Fig. 6. Same as Fig. 5, but with $NiCl_2$ rather than $MnCl_2$.

for the Ni^{++} system than for the Mn^{++} system.

The available experimental data shown in Figures 5 and 6 indicate that, if one neglects various uncertainties which can be evaluated by more experimental work of various kinds, the 1 rigid, 1 soft model for the Ni^{++} cosphere structure is preferred over the 2 soft model. It seems quite likely that these data could also be fit if we assume a 2 soft model for Ni^{++} and a 1 rigid, 1 soft model for Li^{+}, but there is independent evidence that this model is not realistic. Thus, the exchange of water between the solvent and the inner coordination layer (6 water molecules) of $Ni^{++}(aq)$ is the same in solutions of Ni^{++} in 6.5M aqueous $LiClO_4$ as in dilute solutions. /27/ Taken together, the results in this paragraph show, subject to the uncertainties which have been noted, that the hydration layer on Li^{+} is penetrated rather deeply when the lithium ion collides with nickel ion in aqueous solution.

II.5.6. PROSPECTS

On the basis of the work reviewed above we venture to remark upon several different directions of research which promise to improve our knowledge of solute-solute interaction forces in solution and their effect upon measurable quantities.

6.1. Models at a deeper level.

The famous McMillan-Mayer theory/3,28/ gives the functional dependence of the solute-solute pair potentials u_{ij}, which here may be called solvent-averaged potentials, upon the so-called direct potentials

$$U_{ij,1} , U_{ij,2} , \ldots, U_{ij,m} , \ldots$$

for the mutual interaction of solute particles i and j and m solvent particles, now in the absence of all other particles, i.e. in a vacuum. It is natural to think of the direct potential functions as providing a deeper description of the system. Then all of the work described above can be classified as belonging to the second exercise in the sequence /29/

$$\begin{array}{ccccc} \text{direct} & & \text{solvent} & & \text{observable} \\ \text{potentials} & \longrightarrow & \text{averaged} & \longrightarrow & \text{solution} \\ & & \text{potentials} & & \text{properties} \end{array}$$

The problems in contributing to the understanding of solutions via the first exercise are twofold.

1) Even the pairwise components of the $U_{ij,m}$ are poorly known. Moreover, the non-pairwise contributions are quite sure to be important, judging from what is known about the potential of interactions of small clusters of water molecules. /30/

2) Even for the simplest case in which the non-pairwise contributions to the $U_{ij,m}$ are neglected, there are no approximation methods which have been shown to be satisfactory for calculating the solvent-averaged potentials of such a model with the required accuracy.

However, it now seems likely that both of these problems may soon be overcome in view of the current important progress with rather similar problems involving models for liquid water. /31-33/ Those advances of this kind which employ the molecular dynamics approximation method will lead also to the dynamic information corresponding to the solvent-averaged interactions, namely to information about reasonable forms for functions which describe the fluctuating motion $\dot{r}_{ij}$ of two interacting solute particles. Such functions would incorporate both the brownian motion and the hydrodynamic interactions mentioned in §5.

6.2 Higher-order thermodynamic coefficients

Quite recently it has been found that certain higher-order thermodynamic coefficients are rather different for models which otherwise have the same thermodynamic excess functions. /34/ Thus it is found that several models for aqueous $LiCl, MnCl_2$ which give osmotic coefficients ϕ and zeroth-order mixing coefficients w_0 in agreement with experiment also give rather different results for w_1, the first order mixing coefficient for the osmotic pressure. /14/ At ionic strength $I<1M$, w_0 is scarcely measurable at all, but the corresponding coefficients of the enthalpy of mixing, namely h_0 and h_1 /14/ now have been rather accurately measured down to $I<0.1M$ in the laboratory of R. H. Wood. /35/ Thus comparison of model and experimental h_1 coefficients might serve to eliminate some models which otherwise fit the thermodynamic data. Unfortunately, we find that the usual self consistency tests /7,11/ extended to the calculation of w_1 are rather poorly satisfied to the degree where it may be concluded that in the calculation of w_1 we have reached the limit of usefulness of the HNC approximation method. Therefore, developments based on model calculations for higher-order thermodynamic functions will depend upon the application of more powerful statistical-mechanical approximation methods.

6.3. Orientation-dependent and non-pairwise solvent-averaged interactions

Numerous interesting systems have non-spherical solute particles ranging from solutions of OH^- to those of DNA. Unfortunately, approximation methods which are useful for calculating the properties of such models beyond the second virial coefficient level and with controlled accuracy still remain to be developed.

The study of the effects of non-pairwise contributions in the models may be premature, in view of the limited progress in the study of the solvent-averaged pair interactions, except that some information about the non-pairwise effects on observable properties is necessary for reliable estimation of the concentration above which they become too important to neglect.

6.4. Dynamical properties

Most obvious is the need for extension of the work described here to all other dynamical properties of solutions which we may measure. That is, for models which represent the solute-solute interactions in a fairly realistic way, we should be able to calculate the effects of these interactions upon any transport, relaxation, or spectral coefficient of the solutions. While the necessary formal structure of the theory is provided by the linear response theory, there are many interesting problems to be solved before we can calculate any given coefficient with the required accuracy.

REFERENCES

1 H. L. Friedman and C. V. Krishnan in "Water, A Comprehensive Treatise" (Plenum Press, New York, 1973). Vol. 3, F. Franks, editor.

2 H. L. Friedman, Chemistry in Britain, 9, 300 (1973).

3 H. L. Friedman in "Modern Aspects of Electrochemistry", Vol. 6 (Plenum Press, New York, 1971), J. O'M Bockris and B. E. Conway, editors.

4 H. L. Friedman, "Ionic Solution Theory" (Interscience Publishers, Inc., New York, 1962).

5 P. S. Ramanathan and H. L. Friedman, J. Chem. Phys., 54, 1086 (1971).

6 H. L. Friedman, A. Smitherman and R. DeSantis, J. Solution Chem., 2, 59 (1973).

7 P. S. Ramanathan, C. V. Krishnan and H. L. Friedman, J. Solution Chem., 1, 237 (1972).

8 R. W. Gurney, "Ionic Processes in Solution", Dover Publications, Inc., New York, New York, 1953.

9 H. S. Frank, Zeits. Physik. Chem. (Leipzig), 228, 364 (1965).

10 J. C. Rasaiah and H. L. Friedman, J. Chem. Phys., 48, 2742 (1968).

11 J. C. Rasaiah and H. L. Friedman, J. Chem. Phys., 50, 3965 (1969).

12 H. L. Friedman and C. V. Krishnan, J. Solution Chem., 2, 119 (1973).

13 H. L. Friedman, J. Solution Chem., 1, 387, 413 (1972).

14 H. L. Friedman, J. Solution Chem., 1, 419 (1972).

15 H. L. Friedman and P. S. Ramanathan, J. Phys. Chem., 74, 3756 (1970).

16 J. C. Rasaiah and H. L. Friedman, J. Phys. Chem., 72, 3352 (1968).

17 J. C. Rasaiah, J. Chem. Phys., 52, 704 (1970).

18 C. V. Krishnan and H. L. Friedman, J. Solution Chem., 3, 000 (1974).

19 W.-Y. Wen, J. Solution Chem., 2, 253 (1973).

20 H. L. Friedman, Journal de Chimie Physique, (Numero Special, Oct. 1969) p.75.

21 H. L. Friedman, "Effects of Non-Brownian Motion" in "Structure and Transport Properties in Aqueous Solutions", Wiley, New York, 1971. R. A. Horne, editor.

22 L. P. Hwang, C. V. Krishnan and H. L. Friedman, Chemical Physics Letters, 20, 391 (1973).

23 L. P. Hwang, Ph.D. Dissertation, "Theory of NMR Relaxation Probes for Solution Structure," State University of New York at Stony Brook, 1973.

24 N. Bloembergen and L. O. Morgan, J. Chem. Phys., 34, 842 (1961).

25 A. Abragam, Nuclear Magnetic Resonance, (Oxford University Press, New York, 1961), pp.301,302.

26 We are grateful to Dr. R. F. Platford for communicating these w_0 data in advance of publication.

27 (a) S. F. Lincoln, F. Aprile, H. W. Dodgen and J. P. Hunt, Inorg. Chem., 7, 929 (1968).
(b) A. G. Desai, H. W. Dodgen and J. P. Hunt, J. Am. Chem. Soc., 91, 5001 (1969).

28 W. G. McMillan and J. E. Mayer, J. Chem. Phys., 13, 276 (1945).

29 H. L. Friedman, J. Chem. Phys., 32, 1134 (1960).

30 D. Hankins, J. W. Moskowitz and F. H. Stillinger, Chem. Phys. Lett., 4, 527 (1970).

31 A. Rahman and F. H. Stillinger, Jr., J. Chem. Phys., 55, 3336 (1971), 57, 1281 (1972).

32 J. A. Barker and R. O. Watts, Chem. Phys. Letters, 3, 144 (1969).

33 H. Popkie, H. Kistenmacher and E. Clementi, J. Chem. Phys., 59, 1325 (1973).

34 C. V. Krishnan and H. L. Friedman, unpublished studies.

35 R. H. Wood, personal communication.

II.6. THERMODYNAMICS OF MIXTURES OF AQUEOUS ELECTROLYTE SOLUTIONS

A Viewpoint on the Structure of Electrolyte Solutions

by

Yung-Chi Wu

National Bureau of Standards
Washington, D.C.

Abstract

The structural effect of water on aqueous ionic solutions is best interpreted by the concept of ion-solvent cosphere overlapping, developed by Gurney and by Frank. Friedman has advanced a theory which includes the overlapping effect, termed "Gurney potential", superimposed on the Coulomb, core and cavity potentials. Although there is an adjustable parameter in the Gurney potential, the success of fitting experimental data for several 1-1 electrolytes up to 1 molal concentration has proven the existance of the overlapping effect.

The differentiation of the ionic cosphere overlapping effect is a difficult task both experimentally and theoretically. It is the heart of the structural problem, since it is still unknown how the water molecules would be arranged around the central ion and how thick the layer of water molecules would be in the copshere. Experimentally, we can only lump the entire effect into two terms, order-disorder for water molecules and order-producing - order-destroying for ions, then observe how the thermodynamic quantities are influenced by the overlapping effect.

Experimentally, one of the best methods of differentiating the ionic cosphere overlapping effect seems to be the study of mixed electrolyte solutions. The effect is more easily observed in the mixing of the 1-1 charge types with common ions. The reason is that the excess thermodynamic function, Q^E, of mixed two electrolytes, A and B, may be expressed as:

$$\Delta Q^E_{mix} = yQ^E_A + (1-y)Q^E_B + \Delta_m Q^E$$

Where the excess function of mixing, $\Delta_m Q^E$, is a measure of the difference of the excess function of the mixture from those of the pure binary electrolyte solutions. In the common ion mixture, this difference has been shown from the same charge ion interactions. Recently, Robinson, Wood and Reilly (1971) have derived from the chemical equilibrium point of view that the same charge ion interactions are specific. In other words, the Debye-Hückel type of charge effects, for the same charge ions, are very small in the excess free energy of mixings, as observed by Robinson et al. The specificity of the same charge ion interactions appears to rise from the cosphere overlapping effect. In a series of thermodynamic determinations on mixtures of alkali halides the effects are small but indeed distinct. Gurney postulated a rule that the mean activity coefficients, based on the order-disorder character of each ionic cosphere, go "from dissimilar character downward to similar character". The ionic cosphere overlapping effects on the thermodynamics of mixtures appear to parallel the same rule.

II.6.1. Introduction

One aspect of the study of electrolyte solutions is the understanding of the effects of salt upon chemical equilibria and rates of chemical reactions. In both of these chemical phenomena the activity coefficient functions of various electrolytes play an important role. The present knowledge of the behavior of activity coefficients is limited to the long range interactions. There is a lack of clear understanding of the short range interactions. Quantitatively, we know very little about the molecular forces that solvents exert on the ions, such as ion solvent interactions, the energetics of ion solvations; and particularly the significance of the effects of solvent structure on the ions.

In the interpretation of ion solvent interactions, the concept of ionic cosphere overlapping which was developed independently by Gurney 1/ and Frank 2/ - 5/ has been known for some time. They suggested that in ionic solutions, an ion, due to its electrical field is capable of orienting its surrounding water molecules to form a cosphere. The water molecules in the cosphere are structurally different from pure water. As two ions are near each other, the cospheres overlap; the two ions are surrounded by a common portion of water molecules. The molecular forces of attraction and repulsion between two unlike ions are then arising from the degree of order in water molecules adjacent to each ion.

Using this concept Friedman 6/ - 10/ has advanced a theory in which he set up a Gurney potential superimposed on the Coulomb, core and cavity potentials, such that:

$$U_{ij}(r) = Coul_{ij}(r) + Core_{ij}(r) + Cav_{ij}(r) + Gur_{ij}(r) \qquad (1)$$

Where U_{ij} is the pair potential of ions i and j, and is a function of the separation, r, between them.

The Coulomb potential is an attractive potential.

The core potential is a repulsive potential.

The cavity potential accounts for the dielectric change due to the ionic field.

The Gurney potential is expressed as:

$$Gur_{ij}(r) = A_{ij} V_{mu} V_w^{-1} (r_i^* + w_i,\ r_j^* + w_j,\ r) \quad (2)$$

Where A_{ij} is an adjustable parameter

V_w is the molar volume of the solvent

w's are the cosphere thickness

r*'s are the crystal ionic radii

r is the center to center interionic distance

V_{mu} is the mutual volume which is a displaced portion of overlapping.

With the Gurney potential superimposed on the other potentials, Friedman has been able to fit the equation with experimental activity coefficients up to 1 molal for several 1, 1 and 1, 2 electrolytes. Apparently, all other effects which are not accounted for by the Coulomb, core and cavity potentials are lumped into the adjustable parameter, A_{ij}. Since we are still lacking the knowledge of molecular interactions mainly due to the solvent, this becomes necessary. Hopefully, this parameter will lead us to a better understanding of the nature of the problem in the future. Moreover, the success of fitting the equation to the real solution has proven the utilities of the Gurney potential. Furthermore, the Gurney coefficient, A_{ij}, as shown by Friedman, affects very sensitively the final results of the computed osmotic coefficients. In other words, the Gurney potential not only does exist but is also significant. This is to say that the ionic cosphere overlapping effect proposed by Gurney and Frank is indeed important in the behavior of electrolyte solutions. We shall attempt to understand this effect, using as a base the available information on the characteristics of the thermodynamics of mixtures of electrolyte solutions.

II.6.2. Ionic Cosphere Overlapping Effect

In view of the concept of the cosphere and overlapping and the evidence shown to be associated with the effect, the ionic cosphere may be considered as a consequence of the interaction between the ion and its surrounding water molecules. The force of the interaction is weak but it is strong enough to hold the ion and the cosphere water molecules together as an entity. The overlapping concept is that two units overlap and that they arise from the electrostatic interaction of the ions and/or from the molecular forces of the cosphere water molecules. The macroscopic properties of an ionic solution appear to include the cosphere overlapping effect, as well as the electrostatic interaction.

In Friedman's theoretical equation, there are four potential terms on the right hand side. The first two terms, Coulomb and core potentials, may be characterized as electrostatic, and the last two may be considered as contributions from ionic cosphere and overlapping. While the electrostatic effects have been demonstrated by the primitive model, such as the Debye-Hückel theory, the ionic cosphere and overlapping effects are difficult to differentiate from thermodynamic observations. The effects appear to be a structural problem.

The exact nature of the ionic cosphere is still unknown. Questions include: 1) how the water molecules would be arranged around the central ion; 2) how thick the layer of water molecules would be in the cosphere; and 3) what would happen when they overlap. Instead of seeking a direct answer to these questions we may, however, lump the entire effect into two terms, order-disorder for water molecules and order-producing - order-destroying for ions, then observe how the thermodynamic quantities are influenced by the cosphere and overlapping effects.

Experimentally, one of the best methods of differentiating the ionic cosphere overlapping effect seems to be the study of mixed electrolyte solutions. The effect is most easily observed in the mixing of the atomic 1-1 charge type electrolytes with common ions, for the obvious reason of the avoidance of structural complication and the complexity of molecular interaction which may arise from the ion itself. In mixed electrolyte solutions, the excess thermodynamic function is a measure of a secondary effect which cannot be distinguished from a binary solution. It is thought that the secondary effect may be considered as arising from the cosphere overlapping.

II.6.3. Observation on Thermodynamics of Mixture

In mixing a pair of electrolytes, A and B, at constant total concentration, the excess thermodynamic function, Q^E, may be expressed as:

$$\Delta Q^E = y_A Q_A^E + y_B Q_B^E + \Delta_m Q^E \qquad (3)$$

Where Q^E is a general excess thermodynamic quantity of the mixture, such as free energy, heat or entropy

Q_i^E is the same excess thermodynamic quantity of component i.

y_i is the ionic strength fraction of component i.

$\Delta_m Q^E$ is the excess thermodynamic function of mixing

It is this excess function of mixing, $\Delta_m Q^E$, from which we observe three distinctive characteristic types of behavior that cannot be seen in a pure binary electrolyte solution, and these could be considered as contributions from the ionic cosphere overlapping effect.

1. The excess function of mixing is a measure of the difference between that of the mixture and the pure binary electrolyte solutions.

2. In common ion mixtures, the excess function of mixing is independent of the common ion. In other words, the difference arises from the interactions of ions of the same charge.

3. The Debye-Hückel type of electrostatic interactions has

very little influence on the excess function of mixing at the range of concentrations of experimental interest.

Since it appears that the characteristic effect of the excess function of mixing is derived from the interactions of ions of the same charge without the electrostatic interactions, the effect must come from the ionic cosphere overlapping.

In order to substantiate these points further, we observe: the first point is obvious as seen from equation (3). The second point that the excess function of mixing is independent of the common ion*, may be demonstrated by the heats of mixing 11/ listed in Table I.

Table I. Heats of Mixing ($m_A + m_B = 1.0$, $y_A = y_B = 0.5$)

Mixing Pairs	$\Delta_m H$ cal. kg^{-1}*	Common Ion
NaCl - KCl	-9.58	Cl^-
NaBr - KBr	-9.43	Br^-
LiCl - LiBr	0.81	Li^+
NaCl - NaBr	0.79	Na^+
KCl - KBr	0.80	K^+

*The thermochemical calorie (defined) is equal to 4.184J.

These are the experimental data for the heats of mixing with the common ions indicated in the table. It can be seen that a pair of ions of the same charge produce the same amount of heat, independent of whatever the common ion is. It is worth noting that in the last three pairs of mixing, the common ions are Li^+, Na^+ and K^+, of which Li^+ ion differs very much from K^+ ion in terms of size and charge density; yet, the heats of mixing are practically identical. Therefore it is safe to state that the heat effect in a common ion mixture must come from the pair of ions of the same charge.

The third point, on the influence of the Debye-Hückel type of electrostatic interactions, may be further illustrated by the theory developed recently by Robinson, Wood and Reilly 12/. The theory is based on the chemical equilibrium principle that an equilibrium constant is derived from the interactions of a pair of ions of the same charge which may be written as:

$$M^+ + N^+ = MN^{++} \tag{4}$$

$$K_{MN} = (MN^{++})/(M^+)(N^+) \tag{5}$$

These equations indicate a mixture for a pair of electrolytes, MX and NX, with a common ion X^-, where the parentheses denote the concentration or activity of each ion. Robinson, et al., first treat it in the absence of any interaction, in this case, only the concentration is considered; and later they introduce the activity coefficients. They obtain the final equation for the excess free energy of mixing which is shown as follows:

$$\Delta_m G^E/(y_A y_B RTI^2) = K_2 \mathrm{Exp}\left[2A\sqrt{I}/(1 + B\sqrt{I})\right] + K_3 I \tag{6}$$

Where R is the gas constant; T, the kelvin temperature; and I the molal ionic strength. And,

$$K_2 = K_{MM} + K_{NN} - K_{MN} \tag{7}$$

for pairwise interactions in common anion mixture, and

$$K_3 = K_{MMX} + K_{NNX} - K_{MNX} \tag{8}$$

for triplet interactions.

The exponential term of equation (6) is the Debye-Huckel equation, A is the Debye-Hückel constant and B accounts for the effect of the ionic size parameter.

In substantiating the validity of equation (6) Robinson, et al., compare equation (6) with two well known treatments of the excess free energy of mixing of Friedman 13/ & 14/ and of Scatchard 15/, which respectively are:

$$\Delta_m G^E/(y_A y_B RTI^2) = g_o + \text{higher order terms} \tag{9}$$

and $$\Delta_m G^E/(y_A y_B RTI^2) = b^{(0,1)} + 1/2b^{(0,2)}I + \text{higher order terms} \tag{10}$$

They show the identity with Friedman's limiting law. In both of their equations (6 and 9) the Debye-Hückel type of interaction is included. However, as comparing equation (6) to Scatchard's equation (10), the two equations would be identical if the Debye-Huckel constant A in equation (6) is set equal to zero. They then use both their Ks and Scatchard's $b^{(0)}$s to fit their experimental data. The fittings are practically indistinguishable.

The equivalence of equations (6, 9 and 10) on fitting experimental data shows that the Debye-Hückel effect has very little influence on the excess free energy of mixing. However, we may recall that the excess thermodynamic function is a measure of the secondary effect of an interaction. The Debye-Huckel effect being suppressed is understandable. The effect may reappear in the mixing parameters. The following example illustrates this point further.

We observed that for the 1-1, 1-2 mixture $NaCl-Na_2SO_4$ 16/, the difference between the two different fittings is also very small and is within the limit of experimental error. This may be seen from Figure 1.

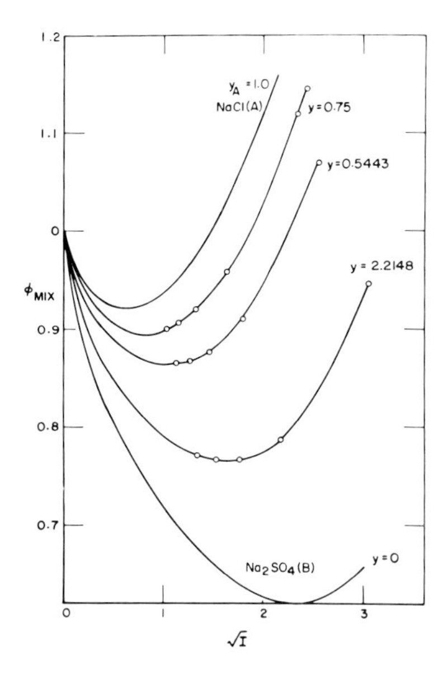

Figure 1. Osmotic Coefficients of the System, $NaCl-Na_2SO_4-H_2O$, as a function of total concentration, $\sqrt{I}$.

The points are experimentally measured. The curves are based on Scatchard's method of computations to extrapolate to zero concentration. The McKay and Perring method 17/ has also been used to obtain the activity coefficient of each component, γ's, and then to back-calculate the osmotic coefficients of the mixture, ϕ_{mix}. At higher concentration (above 0.5 molal ionic strength), the results from both methods are identical. At lower concentrations, they differ slightly, which could not be seen from these curves.

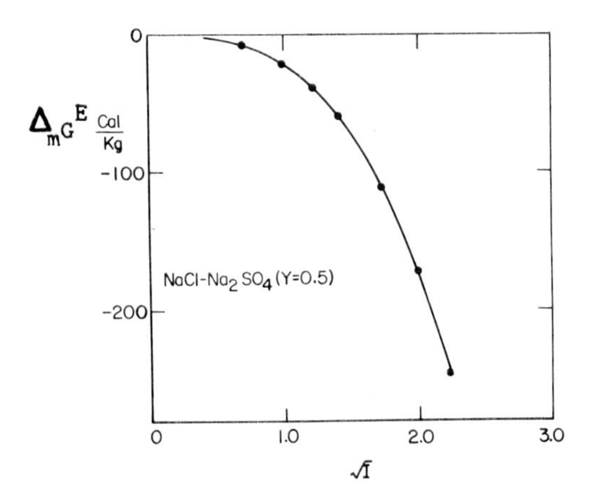

Figure 2. Free energy of mixing, $\Delta_m G^E$, as a function of total concentration, $\sqrt{I}$.

The values of $\Delta_m G^E$, as shown in Figure 2, are also computed using two different methods. One is computed by means of Scatchard's method, the other is by the integration of the partial molal free energy to obtain the excess free energy of the mixture,

$$\Delta G^E = -2RT(1 + y)I \int_0^I [(1-\phi_{mix})/\sqrt{I}]\, d\sqrt{I} \qquad (11)$$

at constant y. Then $\Delta_m G^E$ is obtained from

$$\Delta_m G^E = \Delta G^E - yG_A^E - (1-y)G_B^E \qquad (12)$$

The results from the two computations are practically identical.

It is seen that as far as the excess free energy of mixing is concerned, there is no difference in different fittings, but the para-

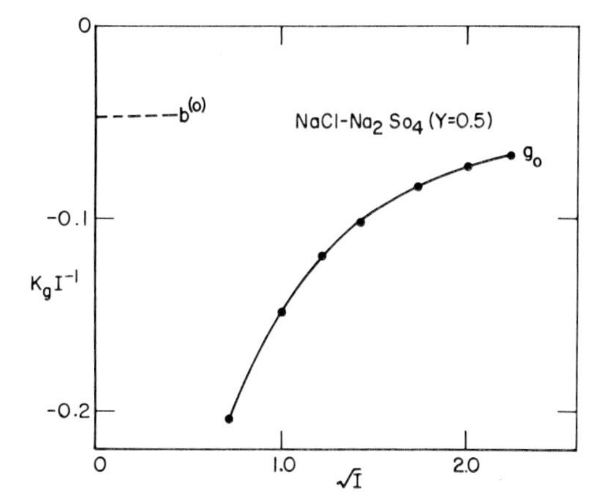

Figure 3. "Mixing coefficients, $K_g = \Delta_m G^E/y(1-y)RTI^2$, in kilogram per molal ionic strength, as a function of total concentration, $\sqrt{I}$."

meters themselves are quite a different matter 18/. This point is illustrated in Figure 3.

The g_o curve appears to go to infinity at zero concentration. Friedman in his "Ionic Solution Theory" 14/ has shown the infinity for g_o at zero concentration. While the dotted line represents Scatchard's $b^{(o)}$. It takes a finite value at zero I.

The fact is that at high dilutions the long range effects are so strong that the parameters diverge dramatically. As the concentration increases, the cosphere overlapping effect becomes predominent. Based on these observations, it is reasonable to believe that the excess function of mixing is a measure of the cosphere overlapping effects, at least for the part of the ions of same charge.

II.6.4. Effects on the Thermodynamics of Mixtures

Up to this point, all we have shown is the existence of the effect of ionic cosphere overlapping and the utility of the Gurney potential, and that the excess function of mixing is a measure of the ionic cosphere overlapping effect. It would be interesting to see how big this effect would be and how specific it is.

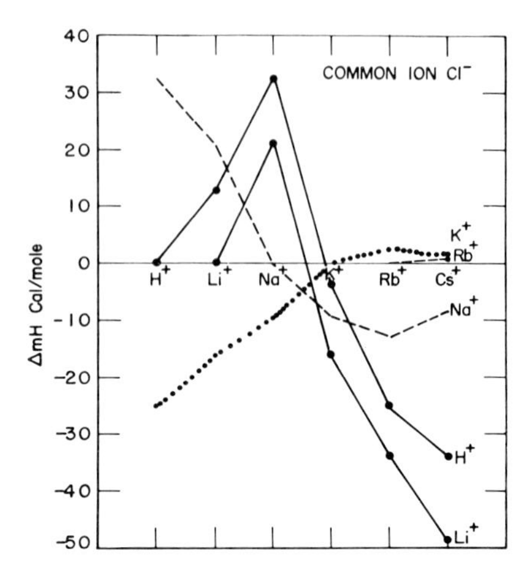

Figure 4. Heats of mixing for the inter-mixings of alkali chlorides and HCl. The abscissa is an arbitrary scale for the increasing order of ionic radii.

We have reported 11/ & 19/ a series of measurements of the heat of mixing for alkali halides mixtures. The results for the chlorides are summarized in Figure 4.

There are several points of interest here:

1. The curve for HCl mixed with the alkali chlorides is parallel and fairly similar to that for LiCl.
2. The other chlorides, each mixed with the rest of the chlorides, are crossed over.
3. If we draw a line to separate the H, Li-and Na-chlorides from the K, Rb-and Cs-chlorides, then we observe that the heats of mixing within each group are positive; and the heats of mixing are negative when the mixing takes place with a pair of salts, one from each group.
4. There is a highest and lowest heat of mixing for all of the chloride mixtures. They are approximately 80 cal apart.

McKay 20/ has calculated the excess free energy of mixing for

Figure 5

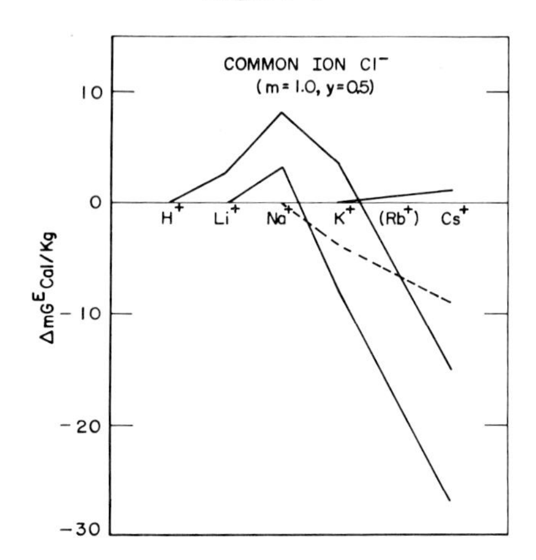

Figure 6

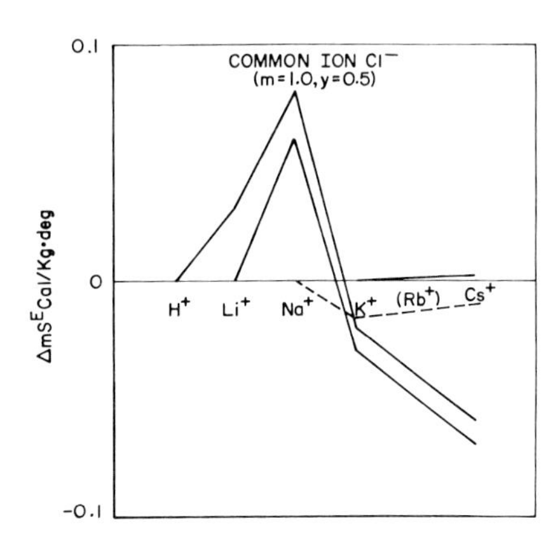

Figure 5. Free energies of mixing for the inter-mixings of alkali chlorides.

Figure 6. Entropies of mixing for the inter-mixings of alkali chlorides.

the same mixtures.* These are shown in Figure 5.

It is seen that the excess free energies of mixing behave similarly to those of heats of mixing. From these two quantities, we may calculate the excess entropy of mixing which is shown in Figure 6.

The excess entropies of mixing are also similar to those of heats of mixing.

Since entropy is a measure of order-disorder, the positive entropy means the final solution of mixture is more random than the linear combination of its components, and that the negative entropy indicates more order.

If we follow the same arguments of Gurney and of Frank that H^+, Li^+ and Na^+ ions are order-producing and that K^+, Rb^+ and Cs^+ ions are order-destroying, it is seen that ions with similar character

*Some of these values were recalculated.

produce more disorder in the mixture and that ions with dissimilar character are making the mixture more orderly as judged from the entropy of mixing.

The fact is that larger entropy is gained for mixing two ions with less similar character, but smaller entropy is gained for mixing those with more similar character. The same trend is observed for entropy loss, i.e., more loss for the most dissimilar and less loss for the least dissimilar.

Gurney observed a rule for the mean activity coefficients which go "from dissimilar character downward to similar character." The thermodynamics of mixtures appear to parallel the same rule. It can be said that the structural ordering effects also go from dissimilar character downward to similar character. The magnitude of the ordering effects follow the degree of dissimilarity. The extrema go to the extremities of the dissimilarity.

II.6.5. Discussion

The experimental facts shown in the last section have been known for some time, however, several interesting questions are posed which remain to be answered. The fact is that the structure of the final solution with a common ion is disordered for mixing two solutions which contain order-producing ions, and also disordered for mixing two order-destroying ions. The structure of the end solution becomes ordered when one solution containing an order-producing ion and another with an order-destroying ion are mixed. The thermodynamic quantities for HCl mixed with the alkali chlorides are parallel and fairly uniform to that for LiCl. With the advancement of Friedman's theory, an attempt is offered here to relate the foregoing phenomena with the cosphere overlapping effects.

Friedman argues that when two ionic cospheres come close to each other, they overlap, the overlapping portion of the cospheres is viewed as the mutual volume. The water molecules in the mutual volume would relax to equilibrium with the normal water in the bulk.

$$H_2O(V_{mu}) \xrightarrow{\Delta E} H_2O(\text{normal}) \tag{12}$$

Energy will be acquired or released in the relaxation process depending upon whether the water molecules in the mutual volume are ordered or disordered relative to those in the normal state.

It is easy to understand why mixing of two structure-makers could gain entropy, for the water molecules in the mutual volume are more ordered than in the normal state, and the process of relaxation increases the entropy. It is more difficult to understand why mixing of two breakers should also result in a small entropy gain. Since the water molecules in the cosphere of an order-destroying ion are supposed to be more random, the mutual volume derived from the ionic cosphere overlapping should contain disordered water molecules; one might suppose that the process would have lost entropy. Yet the observation is the opposite. It seems likely that the B-region in Frank and Wen's model for the cosphere is rather loosely bound, the overlapping effect penetrates into the A region which is frozen as suggested by the model.* In that case, water molecules in the mutual volume possess some degree of order and so cause the entropy increase.

An alternative explanation is that when two cospheres overlap, the water molecules in the cosphere go through rearrangement such as to meet the demand of the electrical field strength of the ions and entropy loss accompanies this process. The degree of ordered rearrangement depends on the degree of dissimilarity of order-disorder characters of the overlapping ions. In this case, the degree of order of water molecules in the relaxing mutual volume also depends on this same factor. The entropy change of these two processes would be nearly cancelled when the two ions closely resemble each other. The net gain of entropy

---

*Frank and Wen [4] have suggested for their model that an ion is surrounded by three concentric regions. The innermost is region A of immobilization of water molecules. The second one, adjacent to A, is region B, in which the water molecules are more random in organization than "normal" and the third, the outermost, is region C which contains normal water.

is apparently due to a larger gain in the relaxation of the mutual volume. For most of the atomic singly charged order-destroying ions the entropy gain in mixing is relatively small.

For the mixings of alkali halides with cations as common ion they show the same kind of small entropy gains as those order-destroying ions produce.

If the foregoing argument is followed, the entropy loss of the rearrangement process in the overlapping cospheres is the greatest for the most dissimilar ions in mixing, and the entropy gain for the relaxation of the mutual volume is relatively small. It may be seen from the two most order-producing ions, H^+ and Li^+, the entropy gain is mainly due to the relaxation of water molecules in the mutual volume. It follows that the entropy change should be solely caused by the most ordered water molecules in the mutual volume, rather than the entropy change involved by those in the rearrangement process; yet, its magnitude is still smaller than those causing entropy loss mainly due to rearrangement. It seems to be right to state that the most dissimilar character of ionic cospheres enhance the orderliness in mixing, as compared to the linear additivity of the two components. All the mixings of the alkali chlorides with chloride as common ion confirm this observation.

There is another interesting characteristic in the figures. The effects of H^+ ion and Li^+ ion in mixings with other ions are parallel and fairly uniform. Based on the foregoing argument for the net gain or net loss of entropy that "more change for the most dissimilar character and less for the least dissimilar character," the curves should have been crossed over as compared with the relation between Li^+ ion to Na^+ ion. Li^+ ion is generally considered to be a stronger structure maker than Na^+ ion. The fact is, however, that when H^+ ion and Li^+ ion, respectively, are mixed with Na^+ ion, the entropy gain is more for the $H^+ - Na^+$ ions mixture than that for the $Li^+ - Na^+$ ions mixture. This indicates that H^+ ion is an even stronger structure-maker than Li^+ ion. It follows therefore, when H^+ ion and Li^+ ion, respectively, mixing with structure-breaking ions, K^+, Rb^+ and Cs^+, each individually, the net loss of entropy of mixing should be in each case larger for H^+ ion than that for Li^+ ion. One might assume the cross-over would have been a natural consequence, in accordance with the argument stated above; however, they are not. The two curves are uniformly parallel instead. The only explanation for such behavior is that the two ions exercise a uniform influence in the rearrangement process, and that the relaxation of water molecules in the mutual volume for H^+ ion mixture is always greater by the same amount, as compared to those for the Li^+ ion in mixture.

II.6.6. Summary

To summarize, it appears that the ionic cosphere and overlapping effects could be revealed by study of the thermodynamics of mixtures of electrolyte solutions.

In common ion mixing an entropy gain seems to be always associated with the relaxation process of the water molecules in the mutual volume to the normal state, and an entropy loss is associated with the rearrangement process of water molecules in the cosphere overlapping. The net gain or net loss of entropy in mixing depends on these two competing processes.

The net gain or net loss of entropy of mixing in the common ion mixture, which is affected by the ions of same charge and their order-disorder characteristics, with the exception of H^+ ion, goes from dissimilar character downward to similar character, following closely the Gurney's rule for mean activity coefficients.

References

1. R.W. Gurney, "Ionic Processes in Solution" (Dover Publications, Inc., New York, 1962)

2. H.S. Frank and A.L. Robinson, J. Chem. Phys. 8, 933 (1940).

3. H.S. Frank and M.W. Evans, J. Chem. Phys. 13, 507 (1945)

4. H.S. Frank and W.Y. Wen, Discussion, Faraday Soc. 24, 133,(1957)

5. H.S. Frank, "Chemical Physics of Ionic Solutions," B.E. Conway and R.G. Barradas, Eds. (John Wiley & Sons, New York, 1965)

6. H.L. Friedman and P.S. Ramanathan, J. Phys. Chem. 74, 3756 (1970)

7. H.L. Friedman, in "Modern Aspects of Electrochemistry," J.O M. Bockris and B.E. Conway, eds. (Plenum Press, New York, 1971)

8. P.S. Ramanathan and H.L. Friedman, J. Chem. Phys. 54, 1086 (1971)

9. P.S. Ramanathan and H.L. Friedman, J. Solution Chem. 1, 237 (1972)

10. H.L. Friedman and C.V. Krishnan, in "Comprehensive Treatise on Water and Aqueous Solutions," F. Franks, ed. (Plenum Press, New York, 1973) Vol. III

11. Y.C. Wu, M.B. Smith and T.F. Young, J. Phys. Chem. 69, 1870 (1965)

12. R.A. Robinson, R.H. Wood and P.J. Reilly, J. Chem. Thermodynamics 3, 461 (1971)

13. H.L. Friedman, J. Chem. Phys. 32, 1134 (1960)

14. H.L. Friedman, "Ionic Solution Theory," (Interscience Publishers, New York, 1962)

15. G. Scatchard, J. Amer. Chem. Soc. 83, 2636 (1961)

16. Y.C. Wu, R.M. Rush and G. Scatchard, J. Phys. Chem. 72, 4048 (1968)

17. H.A.C. McKay and J.K. Perring, Trans. Faraday Soc. 49, 163 (1953)

18. Y.C. Wu, J. Phys. Chem. 74, 3781 (1970)

19. T.F. Young, Y.C. Wu and A.A.Krawetz, Discuss. Faraday Soc. 24, 37 (1957)

20. H.A.C. McKay, Discuss. Faraday Soc. 24, 76 (1957)

III INFRARED METHODS

III.1. THE FAR-INFRARED SPECTRUM OF WATER

Basil Curnutte and Dudley Williams
Physics Department, Kansas State University
Manhattan, Kansas 66506, U.S.A.

ABSTRACT

Studies of the far-infrared spectrum of water provide information concerning the motions of hydrogen-bonded water molecules in the lattice structure of the liquid. A plot of the Lambert absorption coefficient as a function of frequency has one maximum at 680 cm^{-1} associated with a broad absorption band attributed to hindered rotations or librations of H_2O molecules in the lattice and a second maximum near 200 cm^{-1} associated with a narrower band attributed to hindered translations of H_2O molecules in the lattice; an inflection near 2 cm^{-1} in the plot of the Lambert coefficient is attributed to absorption associated with Debye relaxation phenomena. In the present paper we present the values of the indices of refraction and absorption based on recent experimental work and attempt to interpret the results in terms of two recent theories of the structure of water.

The infrared spectrum of the water vapor molecule in its ground electronic state is fairly well understood. In the far-infrared the absorption spectrum is that of an asymmetric rotor and is characterized by irregularly spaced pure rotational lines in the entire spectral range between 800 cm^{-1} and the microwave region. The vibration-rotation spectrum is characterized by fundamentals ν_1 and ν_3 centered at 3651 cm^{-1} and 3756 cm^{-1}, respectively, in the spectral region where molecules containing OH groups have strong absorption bands. The third fundamental ν_2, associated with the bending motion of the molecule, appears at 1595 cm^{-1}. Overtone and combination bands are less intense and appear at higher frequencies. In addition to rotational and vibrational energies, the water vapor molecule has translational energy; at atmospheric pressure and below no observable absorption is associated with changes in translational energy.

III.1.1. THE OBSERVED SPECTRUM

Liquid water is strongly absorbing throughout most of the infrared; the Lambert absorption coefficient $\alpha(\nu)$ in the 3400 cm^{-1} region is nearly 10^8 times $\alpha(\nu)$ in the center of the visible region, where water is most transparent. Because of the strong spectral absorption of water in the infrared, it is difficult to obtain quantitative values of absorption by simple transmission measurements; special techniques must usually be employed (1-6). The general appearance of the water spectrum in the infrared is illustrated by Fig. 1, which gives the spectral transmittance $T(\nu) = \exp[-\alpha(\nu)x]$ as computed for a layer of thickness x = 10μm on the basis of Lambert coefficients obtained from a combination of transmission and reflection measurements. A layer of this thickness is essentially opaque in the 3400 cm^{-1} region; the upper curve in the 3400 cm^{-1} region gives the spectral transmittance of a water layer with a thickness of 1.125μm, much smaller than the wavelength of the radiation in this spectral region. In preparing the curves in Fig. 1 and in most of the subsequent figures, we have used values of $\alpha(\nu)$ based on our own measurements (6) in the 5000-100 cm^{-1} range, on those of Davies *et al*.(7) in the 100-20 cm^{-1} range, and on those listed by Ray (8) for the 20-1 cm^{-1} range.

Because of its proximity to the corresponding bands of water vapor, the strong absorption in the 3400 cm^{-1} region is attributed primarily to the ν_1 and ν_3 fundamentals of the hydrogen-bonded H_2O molecule but also includes some contribution from the overtone $2\nu_2$; the general contours of this composite band are strongly influenced by temperature changes (9). Similarly, the sharp absorption band near 1650 cm^{-1} is attributed to the fundamental ν_2; the ν_2 band shows little change with temperature. In addition to the fundamental bands there are two weaker associational bands in the near infrared portion of the spectrum shown in Fig. 1; these *associational bands* near 2120 and 4000 cm^{-1} are attributed to combinations of far-infrared bands with fundamentals and have no counterparts in the water vapor spectrum.

The far-infrared region of the liquid-water spectrum shown in Fig. 1 is dominated by two major bands. The first of these is an extremely broad, intense band with minimum transmittance near 680 cm^{-1}; this band shifts to 505 cm^{-1} in the spectrum of liquid D_2O (10). Because of this large isotopic shift, this broad band has been attributed to the hindered-rotation or librational motion of the H_2O molecule in the field of its neighbors. The second far-infrared band is much narrower and appears as a shoulder near 200 cm^{-1} on the strong librational band; in D_2O the shoulder band shifts only slightly to lower frequencies (10). This small isotopic shift is consistent with an interpretation of the weaker band in terms of the hindered translational motion of the entire H_2O molecule in the field of its neighbors.

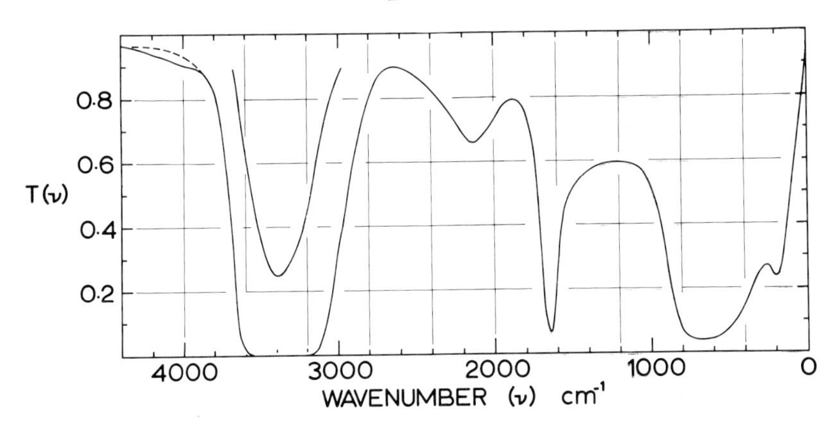

Fig. 1 The spectral transmission of a thin layer of liquid water in the infrared at 25°C.

More detailed information regarding the absorption of water in the far-infrared is presented in Fig. 2, in which we give a plot of the Lambert absorption coefficient as a function of frequency. Absorption in most of this region is associated with the strong librational band with a maximum near 680 cm^{-1}; the general contours of this broad band indicate that it is not a simple band but consists of several overlapping unresolved components. As noted earlier, the librational band has its maximum at 505 cm^{-1} in D_2O; it also has a somewhat smaller half width in D_2O. In an attempt to locate the position of the hindered-translation band, Draegert et al.(10) have arrived at the value of 170 cm^{-1} by attempting to construct a contour of the overlapping librational band; their corresponding value for the hindered-translational band in D_2O is 165 cm^{-1}.

The librational band shifts to lower frequencies and becomes broader with increasing temperature (9,10,11); the peak value of $\alpha(\nu)$ shows little change with temperature. In ice slightly below the melting point the peak of the librational band appears at 830 cm^{-1} and is narrower than the corresponding band in water (11); Zimmerman and Pimentel (12) have located the hindered-translational band at 225 cm^{-1} in the spectrum of ice. In a study of ice at liquid-nitrogen temperatures Bertie, Labbé and Whalley (13) have shown that the libration and hindered-translation bands have several components.

The value of $\alpha(\nu)$ is not small anywhere in the entire range between the ν_2

fundamental and the librational band; as indicated in Fig. 2, the value of $\alpha(\nu)$ at 1200 cm^{-1} is nearly 20 percent of that of $\alpha(\nu)$ at the librational peak. Absorption of water in this interband region varies with temperature (9). The value of $\alpha(\nu)$ decreases rapidly with decreasing frequency for frequencies lower than that of the hindered-translation band; there are no indications of additional absorption peaks. Our own studies (6,10) and the recent study of Zafar et al.(14) in the submillimeter range have given no evidence of an observable infrared peak near 60 cm^{-1}, where Raman (15) and inelastic neutron-scattering studies (16) have revealed a maximum. There is an inflection in the curve in Fig. 2 in the region between 10 cm^{-1} and 1 cm^{-1}; this is associated with the so-called Debye absorption associated with a low-frequency relaxation process (8).

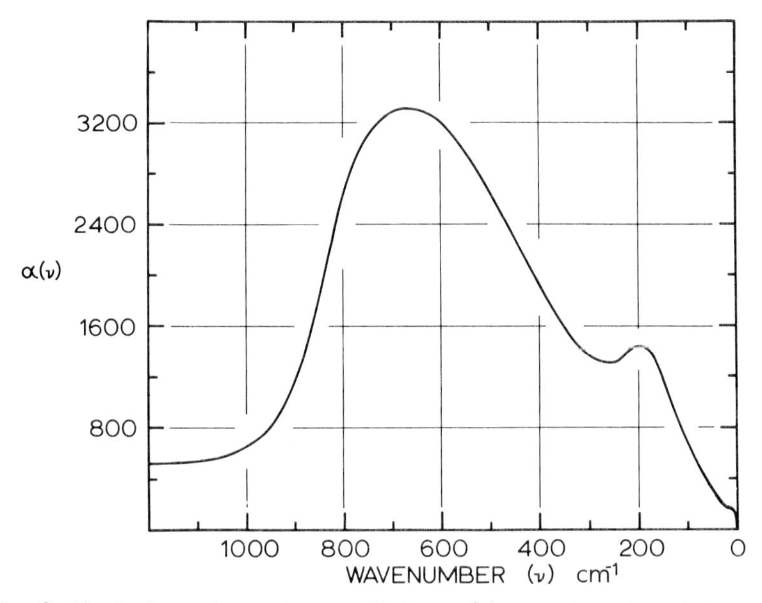

Fig. 2 The Lambert absorption coefficient $\alpha(\nu)$ as a function of frequency in the intermediate and far infrared at 25°C.

Although infrared spectroscopists usually content themselves with a plot of $\alpha(\nu)$ - vs - ν, a complete description of the optical properties of a medium like water actually requires a knowledge of its complex refractive index. The imaginary part of the refractive index $k(\nu) = \lambda\, \alpha(\nu)/4\pi = \alpha(\nu)/4\pi\nu$ gives a measure of absorption per wavelength; thus, even though $\alpha(\nu)$ becomes small in the far-infrared, $k(\nu)$ can remain large. By combining the results of measurements of reflection and absorption, the real part $n(\nu)$ of the refractive index can be determined. In the remote infrared, known as the submillimeter region, both $k(\nu)$ and $n(\nu)$ for water become large and join smoothly with values of these quantities based on dielectric-constant measurements in the microwave and radio frequency regions (8,14). Although for many purposes it is more convenient to give values of optical constants as a function of frequency or wavelength in a log-log plot, we find it desirable for present purposes to give a linear plot of these quantities in the far infrared.

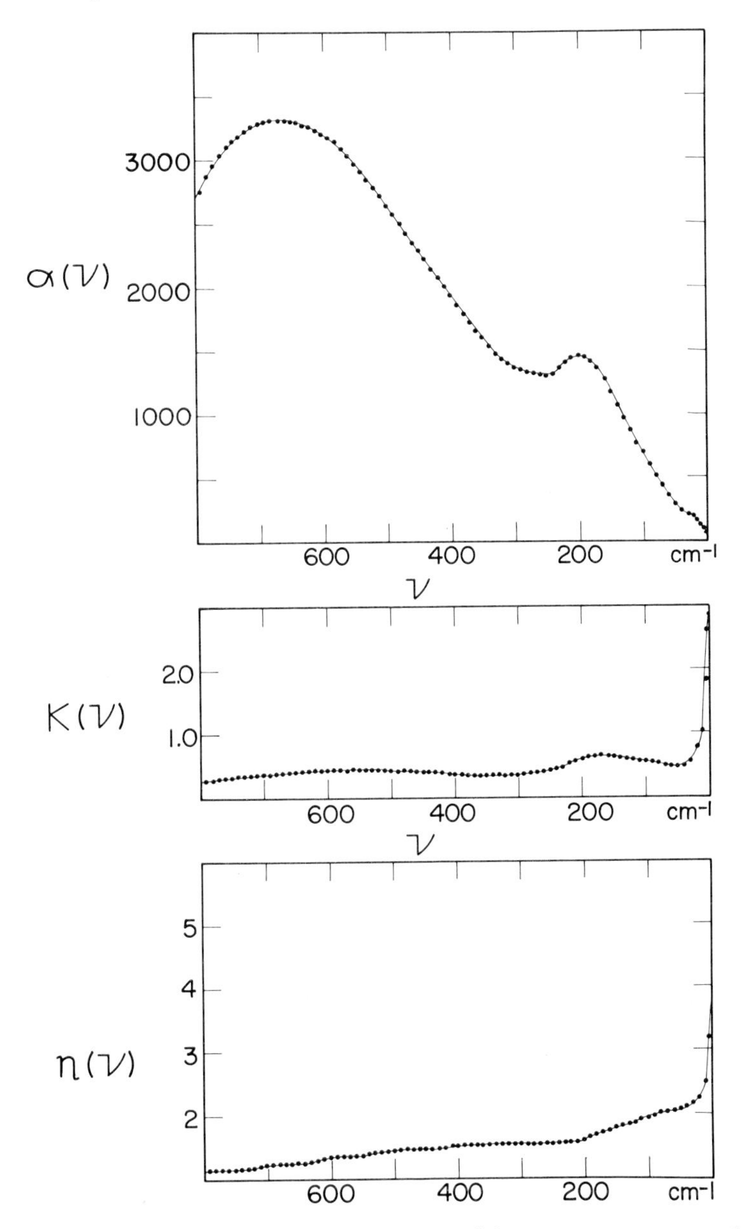

Fig. 3 The Lambert absorption coefficient $\alpha(\nu)$, the imaginary part $k(\nu)$ of the refractive index, and the real part $n(\nu)$ of the refractive index in the far-infrared for liquid water at 25°C.

In Fig. 3 we give plots of $\alpha(\nu)$, $k(\nu)$, and $n(\nu)$ as functions of frequency in the region between 800 cm^{-1} and 1 cm^{-1}. The maximum value of $k(\nu)$ for the broad librational band appears at 570 cm^{-1}, and the corresponding maximum for the narrow hindered translation band is at 170 cm^{-1}; the peak value of k(ν) for the hindered-translation band is nearly 50 percent larger than the peak value of k(ν) for the libration band (6). At frequencies below 170 cm^{-1}, k(ν) remains large and reaches a peak in the vicinity of 1 cm^{-1}; this peak is associated with Debye relaxation phenomena (8). The curve for n(ν) exhibits dispersion features in the vicinities of the libration and hindered-translation bands (6) and extrapolates to a value of approximately 9 at very low frequencies (8).

Although the libration and hindered-translation bands are strong in ice (11), values of k(ν) for ice in the region 100 - 1 cm^{-1} are only 10^{-2} to 10^{-3} times the corresponding values for water (8). A detailed knowledge of k(ν) and n(ν) in the submillimeter region may therefore be extremely important to an understanding of the differences of the properties of water and ice on a molecular scale.

III.1.2. THEORETICAL INTERPRETATION OF THE SPECTRUM

Liquid water consists of a hydrogen-bonded network of H_2O molecule units that remain hydrogen bonded to the same neighbors in the local lattice for appreciable times compared with librational periods. Any calculation of the spectroscopic properties of water involves not only the difficulties of the calculation for a disordered crystal lattice but also the additional difficulties imposed by a constantly changing lattice structure. The presence of strong bands with frequencies close to those of the water-vapor fundamentals indicates that individual H_2O molecules making up the water structure maintain their identities for times that are long compared with their fundamental periods of vibration. All absorption at lower frequencies must be associated with lattice vibrations of a system containing these units bound together in various ways.

We shall attempt to interpret the observed far-infrared spectrum in terms of two quite different theories. The first of these, developed by Curnutte and his students (17,18), attempts to interpret the spectrum by means of a model involving parameters fitted to the observed major spectral features of water at 25°C; the theory is then tested by applying it without changes in parameters to water at other temperatures and to ice. The second theory, developed by Rahman and Stillinger (19), involves a molecular-dynamics study of a set of rigid H_2O molecular models subject to interaction potentials consistent with measured macroscopic thermodynamic properties of water. By a computer study of the motions of a set of these H_2O units, these authors were able to obtain auto-correlation functions of various physical quantities and to interpret these functions in terms of lattice frequencies; since infrared absorption involves a change of the electric dipole moment of the absorber, all characteristic lattice motions do not necessarily produce features in the far-infrared spectrum. Fortunately, the two quite different theories to be discussed do not lead to basically conflicting conclusions.

A. THE LOCAL LATTICE MODEL THEORY

In their initial study Bryan and Curnutte (17) considered a single H_2O molecule held in position in a rigid tetrahedral cage of neighboring molecules by forces given by the Lippincott-Schroeder potential, which had been developed (20,21) to describe intermolecular hydrogen bonding, and by an electrostatic interaction between the H_2O molecule's dipole moment and the electric intensity at the site of the molecule. Such a hydrogen-bonded molecule has three fundamental librational modes and three fundamental hindered-translational modes of oscillation in addition to its internal vibrational modes. Starting with the most probable O-O distance as determined by x-ray studies (22) and with the peak librational and hindered-translational frequencies listed by Walrafen (15), Bryan and Curnutte adopted four empirical parameters needed to give their tetrahedrally hydrogen-bonded molecule the observed peak frequencies. After selection of the parameters they then used Monte Carlo calculations to determine the spectral effects produced by the distribution of O-O distances as determined by x-ray diffraction; they found that variation of the distance between the oxygen atom of the central molecule and the oxygen atoms in the rigid cage over the observed range established in x-ray studies resulted in a broadening of each of the librational bands by approximately 220 cm^{-1} and each of the hindered-translational bands by approximately 120 cm^{-1}. In view of the relaxation time $\tau \simeq 10^{-12}$ seconds given by dielectric-constant studies (23), the lifetime of a molecular cage is limited, and this leads to a superposed uncertainty broadening $\Delta\nu \approx 1/c\tau$ of the order of 35 cm^{-1}.

A plot of the results of the Monte Carlo calculations gave a frequency spectrum that compared favorably in certain respects with the profiles of the observed Raman and infrared spectra. However, the introduction of random O-O distances destroys the original tetrahedral symmetry at the site of the central molecule so that little can be said regarding definite selection rules or spectral intensities. After the final selection of parameters for water at 25°C, the Bryan model was then employed to produce spectra of water at other temperatures, the spectrum of liquid D_2O, and the spectrum of ice with generally satisfactory results in the vicinity of the observed librational and hindered-translational bands from O-O spacings given by x-ray studies. The frequencies of the lattice modes given by the model for water at 25°C are listed in the first column of Table I; the frequencies listed represent peak values in the spectrum of a single molecule connected by four tetrahedral hydrogen bonds to neighbors in a rigid cage; the O-O distances are randomly spaced over the range established by x-ray studies.

Not all H_2O molecules in liquid water have complete tetrahedral hydrogen bonding; thus it becomes necessary to estimate the spectral effects of incomplete bonding. The effects of incomplete bonding calculated for a central H_2O molecule in a rigid cage of neighbors on the basis of most probable O-O distances are listed in the second and third columns of Table I. The second column lists the frequencies to be expected from a molecule with one hydrogen bond to its oxygen atom broken; the third column lists the frequencies to be expected for a molecule with a single unbonded hydrogen. The additional effects introduced by breaking a single hydrogen bond include new hindered-translational modes at 100 cm^{-1} and new librational modes at 350 and 400 cm^{-1} as indicated in the table. Incomplete intermolecular bonding thus results in the general lowering of lattice frequencies.

TABLE I

Calculated Frequencies in the Lattice Spectrum of Water at 25°C

Single Molecule in a Rigid Cage			Five-Molecule Unit in a Rigid Cage
Tetrahedral Bonding	Free O	Free H	(Tetrahedral Bonding)
	Librational Frequencies (cm^{-1})		
450	475	350	450,450,450,452,452
551	575	450	553,553,553,554,554
722	725	625	722,722,722,723,723
	Translational Frequencies (cm^{-1})		
165	100	200	77,217,164,164,164
165	175	125	71,219,160,163,165
166	100	125	75,219,162,165,165

Recognizing that the assumption of a completely rigid cage of nearest neighbors was somewhat unrealistic, Bandekar and Curnutte (18) extended the general method to include a central molecule in a non-rigid cage of nearest neighbors -- *i.e.*, a hydrogen bonded five-molecule unit -- connected by hydrogen bonds to a rigid cage of next-nearest neighbors. A normal-coordinate analysis of the five-molecule unit yielded a set of frequencies associated with its own fundamental vibrational modes plus additional frequencies associated with hindered rotational and translational motions of the entire five-molecule unit in the rigid cage of surrounding molecules. With little change in the Bryan parameters, Bandekar obtained the sets of frequencies listed in the final column of Table I. His treatment yielded a set of 5 frequencies close to 450 cm^{-1} that can be associated with the original 450 cm^{-1} librational frequency of the single H_2O molecule in the Bryan study; a set of 5 frequencies near 555 cm^{-1} associated with the original 551 cm^{-1} libration; and a set of 5 frequencies near 720 cm^{-1} associated with the original 722 cm^{-1} libration. In the 150-175 cm^{-1} frequency range of hindered translation for the single H_2O unit in the Bryan study, the Bandekar treatment gives 9 frequencies in the narrow range 160-165 cm^{-1}; in addition, there are 3 frequencies between 71 and 77 cm^{-1} that are associated with hindered translation of the entire five-molecule unit in its rigid cage; finally, there are 3 new frequencies in the range 217-219 cm^{-1} that can be classified as out-of-phase translatory motions of parts of the five-molecule unit within its rigid cage. Absorption bands centered at the listed frequencies are broadened by amounts comparable with those listed for the earlier single-molecule model.

This general type of analysis could be extended to include a larger non-rigid group of molecules consisting of the central H_2O molecule with its nearest and next-nearest neighbors inside a still larger rigid cage. However, even with the five-molecule model, the computed lowest frequencies are comparable with the frequency of dielectric relaxation; furthermore, the widths of the resulting absorption bands are comparable with the band frequencies themselves. Thus, particularly in view of the amounts of computer time involved in taking account of random O-O spacings, the wisdom of extending calculations of this type is questionable.

In summary: The calculations based on this model, which was parameterized to account for the major libration and hindered-translation bands observed in the Raman spectrum of water at 25°C, also accounts satisfactorily for the variation of these bands with temperature, for the corresponding bands observed in ice, and for the corresponding bands appearing in the spectrum of liquid D_2O. The calculations further predict absorption at frequencies lower than that of the observed hindered-translation peak and at frequencies in the region between the libration and hindered-translation peaks; infrared absorption is indeed observed in these regions. In order to account for the observed general absorption in the interband region between the ν_2 fundamental and the libration bands, it would probably be necessary to invoke overtones and combinations of the major lattice bands; the possibility of difference bands involving the ν_2 fundamental and certain lattice modes might also be considered.

B. THE MOLECULAR-DYNAMICS STUDY OF RAHMAN AND STILLINGER

Turning now to the molecular dynamics study of Rahman and Stillinger (RS), we must first emphasize the basic difference between their treatment and the spectroscopic treatment of Curnutte and his colleagues, who introduced certain parameters to account for major spectral features previously observed; RS introduced no fitted parameters to account for known spectral features in the far infrared or in the low-frequency Raman spectrum. Thus, in considering the results of their theory we should not expect quantitative reproduction of the observed spectrum but should look for qualitative general agreement between predicted and observed spectral features. Closer quantitative agreement could probably be accomplished by the introduction of fitted parameters.

The RS procedure involved a computer study of the dynamical behavior of 216 rigid H_2O molecular models having a number density corresponding to that of water and subject to a rather simple intermolecular potential proposed by Ben-Naim and Stillinger (24). The temporal evolution of this system of rigid molecular units was studied by the examination and analysis of successive molecular arrangements generated by the computer. By analysis of these computed molecular arrangements it was possible to obtain auto-correlation functions for various properties of the system; Fourier transforms of the auto-correlation function yield the prediction of the frequency spectrum associated with the properties involved. The Fourier transform plots present a quantity related to the square of the amplitude of oscillation per unit frequency interval as a function of frequency; for want of a better term we shall refer to the ordinates of these plots as spectral density. If electromagnetic radiation were involved, the Fourier transform plot would represent what the electrical engineer calls a power spectrum.

In an analysis of the diffusive motions of the molecular centers of mass, RS have obtained a center-of-mass velocity auto-correlation function; the Fourier transform gives a spectrum that can be compared in a general way with what we have called the hindered-translation spectrum. The plot of spectral density is essentially flat from very low frequencies to approximately 212 cm^{-1}; beyond 212 cm^{-1} spectral density decreases rapidly with increasing frequency. Superposed on the flat portion of the spectral curve is an additional fairly sharp peak at 50 cm^{-1}; the peak height is 40 percent higher than that of the flat portion of the spectrum. The sharp peak at 50 cm^{-1} may well be related to the peak observed near 60 cm^{-1} in Raman (15) and inelastic neutron scattering studies (16). By itself the Fourier transform of the center-of-mass velocity auto-correlation function does not predict the absorption peak observed near 200 cm^{-1}.

A more favorable direct comparison with experimental results exists in the librational region, where there is experimental evidence of three bands. The spectrum obtained from the angular-velocity correlation functions for angular velocity about the three principal axes of inertia of the monomer has well separated peaks in spectral density at 395, 420, and 847 cm^{-1}. The corresponding spectrum for total angular momentum covers a broad spectral region extending from near zero to 1400 cm^{-1} with evidence of component peaks at 424, 636, and 850 cm^{-1}. The really important results given by these spectra are that the spectral density involved in the rotational motion of the molecular units covers an extremely broad range of frequencies and that three separate peaks are predicted.

The RS result most directly related to the absorption spectrum is given by a plot of spectral density as a function of frequency that is based on the proton total velocity auto-correlation function. This composite spectrum combines features of the center-of-mass velocity spectrum with a spectrum obtained from the velocity auto-correlation function for proton motion relative to the centers of mass of the H_2O monomers. The plot of spectral density is dominated by a broad band with a maximum at 425 cm^{-1} and with inflection points at 636, 850, and 1110 cm^{-1} in the high frequency wing, which exhibits appreciable spectral density for frequencies as high as 1300 cm^{-1}. A peak at 212 cm^{-1} appears as a low-frequency shoulder of the main band; a smaller additional sharp peak appears at 53 cm^{-1}; and the spectral density is non-zero at very low frequencies. Although the characteristic peak and inflection frequencies are different from the frequencies observed experimentally, the major portion of the broad band coincides with the broad librational band observed in infrared absorption; the low-frequency shoulder band at 212 cm^{-1} corresponds to the hindered-translation band appearing as a shoulder near 200 cm^{-1} in the plot of observed $\alpha(\nu)$ - vs - ν given in Fig. 2. The peak at 53 cm^{-1} may well correspond to the 60 cm^{-1} peak observed by Raman and neutron-scattering techniques.

III.1.3. DISCUSSION OF RESULTS

The theoretical plots of Curnutte and his students, which were based initially on Raman results, and those of RS based on a more general study of the molecular dynamics of rigid H_2O units subject to a particular form of intermolecular potential show relative spectral density as a function of frequency. It is not necessarily true that all features in these plots will appear in the infrared absorption spectrum, since absorption occurs only when changes in dipole moment are produced by radiation of the appropriate frequency. For example, rotational motion of an isolated H_2O molecule about its symmetry axis would not be infrared active since no dipole moment changes occur in the course of the motion. However, such a rotational motion of a molecule surrounded by neighbors could become infrared active as a result of dipole moments induced, primarily as a result of changes in hydrogen bonding, in the polarizable neighboring molecules. Any detailed theory of the absorption of infrared radiation should take account of these and similar processes. Such processes are not considered in detail in the formulation of the Curnutte theory and are prohibited in the RS treatment by the assumption of unpolarizable rigid molecules.

A related phenomenon known as translational absorption has been suggested by Litovitz (25) to account for a portion of the observed infrared absorption in the extreme infrared. This type of absorption, which was discovered by Welsh and his collaborators (26), can occur when two colliding polarizable molecules produce a collision pair with a resultant dipole moment capable of interacting with incident radiation; such a collision pair can absorb a quantum of radiant energy. When such an absorption occurs, the members of the collision pair separate with greater translational energy than they had prior to collision; since the energy absorbed in this non-periodic process goes immediately into translational energy, the process is termed translational absorption. Typical translational absorption bands are broad and nearly structureless. Since translational absorption bands for colliding H_2O molecules in liquid water may occur in the remote infrared, they could well account in part for the almost continuous absorption observed in the 100-20 cm^{-1} range by Davies *et al.*(7).

It is interesting to speculate on the nature of the absorption processes involved in the rapid rise of the $k(\nu)$ curve in Fig. 3 at frequencies below 10 cm^{-1}. Little detailed information on the subject is given by the RS curves because of the long total observation times that would be required to give valid estimates of spectral density at extremely low frequencies. The RS molecular-dynamics study does, however, provide some estimates of the logarithmic decay of librational motion that occurs with a time constant comparable with measured time constants for dielectric relaxation. Further detailed experimental and theoretical studies in the millimeter and submillimeter regions should prove to be of importance to an understanding of the liquid structure.

As indicated earlier, the RS spectrum has non-zero spectral density at frequencies considerably higher than those of the main librational peaks. Infrared absorption has been observed (9) in the entire interband region between the librational band and the ν_2 fundamental. Since the observed absorption extends to frequencies higher than those predicted by the RS plot, it seems likely that for the 1300 cm^{-1} region the observed absorption may indeed result in part from a difference band involving the ν_2 fundamental and lattice modes. If the rigid H_2O units in the RS scheme could be replaced with units capable of vibration, this suggestion could be checked. At the expense of greatly increased computation time other intermolecular potentials might be used in the RS scheme to bring the predicted spectrum into closer quantitative agreement with Raman and infrared observations.

The authors should like to express their thanks to the U. S. Office of Naval Research and to the Kansas State Agricultural Experiment Station for support of much of the research reported here.

REFERENCES

1. L. Pontier and C. Dechambenoy, Ann. Geophys. 21, 462 (1965) ibid. 22, 633 (1966).

2. V. M. Zolatarev, B. A. Mikhailov, L. I. Aperovich, and S. I. Popov, Optika Spectrosk. 27, 790 (1969). [Translation: Optics Spectrosc. 27, 430 (1969).

3. M. R. Querry, B. Curnutte, and D. Williams, J. Opt. Soc. Am. 59, 1299 (1969).

4. A. N. Rusk, D. Williams, and M. R. Querry, J. Opt. Soc. Am. 61, 895 (1971).

5. C. W. Robertson and D. Williams, J. Opt. Soc. Am. 61, 1316 (1971).

6. C. W. Robertson, B. Curnutte, and D. Williams, Mol. Phys. 26, 183 (1973).

7. M. Davies, G. W. F. Pardoe, J. Chamberlain, and H. A. Gebbie, Trans. Faraday Soc. 66, 273 (1970).

8. P. S. Ray, Appl. Opt. 11, 1836 (1972).

9. G. M. Hale, M. R. Querry, A. N. Rusk, and D. Williams, J. Opt. Soc. Am. 62, 1103 (1972).

10. D. A. Draegert, N. W. B. Stone, B. Curnutte, and D. Williams, J. Opt. Soc. Am. 56, 64 (1966).

11. J. W. Schaaf and D. Williams, J. Opt. Soc. Am. 63, 726 (1973).

12. R. Zimmerman and G. C. Pimentel, Proc. Intern. Meeting Mol. Spectry., Bologna, 2, 726 (1962).

13. J. E. Bertie, H. J. Labbe´, and E. Whalley, J. Chem. Phys. 50, 4501 (1969).

14. M. S. Zafar, J. B. Hasted, and J. Chamberlain, Nature (Phys. Sci.) 243, 106 (1973).

15. G. E. Walrafen, J. Chem. Phys. 40, 3249 (1964).

16. J. O. Burgman, J. Sciensci, and K. Skold, Phys. Rev. 170, 808 (1968).

17. J. B. Bryan and B. Curnutte, J. Mol. Spectry. 41, 512 (1972), and J. B. Bryan, Doctoral Dissertation, Kansas State University (1969).

18. J. Bandekar, Doctoral Dissertation, Kansas State University (1973).

19. A. Rahman and F. H. Stillinger, J. Chem. Phys. 55, 3336 (1971).

20. E. R. Lippincott and R. Schroeder, J. Chem. Phys. 23, 1099 (1955).

21. R. Schroeder and E. R. Lippincott, J. Phys. Chem. 61, 921 (1957).

22. A. H. Narten, M. D. Danford, and H. A. Levy, ORNL-3997, UC-4 Chemistry, Oak Ridge National Laboratory (1966).

23. D. Eisenberg and W. Kauzmann, The Structure and Properties of Water (Oxford University Press, 1969), pp. 206-208.

24. A. Ben-Naim and F. H. Stillinger, "Aspects of the Statistical-Mechanical Theory of Water" in "Structure and Transport Processes in Water and Aqueous Solutions", edited by R. H. Horne (Wiley-Interscience, New York, In Press.)

25. T. A. Litowitz, Comment at Marburg Symposium (1973).

26. Z. J. Kiss and H. L. Welsh, Phys. Rev. Letters 2, 166 (1959).

27. Z. J. Kiss, H. P. Gush, and H. L. Welsh, Can. J. Phys. 37, 362 (1959).

III.2. INFRARED FUNDAMENTAL REGION

Werner A. P. Luck
Physikalische Chemie Universität Marburg
D 3550 Marburg/Lahn, West-Germany

ABSTRACT

I.R. spectroscopy provides some of the best criteria for the quantitative study of H-bonding. The fundamental stretching bands of H_2O, D_2O or HDO as a function of temperature show that with increasing T the number of H-bonds decreases. Quantitative measurements in the fundamental region are difficult due to band-overlapping and also some instrumental problems. However, the fundamental vibrations give important information on the angle and distance-dependence of the H-bond interaction. The stretching vibrations give indications of the existence of different bands due to free OH groups, and energetically unfavoured and favoured H-bonds. These bands are consistent with the Raman experiments and the overtone spectra.
The comparison of the alcohol solution spectra with the spectra of pure alcohols shows a smaller content of non H-bonded OH groups in pure liquids than most theories claim.
Spectra of aqueous electrolyte solutions give an ion series similar to the lyotropic ions series. The latter is attributed to changes of the water structure.

III. 2.1. INTRODUCTION

Spectra are the language of atoms and molecules so that if the problem of water structure is soluble the spectra should provide the answer. We thus need to be able to interpret the spectra. The interpretation of the intermolecular vibrations in the far I.R. is only beginning to be understood. Intermolecular interactions other than H-bonding interactions only slightly disturb fundamental vibrations. There are few papers on the theory of these effects and therefore we need a lot of experimental data to identify the intermolecular effects on spectra. The interpretation of H-bond effects on fundamental vibrations has some difficulties due to band-overlapping. This region gives excellent qualitative information on H-bond bands. Quantitative measurements of the H-bond state are much more precise with overtone data (see chapter: I.R. overtone region)/1//2/. Thus it is not surprising that the literature of liquid-spectra is not uniform. Some doubts have been expressed concerning I.R. spectroscopic methods; therefore a review on the I.R. methods has also to refute this published criticism.

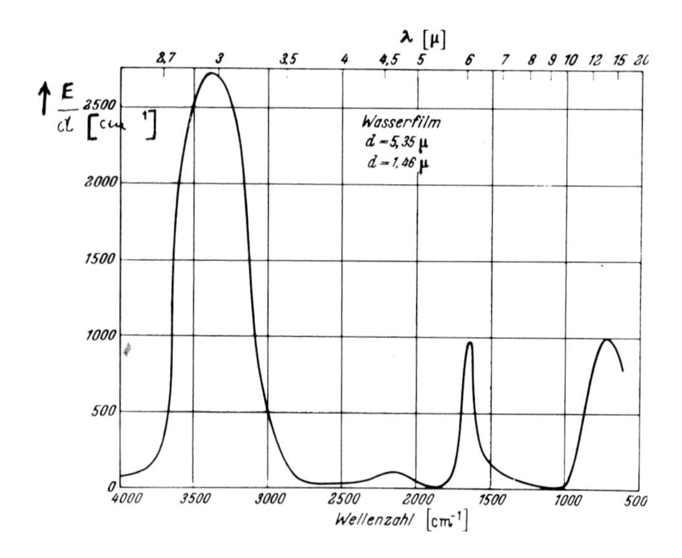

Figure 1. Fundamental absorptions of liquid water at room T.

The fundamental absorption bands of liquid water are shown in fig. 1. The frequencies in the vapour state are given in table 1. /3/

Fundamental frequencies of H_2O in vapour state

OH-stretching	H_2O cm^{-1}	CH_3OH	D_2O cm^{-1}	HOD cm^{-1}
ν_1	3651,7	3682	2666	2719
ν_3	3755,8		2789	
OH-bending ν_2	1595	1340	1178,7	1402

Fig. 2 shows the normal modes of H_2O.

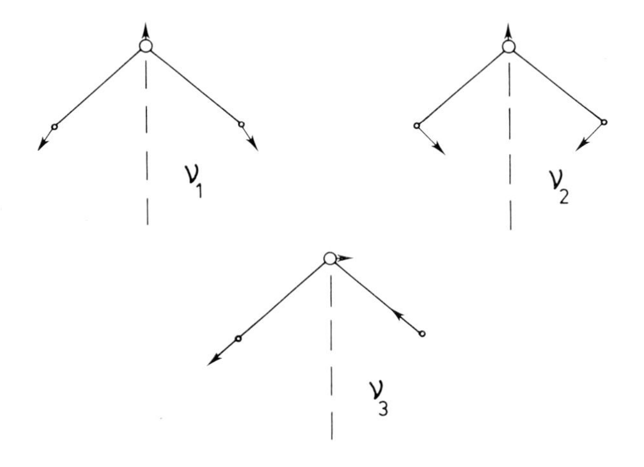

Figure 2. Normal modes of a water molecule, ν_1 and ν_3 stretching, ν_2 bending mode.

In the far I.R. are some intermolecular vibrations (see chapter Far I.R.). The band 2100 cm^{-1} (fig.1) is only observed in liquid water. It is a combination band of the bending and a libration mode.

III. 2.2. SOLUTION SPECTRA

Extinction coefficients ε in I.R. spectra are not very sensitive to non-polar intermolecular interactions /1/ /4/. For example ε of cyclohexane in solutions of CCl_4 does not change until concentrations of 100 ml/l and changes at about 10% to cyclohexane in bulk (Fig.3).

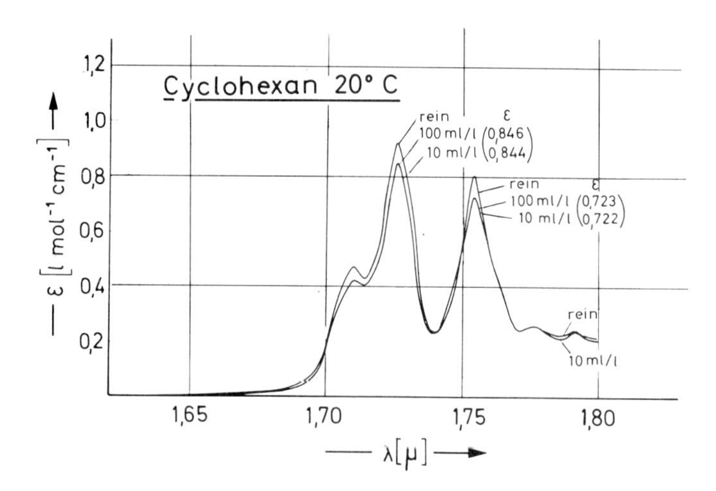

Figure 3. Extinction coefficients of cyclohexane pure in CCl_4 solution change only slightly.

But in systems with H-bonds such as OH or NH valence vibrations one observes a large change in the ε -values. (Fig.4). For example the band of non-H-bonded OH groups of CH_3OH (3642 cm^{-1}) or C_2H_5OH almost disappears in pure methanol /1/ /5/. With increasing concentration (increase of the content of H-bonds) a broad and very intense band with a maximum at 3340 cm^{-1} appears. The CH vibrations (2900-3000 cm^{-1}) change a little with concentration by overlapping with the H-bond band. This disappearance of the vibration band of the non-H-bonded OH groups is one of the most sensitive indicators for the existence of H-bonds /1/ /6/. Indeed perhaps this provides the best definition of a H-bond which one can give. The $\int \varepsilon d\nu$ or ε_{max} of this band gives a quantitative measure of the concentration of non-H-bonded OH groups and naturally there are very many papers devoted to study the H-bonds in solutions by this method. The equilibrium constants obtained by I.R. solution spectroscopy are the best which physical chemistry knows /7/

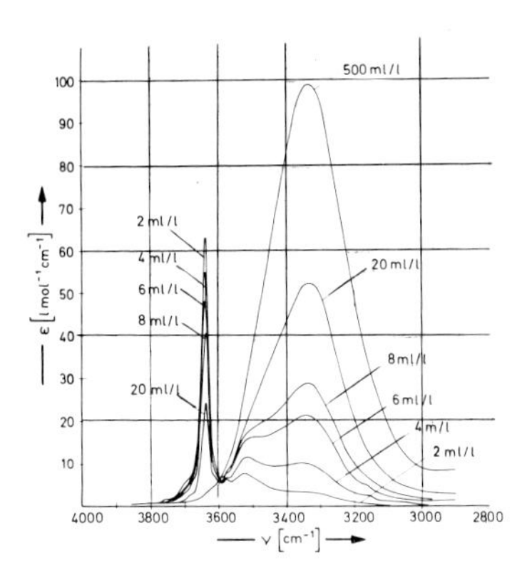

Figure 4. νOH fundamental absorption of CD_3OH in CCl_4 $30^{\circ}C$.

The content of monomers MO or the degree of dissociation of the H-bonds can be determined quantitatively by this I.R. technique.

$$\alpha = \frac{MO}{C} = \frac{\varepsilon_{max}}{\varepsilon_0} = \frac{[\int \varepsilon d\nu]_C}{[\int \varepsilon d\nu]_0} \tag{1}$$

C = concentration by weight

index 0 means limit-values for $C \to 0$.

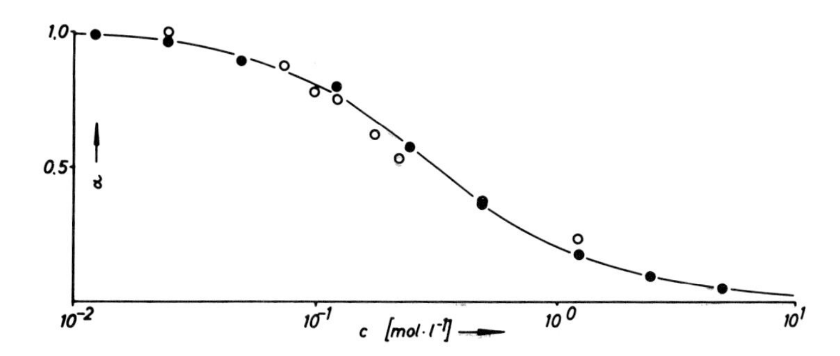

Figure 5. The content α of free OH of CH_3OH in CCl_4-solutions at different concentrations C at $30^{\circ}C$, determined by the fundamental band (2.75 μ) ○ and the first overtone (1.4 μ) ● . The α-values extrapolated to pure CH_3OH contradicts the most theories of H-bonded liquids.

Fig. 5 gives the concentration dependence of α of methanol in CCl_4, determined from the fundamental OH vibration and by the first overtone /1/. This result shows that the content of non-H-bonded OH groups in pure methanol should be of the order of 1%. /8/. Comparing this with water and remembering that the H-bond energy of both molecules per OH bond is similar we can extrapolate that in pure water at room temperature the content of non-H-bonded OH groups should be comparably small. This is in contradiction to most theories of water. The old paper by Eucken /9/ would demand 70% non-bonded-OH groups in water at room temperature and the Némethy-Scheraga /10/ treatment proposes about 50% non-bonded OH groups. Both statements have thus to be greatly doubted by these I.R. measurements on alcohol solutions.

III.2.3. HOD-SPECTRA

Fishman /11/ 1961 has determined the T-dependence of the fundamental vibration of pure methanol to 270°C. His measurements show that the band maximum moves from the frequency of H-bonded species at low T to frequencies of non H-bonded species at high T. Saumagne and Josien have done measurements with water to the critical temperature T_c/12/13/.

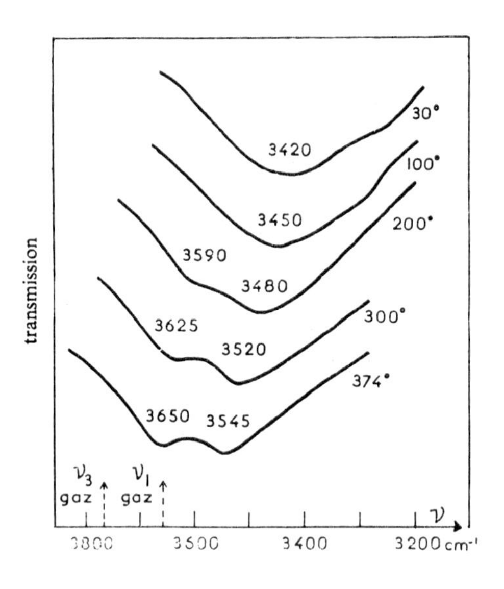

Figure 6. Infra-red absorption of liquid water in the hydroxyl stretching region at different temperatures./12/13/

(Fig. 6) This work shows a shift of the band maxima towards higher frequencies - the region of free OH - with increasing T. These spectra can only be discussed qualitatively because the exact pressure is not known and the transmission only is given. But they show that the intensity ratio of ν_1 and ν_3 is dependant on association. Mohr, Wilk and Barrow /14/ suppose similar indications on the intensity change of ν_1/ν_3 on solution spectra of H_2O in CCl_4.

The HOD spectra are more convenient for quantitative discussions since one can separately observe the OH and the OD (2720 cm^{-1}) vibration. The HOD molecule is measured in D_2O. This allows a cell length of 35 μ, which can more easily be determined than the cell lenth of 1-5 μ required for H_2O spectra. Hartmann /15/ and Senior and Verrall /16/ have studied the fundamental stretching of HOD under saturation conditions to 100°C. Both groups observe a decrease of intensity at

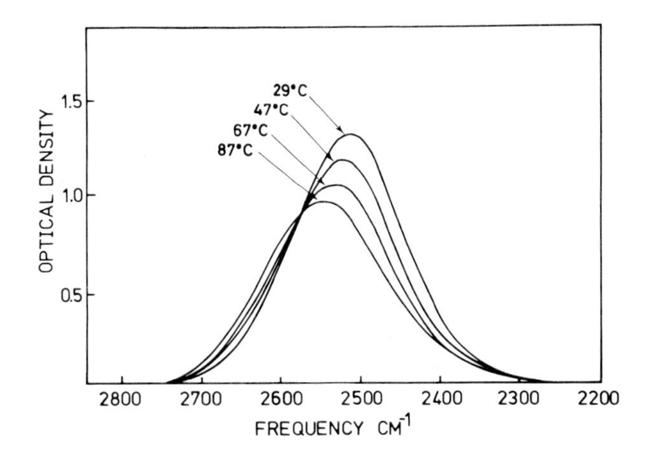

Figure 7. Plot of optical density vs. frequency for the ir absorption of HOD at 2500 cm^{-1} for various temperatures.

lower wavenumbers (region of H-bonds) and an increase of intensity at higher wavenumbers with increasing T (region of non or weaker H-bonded OH groups) Senior and Verrall suppose that the I.R. HOD band is consistent with two T-dependant bands with maximum at 2525 cm^{-1} and 2650 cm^{-1} /16/ Fig. 7). This points to an decrease of H-bonds and an increase of free or weaker bonded OH groups with higher T. These experiments agree very well with similar Raman measurements of Walrafen /17/ /18/ (Fig. 8). Walrafen has shown by a curve analyser that this observation can be understood by the assumption of the overlapping of different bands with the main components at 2520 and 2650 cm^{-1}. Franck and

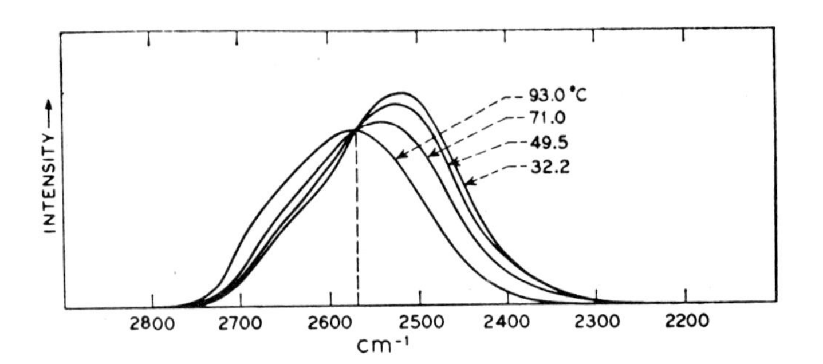

Figure 8. Quantitative Raman intensities as functions of frequency corresponding to a 6.2 M solution of D_2O in H_2O at temperatures from 32.2° to 93.0°C. The intensity scale is arbitrary. /17/.

Roth have measured the HOD spectra at different T and different densities /19/ /20/. Fig. 9 shows how the precision of fundamental measurements suffer: The zero line is a function of wavenumbers, the instru-

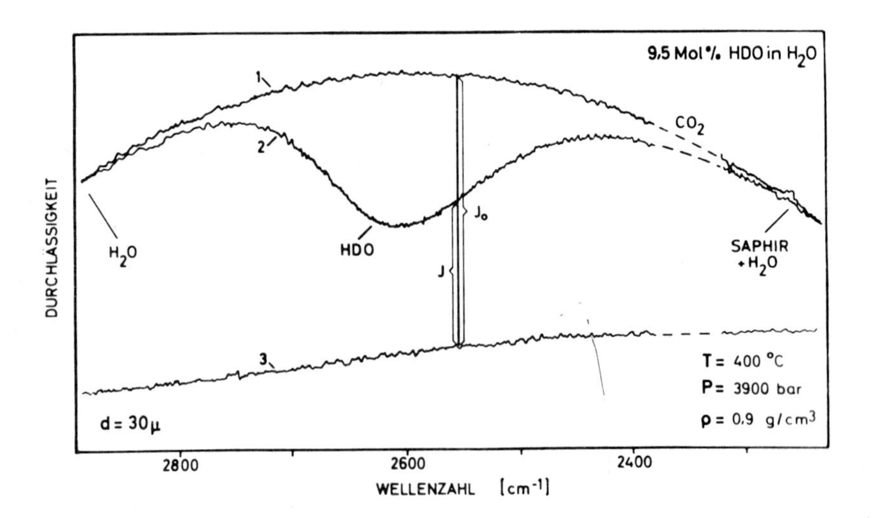

Figure 9. One set of original curves needed to determine the HDO fundamental band. /19/.

ments gives transmittance values, there is a background of H_2O and CO_2. These authors have made no measurements under saturation conditions. As comparison fig.10 gives original curves of the optical den-

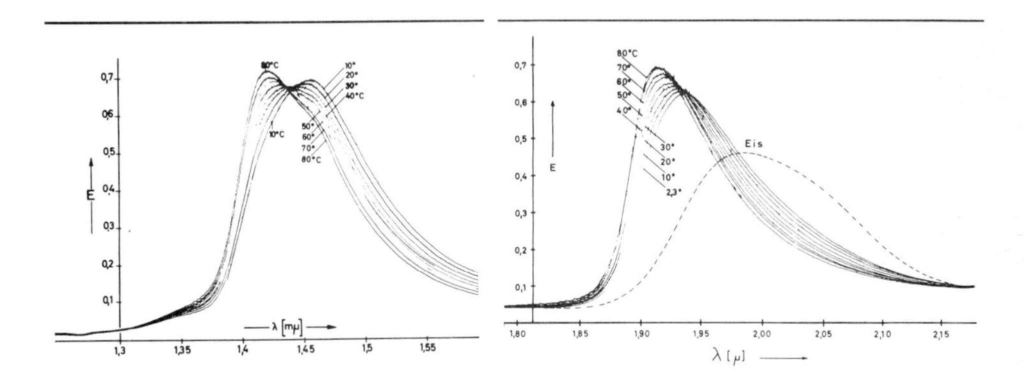

Fig.10 Set of original curves of the H_2O first overtone band at different T and combination band $\nu_{1,2} + \nu_3$ /21/.

sity of the H_2O first overtone. These curves have no background and have to be corrected only by the density to get directly the extinction coefficients. Fig.11 shows measurements of Franck and Roth at different densities. With increasing T the spectra change according to a system with higher content of non H-bonded molecules. But the high density induces higher pressure at higher T and induces a higher degree of H-bonded molecules than at saturation conditions and brings complications in addition.

Reducing the density at T_c the critical density changes the spectrum nearer to the free OH state. The wavelength maximum at critical conditions of HOD is 2650 cm^{-1}, a wavenumber which Senior-Verrall (S-V) /16/ and Walrafen (W) /17, 18, 22/ have indirectly concluded from the band analysis of their low T spectra. This result and the approximate isosbestic point in the spectra of S-V and W. gives an indication that water can approximately be described with a two species model like Roentgen has done (details see chapter overtone).
By density reductions to 0.01 g/cm^3 one observes the rotational structure with three branches (maxima: 2820, 2718 and 2600 cm^{-1}). A quantitative analysis of bands with rotational fine structure and the

transition to a continuum band is very complicated /23/.

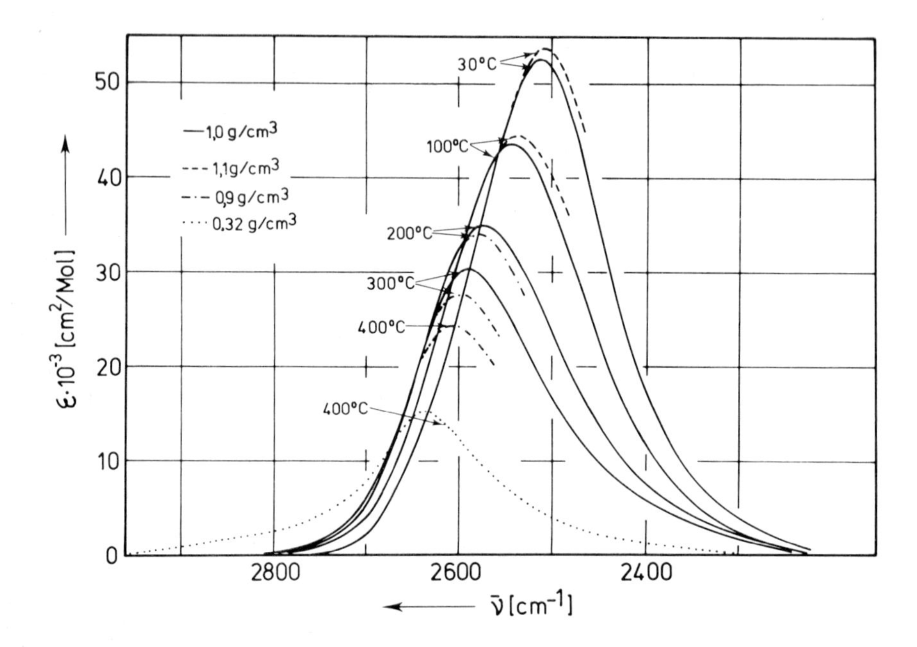

Fig.11 OD fundamental vibration of HOD /19, 20/

The consensus of all the papers on the HOD fundamental stretching vibration leads to the statement that with an increase of T one observes a decrease of the content of H-bonds. A quantitative analysis of the H-bond content has yet to be made and is difficult due to the high intensity of the H-bond bands and the overlapping effects with the weaker free OH-band. (But see the overtone chapter).

Walrafen obtained similar results (fig.8) with Raman measurements on the fundamental vibration of H_2O /24/ at 30 - 93 ^{o}C. He showed that the profile of this Raman band (the fundamental stretching vibration) could be separated into four Gaussian components with the frequencies: 3247, 3435, 3535, 3622 cm^{-1}. He also /25/ showed that I.R. fundamental band of H_2O could be separated by a curve analysis into four components at 3240 (310), 3435 (260), 3540 (150), 3620 (140) cm^{-1} /25/. In brackets are given the half-width in cm^{-1}. The half-widths are in agreement with the experience that they increase with H-bonds. Distinct peaks at 3620 and 3200 cm^{-1} of the H_2O I.R. stretching band have been observed by Crawford and Frech /26/. If we compare the I.R. results of Walrafen with the methanol spectrum in fig.2 we can see that the curve analyser fits data with one band in the region of free OH, one with in the region of dimers (H-bonds with unfavoured angles) and two bands in the region of linear H-bonds. Lindner has made experiments of the HOD Raman band in a wide region of T and density (till 400 °C) /53/. He established Walrafens result of the possibility to analyse the Raman bands with 3 overlapped bands /53/. Lindner fit his data for all T and all densities with the sum of the three bands:

1. 2630 cm^{-1} free OH band, maximum not T-dependent
2. 2510 - 2620 cm^{-1} (H-bond band, maximum f(T).)
3. 2300 - 2550 cm^{-1} (H-bond band, maximum f(T).)

The T-independence of the maximum of the free OH band indicates the existence of a well defined species the non H-bonded OH groups. The T-dependence of the H-bond bands would agree with the assumption of a distribution on different H-bonds. But one has to stress that the curve analysis may be possible in different ways. This is not the case for the curve analysis of the overtone band with the iceband and the overcritical extinction coefficients (see overtone chapter).

Naturally the free OH groups in this nomenclature cannot be gas-like species. We know from solution spectra that there are solvent effects in non-polar solvents also. Free-like molecules could only mean free in the environment of the polar solvent H_2O. But the differences between the large effects of H-bonds and these solvent effects can be clearly demonstrated by a fig. given by Franck and Buback /27/ (fig.12). This fig. differs between the large frequency shifts induced by the

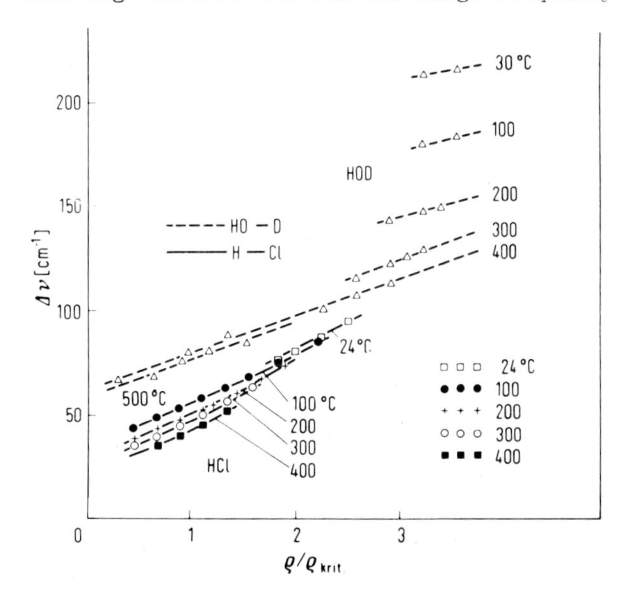

Fig.12 $\Delta\nu$ cm^{-1} = $\nu_{vapour} - \nu_{liquid}$ of the maximum of the fundamental band of HCl and OD of HOD as function T and the reduced density /27/.

strong H-bonds of water and the smaller frequency shifts induced by smaller dipole effects of HCl. Roth /20/ has given a similar diagram including the methanol results. Ethanol has similar large frequency effects like water. The frequency shift of HCl and the general (non H-bond induced effect) of water and ethanol includes effects during the change from the gas spectrum with rotation fine structure to the broad liquid spectra.

The fundamental I.R. results have not been interpreted quantitatively in relation to the H-bond content. But the fig.12 shows that one could try to calculate the extra effects of H-bonds, assuming the overcritical state as standard for the liquid state of water without H-bonds. Fig.12 shows clearly that the large effects of H-bonds on $\Delta\nu$ can be neglected for $T > T_c$. For these T the behaviour of water and HCl become similar.

III. 2.4. ELECTROLYTE SOLUTIONS

The spectra of aqueous electrolyte solutions also give indications of different water species. The ClO_4^- ion induces a change in the water spectrum similar to that of an increase of T. Keçki /28/ has shown that increasing ClO_4^- - concentration induces an intensity change of two different components of the HOD OD stretching band (fig.13). The two components have maxima at 2510 cm^{-1} and 2620 cm^{-1}. These two com-

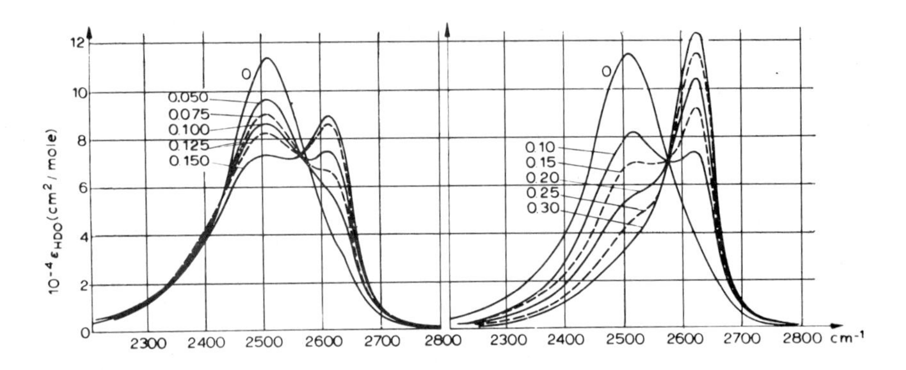

Fig.13 OD band of HOD in solutions of $Mg(ClO_4)_2$ (left) and $Ca(ClO_4)_2$ (right). Numbers indicate the $C_{ClO_4^-}/C_{H_2O}$.
C_{H_2O}: concentration of all isostopic forms of water.
ClO_4^- induces an increase of the free OH-band (2620 cm^{-1}) and a decrease of H-bond band (2510 cm^{-1}) /28/.

ponents are related to the two components proposed by Senior-Verrall or to the main-components given by Walrafen. This is further indication that the T-dependant state of liquid water can be described in the first approximation with a two state model: bonded and non H-bonded OH groups. Similar results are obtained by Brink and Falk /29/ with $NaClO_4$, $Mg(ClO_4)_2$ and $NaBF_4$ +). They observe two components: one component maximum 2630 cm^{-1}, $\Delta\nu_{1/2} \approx 50$ cm^{-1}, second component maximum 2503 cm^{-1}, $\Delta\nu_{1/2} \approx 160 - 200$ cm^{-1}. Brink and Falk /29/ discuss the 2503 cm^{-1} band as band of H-bonded water molecules and the 2630 cm^{-1} band as water molecules weakly bonded to the ClO_4^- - ions. But one can also get a consistent picture with the spectra of pure water, if one remembers that the 2630 cm^{-1} band is similar to the high frequency band in the interpretation of Senior-Verrall and Walrafen and the bands observed by Franck - Roth for high T in pure water. Especially one can stress the similarity of the change by a factor of 2 of the half-widths observed by Brink - Falk and Kecki and the half-widths of the components of the pure water band proposed by Walrafen.

Generally it is observed that anions induce much larger changes in water spectra than cations (see chapter overtone region). To a first approximation the ions alter the water spectra similar to that brought about by a change ot T on pure water. Fig.14 gives the frequency shift of the OH stretching vibration of HOD as a function of the salt con-

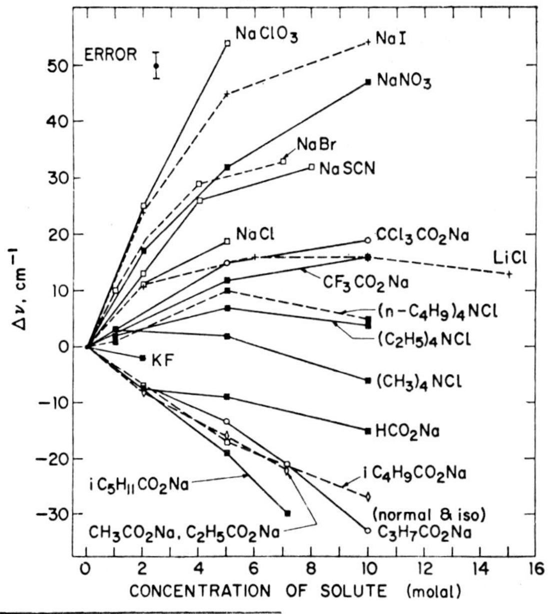

Fig.14 $\Delta\nu$ is the shift in frequency of the νO-H (HDO in D_2O salt solution) band from the νO-H (HDO in pure D_2O) band plotted vs. the molal concentration of dissolved salt. /15/

+) I.R. spectra of aqueous solutions other ions and hydrates see /55/

centration as observed by Hartmann /15/.

There are two groups of ions: the first induces a positive $\Delta\nu$ as that of heating pure water; the second group gives a negative $\Delta\nu$ similar to that of cooling pure water. These two groups can be observed by different methods for example with NMR chemical shifts /30, 15/. This is the known structure-breaker and structure-maker effect in a rough heuristic nomenclature. The ion effects on the fundamental bands are similar to these on the overtone bands. In the overtone region they can be determined more precisely and they will be discussed in more detail in the overtone chapter.

A review of the I.R. spectra of aqueous ion-solutions is given by Verrall /31/. There is a review of the relation between the spectroscopic results and the properties of electrolyte solutions, especially the relation between the spectroscopic ion series and the lyotropic ion series /21/.

III. 2.5. THE H-BOND BANDS

All solution spectra of alcohols show at low or medium concentration a medium band at about 2.87 µ or 3510 cm^{-1} (fig.4). Barnes, Hallam, and Jones have shown that this observation is valid for fluoroalcohols too /32/. This band cannot be observed at higher concentrations as it merges into one broad association band centred ca 3340 cm^{-1}. This small intermediate band is related to a minimum of the dipole moment in the same concentration region /33, 34, 35/ (fig.15). A minimum in dipole moment indicates the presence of cyclic structures. The mean size m of aggregates Ass can be estimated spectroscopically /36, 1/:

$$m\,Mo \qquad Ass \tag{2}$$

$$K = \frac{Ass}{Mo^m} = \frac{C - Mo}{Mo^m} \tag{3}$$

C: Total concentration

$$m \log Mo = \log(C - Mo) + \text{const} \tag{4}$$

Fig.16 shows the part of eq.(4) for methanol. The intermediate band at 3520 cm^{-1} dominates at room T until concentrations of 0.1 mol/l ($x_{CH_3OH} \approx 0.01$). A peak of the main H-bond band (3340 cm^{-1}) can be

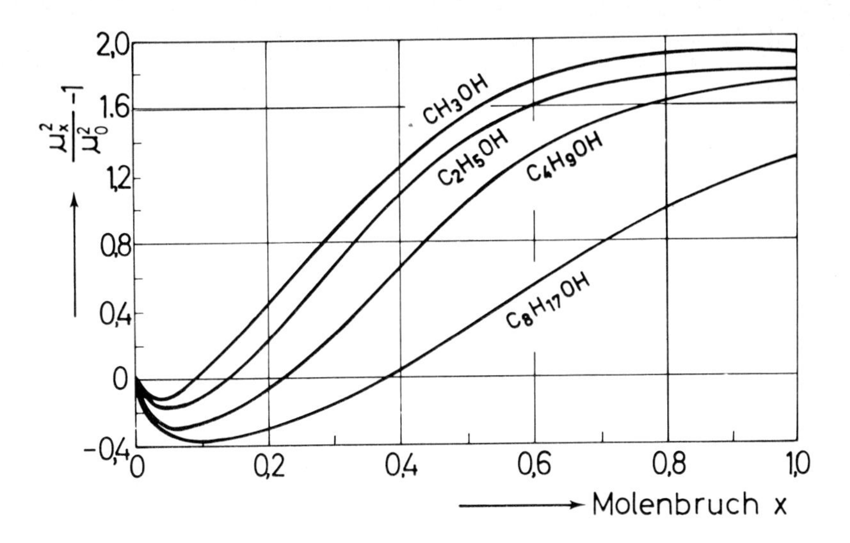

Fig.15 The dipolemoment minimum of alcohols in CCl_4-solutions as function of alcohol-concentration (indicates cyclic structures).

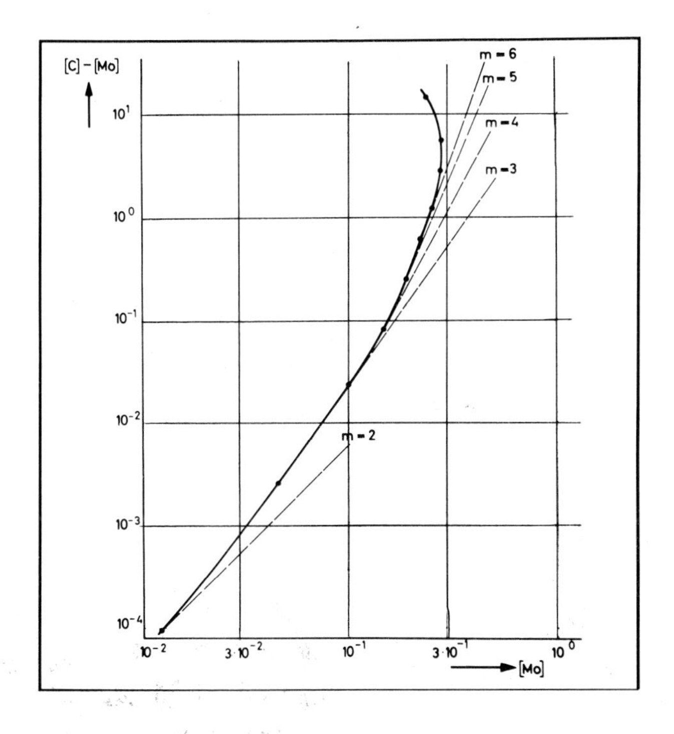

Fig.16

Determination of the mean aggregation number m of methanol in CCl_4-solution 20 °C

observed in this spectral region till 0.2 mol/l (x_{CH_3OH} = 0.02). The dipole moment minimum is observed at about x_{CH_3OH} = 0.02. In this region the mean association number is 3 (fig.16). Both methods are thus consistent with the assumption that in this region there exist small cyclic H-bond structures (dimers and trimers). The interaction energy of H-bonds has a maximum if the angle between the proton axis and the axis of the lone pair orbital is zero /1, 36, 37, 38/. Cyclic dimers and trimers have unfavoured angles. That means the interaction energy is smaller than in the case of linear H-bonds. In H-bond spectroscopy /1/ there exists the Badger-Bauer rule which indicates that in similar molecules the frequency shift $\Delta\nu$ of the H-bond band in comparison to the maximum of the free OH vibration is proportional to the H-bond interaction energy /1, 39/. For methanol in CCl_4 one observes $\Delta\nu$(dimer) = 130 cm^{-1}; $\Delta\nu$(linear H-bond) = 300 cm^{-1}

$$\frac{\Delta\nu\ (\text{dimer})}{\Delta\nu(\text{linear})} = 0.45 \approx \frac{\Delta H(\text{dimer})}{\Delta H(\text{linear})} \qquad (5)$$

This result would mean that the two H-bonds in a dimer would induce a similar interaction energy to one linear bond. This result shows why a cyclic dimer is energetically not unfavoured to a linear bond whether the angle is energetically unfavoured. This "dimer-band" can more readily be observed with the matrix technique (s.page 285). There it is discussed in terms primarily of an open end dimer though with the presence also of a cyclic species. There is no argument however why an open end dimer should have only half the $\Delta\nu$ of the linear multimer. The frequency shift $\Delta\nu$ = 30 cm^{-1} of the OH asymmetric stretching band between water (3651,7 cm^{-1}) and methanol (3682 cm^{-1}) is so small that the argument that a linear dimer $\Delta\nu$ = 130 cm^{-1} should have such strong effect in comparison to a linear multimer $\Delta\nu$ = 300 cm^{-1} cannot in our view be understood. But the angle dependence of H-bond energy could provide a good interpretation if one assumes a cyclic dimer consistent with the dipole-moment measurements. A third argument for the existence of a cyclic dimer is the similar angle dependence of H-bond energy determined by equilibrium constants of the association of oximes and the $\Delta\nu$-values of the intramolecular H-bonds of diols /36, 37/.

A further indication of the influence of H-bond angles on the water spectra are the spectra of different ice-modifications /40/. These modifications have similar structures /41/ which differ on the angles of the H-bond angles. The spectra show different ice types differing mainly on the frequency shifts to the free OH vibration corresponding to the H-bond angles.

The matrix spectra (see chapter matrix spectra) of water can be understood also from the view that one observes different H-bond species with different angles. The angle dependence of the H-bond energy obtained by this interpretation of the matrix spectra is consistent with the other results discussed in this chapter.

The energy of the H-bond interaction depends also on the OH---O distance /42/. Wall and Hornig have given an experimental function of the ν of the OH fundamental in dependence of r /43/. But in their diagram are collected data from very different molecules with different acidity and basicity and the acidity of the proton and the basicity of the lone pair electrons, also determines the H-bond interaction /6, 44, 45, 46/. But we have found that the maximum of the ice band in the overtone region as a function of T (that means on density and r too) depends on a similar function to the Wall-Hornig function. Our data can be described by a dependence: $\Delta\nu \approx 1/r^{14}$. This may be related to a polarisation effect of the dipole-moment. Compare the calculated dipole-moment of H_2O in ice of 2.6 Debye (Coulson-Eisenberg /47/). For large distance r > 3 Å the Wall and Hornig review gives a function: $\Delta\nu \approx 1/r^3$. In a smaller r-region there is an approximation: $\Delta\nu \approx A-B\cdot r$ /6, 48/. Bellamy and Owen /9/ gave a theoretical relation:

$$\Delta\nu\ (cm^{-1}) = 50\ \left[(d/r)^{12} - (d/r)^{6}\right] \qquad (6)$$

d: gas kinetic collision diameter.

The result of both effects is: H-bond energy therefore depends on H-bond angle and O...O distance.

The molecule

$C_6H_5-CH_2-CH(CH_3)OH$

gives an example that one can differentiate the following species /1/ by I.R. methods:

non-H-bonded	3622 cm^{-1}
intramolecular H-bond to π-electrons	3600 cm^{-1}
intermolecular H-bond to π-electrons	3555 cm^{-1}
intermolecular H-bond (unfavourable angle)	3480 cm^{-1}
intermolecular H-bond (linear)	3360 cm^{-1}

All these observations suggest that in liquids there are different H-bonds with different angles and different distances. But the results of the matrix technique show that there is a energetically prefered small number of arrangements. The data on the angle dependence of H-bonds show that if there would be a continuum of angles one would expect the preference of some frequencies of the H-bond stretching band and that there are some energy minima of a small number of cyclic arrangements /38, 1/. That would mean that the distribution of angles and the distribution of H-bond energy has some maxima.

The assumption of a linear dimer with half the $\Delta\nu$-values than linear multimers would have an other difficulty. There are many indications that the frequency of a non H-bonded OH group does not change much if the free electron pair, or the second OH group in the case of water, are H-bonded.

1. Diols in dilute CCl_4-solutions show two OH-bonds: one of a free $OH\nu_F$ and one of the second OH which is intramolecular H-bonded /1, 36, 37/ ν_H. There is a linear relationship $\Delta\nu_F$ (cm^{-1}) = $0.08\ \Delta\nu_H$.

   $\Delta\nu_F$: frequency shift of ν_F compared with a free molecule
   $\Delta\nu_H$: frequency shift of ν_H compared with a free molecule

2. In dilute solutions: CCl_4 - H_2O - acceptors Mohr et.al. /4/ have observed two different OH stretching bands to one of a free OH group and one of an intermolecular H-bonded OH group water-acceptor (fig.17). The frequency ν_F of the free OH gives a similar linear relation (fig.17) for the water molecule /1, 36, 37/

$$\Delta\nu_F\ (cm^{-1}) = 0.075 \cdot \Delta\nu_H$$

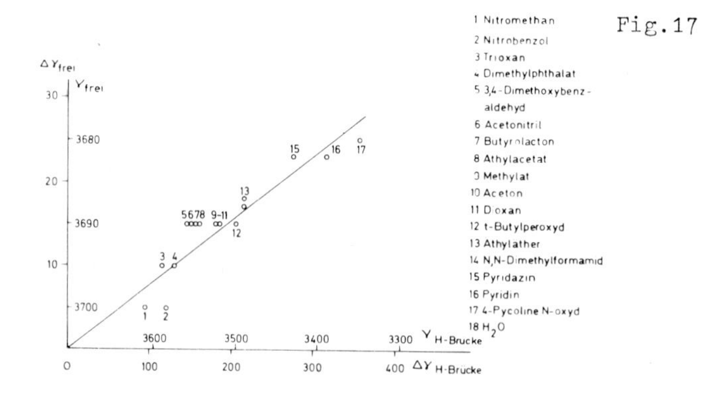

Fig.17 Dilute solutions of H_2O and different acceptors in CCl_4 /14/. Frequency of the free OH depends linearly on the frequency shift of the second H-bonded OH group /1, 36, 37/.

3. With the matrix technique H_2O in solid N_2 or solid Ar Tursi and Nixon /50/ and Neikes and Luck /51, 38/ have observed a splitting of the free OH if different H-bonded species appear. We believe this is the same effect than the diols or H_2O-acceptors give in solution. One can suppose that every type of H-bonded species with a certain H-bond interaction gives a small shift of the second free OH group. In this case, we ontained a linear relation:

$$\Delta\nu_F \; cm^{-1} = 0.077 \; \Delta\nu_H$$

The conformity of these three different experiments adds strength that this interpretation is correct. This casts doubt on the interpretation of linear water dimers with their own H-bond bands. In addition we can expect that this small effect on the free OH band of liquid water cannot be observed exactly because in liquids the half width are too large.

III. 2.6. COMPARISON FUNDAMENTAL AND OVERTONE TECHNIQUE

The efficient I.R. solution method /1/ to study the H-bond state cannot be applied to water since there is no solvent for water with sufficient solubility without H-bond interaction. Therefore we have tried to get information on water by studying the T-dependence of the water spectra. This method can be adjusted by studying pure alcohols and comparing these results with the solution measurements /1/. This work has shown that there are large T-effects on the I.R. spectra of pure alcohols. But there are some other environmental effects on alcohols in CCl_4 or pure liquids also. These effects are smaller than the H-bonded effect. All these different environmental effects on the alcohol spectra change the band profile but it was found /5/ that the $\int \varepsilon d\nu$ of a whole band was constant even with considerable changes in the band profiles (page 27o). This constancy provides the possibility for studying the main effects without the knowledge of the partition functions of different interactions. Independant from the disturbance of bands by small interactions like dispersion forces we can alter the band of non H-bonded OH groups and bands of H-bonded OH groups of alcohols. The overtone band of non H-bonded state of CH_3OH and C_2H_5OH increases with T till 350 °C under saturation conditions /5/. Above this T the intensity becomes constant. From this evidence we have assumed that above this T all OH groups are not H-bonded and if this reasonable probable assumption is accepted we can for all T give the values of non H-bonded OH groups /5,1/. The result that at room T this content is at about 2% is in excellent agreement with the solution experiments (fig.5), thus we have verified that we have an efficient method for studying the H-bond state in a pure liquid. Roth /20/ has studied the fundamental stretching vibration of pure ethanol up to the T_c. At 20 ^{o}C the band maximum has the frequency of 3320 cm^{-1} (linear H-bond band). The intensity of this band decreases with increasing T. Under critical conditions the frequency of the band maximum is 3580 cm^{-1} (dimer band). This agrees with the overtone result that a number of H-bonds exist in alcohols above T_c, these disappear above 370 ^{o}C, the critical region of water. From our viewpoint the magnitude of T_c of water is mainly determined by the equilibrium of H-bonds, but the T_c of alcohols depends on this equilibrium and on the dispersion forces also.

We have to stress that the quantitative results on alcohols were mainly obtained in the overtone region. Lescombe, Latimer and Rodebush and Mecke utilised the overtone region to determine the H-bonds in solutions in the twenties and thirties. But this excellent region has been virtually neglected since the development of recording I.R. instruments which made the fundamental region so readily accessible. Some doubts have been expressed on the results for pure liquids given from the point of view of the fundamental region. But there are some very important differences between the two techniques. First the photometric accuracy of near-I.R. instruments is far higher than that of mid-I.R. instruments. Last but not least there is a fundamental difference in H-bond spectral features for the two different regions. In the fundamental region the intensity of the H-bond band is higher by a factor of 20 or more than the intensity of the band of free OH-groups (fig.4) /1/. In overtone regions $\int \varepsilon d\nu$ of the free and bonded OH-bands are nearly the same (fig.18). In addition the frequency shift between

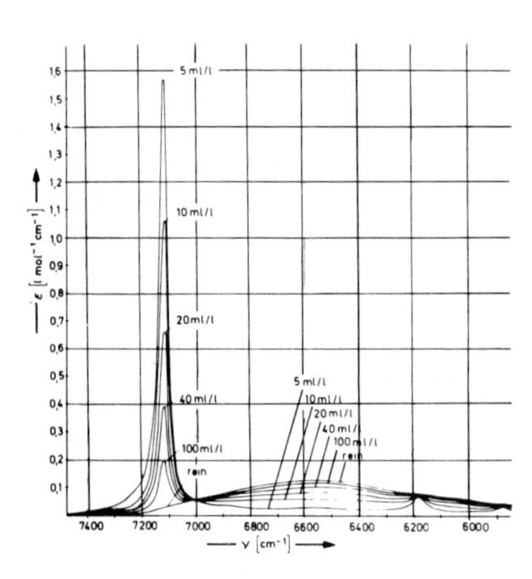

Fig.18 Concentration dependence of the first overtone of the OH stretching of CD_3OH in CCl_4, 20 ^{o}C (compare with fig.4).

these two bands is proportional to the number of the overtone. In the liquid state there is a broading effect on the OH bands. Therefore the quantitative analysis of H-bond state is impeded by this overlapping effect of free and H-bond bands. But this difficulty is much greater in the fundamental region than in overtone region. The result is that the fundamental region is efficient for studying H-bond bands qualitatively from the frequency shifts. Whereas the overtone region is efficient for quantitative work especially for the determination of the content of non H-bonded groups. The Raman measurements have similar difficulties to the fundamental I.R. region. Overtones are usually very weak in the Raman and thus measurements with the Raman effect are very limited. In addition quantitative intensity determinations in the Raman are inherently difficult.

The determination of H-bonds by studying the H-bond bands is complicated because we have to taken in account in liquids a distribution of H-bonds with different angles and different distances. Fermi resonance effects often give further complications. In the overtone region:

1. We can study different overtone bands of the free OH.
2. The half-width of the free OH band is smaller than the half width of the H-bond band;
3. The Fermi resonance effect is more important between two bands if one band is a fundamental band /52/.

For these three reasons our overtone technique has the advantage that the Fermi resonance effect can disturb quantitative results much less.

The possibility to identify or to determine quantitatively two bands depends on three factors: half-width $\Delta\nu_{1/2}$, frequency distance $\Delta\nu$ and intensity ratio /1, 46, 54/. Fig.19 gives a theoretical example: If we add two bands of two components with different concentrations, there

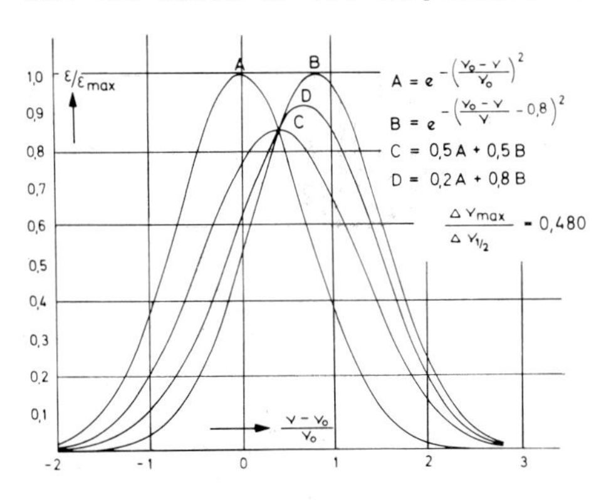

Fig.19 Calculated overlapping of two bands of Gaussian profiles with $\Delta\nu = 0.48\,\Delta\nu_{1/2}$. A mixtures of two of these bands give no indication whether there are two bands or not.

may be concentration regions in which only one band seems to exist. If two bands overlap with equal intensity one can recognize only two bands if $\Delta\nu/\Delta\nu_{1/2} > 0.85$ /53/. This situation can worse if two bands with different intensity overlap.

The quotient $\Delta\nu/\Delta\nu_{1/2}$ is more unfavourable for the fundamental OH stretching of water than for the overtone. Further the intensity of the free OH-band and the H-bond bands are more favourable for the overtone band than for the fundamental band. Fig.20 simulates the conditions of the HOD fundamental and the first overtone band. Fig. 20

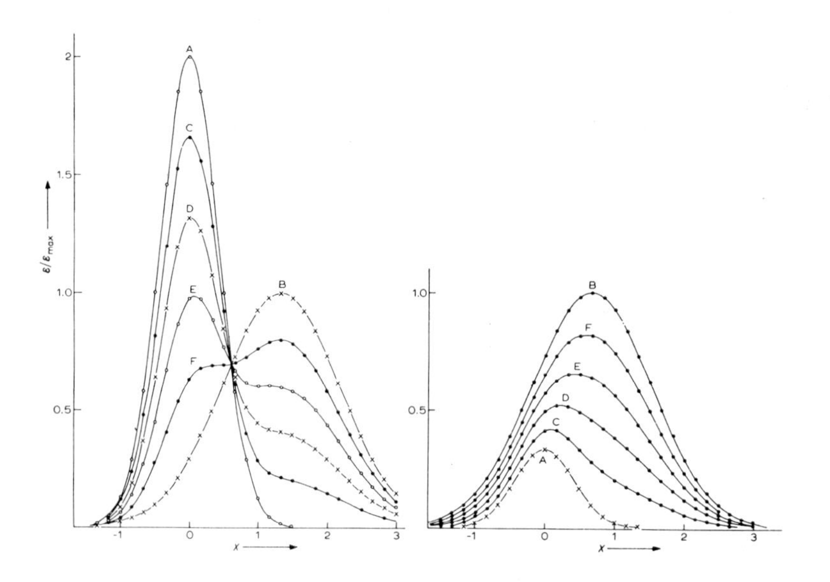

Fig.20 Simulated overlapping of the HOD stretching at different T
a) first overtone, b) fundamental
$\Delta\nu$, $\Delta\nu_{1/2}$ and intensity correspond to the experimental values.
a) $A = 2\exp(-4x^2)$ $\quad B = \exp - (x - 1,33)^2$
b) $A = 0.33\exp(-4x^2)$ $\quad B = \exp - (x - 0.66)^2$
$C = 0.8\,A + 0.2\,B$ $\quad D = 0.6\,A + 0.4\,B$
$E = 0.4\,A + 0.6\,B$ $\quad F = 0.2\,A + 0.8\,B$

shows clearly that one can make statements on the H-bond content in HOD easier by studying the overtone spectra. The fundamental band gives but qualitative support for the overtone results. Fig.20 provides one consistent picture of the H-bond state with both fundamental and overtone spectra and shows that apparant discrepancies between these two region can be cancelled out.

From this view the fundamental region is mainly important for the problem of the structure of liquid water for two reasons:

1. It aids the interpretation of the H-bond bands in the overtone region.
2. The matrix technique gives indications of a limited number of H-bonded species and it gives information on the angle dependence of the H-bond energy of water.

REFERENCES

1 W.A.P.Luck in: Water, a Comprehensive Treatise, editor F.Franks, Vol.2, p.235 Plenum Press, New York 1973

2 W.A.P.Luck and W.Ditter, J.mol.structure 1, 261 (1967/68)

3 G.Herzberg: Molecular Spectra and Molecular Structure, Vol.II, p. 489, von Nostrand Co., Princeton 1953

4 W.A.P.Luck, Habilitationsschrift Heidelberg 1968

5 W.A.P.Luck and W.Ditter, Ber.d.Bunsenges. 72, 365 (1968)

6 C.Pimental et.al. McClellan "Hydrogen Bond" Freeman and Co., San Francisco 1960

7 W.A.P.Luck, Z.Naturfosch. 23b, 152 (1968)

8 W.A.P.Luck, Naturwissenschaften 52, 25, 49 (1965)

9 A.Eucken, Ztschr.Elektrochem. 52, 264 (1948)

10 G.Némethy and H.A.Scheraga, J.Chem.Phys. 36, 3382 (1962)

11 E.Fishman, J.Phys.Chem. 63, 2204 (1961)

12 P.Saumagne, Thèses Bordeaux 1961
P.Saumagne and M.L.Josien, Bull.Soc.Chim. 1938, 813

13 M.L.Josien, Disc.Far.Soc. 43, 142 (1967)

14 S.C.Mohr, W.D.Wilk and G.M.Barrow, J.Am.Chem.Soc. 87, 3048 (1965)

15 K.A.Hartmann, J.Phys.Chem. 70, 270 (1966)

16 W.A.Senior and R.E.Verrall, J.Phys.Chem. 73, 4242 (1969)

17 G.E.Walrafen, J.Chem.Phys. 48, 244 (1968)

18 G.E.Walrafen and F.Franks: Water, Vol.1, p. 151, Plenum Press, New York 1972

19 E.U.Franck and K.Roth, Disc.Far.Soc. 43, 108 (1967)

20 K.Roth, Dissertation Universität Karlsruhe 1969

21 Fortschr.chem. Forschung 4, 653 (1964)

22 G.E.Walrafen: Hydrogen-bonded Solvent Systems, ed. A.K.Corrington and Pitones, Taylor and Francis, London 1968

23 W.A.P.Luck, Ztschr.Naturforsch. Nr. 2 and Nr.4, 1951

24 G.E.Walrafen, J.Chem.Phys. 47, 120 (1967)

25 G.E.Walrafen, J.Chem.Phys. 47, 114 (1967)

26 R.E.Frech, Thesis University of Minnesota 1968

27 M.Buback and E.U.Franck, Ber.Bunsenges. 75, 38 (1971)

28 Z.Kęcki, Adv.Molec.Relaxation Processes 5, 137 (1973)

29 G.Brink and M.Falk, Can.J.Chemistry 48, 3019 (1970)

30 J.N.Shoorley and B.S.Alder, J.Chem.Phys. 23, 805 (1955)

31 R.E.Verrall: Water a Comprehensive Treatise, ed.F.Franks, Plenum Press, New York 1973, Vol.3, p. 211

32 A.J.Barnes,H.E.Hallam and D.Jones, Proc.R.Soc.Lond. A 335, 97 (1973)

33 R.Mecke, Ztschr.Elektrochem. 52, 280 (1948)

34 R.Mecke, A.Reuter and R.L.Schupp, Ztschr.Naturforsch. 4a, 182 (1949)

35 R.Mecke and A.Reuter, Ztschr.Naturforsch. 4a, 317 (1949)

36 W.A.P.Luck, Naturwissenschaften 52, 25, 49 (1965)

37 W.A.P.Luck, Naturwissenschaften 54, 601 (1967)

38 W.A.P.Luck: Recent Progress in Hydrogen Bonds, "Stereochemistry of Hydrogen Bonds", ed. P.Schuster, C.Sandorfy and G.Zundel, North Holland publ. Amsterdam, in press

39 H.E.Hallam: Infrared Spectroscopy and Molecular Structure, ed. M. Davies, Chapter XII, Elsevier 1963

40 J.E.Bertie and E.Whalley, J.Chem.Phys. 40, 1646 (1964)

41 N.H.Fletcher: The Chamical Physics of Ice, Cambridge University press 1970

42 H.Ratajczak and W.S.Orville-Thomas, J.mol.struct. 1, 449 (1968)

43 T.T.Wall and D.F.Gornig, J.Chem.Phys. 43, 2079 (1965)

44 L.J.Bellamy and R.C.Pace, Spectrochim.Acta 25, 319 (1969)

45 L.J.Bellamy, H.E.Hallam and R.I.Williams, Trans.Far.Soc. 54, 1120 (1958)

46 W.A.P.Luck, J.mol.structure 1, 339 (1967/68)

47 C.A.Coulson and D.Eisenberg, Proc.Roy.Soc. 291A, 445 454 (1966)

48 R.C.Lord and R.E.Merrifield, J.Chem.Phys. 21, 166 (1953)

49 L.J.Bellamy and A.J.Owen, Spectrochim.Acta 25A, 319,329 (1969)

50 A.J.Tursi and E.R.Nixon, J.Chem.Phys. 52, 1521 (1970)

51 Th.Neikes, Diplomarbeit Marburg 1974

52 H.A.Stuart: Molekülstruktur, 3.Aufl. p.474, Springer Verlag, Heidelberg 1967

53 H.A.Lindner, Thesis Karlsruhe 1970

54 W.A.P.Luck, Disc.Far.Soc. 43, 115, 133 (1967)

55 G.Brink and M.Falk, Spectrochim.Acta 27A, 1811 (1971) ($Ba(No_2)_2H_2O$)
V.Seidl, O.Knop and M.Falk, Can.J.Chem. 47, 1361 (1969) ($CaSO_4 2H_2O$)
H.R.Wyss and M.Falk, Can.J.Chem. 48, 607 (1970) (NaCl)
G.Brink and M.Falk, Can.J.Chem. 48, 2098 (1970) ($NaClO_4\ H_2O$; $LiClO_4\ 3H_2O$; $BaClO\ 3H_2O$)
M.Falk, Can.J.Chem. 49, 1137 (1971) ($((CH_3)_3\ N\ 101/4\ H_2O$)
M.Holzbecher, O.Knop and M.Falk, Can.J.Chem. 49, 1413 (1971) ($Na_2[Fe(CN)_5\ NO]2H_2O$)

III.3. INFRARED OVERTONE REGION

Werner A.P. Luck
Physikalische Chemie, Universität Marburg
D 3550 Marburg/Lahn, West-Germany

ABSTRACT

The temperature dependence of the overtone spectra of fluid water in comparison with that of solutions gives a good criterion for the determination of the content O_F of non H-bonded OH groups. $O_F = f(T)$ is given from 0° - 400 ^oC for saturation conditions. This function is the simplest approximation to describe the properties of liquid water. It is shown that: heat of melting, heat of vaporisation, heat content, specific heat, density, surface energy, dielectric constant, NMR data and magnetic susceptibility can be understood as a function of T from the view of this simple experimental O_F function.
It is shown: for T < 200 ^oC under saturation conditions there are networks of large numbers of molecules H-bonded. The mean size of aggregates is estimated. Indications on smaller aggregates can only be recognized for 200 $^oC < T < T_c$.
The spectroscopic experiences to get this O_F function are reported. The consistency of the overtone results with fundamental and Raman spectra is discussed.
The structure temperatures T_{str} of aqueous electrolyte solutions - defined by Bernal and Fowler - are given quantitatively by overtone spectra. T_{str} yields an ion series. This series is identically with the lyotropic ion series. This series is now understood as effect on the water structure.
Determinations of O_F and much more T_{str} also are approximation methods. All who dislike approximations can accept from overtone spectra semiquantitatively that the content of non H-bonded OH groups increases with increasing temperature and that ions also change this content.

III.3.1 INTRODUCTION

The determination of the concentration dependence of the overtone bands of non H-bonded OH or NH groups is one of the most powerful methods to determine the H-bond equilibrium (see fig. 5 in chapter fundamental region, page 225). The advantage of the overtone spectra for this technique in comparison to the more familiar fundamental region are /1/:

1. The extinction coefficients of the maxima of non H-bonded groups are larger than those of H-bonded species.

2. The frequency shift $\Delta\nu$ between the band maximum ν_F of non H-bonded groups and the ν_H of bonded ones are large so that overlapping effects are small /2/.

3. The photometric accuracy of vis-near I.R.-instruments is better than that of mid-I.R. spectrometers.

4. The cell thickness required is in the range 1 mm - 10 cm compared with some μ in the fundamental region.

5. There are near I.R. instruments with cell compartments behind the monochromator; this arrangement cannot be constructed with mid-I.R. instruments.

6. From the view of point 4 and 5 the construction of thermostatted cells is easier.

7. The zero line is thus better than in the fundamental spectra and allows high quality quantitave measurements to be made.

The comparison of spectra in inert solvents with pure liquids shows that in pure liquids with OH groups the content of non H-bonded groups increases with increasing T. We demonstrate this on the first overtone of the OH-stretching. Fig.1 (compare with fig. 18 on page 240) ν_F and the half-width $\Delta\nu_{1/2}$ of spectra in inert solvents and of liquid spectra at higher T are also similar if the molecules have non-polar groups. Compare in fig.2 the spectra of liquid cyclohexanol with the CCl_4-solution spectra (dotted) +). The maximum of the band of liquid phenol at

+) The concentration dependence of the solution spectra see /2/.

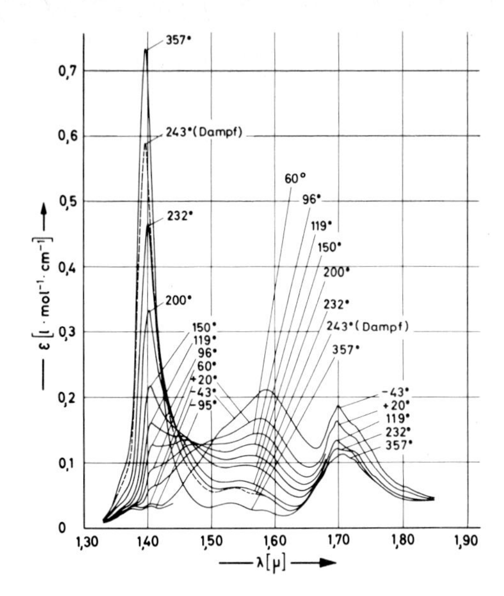

Fig.1 The T-dependence of the extinction coefficient ε of the first overtone of pure methanol. The band of non H-bonded OH at 1.6 μ increases with T /3/.

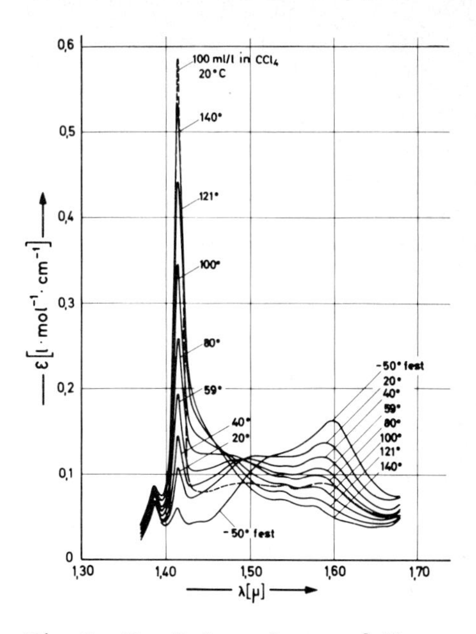

Fig.2 The T-dependence of the first overtone of pure cyclohexanol /2/. Dotted solution spectra 20 °C 100 ml/l cyclohexanol in CCl_4.

higher T has a frequency shift of ν_F in comparison with the spectrum in CCl_4 because there is an intermolecular H-bond interaction to the π-electrons of the phenyl-groups (fig.3). This induces a $\Delta\nu$ and a change of $\Delta\nu_{1/2}$. $\Delta\nu$ decreases with T because this H-bond to the π - electrons also decreases with T. Methanol shows a different $\Delta\nu_{1/2}$ in the pure liquid compared with CCl_4 solutions. The cause is the different environment in CCl_4 and in CH_3OH. Usually the $\Delta\nu$ of stretching bands are negative and proportional to the interaction energy (Badger-Bauer-rule) (see in fig.4 the comparison between CCl_4 and CS_2 as solvent). In fig.4 are also plotted the spectra of CH_3OH at T_c and different densities. ν_F of liquid methanol is higher than in CCl_4-solution. This effect parallels the rotational effects of methanol in pure liquid and CCl_4 solution. It is well known that small molecules such as HCl, NH_3 and H_2O exhibit rotational wings in non-polar solvents. However we find that CH_3OH exhibits a sharp ν_{OH} and $2\nu_{OH}$ in CCl_4, which clearly indicates a lack of rotation. Water shows only a very small difference of $\Delta\nu$ in pure water at T_c (7163 cm^{-1}) and in CCl_4-solution at room T (7158 cm^{-1}).

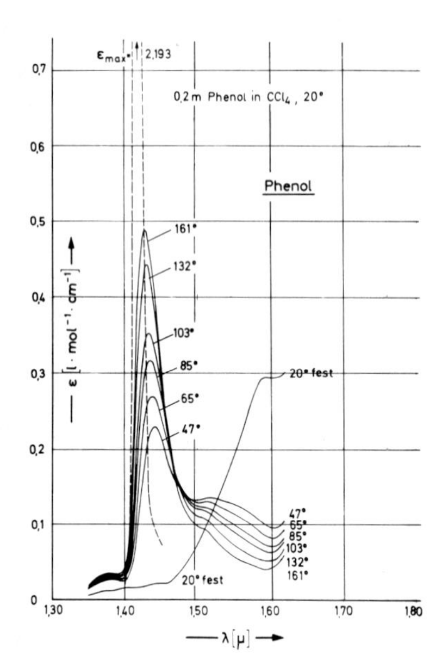

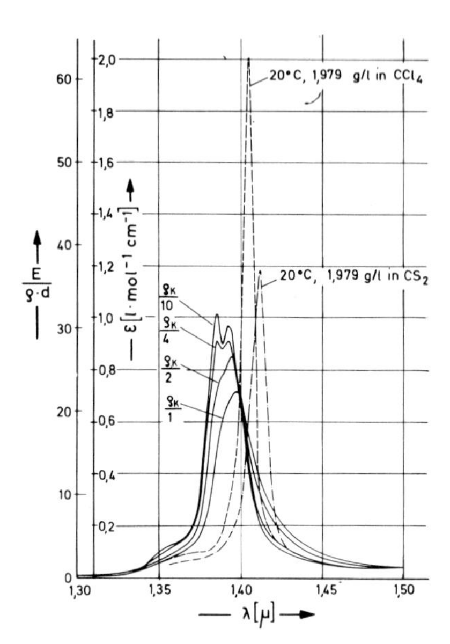

Fig.3 The T-dependence of the first overtone of pure phenol. Dotted 0.2 m phenol in CCl_4 20 ^oC. The frequency difference between dilute solutions and pure phenol is induced by intermolecular H-bonds to π-electrons.

Fig.4 Full lines: first overtone of OH stretching at T_c and different densities (units ζ_k: critical density). Dotted line: dilute solutions of CH_3OH in CCl_4 and CS_2 at 20 ^oC) /3/.

All our experience shows that in H-bonded liquids the content of non H-bonded OH or NH groups can be determined quantitatively. The details of the quantitative determination will be given in the next section. Bellamy, Hallam and Williams /4/ have shown with the OH fundamental vibrations that the H-bond interaction of water to a wide variety of different acceptors is very similar to the interaction of methanol to the same acceptors. This shows that the I.R. spectra of water can sensibly be compared with the spectra of methanol. The comparison of the spectra of pure methanol with the well-studied solution spectra of methanol therefore support the interpretations made of the spectra of pure water. The data /4/ also show that in the more polar media ν_1 and ν_3 of H_2O merge into one broad band, which is in line with the same results for pure liquid H_2O.

III.3.2 THE OVERTONE BANDS OF WATER

Fig.5 shows the general appearance of the T-dependence of the water overtone region on a logarithmic scale. We expect overtones of the

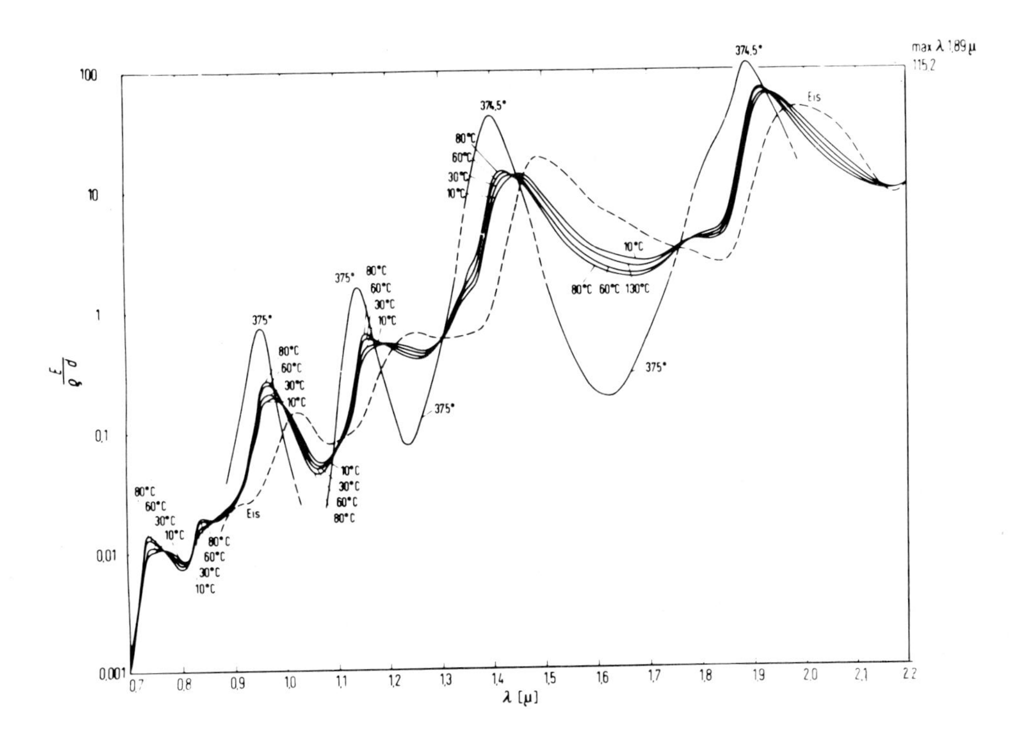

Fig.5 T-dependence of the overtone bands of water. Dotted: ice-spectrum.

asymmetric (ν_1 = 3651 cm^{-1}) and the symmetric (ν_3 = 3755,8 cm^{-1}) stretching vibrations and combination bands of these stretching and the bending vibration (ν_2 = 1595 cm^{-1}). The simplest interpretation of the different bands is given in Table 1.

Table 1

Interpretation of the water overtone spectra in the vapour state /5/

	ν cm^{-1}	λ μ
$\nu_{1.3} + \nu_2$	5332	1.875
$2\ \nu_{1.3}$	7251	1.379
$2\ \nu_{1.3} + \nu_2$	8807	1.135
$3\ \nu_{1.3}$	10613	0.9422
$3\ \nu_{1.3} + \nu_2$	12151	0.824
$4\ \nu_{1.3}$	13831	0.723

This interpretation does not differentiate between individual bands of the ν_1 and ν_3, because they are not separated in the vapour state and in the pure liquid state one observes only one band. Therefore Greinacher, Lüttke and Mecke /5/ use the nomenclature $\nu_{1,2}$ to demonstrate that one cannot distinguish between the two band-systems. In addition we observed with a 10 cm cell in the vapour state under saturation conditions at 119°C a very weak band at 1.48 µ (fig.6)

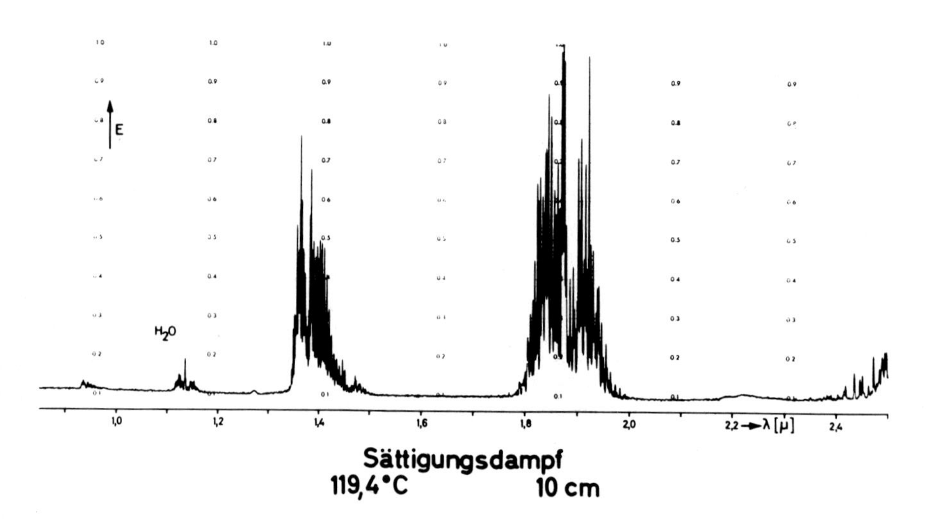

Fig.6 H_2O vapour, 119 °C saturation density, 10 cm

This may be a H-bond band.

In the liquid state (fig.5) all the bands are shifted to longer wavelength. In addition there appear two small bands at 1.78 μ and 1.35 μ . These could be: $\nu_{1.3} + \nu_2 + \nu_L \approx 5600\ cm^{-1}$ (1.78 μ); $2\nu_{1.3} + \nu_2 + \nu_L \approx 7420\ cm^{-1}$ (1.35μ). They correspond to the 4.75 μ band, which is a combination of the bending ν_2 and the libration band ν_L. This 4.75 μ band also appears only in liquid water.

III.3.3 QUANTITATIVE INTERPRETATION OF THE H-BOND CONTENT

There is no solvent which water dissolves in a sufficient concentration without H-bond interaction. Therefore a comparison with the excellent solution method of overtone spectroscopy cannot help to determine the H-bond content in water. This we have done with methanol and ethanol to test the interpretation of spectra of pure liquids /3, 1/ (see page 224). The H-bonds of water can only be observed with spectra at different T. Falk and Ford have expressed doubts whether overtone spectra could give quantitative data because they are so many combination bands /6/. But their qualitative objection is not valid quantitatively. It is clear from his theoretical spectral-line diagram (see fig.7) that a number of bands are so weak that they can neglected +). Furthermore their theoretical bands clearly show that the frequencies of the different H_2O-bands of the non H-bonded groups are located in certain frequency regions i.e. there are substantial

+) It is clear from his diagram (fig.7) that one has to stress that the intensity of I.R. bands decreases about a factor of ten in going from the fundamental to the first overtone or by a change from the first overtone to the second and so on. If the collection of possible bands given by Falk and Ford /6/ are compared: combinations of fundamentals with the second overtone; first overtone with the third; combinations of the first overtone with the fourth overtone would mean intensity differences by factors about: 100 and 1000. Such effects are mainly with in the limits of error of the quantitative spectroscopy and cannot give arguments against this method.

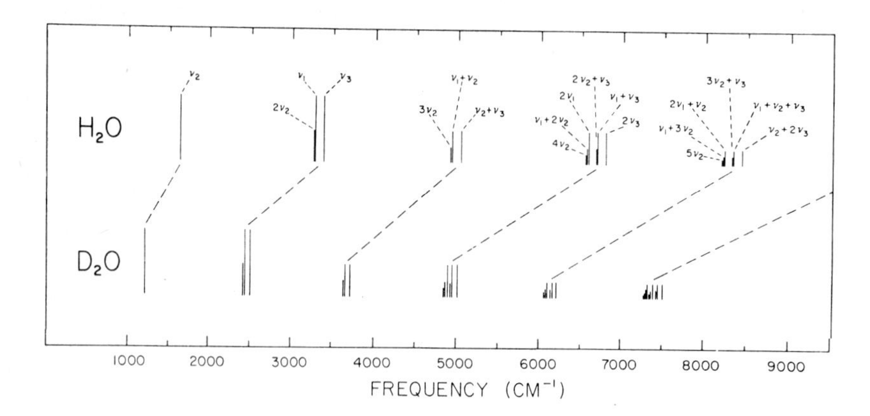

Fig.7 Computed overtone and combination bands, Falk and Ford /6/.

gaps which are the regions in which the linear H-bonded OH-groups absorb (also see fig.5). The quantitative method for the determination of the H-bond content is however also valid even if there are some bands overlapping. The method is a standard analytical procedure and needs only the extinction coefficients of the different molecular states. For example if there are two species F and H with extinction coefficients ε_F and ε_H at one wavelength, one can determine the fraction O_F of species F from the optical density E:

$$E = (\varepsilon_F \cdot C_F + \varepsilon_H \cdot C_H)\, Z = \varepsilon_C \cdot C \cdot Z \tag{1}$$

Z = cell length

$C = C_F + C_H$ total concentration

$$E = \left[\varepsilon_F \cdot O_F + \varepsilon_H (1 - O_F)\right] C \cdot Z = \left[(\varepsilon_F - \varepsilon_H) O_F + \varepsilon_H\right] C \cdot Z \tag{2}$$

$$O_F = \frac{C_F}{C_F + C_H} \tag{3}$$

If ε_F, ε_H and C are known, O_F can be determined independently if at the required wavelength there aborbs one band or many. If the absorption is due to many different bands 1, 2, 3, ... i we have

$$\varepsilon_F = \varepsilon_{F_1} + \varepsilon_{F_2} + \varepsilon_{F_3} + \ldots\ \varepsilon_{F_i} \tag{4}$$

$$\varepsilon_H = \varepsilon_{H_1} + \varepsilon_{H_2} + \varepsilon_{H_3} + \ldots\ \varepsilon_{H_i} \tag{5}$$

This statement is also true if there is any Fermi-resonance at this wavelength provided that the intensity increase due to Fermi-resonance is linearly proportional to the concentration. Thus the main doubts expressed about the overtone method are invalid. The argument is well established experimentally by the fact that with our method four different overtone bands give identical values of the H-bond content /7/. If the overlapping of different bands could interfere this effect could not be the same at four different overtones. Overlapping effects should especially be different in HOD spectra. On the first overtone of HOD stretching however we have found the same H-bond content as we did on the four different H_2O-bands /8/. This is a further evidence that the overtone method gives excellent results.

The problem of the method, as indeed also with the fundamental or any other spectroscopic region, is the determination of the extinction coefficients. There are three possibilities to get these values.

1. We tried to use our knowledge of the H-bond content $(1 - O_F)$ at one T /9/. If one value $1 - O_F$ or O_F is known the extinction coefficient can be calculated. The estimate of O_F at one T can be evaluated from calorimetric data. Pauling /10/ compared the heat of melting of ice $\Delta H_m = 1.43$ kcal/mol with the H-bond energy $\Delta H_H = 4.5$ kcal/mol per OH-group:

$$O_F = \frac{\Delta H_m}{2 \cdot \Delta H_H} = \frac{1.43}{2 \cdot 4.5} = 0.16 \qquad (6)$$

Fox and Martin /11/ assumed that the fraction of H-bonds on ΔH_M is the same as on the heat of sublimation ΔH_S.

$$O_F = \frac{\Delta H_M}{\Delta H_S} = \frac{1.43}{11.6} = 0.12 \qquad (7)$$

Pauling's approximation can be corrected with an estimate of the influence of the dispersion forces on ΔH_M /12/. The heat of melting of H_2S can assist as this estimate. A better value of O_F would be:

$$O_F = \frac{\Delta H_M\,(H_2O) - \Delta H_M\,(H_2S)}{\Delta H_H} = \frac{1.43 - 0.57}{2 \cdot 4.5} = 0.096 \qquad (8)$$

Pauling's early value of ΔH_H seems to be a little too large. We would prefer ΔH_H = 4 kcal/mol (see page 265). With this value eq. (8) would give $O_F \approx 0.1$. The last two estimations agree very well with the result of Haggis, Hasted and Buchanan (H-H-B) /13/. They tried from a rough theory of the T-dependence of the dielectric constant ε to estimate O_F from the heat of vaporisation (fig.8).

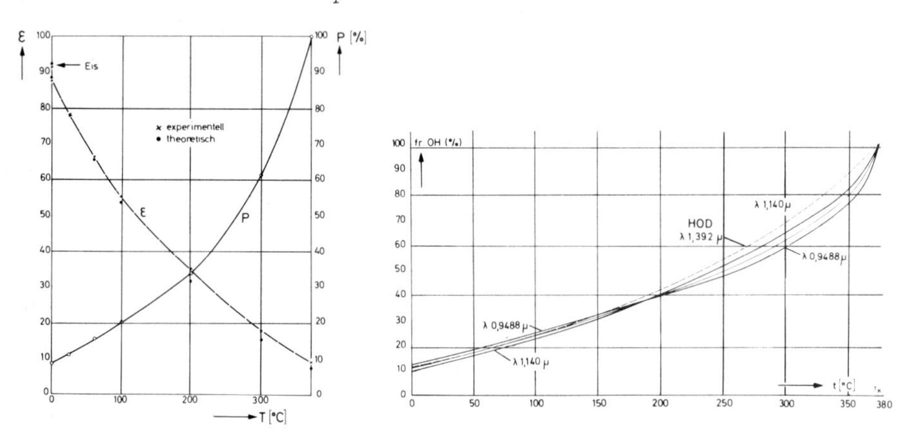

Fig.8 P: estimated values of O_F in percent: content of non H-bonded OH

ε: dielectric constant calculated with P (Haggis, Hasted, Buchanan /13/.

Fig.9 Content O_F, free OH in percent, determined by two H_2O bands (λ_{max} = 0.9488 μ and λ_{max} = 1.14 μ) and one HOD band (λ_{max} = 1.39 μ). The curve between the two H_2O bands is the mean value.

To get conformity between their theory and the experimental ε-values they have to assume O_F (0 °C) = 0.09 (fig.9). If one accepts this approximation value of 0.09 we get a good agreement between the spectroscopic determined O_F-values till 100 °C /7/ and the values got by H-H-B /13/.

2. Buijs and Choppin /14/ tried to estimate the optical density E at the three wavelengths 1.16 μ; 1.2 μ and 1.25 μ as characteristic for species with two H-bonds C_2, with one H-bond C_1 and without any H-bonds C_o. They took the equations:

$$\varepsilon^0_{1.16}C_0 + \varepsilon^1_{1.16}C_1 + \varepsilon^2_{1.16}C_2 = E_{1.16}$$

$$\varepsilon^0_{1.20}C_0 + \varepsilon^1_{1.20}C_1 + \varepsilon^2_{1.20}C_2 = E_{1.20}$$

$$\varepsilon^0_{1.25}C_0 + \varepsilon^1_{1.25}C_1 + \varepsilon^2_{1.25}C_2 = E_{1.25}$$

$$C_0 + C_1 + C_2 = 1 \qquad (9)$$

They took optical densities at different T to calculate the six unknown variables C_i and ε^i_i by computer. They found a good agreement with the cluster theory of Némethy and Scheraga /15/. We would doubt whether the wavelength 1.20 μ is characteristic for species with one H-bond but the misfortune of the system of equations (9) is that the given solutions are not sound. These are non-linear equations which have infinite solutions /16/. If one starts the computer calculation with a theoretical assumption and looks for the correct solution by the method of smallest errors, the computer stops at the solution from the infinite possibilities which is the nearest to the start point /16/. This method is fine to demonstrate that a model is a possible one but is no method to demonstrate that the theory is the correct one.

3. To obtain information on the H-bond contents is difficult because there could be a continuum of different H-bond interactions or a distribution of different H-bond species as the matrix technique demonstrates (page 285). Therefore the best way to get quantitative data from the I.R. method is to restrict the data to the non H-bonded species. If there is a continuum of H-bond species or a distribution of a finite number of different H-bonds there is certainly a defined species: non H-bonded molecules. The results /3/ that the overtone spectra of alcohols and the frequencies of water at $T > T_c$ are in agreement with solution spectra supports this approach.

To get the extinction coefficients of the non H-bonded band we have determined the overtone spectra up to 400 °C under saturation conditions. The optical density of water at $T > T_c$ and critical density ρ_c is constant, This result gave us the possibility at $T > T_c$ and $\rho = \rho_c$ to take as the standard state for liquid water without H-bonds and eliminate the complicated transition to the rotation structure of the vapour spectra. The spectral profiles are sensitive to environmental interaction therefore the vapour state cannot be taken as the standard state for liquids or solutions. The results of Franck and Buback /17/ (fig.12page 231) show that the spectra seem to differ between H-bond interactions and other smaller interactions. To reduce the overlapping effects we tried to estimate the content O_F of non H-bonded molecules at the wavelength maximum at T_c (fig.5). The results of two H_2O bands and one HOD band are given in fig.9 (page 266). The most surprising result has been /18/ that with this independent method O_F (0 ^oC) = 0.09 - 0.1 we get excellent agreement with the first rough method /7/ with the calorimetric estimations /12/ and with the results of H-H-B /13/. The first overtone band of HOD is more isolated than the H_2O band because HOD has not the complication of overlapping ν_1 and ν_3. However the result obtained by this HOD band /8/ is in good agreement with the H_2O-bands and gives this third method a higher degree of confidence. In the next but one section we will give further spectroscopic tests of the quality of this result.- We have to stress the determination of O_F is only an approximation.The real values may be a little lower.

III.3.4.CONSISTENCY OF THE SPECTROSCOPICALLY DETERMINED O_F WITH PROPERTIES OF WATER.

It is clear that the third method given in the previous section to determine O_F has the character of a simplified model. The proposition of a model in science is to describe the properties of matter and to forecast its behaviour. Therefore at first we will look whether the result of fig.9 is in agreement with the properties of liquid water.

1. Heat Content.

a) ΔO_F during melting is consistent with ΔH_S.

b) The main interest on water concerns the region 0° to 100 ^oC. The heat content of water by the influence of the liquid state increases in this T-interval by:

$$\Delta H = (C_p\ (liq) - C_p\ (vapour))\Delta T = (18\ cal/mol^o - 9\ cal/mol^o)\ 100$$
$$= 1\ kcal/mol \qquad (10)$$

That means: If this ΔH is induced mainly by H-bonds, about 1/8 of the total H-bonds are opened during heating from 0° to 100 ^oC which is in agreement with the spectroscopic result which is 0.12 in O_F.

c) The anomalous T-increase of the specific heat C of liquid water up till T_c can be described quantitatively /19/ from the experimental O_F-values by the eq.:

$$C = C_{vap} + \left(\frac{\partial O_F}{\partial T}\right) \Delta H_H + 2\,(1 - O_F)\,R \qquad (11)$$

C_{vap} = specific heat of H_2O vapour in the ideal gas state as measure of the intramolecular degrees of freedom.

R = gas constant.

The last term gives an approximation of the general interaction factor for all liquids. One can show /19, 20/ that this fraction of the specific heat due to non-polar interactions in the liquid state can be given approximately by the number of first nearest neighbours times R/2. For water this number can be taken as the experimental function $(1 - O_F)$ of the H-bonded ordered areas which have the coordination numbers not far from 4 (compare the C_{2v} symmetry determined by Raman spectra). This last term in eq.10 is a rough approximation, but this term is small compared with $\left(\frac{\partial O_F}{\partial T}\right) \Delta H_H$ and can be neglected in the majority of the T-region. Eq.10 needs no adjusted parameters O_F and $\frac{\partial O_F}{\partial T}$ is given by fig.9. It gives a quantitative description of the specific heat as f(T) within the error limits of the spectroscopic method. The maximum of the specific heat at T_c changes with pressure p. Some semi-quantitative experiments in the overtone region have shown that the pressure dependence of the O_F-values parallels the change of the maximum of the specific heat with p. Eq.10 is able to describe the specific heat up till T_c. The specific heat minimum of liquid water at 38 °C is also given by eq.10.+) Walrafen has given a more complicated formula to fix the specific heat $0\,^{\circ}C < T < 100\,^{\circ}C$ /21/. He assumes three types of molecules with two, three and four H-bonds and two constants. He adjusted the content of the different species by a computer method with reference to his Raman measurements of the intermolecular Raman bands. But one has to stress the present Raman data give no absolute values for the H-bond content because the intensity coefficients of the different H-bond states are as yet unknown and the change of the Raman spectra during melting is as yet unknown. We imply that the spectra mainly give information on the bonded and non H-bonded OH groups. The interpretation of H-bond bands of different species seems to be complicated. Therefore we compare only the content of the non H-bonded OH groups given by the Walrafen formula at 0 °C is $O_F \approx 0.075$,

+) /48/

this value and the values to 100 °C agree well with our overtone results, which are absolute values without any extrapolation.

2. Heat of Vaporisation.

The heat of vaporisation L_V at constant volume V can be represented /19/ by:

$$L_V = W_{real} + (1 - O_F)\Delta H_S + O_F W - 2(1 - O_F) RT \qquad (12)$$

W_{real}: intermolecular interaction in the saturated vapour state, determined by the experimental heat content of the vapour minus the heat content of H_2O-vapour in the ideal gas state.

ΔH_S : total intermolecular interaction in the liquid state of H-bonded OH groups, determined by the sublimation energy.

W : intermolecular interaction energy of non H-bonded type, determined by $\Delta H_S - 2\Delta H_H = 3.6$ kcal/mol and calculated by Briegleb /22/.

The last term is the heat content of the intermolecular degrees of freedom (see eq. (10)).
Eq.11 also has no adjusted constants and is determined only by experimental data: O_F, W_{real}, ΔH_S and ΔH_H. It is quantitatively in agreement with the experimental L_V-values /19/ (fig.10).

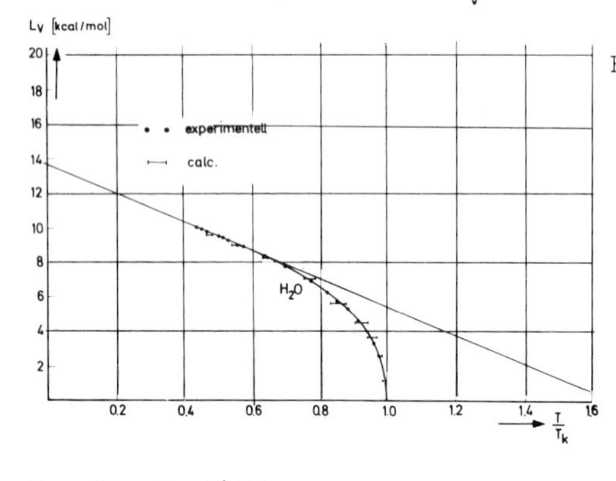

Fig.10 Heat of vaporisation, L_V, of water under saturation conditions.
O: experimental values; calculated by eq.(11)

3. The density

The density ρ of liquid water included the anomalous maximum at 4 °C can be described /19/ by eq.(12)

$$\frac{1}{\rho} = (1 - O_F) V_B + O_F V_F \qquad (13)$$

V_B: adjusted partial molar volume of H-bonded OH groups, taken as ice like volume.

V_F: adjusted partial molar volume of non H-bonded OH groups.

V_B and V_F are taken as linear T-functions. The expansion coefficient of V_B is taken as ice like. Eq.(12) gives the T-function of ρ till 350^oC /19/. Above 350^oC are some deviations from the experimental values. The reason is that for all liquids a linear coefficient of thermal expansion is not valid near T_c.

4. Surface Energy.

The surface tension σ of water is anomalously large compared with other molecules. The main reason is the small size of the H_2O molecules and the large number of molecules in 1 cm^2 surface of water. For structure discussions one should calculate the molar surface tension:

$$\sigma_M = \sigma \cdot N_A^{1/3} V^{2/3}$$

N_A : Avogadro number, V : molar volume

σ_m is also anomalously large compared with alcohols, but smaller than large molecules such as soap-like molecules.

σ_M is a free enthalpy in the nomenclature of thermodynamics, (determination in a isothermic experiment with heat contact with the environment) /20/. The surface energy H_σ is given by the Gibbs-Helmholtz equation:

$$H_\sigma = \sigma_M - T \left(\frac{\partial \sigma_M}{\partial T}\right) \qquad (14)$$

H_σ is T-independent for all non-strong polar liquids for $T < 0.9\ T_c$ /20/, but this is not valid for strongly H-bond liquids /19/. In this case H_σ has a maximum as a function of T. The absolute value of H_σ and the T-increase of water can be described /19/ by the eq.(15).

$$H_\sigma = [(1 - O_F)\ \Delta H_S + O_F \bar{W}]_{in} - [(1 - O_F)\Delta H_S + O_F \bar{W}]_{surf} \qquad (15)$$

This eq. is obtained from eq.(12). The indices <u>in</u> and <u>surf</u> mean the values in the bulk of the liquid and at the surface. The errors of (15) are higher than those of the eq.(10), (11) and (12) because it involves the difference of two terms of similar magnitude. But the absolute value of H_σ at room T H_σ = 1.6 kcal/mol is got by (15) and the increase of H_σ with T too.

5. NMR-Data.

The chemical shift of the NMR signal of a liquid depends on the content of H-bonds. Glasel has shown (page 428) that the proton chemical shift of water increases nearly linearly from 0° - 100 °C by about 0.9 ppm (see fig.1 page 428). It changes from the liquid state at 0°C to the vapour state at 100 °C about 4.8 ppm. There are no data on the chemical shift during melting. But if we assume from our discussion of the calorimatric data (page 255) that during melting the H-bond content decreases about 10% and the change of the H-bond content from 0°C liquid to the vapour state changes about 90%, the NMR data would give the approximation: the H-bond content during heating water from 0°C - 100 °C would change about 17.0% in the order of magnitude with the overtone data.

The O^{17}-chemical shift changes from 0°C to 100 °C by about 5 ppm and from 0°C liquid state to 100 °C vapour state ca. 38 ppm (fig.2 page 428). Assuming that the last change means 90% change of the H-bond content the change from 0°C to 100 °C in the liquid state would mean a change of O_F about 12%. This agrees with fig.9. The change of the O^{17}-chemical shift in the range 0° - 100 °C is linear. Thus the H^1 and O^{17}-chemical shift data are in agreement with the O_F-data from overtone spectroscopy.

6. Cluster Model.

Némethy and Scheraga have calculated the free energy using a cluster model /15/. They obtained a certain concentration of monomers and large clusters which leads to an O_F value of 0.47 and a mole fraction of monomer = 0.25 at 0°C. Both these values are in contradiction to fig.9 and to the calorimetric indications. Neither the monomer content nor the cluster size can be determined directly spectrocopically. However if one assumes the cluster model then one can obtain an estimate of the mean cluster size. If we assume in accordance with the C_{2v} symmetry of the Raman data /21/ a tridymite-like arrangement of the H-bonded and we assume that the non H-bonded OH groups are mainly at the surface of the cluster of H-bonded molecules we can calculate the mean numbers of molecules which are H-bonded. This assumption is in agreement with the theory of cooperative mechanism in ordered liquids.

The angle dependence of the H-bond energy (page 296 fig. 5) gives a good interpretation of a cooperative mechanism. If there is any hole defect in a tridymite-like arrangement of water molecules, instead of six-membered rings, five-membered rings of H_2O molecules will be formed. A five-membered ring has unfavoured H-bond angles and thus a smaller H-bond energy. The probability of getting a second hole defect is more likely in a five-membered ring than in an undisturbed area. From this view point the concentrations of defects in liquid water can be understood. Fig.11 gives the estimation of the dimensions of the area of H-bonded molecules from the overtone O_F-values. Bearing in mind these very idealized assumptions we should further stress that there is no direct indication of clusters in the overtone spectra. Fig.11 is only given to show whether the overtone spectra could be reconciled with a cluster model.

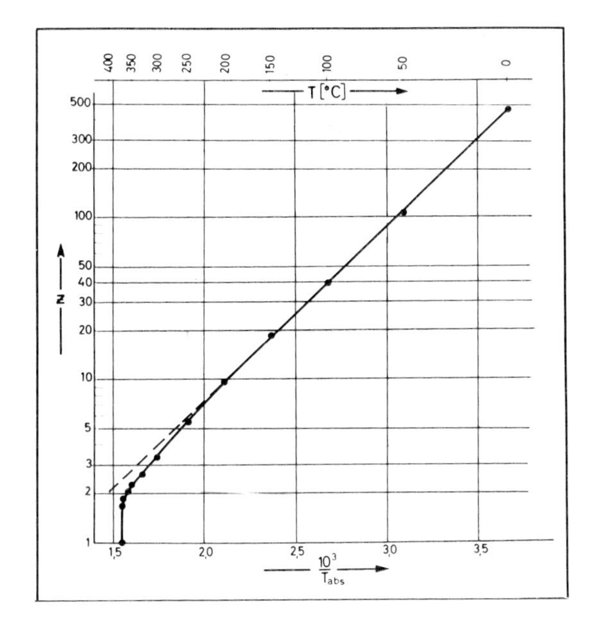

Fig.11 Estimation of the mean size of the system of H-bonded water molecules versus 1/T /23/.

This model (fig.11) can be supported with the frequencies of the maxima of the non H-bonded OH. The matrix technique and other experiments (see page 238) have shown that the frequency of one non H-bonded OH group of water change a little if second OH group or if one of its lone pairs is involved in a H-bonding interaction. We would expect indications of this small effect on the free OH band of liquid water also. Fig.12 gives the frequency of the maximum of water bands as f(T). All four overtone bands have their own frequency scale. The zero points are the maximum of the related ice-band, as the standard of the linear H-bond state. But we habe to stress that we cannot be certain that the maximum of the ice-band belongs to the free OH band under discussion because there are possibilities of other weak sum or difference bands

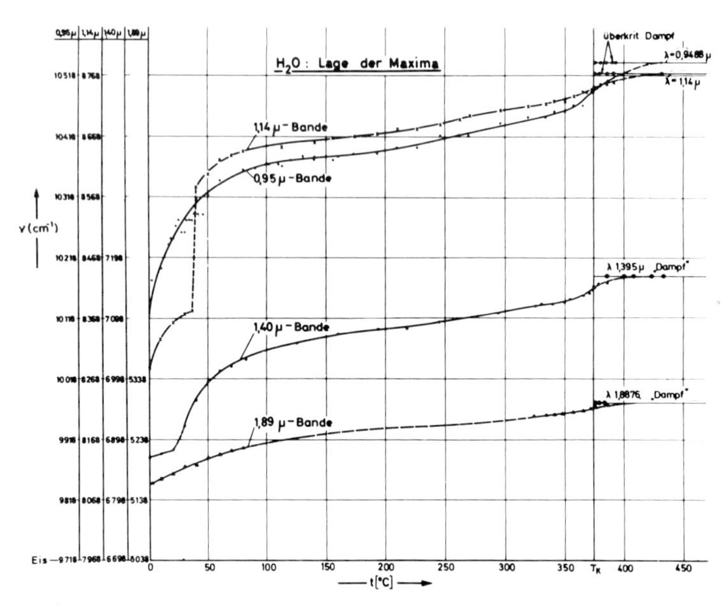

Fig.12 Frequencies of the maximum of the water overtone bands versus T

with their own ice bands. But all four bands in fig.12 show clearly that the maximum to 200 °C depends on the overlapping of a band of non H-bonded species and one or more bands of Hbonds. The frequencies in this T-region depends on the intensities and on the frequency shifts of these bands and can be different for the different overtone bands. But for T > 200 °C the overlapping effect with H-bond bands may be smaller and the more or less isolated non H-bonded molecules dominate. Similar effects can be observed on the more isolated HOD band (fig.13).

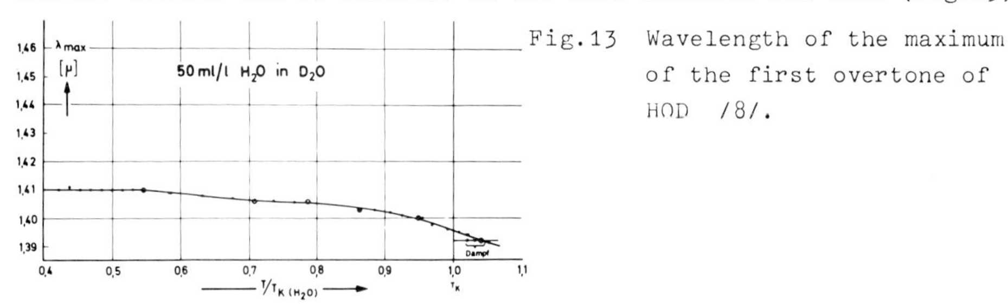

Fig.13 Wavelength of the maximum of the first overtone of HOD /8/.

In this case overlapping effects with other bands are smaller. But there is also a small effect of overlapping on the frequency of the HOD maximum of the free OH to T = 210 °C or T/T_k = 0.75. At the same T the aggregate sizes in the rough cluster model begin to get smaller than 8 (fig.11). That means for T > 210 °C equilibria of smaller aggregates dominate and the size of small aggregates decreases with T. From the view of the matrix isolated spectra one should expect if this is true small effects on the free OH freqnency depending on the aggregate size. Experimentally the matrix technique in the fundamental vibration gives changes of this frequency at a maximum of 49 cm^{-1} (second OH group belongs to an aggregate larger than 4 probably 6) and for a dimer $\Delta\nu \approx 15$ $cm^{-1} \approx \frac{1}{3} \Delta\nu_{max}$ /1/. Table two gives the observed frequency shifts from 210 °C - T_c and $\Delta\nu$ at T_c.

Table 2

Frequency shifts of the different free OH bands.

Band type	$\lambda_F(\mu)$	$\Delta\nu(210^o - T_c)$ cm^{-1}	$\Delta\nu(T_c)$ cm^{-1}
H_2O liquid			
$\nu_{1.3} + \nu_2$	1.8876	40	15
$2\nu_{1.3}$	1.395	85	20
$2\nu_{1.3} + \nu_2$	1.14	95	25
$3\nu_{1.3}$ +	0.9488	135	45
HOD liquid			
$2\nu_1$	1.392	70	20
H_2O matrix Ar /45/	2.7	49	14
H_2O matrix N_2 /1/	2.7	41	11

Taking into account that all frequency shifts are proportional to the number of the overtone and that experiments gives only a small frequency shift, the spectra of liquids are in excellent agreement with the matrix results. Secondly that means that the spectra are consistent with the conclusion that till 210 °C in water under saturation conditions higher aggregates dominate and above 210 °C we have to taken into account the equilibrium of smaller aggregates with fewer molecules than six.

7. H-bond Energy.

From the view of the frequency shift-of the water spectra comparing ice spectra and spectra of water in the liquid - like state

$T > T_c$ with the similar shift of the alcohol spectra - the H-bond interaction energy of one linear optimal H-bond of water should be similar to the H-bond interaction energy of alcohols. This energy of alcohols is more readily determined and spectroscopically there is general agreement that the value in non-polar solution is about 4 kcal/mol /24/ (compare chapter II, 1; Onsager). This value is in agreement with the calculated dispersion energy of 3.6 kcal/mol of water /22/ and the sublimation energy of 11.6 kcal/mol = 2·4 + 3.6 kcal/mol. Similar value 4.2 kcal/mol has been obtained by Vand and Senior by their statistical thermodynamics calculations /25/ viz: ΔH_H = 4.2 kcal/mol bonds. In the literature there are some values quoted of $\Delta H_H \approx$ 2.5 kcal/mol. For example Walrafen /21/ derived this value from his Raman data. He assumes an equilibrium:

$$OH_{free} \rightleftharpoons OH_{bond} \tag{16}$$

$$K = \frac{C_{OH,bond}}{C_{OH,free}} \approx \frac{\text{Intensity bonded band}}{\text{Intensity non-bonded band}} \tag{17}$$

Walrafen plotted the intensities of two different Raman bands as $f(\frac{1}{T})$. We can evaluate two concentrations approximately from our overtone spectra especially from the HOD data. Plotting this quotient $\frac{\varepsilon_{H,bond}}{\varepsilon_{free}}$ of eq.(17) as a function of $\frac{1}{T}$ we obtain similar energies, ca.2.1 kcal/mol (fig.14). Worley and Klotz found with eq.(17) and the HOD overtone spectra 2.37 kcal/mol /30/. However we have some doubts on the correctness of this assumption of eq.(16).

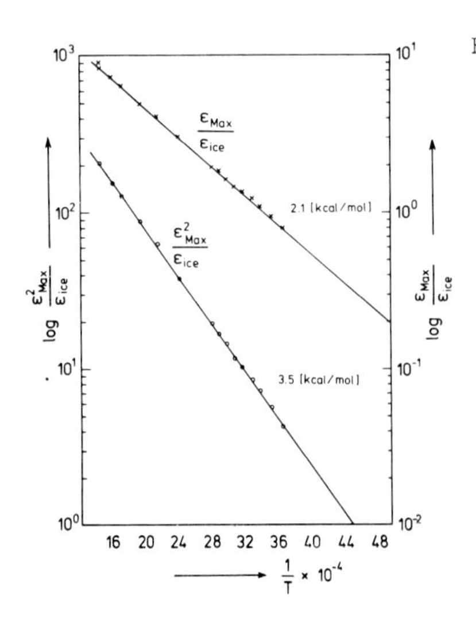

Fig.14 Quotients of the extintions coefficients of the free OH maximum and the bonded HOD as f(1/T). This plot cannot differentiate between the mechanism eq.(16) or (18).

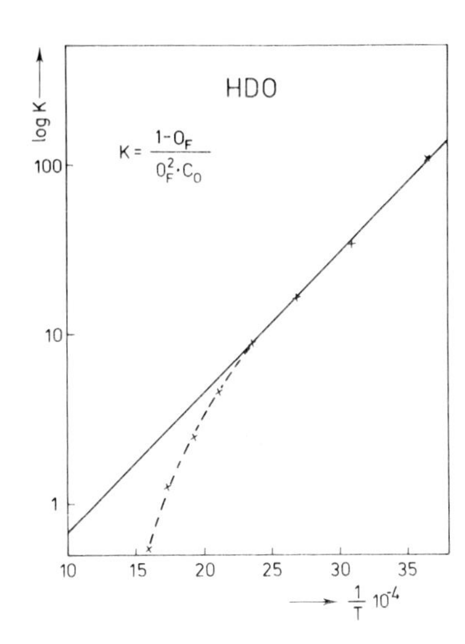

Fig.15 Plot of $(1 - O_F)/O_F\ c^2$ versus 1/T.

We think more exactly we should write:

$$OH_{free} + \theta_{free} \rightleftharpoons OH_{bond} \tag{18}$$

$$K = \frac{C_{OH,bond}}{C_{OH,free} \cdot C_{\theta,free}} \tag{19}$$

θ_{free}: Free electron pairs, as acceptor of H-bonds.

If we calculate the H-bonds in liquid water from an equilibrium, also it would be better to write (see chapter II, 1, Onsager):

$$K = \frac{C_{OH,bond}}{C^2_{OH,free}} = \frac{(1 - O_F)}{O_F^2\ C_O} \tag{20}$$

Because in water : $\quad C_{\theta,free} = C_{OH,free}$ (21)

C_O: molar concentration of water

If we plot ln K values of eq.(20) as $f(\frac{1}{T})$ we get a straight line for $T < 200\ ^oC$. ($\varepsilon_H/\varepsilon^2$ gives a straight line too with ΔH = 3.5 kcal/mol. This effect cannot decide if eq,(19) or eq.(16) is correct (fig.14)). This is in agreement with the discussion of table 2 and fig.11 that above this T we have to taken in to account a coupled equilibrium of small aggregates. In this region eq.(19) would not be valid +). The slope of fig.15 for $T > 200\ ^oC$ gives an energy of 3.8 kcal/mol. This value is in good agreement with the calorimetric values and with the discussion given at the beginning of this chapter.

+) For alcohols is: $C_{\theta_{free}} = (1 - O_F) + 2\ O_F\ C_O$

$$K = \frac{(1 - O_F)}{(O_F + O_F)\ C_O}$$

$\ln K = f(\frac{1}{T})$ gives a straight line also till 200 °C, for $T > 200\ ^oC$ there are similar deviations like in fig.15

8. Dielectric Constant.

The overtone results on O_F agree with the first attempt to describe the T-dependence of the dielectric constant ε by Haggis-Hasted and Buchanan (see fig.8 page 256) /13/. We have plotted (fig.16) the data of ε given in the new review of by Quist and Marshall /26/. The dotted line gives the values under saturation conditions. Fig.16 shows

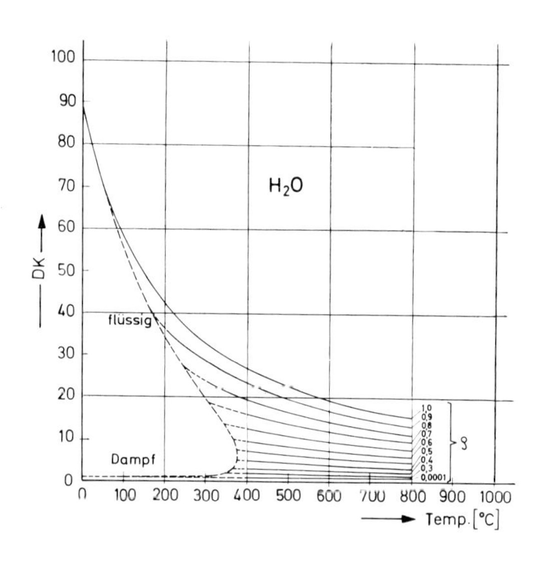

Fig.16 The dielectric constant of water calculated from the data given by Quist and Marshall /26/.

that ε at the critical temperature is near the value of the isolated molecules $\varepsilon_\infty = 3.2$ and near the minimum of $\varepsilon = 3$ of ice doped with HF (page 422). In this HF-contaminated ice the two different orientation mechanisms are cancelled /27/. The fig.16 shows that higher density than the saturation density induces an effect like the decreasing of T, one can assume that this induces higher H-bond contents by higher concentrations. This is in accord with the I.R. experiments given by Franck and Roth /28/ fig.17 shows some experiments on the 1.14 µ overtone band at higher densities +). The higher densities induces a lower content of O_F. Com-

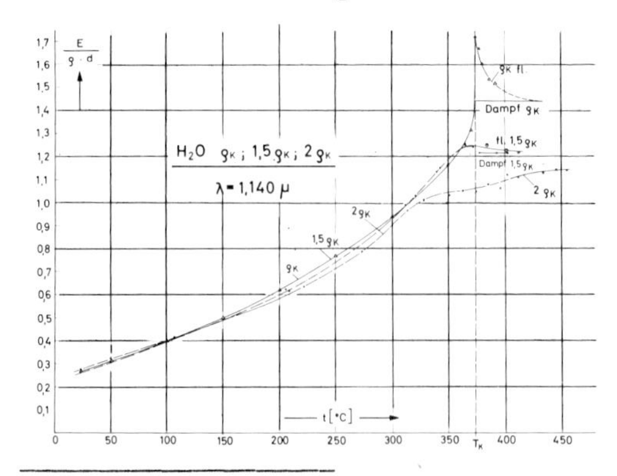

Fig.17 Extinction coefficients of the 1.14 µ free OH water band at different densities.

+) The experimental maximum of E/d at ρ_k in the liquid state is induced by density gradient effects /29/.

paring figs.16 and 17 one can recognize that the content of free OH at 400 ^{o} $\rho = 1.5 \cdot \rho_c$ is nearly the same as at 358 ^{o}C at saturation density ρ_s. The dielectric constant at 400° at $\rho = 1.5\ \rho_c$ is identical to the saturation density value at 355 ^{o}C also. Just as good is the agreement of the temperature of similar free OH content fig. 17 at $\rho = 2\rho_c$ (T = 400 ^{o}C is adequate for 335°, ρ_s; T = 450 ^{o}C adequate for 345 ^{o}C ρ_s) ε at $\rho = 2\rho_c$ and T = 400 ^{o}C adequate to ε at 335 ^{o}C ρ_s, $\rho = 2\rho_c$ and T = 450 ^{o}C adequate to 345 ^{o}C, ρ_s. This discussion shows that the content of free OH determines by overtone I.R. spectroscopy and the data of ε are also consistent.

9. Magnetic Susceptibility.

Cini and Torrini / 26a / found a nearly linear increase of the magnetic susceptibility χ of water from 0°C - 80 ^{o}C. They discuss this in the view of the statistic theory of water given by Vand and Senior /25/ at first approximation Cini and Torrini need a equation

$$\frac{\chi(T)}{\chi(20^{o}C)} = a + b\ x \qquad (22)$$

x: the number of H-bonds per mol.
It is easier however to compare this linear relationship with the approximate linear increase of O_F (fig.9)

III.3.5. FURTHER SPECTROSCOPIC EVIDENCE OF THE APPROXIMATION OF THE CONTENT OF FREE OH

a) General view of band profiles and band areas. The more exact experiment would be to determine the distribution function of different H-bonds or arrangements of water molecules. This might be done by the I.R. method by detailed analysis of the band profiles. But this field of work is only in its early development. We feel therefore that we should try an approximate approach before this exact analysis becomes available. There is one spectroscopic effect which makes it easier to give approximate statements. In the absence of H-bond the interaction $\int \varepsilon d\nu$ of aclohols integrated on a band is relatively insensitive to environment effects. For example fig.4 has given strong effects of rotation on the band profiles of the first overtone of the CH_3OH OH stretching. But even during such large change in bond profiles $\int \varepsilon d\nu$ is nearly constant (see table 3).

Table 3

$\int \varepsilon d\nu$ of the first overtone band of a free OH of CH_3OH /3/

sensity (ζ/ζ_k)	T (^oC)	$\int \varepsilon d\nu$ area-unite	density (ζ/ζ_k)	T (^oC)	$\int \varepsilon d\nu$ area-unite
0.1	314	1040	0.024	174	1145
0.25	307	1097	0.19	201	1150
0.5	310	1066	0.32	220	1100
1	314	1044	(1)	(243)	(875)
			1	377	1020
solution					
2 g/l CCl_4	20	1066			
2 g/l CS_2	20	930			

$\int \varepsilon d\nu$ is constant if there are most of the molecules in a non H-bonded state. This integral is not constant if there are some H-bonds ($T = 243\ ^oC$, $\zeta = \zeta_k$ in table 3). This result shows that $\int \varepsilon d\nu$ gives a good indication of the total content of free OH summarized for all different environment effects. But $\int \varepsilon d\nu$ is sensitive to H-bond interactions.

From this view we have repeated our method to determine O_F by evaluation of the band areas. From the end of short wavelength to the maximum. In this way we reduce difficulties of the overlapping effect with H-bond bands at the longer wavelength side of the free OH-bands (fig.18). The result provided a verification of the simpler method to determine O_F with the extinction coefficient (fig.9) at the maximum of the free OH band at T_c. The values of $2\nu_{1.3} + \nu_2$ of H_2O and D_2O and $2\nu_1$ of HDO which can be determined the most exactly of the overtone bands are in agreement with each other and with the simple method of fig.9. This is a further argument for the quality of this method. Fig.18 and 19 give a criteria of the limits of errors by overlapping effects.

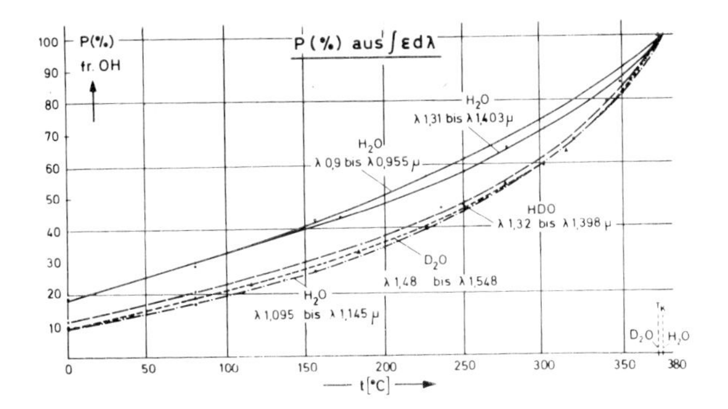

Fig. 18

Content O_F of free OH in percent P determined by the $\int \varepsilon d\nu$ of different H_2O or HOD overtone bands.

b) <u>Band Analysis.</u> The attempt to calculate the unknown concentrations of different species and unknown extinktion coefficients gives an non-linear system of equations and infinite solutions /16/. This objection is also valid for the attempt to analyse the band profile of overlapping bands with an analogue computer for example the Du-Pont curve analyser. But the number of possible solutions is reduced if any knowledge on extinctions coefficients or concentrations can be obtained or assumed. Therefore we have tried to analyse the HOD and H_2O spectra with the assumption that the non H-bonded groups absorb mainly with the extinction coefficients of the supercritical state and the linear H-bonded OH groups with the extinction coefficients of ice /8/. In this case we would have a linear system of equations with definite solutions. But the curve analyser shows that this attempt is not possible exactly. We need a minimum of three different bands. If the ice-like and the overcritical bands are used this third band should have a maximum in between these two bands. This result is in agreement with the matrix technique and that for alcohol solutions (page 225). Both methods give the information that there is a third band of a dimer-like species. Naturally this band does not necessarily belong to a dimer. The angle dependence of the frequency of different H-bond species (fig. 5 page 296) shows that the "dimers", "trimers" and "tetramers" have similar absorption maxima. Therefore a broad medium band could indicate the existence of different small aggregates too.

The band of the free OH absorption could be better adjusted if one assumes two similar bands with a frequency shift of about 100 cm^{-1}. But all fine details of the spectra at every T till 400°C under saturation conditions can be fixed quantitatively by the curve analyser with these four bands /8/. Fig.19 gives examples of the required components of the curve analyser to fix the experimental bands. This result means we could suggest that liquid water can be described with a simple model of three different types of OH groups: non H-bonded, linear-like bonded and bonded with medium favoured angles. In addition in the second approximation there may be distributions of different angles and so on. Fig.19 gives the result of the percentages of these three types of OH groups evaluated by this curve analysis. In this method there are no possibilities for subjective assumptions. In general the curve analyser has infinite solutions when deconvoluting spectra into the various overlapping bands. But in our case we can reduce the number of parameters by taking the ice-spectra and the overcritical spectra as a basis. The curve analyser then shows that only a third band is necessary for a quantitative description of all fine details of the water bands at all T. The frequency of the maximum of this band is similar to that of the dimer band of alcohol solutions. It is also in agreement with the water dimer band found in the matrix technique. If is further corroborated with the result of the ultrasonic method (page 523) that the properties of water can

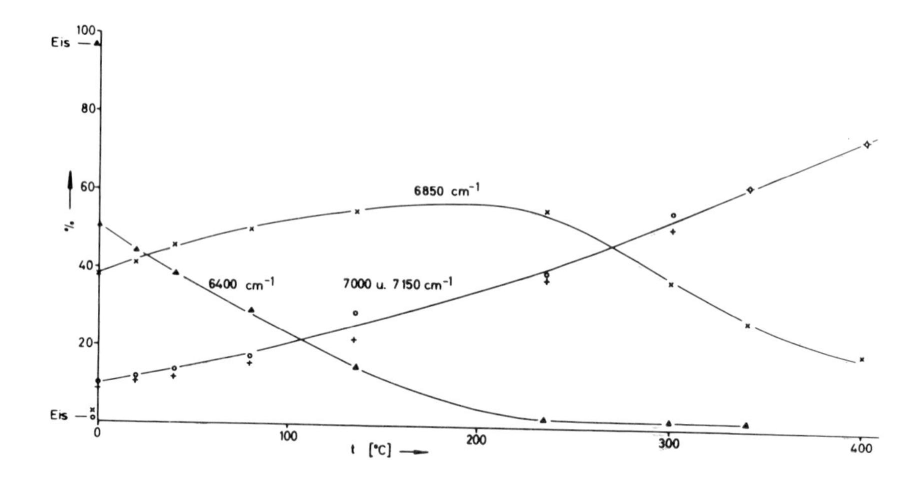

Fig.19 Curve analysis of the HOD band. Percentages of the three needed bands: free OH band 7000 and 7150 cm^{-1}, band of energetically unfavoured H-bonds 6850 cm^{-1} and band of linear H-bonds 6400 cm^{-1}. *)

*) Our method cannot exactly decide how large is the rest of unfavoured H- bonds at T_c or how strong the band profiles deviate from the Gaussian type.

be described with a minimum of three different water species. Our result of the curve analyser is unequivocal since we have fixed the data with a minimum of bands. This third band is broad. Every broad curve of the Gaussian type can be split up into a sum of Gaussian curves. Therefore the curve analysis of the spectroscopic results cannot decide if there are only two types of H-bonded species (ice-like and alcoholic dimer-like arrangements) or whether there are distribution function of different H-bonded arrangements. But on the other hand not every sum of Gaussian curves gives one resultant Gaussian curve. From the view of the continuum model one could not expect a priori that the curve analysis of the overtone data in agreement with the Raman data would yield this simple result of a sum of the free OH-band, the linear H-bond band (ice) and one of angle unfavoured H-bonds.

c) Isosbestic Points. If there are two species with two different absorption bands which overlap slightly the spectra should give a sharp isosbestic point, a point for all spectra at any concentration quotient. Walrafen has stressed that his Raman results show isosbestic points. In the T-region of his measurements up till 100 °C we find in the overtone region not sharp isosbestic but nearly isosbestic points (fig.20 and fig.21). Isosbestic points are also observed on the HOD overtone spectra in a T-region of 7 °C till 61 °C by Worley and Klotz /30/. The results of these authors agree with our measurement enlarged in a more expended T-region. The ice curves do not cut the isosbestic point. But this result is dubious because of the uncertainties in the quantitative measurement of solid-state spectra. The deviation of the ice band can be caused by: experimental difficulties, change of the extinction coefficients in the solid state or an influence of a cooperative mechanism. But also at high T the absorption bands do not cut the "isosbestic points". This result we have to expect from the curve analyser which needs three different bands. But the concentration of this third medium "species" does not change much in the T-region up to 100 °C. Therefore the apparent isosbestic points are in accord with the curve analyser result.

The distinct deviation from the isosbestic point at higher T may be an influence of the third band and of the frequency shift of the non H-bonded band by different H-bonds of the second OH group. (Compare the matrix results).

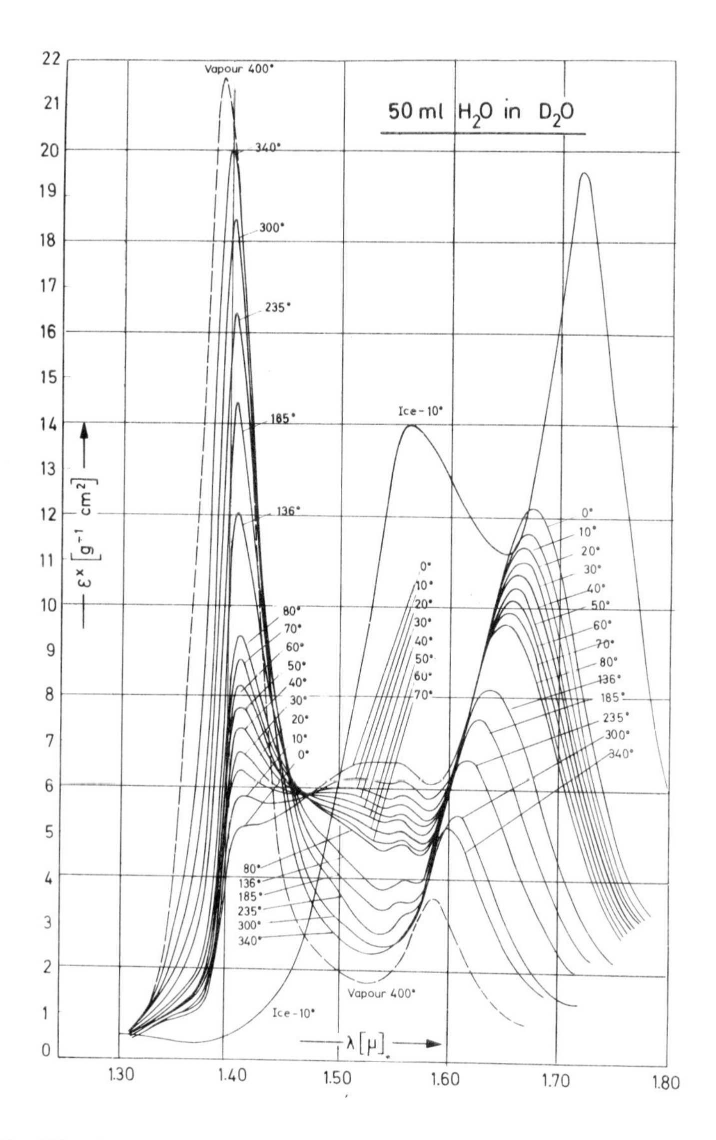

Fig.20 First overtone of HOD versus T, 50 ml H_2O/1 D_2O; D_2O absorption eliminated /8/.

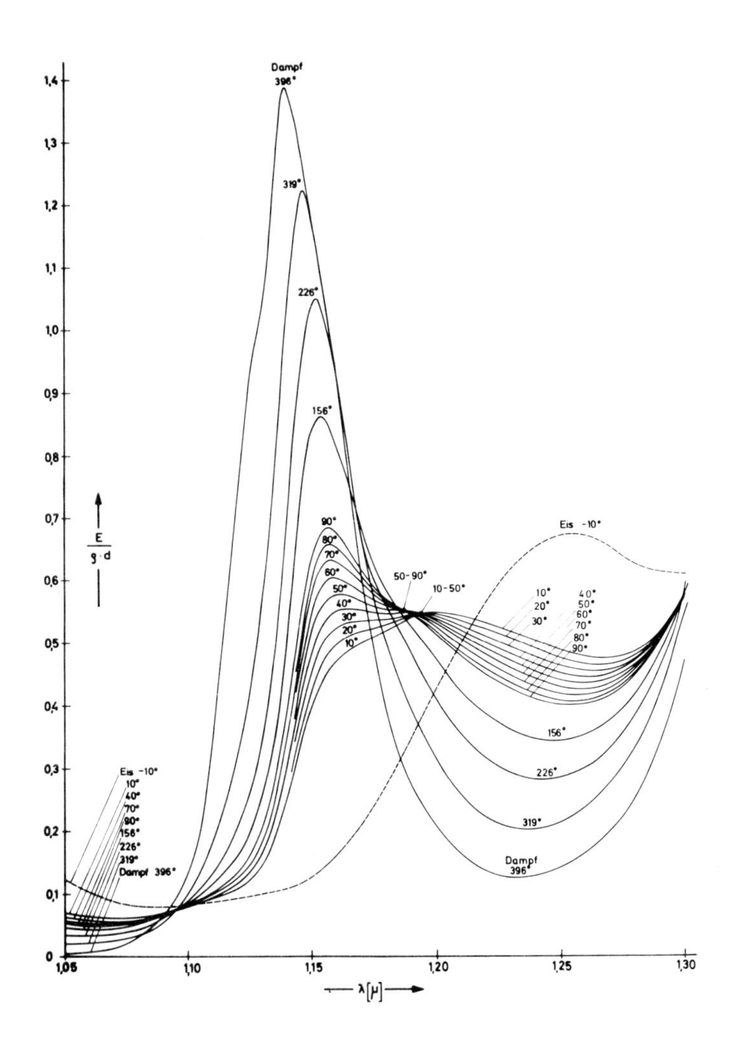

Fig.21 Temperature variation of the 1.15 ν combination band of H_2O

III.3.6. DISCUSSION

Continuum model /31/ or mixture model /32/ this is the question? We stress that our results are not contratictionary to both extremes of authors. But our results are more compatible to a model that gives an approximation with a distinct number of parameters $^{+)}$. Falk was one of the most active attorneys for a continuum model /6/. Therefore one should discuss his arguments. At first he made experiments on the fundamental HOD vibration but only in a small T-region and gave his results in transmittance which should not be done if one is discussing band profiles. His arguments that overtone spectra consist of a "band cluster" and cannot give good data is ruled out by experiments with 4 different H_2O, 2 different D_2O and one HOD overtone band with similar results. In addition he did not discuss that any spectroscopic method is still valid if there is band overlapping provided one determines the extinction coefficients (page). This we have done with spectra at 400°C and of ice. His argument that the change of intensity of ν_1 and ν_3 with H-bonds disturbs the H-bond spectroscopy is ruled out by measurement on HOD and the argument if this change would be linearly proportional to the concentration can be taken into account. His argument that Fermi resonance disturbs the method is ruled out by the experiments on many bands and by the argument that the Fermi resonance effect is weak for combinations of overtones /33/. The overtones of the bending modes are weak and the bending absorptions do not change greatly by H-bonds. His argument of the broadening effects by intermolecular coupling can be ruled out by experiments on $\int \varepsilon d\nu$ because this integral is not sensitive to these effects. We agree with Falk and Ford /6/ that the fundamental vibrations give little information on the H-bonded state. But together with the much better overtone results the fundamental results are consistent with the overtone spectra (see chapter fundamental region page 254).

Numerous authors have cast doubts on spectroscopic results in general. These are summarized for example by Kell /31/. Much of these criticisms are ill-founded and can be countered as follows:

1. The spectra are not put into relation with ice spectra. This is done in our work.

+) In the known paper of Stillinger and Rahman is given a statement that the calculated distributions of molecules with 0,1,2,3..bonds conflicts with two-state theories. But this is not true for a theory with bonded and unbonded OH groups and the experimentally O_F, especially this is valid for the calculated high density of 1 by St. and R.

2. The band analysis into components is non-unique. We are able to reduce the number of parameters by taking ice-like and overcritical spectra as standards and this analysis improved.

3. One should take into account the distance and angle dependence of the H-bond bands. This is discussed in our work. The x-ray experiments show that the distance does not change much under saturation conditions. In non-polar liquids too the main effect of density is the decrease of the number of first next neighbours and not the distance of first next neighbours (compare x-ray results on argon /34/).

4. Kell /31/ has not taken into account the differences between the three spectral methods: overtone, fundamental- and Raman spectroscopy.
   We believe a lot of arguments show that overtone spectroscopy on pure water shows quantitatively an increase of non H-bonded OH groups with T. Some misunderstandings on the overtone method may be induced because the different influence of overlapping effects in fundamental and overtone region is not known /35/

In the addendum to their paper Falk and Ford feel that there are several inconsistencies in the recent estimates of different O_F values. However he appears to be unaware that he is comparing calculations of different qualities. For example Hartmann /36/ has estimated O_F (0^oC) = 0.07. But this value was estimated with the HOD fundamental spectrum as rough approximation with extinction coefficients of alcohol-solutions in CCl_4. Walrafen has estimated O_F (0^oC) = 0.105. But the Raman effect gives no absolute values. Walrafen /37/ extrapolates Raman data together with a calculation of thermodynamic values. Falk and Ford do not discuss the values of the overtone experiments and it seems they have not recognized that only this method gives direct values. Nevertheless these three values are in good agreement taken in account the estimations of the other authors. These three values obtaines with independent methods agree well with the earlier estimation geiven by Haggis-Hasted and Buchanan /13/.- We have to stress that the spectroscopic method below 200 ^oC at saturation conditions do not give any indications on monomers, only on free OH groups. But this is in agreement with the low vapour pressure of a low molecular weight molecule. One could expect that the quotient of the number of monomers to the number of H-bonded molecules in the liquid should be in the same order as the quotient of molecule-numbers per ccm in the saturated vapour state to the number of molecules per ccm in the liquid state. This quotient is for water at 50 ^{o}C, 10^{-4} and at 200 ^{o}C, 10^{-2}. Light scattering experiments of water disagree also with the assumption of a large volume fraction of molecule-sized vacancies /47/.

III.3.7. ELECTROLYTE SOLUTIONS.

From the view of viscosity measurements Bernal and Fowler /37/ concluded that electrolyte admixture changes the structure of water to a first approximation similar to a T change. They defined the terminology structure temperature T_{str} as that T at which pure water would have effectively the same inner structure and therefore the same viscosity, same x-ray scattering, same Raman spectra etc. /37/. Bernal and Fowler gave no quantitative values of their concept. Suhrmann and Breyer have tried to quantify this using the optical density at the maxima of the overtone bands /38/. Ganz has made similar attempts with the frequencies of the overtone maxima as a function of electrolyte content /39/. We believe the best method to give the quantitative content of species by spectroscopy is the optical density at a fixed frequency /40/. The best frequencies to use are the band maxima. Therefore we have determined the T_{str} of different electrolyte solutions at the wavelength of the maxima of the free OH band at 200 ^{o}C /40, 41/. In this way we have a measure of the non-bonded OH groups in an environment of the liquid-like structure at lower T. This gives a measure of the free OH groups in a state where only one OH group is non H-bonded.
The main results of this technique are /40, 41/:

1. The spectra are much more insensitive to electrolyte admixtures than the solubility data of water are. We need concentrations of 0.5 mol/l for good spectroscopic measurements.

2. The main effect on T_{str} are given by anions. The cations have a less effect. This observation and the insensitivity agree with NMR data /42/.

3. The effect of anions is determined by e^2/r. e: charge and r: radius of the ions.

4. The difference of the solution temperature T and T_{str} depends on the position of every ion in an series and on T (fig.22).

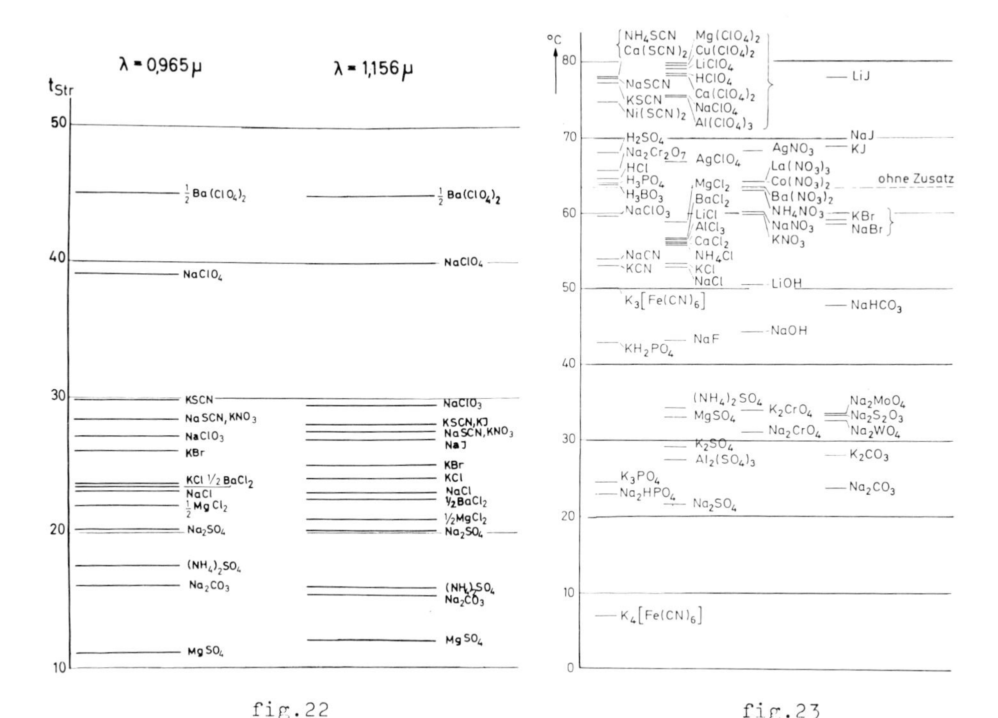

fig.22 fig.23

Fig.22 Structure temperature determined by 0.965 μ and 1.156 μ I.R. band at 20 °C, C_{anion} = 1 mol/l.

Fig.23 Turbidity temperature T_T of aqueous solutions of iso-C_8H_{17}–⟨phenylene⟩–O-$(CH_2 - CH_2)_9$ – OH at 64 °C is changed by salts (C = 0.5 mol/l) by a similar ion series which parallels changes in the I.R. Spectra (fig.22).

The ion series obtained with the overtone technique is identical with the Hofmeister or lyotropic ion series of colloid chemistry /43/. Therefore the cause of the pyotropic ion series can be confirmed to the change of water structure. The lyotropic ion series can be easily determined by measurements of the turbidity point T_T of aqueous solutions of p-iso-octylphenol which contains a chain of 9 ethylenoxide groups. This compound is soluble in water at low T and these solutions are separated in two phases above the turbidity temperature T_T (fig.23). The T_T ion series is similar to the ion series obtained by T_{str}.

5. Ions heading the lyotropic ion series induce more free OH than pure water would have. That means water becomes more hydrophilic through these additions. These ions are called structure breakers. Such salts can induce salting-in effects of organic compounds. Especially compounds with ether groups give salting- in effects. Ions at lower end of the ion series induce less free OH. That means water becomes more hydrophobic. These salt groups induce mainly salting-out effects on gases or organic compunds: they are called structure-makers.

One result of the spectroscopic experiments on electrolyte solutions is not easy to understand. This is the relatively small effect of ions on the water spectra but relatively large effects of ions on solubility of gases or organic compounds in water /12/. If one assumes in a heuristic model the salt-out effect is induced because the water molecules around every ion are fixed in the strong Coulomb field, one can estimate the number of water molecules around every ion which do not solvate organic molecules. The result is these secondary hydration numbers are strongly dependant on the ion-concentration /12/. For example Na_2SO_4 gives the following secondary hydration (HZ) numbers determined by this solubility model at 20 $^{\circ}$C:

concentration (mol/l)	0.15	0.5	0.75	1
HZ	275	90	55	28

The spectroscopic method at concentrations of 0.5 mol/l gives a change of O_F of the order of 1%. But if one recalls that at room T a small O_F means that a large number of water molecules are H-bonded with the assumption that the free OH are arranged in fissure plains and are not statistically distributed, a change of the numbers of H-bonded molecule would not change the content of free OH significantly (fig.24).

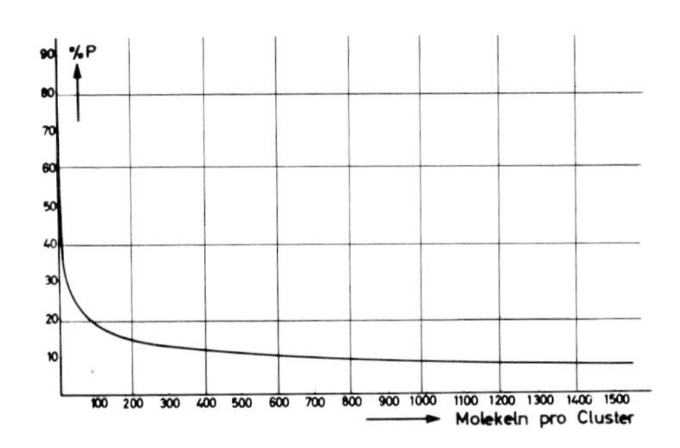

Fig.24 Percentage P of free OH in tridymite-like water aggregates versus the mean molecule number per aggregate.

If we also remember that O_F at room T is nearly 0.1, then in an ideal model a change of O_F to 0.11 would mean that the number of H-bonded molecules would change by about 100 molecules. With this model the small ion effects on spectra can be understood (fig.23 and 24). If every ion fixes a certain amount of ions by Coulomb forces and outside these secondary areas would be arranged the normal area of H-bonded molecules both effects are consistent: the small spectroscopic effects and the large solubility effects. If this model is right then would follow that the ion effects on the spectra should be larger at higher T. Because at higher T the O_F-values decrease that means the expansion of H-bonded molecules would be smaller (fig.24). The change of strongly orientated molecules by the Coulomb forces of the ions and by the H-bonds at higher T should induce a higher change of O_F. This prediction we have verified by spectroscopic determinations of T_{str} as a function of T. Fig.25 shows that the differences T_{str} - T are becoming larger with increasing T for structure maker. Starting with the results of the spectroscopic experiments we have predicted that the salt-out effects should decrease with increase of T. This prediction we have verified also by determination of the partition coefficients of cresol or phenol between cyclohexane and aqueous electrolyte solutions.

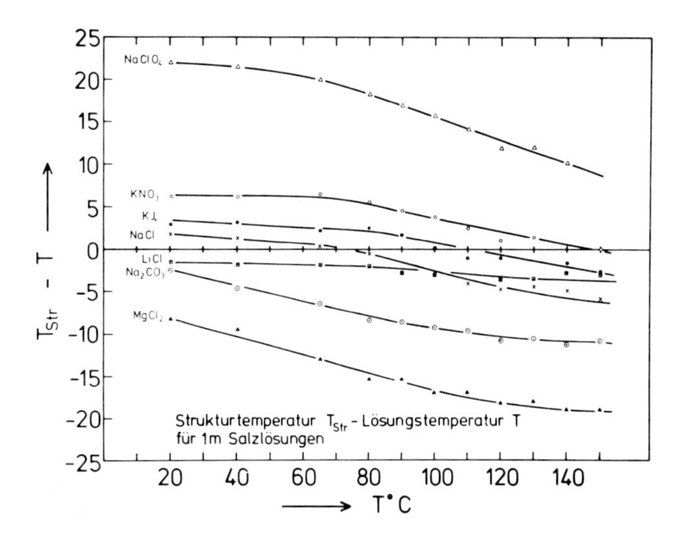

Fig.25 The differences T_{str} - T increase with the solution T for structure makers. The structure breaker effects decreases with T.

Determinations of the partial molar volume in aqueous salt solutions by density measurements gives values which mainly are smaller than the molar volume of pure water. That means the electrostriction effect of ions dominates. But the ion series obtained yield by the partial molar volume in ion-solutions is similar to the spectroscopic determined ion-series too. Structure-breakers give smaller changes of volumes than structure-makers. The dominating effect of the anions can be demonstrated by the partial molar volume in $MgCl_2$ solutions. If one plots the data against the anion concentration the effect of $MgCl_2$ on water is similar to the effect of NaCl. From partial molar effects one can conclude that we have to differentiate between ion effects on the first or the nearest hydration shell (dominates on v-effects) and on large distances (dominates of spectra). One can assume that large anions like ClO_4^-, I^- of SCN^- give in the nearest neighbourhood a more densed package of water molecules. But at larger distances of the ions the arrangement of water molecules is disturbed so that the overall effect gives a smaller expansion area of the H-bonded water molecules and a larger content of free OH groups.

All discussions of the area of H-bonded molecules on the basis of the spectroscopic results refered only to small time intervals (vibrational transition times are of about 10^{-14} sec). Therefore all conclusions are related only to these small time-intervals. This can be demonstrated by measurements of aqueous solutions of acids or bases. In this case we have observed a broadening effect of the water bands /40, 12/. This indicates that in such solutions the dynamics of the H-bonds is different. This effect is insensitive to T-change. Zundel has also observed this broadening effect of acids on the fundamental bands /46/. He assumes that a tunnel effect induces in acids a quick change of the positions of the protons. Janotscheck has presented a theory of this effect at this meeting. (page 625).
We have to stress determinations of T_{str} are only rough approximations. Zukovskij discusses special effects of ions on the H-bond bands by this method as detailed ion effects on the water structure. /49/

REFERENCES

1 W.A.P.Luck: Infrared Studies of Hydrogen Bonding in Pure Liquids Solutions in: Water a Comprehensive Treatise, F.Franks Editor, Vol.2, p. 235-320, Plenum Press, New York, 1973.

2 W.A.P.Luck and W.Ditter, J.mol.Structure 1, 261 (1967).

3 W.A.P.Luck and W.Ditter, Ber.Bunsenges. 72, 365 (1968).

4 L.-J. Bellamy, H.E.Hallam and R.L.Williams, Trans.Faraday Soc. 54 1120 (1958).

5 E.Greinacher, W.L.Lüttke and R.Mecke, Ztschr.Elektrochem. 59, 23 (1955).

6 M.Falk and T.A.Ford, Canad.J.Chem. 44, 1699 (1966).

7 W.A.P.Luck, Ber. Bunsenges. 69, 626 (1965).

8 W.A.P.Luck and W.Ditter, Ztschr. Naturforschg. 24b, 482 (1969).

9 W.A.P.Luck, Ber. Bunsenges. 67, 186 (1963).

10 L.Pauling, Nature of the Chemical Bond, Oxford Univ. Press 1960.

11 J.Fox and A.E.Martin, Proc. Roy Soc. A 174, 23 (1940).

12 W.A.P.Luck, Fortschr. chem. Forschung 4, 693 (1964).

13 G.H.Haggis, J.B.Hasted and J.Buchanan, J. Chem. Phys. 20, 1452 (1952).

14 K.Buijs and G.R.Choppin, J. Chem. Phys. 39, 2035 (1963).

15 G.Némethy and H.A.Scheraga, J.Chem. Phys. 36, 3382 (1962).

16 G.Böttger, H.Harders and W.A.P.Luck, J.Phys. Chem. 71, 459 (1967).

17 M.Buback and E.U.Franck, Ber. Bunsenges. 75, 38 (1971).

18 W.A.P.Luck, Ber.Bunsenges. 70, 1113 (1966).

19 W.A.P.Luck, Discuss. Faraday Soc. 43, 115 (1967).

20 W.A.P.Luck and W. Ditter, Tetrahedron 27, 201 (1971).

21 G.E.Walrafen: Raman and I.R. Investigations of Water Structure in: Water a Comprehensive Treatise, E. Franks Editor, page 151-215, Plenum Press, New York, 1972.

22 G.Briegleb: Zwischenmolekulare Kräfte und Molekülstruktur, page 46, Enke Verlag, Stuttgart, 1937.

23 W.A.P.Luck, Informationsdienst, Arbeitsgem. für Pharmaz. Verfahrenstechnik e.V., 16, 127 (1970).

24 G.C.Pimentel and A.L.McClellan: The Hydrogen Bond, Reinhold Freeman Co., San Francisco 1960.

25 V.Vand and W.A.Senior, J.Chem. Phys. 43, 1869 (1965).
W.A.Senior and V.Vand, J.Chem. Phys. 43, 1873 (1965).
V.Vand and W.A.Senior, J.Chem. Phys. 43, 1878 (1965).

26 A.S.Quist and W.L.Marshall, J. phys. Chem. 69, 3167 (1965).

26a R. Cini and M. Torrini, J.Chem. Phys. 49 ,2826 (1968).

27 C.Jaccard, Ann. New York Ac. Science 125, 390 (1965); C.Jaccard Helv. Phys. Acta 32, 89 (1959);

28 E.U.Franck and K.Roth, Discuss. Faraday Soc. 43, 108 (1967).

29 W.A.P.Luck and W.Ditter, Ber. Bunsenges. 70, 1113 (1966).

30 J.D.Worley and I.M.Klotz, J. Chem. Phys. 45, 2868 (1966).

31 G.S.Kell: Water and Aqueous Solutions, editor R.A.Home, Wiley Intersc. 1972, page 331.

32 C.M.Davies and J.Jarzynski: Water and Aqueous Solutions, editor R.A.Home, Wiley Intersc. 1972, page 377.

33 H.A.Stuart: Molekülstruktur, 3.Aufl. page 474, Springer Verlag, Heidelberg 1967.

34 A.Eisenstein and N.S.Gingrich, Phys. Rev. 62, 261 (1942). N.S.Gingrich, Rev. mol. Phys. 15, 90 (1943).

35 W.A.P.Luck and W.Ditter, J. mol. struct. 1, 339 (1967).

36 K.A.Hartmann, J. Phys. Chem. 70, 270 (1966).

37 J.D. Bernal and R.H. Fowler, J. Chem. Phys. 1, 516 (1933).

38 R.Suhrmann and F.Breyer, Z. phys. Chem. B 20, 17 (1933) R.Suhrmann and F.Breyer, Z. phys. Chem. B 23, 193 (1933).

39 E.Ganz, Ann. Phys. 26, 331 (1936); 28, 445 (1937); E.Ganz, Z. phys. Chem. 33, 163 (1936); 35, 1 (1937).

40 W.A.P.Luck, Ber. Bunsenges. 69, 69 (1965).

41 W.A.P.Luck: Proc. 4.Internat. Symposium on Fresh Water from the Sea, Heidelberg 1973, Vol. 4, p. 531.

42 J.N.Shoorley and B.J.Alder, J. Chem. Phys. 23, 805 (1955).

43 F.Hofmeister, Arch. exp. Path. Pharm. 25, 295 (1890); W.B.Hardy, Z. Phys. Chem. 33, 391 (1900).

44 J. Th. G. Overbeek and H. R. Kruyt, Colloid Science I, p. 307, Elsevier Amsterdam 1952.

45 Th. Neikes, Diplomarbeit, Marburg 1974.

46 G.Zundel, Hydratation and Intermolecular Interaction, Academic Press, New York, 1969.

47 K.S. Mysels, J.Am. Chem. Soc. 86, 3503 (1964).

48 W.A.P.Luck, Adv. Mol. Rel. Proc. 3, 321 (1972).

49 A.P. Zukovskii and W.A. P. Luck . in"Water in biological Systems." Vol. 4. in press Leningrad, editors M.F. Vuks and A.I.Sidorova.

III.4. INFRARED MATRIX ISOLATION STUDIES OF WATER

H. E. Hallam

Department of Chemistry, University College of Swansea,
Singleton Park, Swansea SA2 8PP, Wales.

ABSTRACT

The advantages of cryogenic infrared spectrocopy for the investigation of molecular association is discussed. H_2O and D_2O have been studied in a variety of supports, Ar, Kr, Xe, N_2, CO and D_2, all of which are critically reviewed and compared with similar studies for methanol. The structure of the dimer is interpreted as being open-chain in a N_2 matrix and cyclic in D_2.

III.4.1. INTRODUCTION

The matrix-isolation technique involves the rapid condensation of a gaseous mixture of an absorbing species (A) and a diluent gas (M), usually a noble gas, at a temperature sufficiently low (4 to 20°K) to prevent diffusion of the 'guest' species. The technique was originally developed by G. C. Pimentel [1] and his associates as a means of studying unstable molecular species. With high matrix-absorber ratios the solute may be expected to be 'isolated' in the rigid inert matrix and thus highly reactive species can be preserved for leisurely spectroscopic study [2].

Recent advances in cryogenic technology [3] have led to the increasing application of the matrix-isolation technique to stable molecules [4] in order to eliminate some of the spectral complexity arising from rotational interaction and association effects. The method can also be used to study rotational motion in the solid state and molecular aggregation under controlled conditions. Cryogenic matrix-isolation spectroscopy is thus of considerable interest to the physical chemist and the solid-state physicist. This paper reviews all the published, and some unpublished, infrared solid-matrix studies of water, and relates them to liquid solution studies of water and methanol.

III.4.2. MATRIX ENVIRONMENTAL EFFECTS

The basis of the utility of the technique lies in the fact that the I.R. spectrum of a matrix-isolated molecule (i.e. a monomer) consists of sharp, purely vibrational absorptions. It is clear, however, that the environment, even of noble gas molecules, must perturb the energy levels of the trapped species and modify their spectral features. The spectral changes which are produced in solid solutions generally resemble those which accompany the dissolution of molecules in liquid solvents [5], but there are several additional effects which arise due to the matrix environment. At present we can identify six general effects [4] all of which may contribute to modify the shape, intensity and position of a band and, in many cases, to cause a single vibrational mode to have a multiplet structure. These effects are:

(i) Multiple trapping sites - It is assumed that a guest species is usually trapped in a substitutional site formed by the removal of one or more matrix molecules, but there are also interstitial sites which might accommodate small solute species. Other sites may include dislocation sites. The intermolecular forces between matrix and absorber molecules will be different for different sites (Fig. 1) and the resulting different perturbations of the energy levels may lead to multiple frequencies.

(ii) Molecular rotation - Most trapped species will be so tightly held as to prevent rotation and at cryogenic temperatures the pure vibrational modes will appear as very sharp absorptions, with band half widths typically about 1 to 2 cm^{-1}. This is inherently the great advantage of M.I. spectra. However, in a noble gas matrix possessing a sufficiently large cavity, it might be expected that the rotational levels of a very small guest molecule would be only slightly perturbed and thus rotation be relatively unhindered. Rotational features are identified by reversible intensity changes on temperature cycling.

(iii) Medium effect or matrix shift - Under conditions of perfect isolation in any site the guest molecule is subject only to solute-matrix interactions. These will perturb the vibrational levels of the solute and will be reflected in a frequency shift, $\Delta\nu = (\nu_{matrix} - \nu_{gas})$, which is analogous to the well-known solvent shift [5] and for stretching modes, is usually to lower frequency.

(iv) Molecular aggregation - Studies must be carried out at high M/A ratios, usually 1000, to achieve good isolation. At low M/A ratios molecular aggregates may be formed and trapped in addition to monomers (Fig. 1), a phenomenon which will be greatest for solutes capable of hydrogen-bonding. The technique offers an excellent method for investigating molecular association in a controlled manner. The multiple features due to self-association are identified from their concentration dependence and from "warm-up" experiments in which monomers diffuse to form dimers and higher multimers. The sharpness of the spectra of the multimer species frequently allow interpretations to be made of their geometry.

(v) Coupling with lattice vibrations - In the solid state additional features may arise due to coupling of internal vibrational modes with external lattice modes.

(vi) Phonon bands - A trapped impurity species will disturb the host lattice symmetry and may activate otherwise inactive lattice modes which appear as the so-called phonon bands. For example solid Ar has phonon band maxima at 42 cm^{-1} and 64 cm^{-1} and the presence of an impurity activates a band at 73 cm^{-1}.

III.4.3. ENVIRONMENTAL EFFECTS ON THE ν(OH) STRETCHING FREQUENCIES OF METHANOL

The improvement of the Ar matrix spectra over the spectra in all other phases is clearly shown in Fig. 2. In the vapour phase, difficulties arise from the rotational fine structure of the monomer vibrational band, although bands can be discerned within the rotational envelope, due to dimers and higher multimers. In the pure liquid or solid phases, the OH modes give bands of enormous breadth due to the presence of large multimeric species. In solution in non-polar solvents at ambient temperatures some control of the species is obtained by varying the concentration; at extreme dilution the monomer spectrum is obtained (Fig. 2g, $\Delta\nu_{\frac{1}{2}}$ 22 cm^{-1}), whereas at high concentrations spectra approximating to the pure liquid are obtained. On progressive dilution the broad multimer absorption gradually shows indications of a shoulder on the high frequency side, which gives rise to a poorly defined dimer band at ca 3500 cm^{-1}, before the monomer species becomes dominant. The breadth of the bands, however, makes it impossible to locate bands due to multimeric species

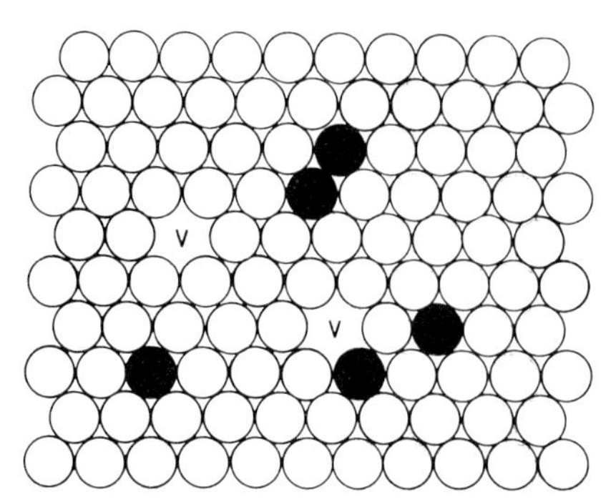

Fig. 1 Different environments of matrix trapped monomers caused by vacancies and next-nearest neighbours.

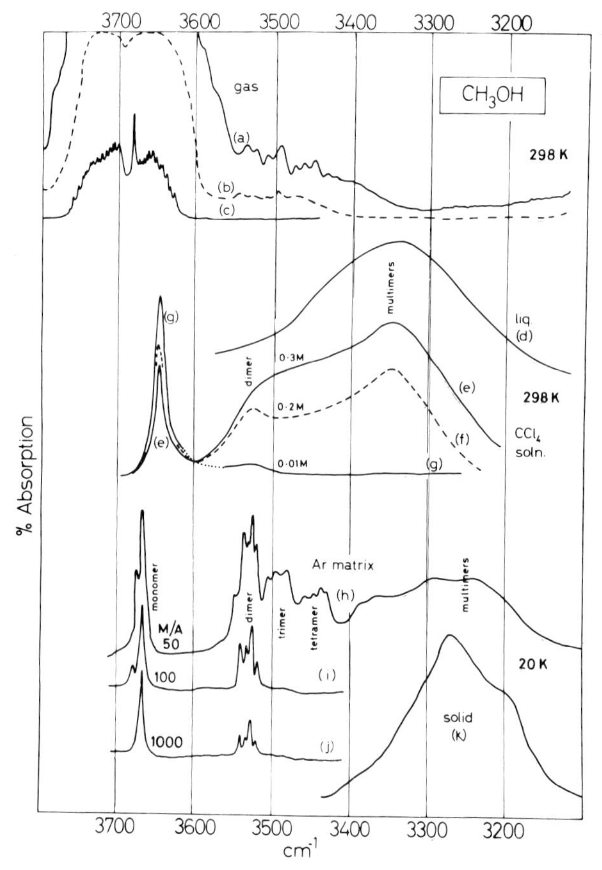

Fig. 2 Environmental effects on the infrared spectra of ν(OH) stretching region of methanol.

intermediate between dimer and high polymer.

In Ar matrices at 20 K a series of bands are observed [6] as the M/A ratio is lowered, which, from their concentration dependence may be assigned to monomer (Fig. 1j, $\Delta\nu_{\frac{1}{2}} = 5\ cm^{-1}$), dimer, trimer, tetramer and high multimer species. In addition to these absorptions, a band is observed on the high frequency side of the monomer band which increases in intensity as the multimers increase. It is interpreted as being due to the free terminal OH group of an open-chain

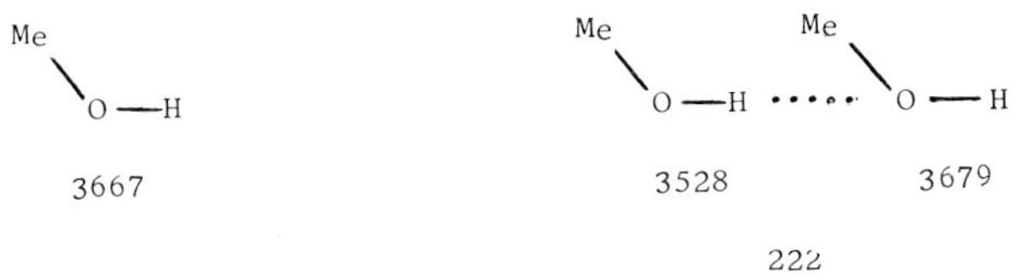

Me–O—H (3667)

Me–O—H ····· Me–O—H (3528, 3679; 222)

dimer (and multimer). This is confirmed by far I.R. data [6]. A cyclic dimer should show an intense hydrogen-bonded stretching mode, but only a weak deformation mode, whereas an open-chain dimer should show stretching and deformation modes of comparable intensity. Two strong bands, at 222 cm^{-1} and 116 cm^{-1}, are observed and the low deuteration shift (to 213 cm^{-1}) of the stretching mode lends further support for an open-chain structure for the dimer. The multiplet structure of the dimer, trimer and tetramer absorptions may arise from different angular orientations of the hydrogen bonds, which in the extreme will involve cyclic species, possibly stabilised in the low temperature matrix.

III.4.4. MATRIX STUDIES OF WATER

H_2O and D_2O have been extensively studied by I.R. cryospectroscopy in a variety of matrix supports:

Ar, Kr, Xe, [7 - 11]; N_2, [12 - 15]; CO [15], and D_2 [16].

The question of molecular rotation in solids is an old one and many molecules have been thought to be rotating freely in at least one crystal phase. More sophisticated experimentation has however over the years disproved many of these and in only a few pure systems, such as crystalline H_2, is there strong evidence for quantized rotation. However, substitutional cavities in noble gas crystals have 12 spherical neighbours and are highly symmetric, so that barriers to rotation for small molecules trapped in such sites (Fig. 3) should be low. It has been established [7 - 11] that the trapped H_2O monomers (Fig.3) undergo esentially free rotation in noble-gas matrices, rotational assignments being made of the prominent fine-structure peaks in the ν_3 (antisymmetric stretch) and ν_2 (deformation) bands. The time dependence of these features at 4 K show [8, 9, 11] intensity changes over a period of a few hours which are consistent with a nuclear-spin-species conversion process and fully confirms the rotational assignment. On the other hand, in N_2, D_2 and CO matrices

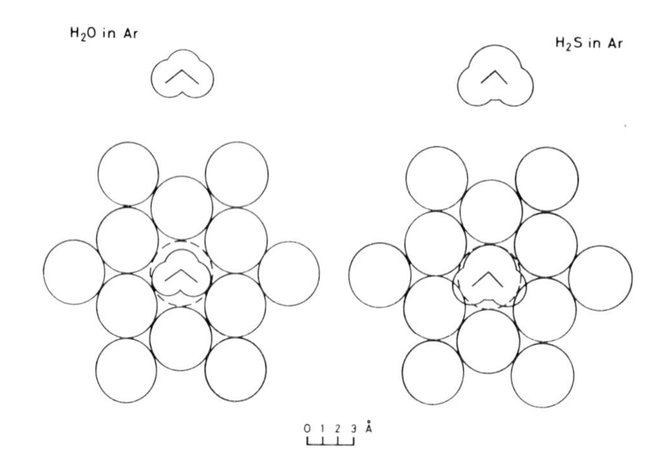

Fig.3

Comparison of volume requirements for H_2O and H_2S trapped in a solid argon matrix.

which provide "cylindrical" substitutional sites, the evidence favours rigidly held monomers. Of more concern to this Symposium are the dimer absorptions and their assignment in terms of the structure (s) of the dimer species. The three possible dimer structures and the number of I.R.-active ν(OH) modes are: (I) open-chain (4); (II) cyclic (2); and (III) bifurcated (4):

I (open-chain: H–O–H···OH_2) II (cyclic) III (bifurcated)

For a small guest molecule such as H_2O the aggregation behaviour is best studied in a host matrix such as N_2 which eliminates the complexity of rotational features and three near-I.R. studies have been made of water suspended in solid N_2. The early study of Van Thiel, Becker and Pimentel [12] of H_2O at 20 K has been extended by Harvey and Shurvell [13] to D_2O at 5 K, and more recently by Tursi and Nixon [14] to H_2O, D_2O and HDO over the temperature range 4 to 20 K and a concentration range M/A = 40 to 400. The absorptions show three distinguishable types of concentration behaviour. Those which exhibit a steady increase in intensity with increasing M/A are due to monomers; the peaks whose intensities pass through a slight maximum at M/A $\simeq$ 70, arise from the first multimer to be formed, assumed to be dimers, and those with continuously decreasing intensity over this concentration range are attributed to higher multimers. For the dimeric species Van Thiel et al. [12] suggested a cyclic structure rather than an open or bifurcated one, their suggestion being based upon a comparison of the number of observed dimer absorptions with the numbers of infrared active fundamentals predicted for the several plausible structures and upon the frequency pattern. They found just two absorptions in the

OH stretching region (3691 and 3546 cm^{-1}) and one in the bending region (1620 cm^{-1}) definitely attributable to the dimer. The possibility that a second weaker band in the bending region (1615 cm^{-1}) might also be due to the dimer led them to speculate further that the cyclic structure might be nonplanar and thus without a centre of symmetry. In the second study, Harvey and Shurvell assign three broad bands (2725, 2598 and 1203 cm^{-1}) to the D_2O dimer but do not comment on the structure of the dimer. They attribute several of sharper multiple peaks to nearly freely rotating D_2O monomers in N_2 as in the noble gases. Tursi and Nixon, however, have conclusively shown that the frequencies do not fit the pattern expected for a freely rotating molecule. Also, in contrast to the case of H_2O in Ar for which Hopkins et al. [11] detected changes of 10 to 20% in the intensities of rotational lines due to nuclear-spin conversion, Tursi and Nixon observed a slight sharpening of the peaks and a concurrent slight decrease in intensities on lowering the temperature from 20 to 4K but no further change with time at 4K.

In contrast to the results of Van Thiel et al, Tursi and Nixon identify with some confidence <u>five</u> fundamentals, and possibly a sixth, both for the H_2O and the D_2O dimer (Table I). They thus eliminate a centre of inversion in the dimer structure. The pattern of the frequencies is significant and bear an interesting

TABLE I

Monomer and Dimer Frequencies (cm^{-1}) of H_2O and D_2O in N_2 at 20 K [14]

Assignment	Monomer H_2O	Monomer D_2O	Monomer HDO	
ν_3 antisymmetric stretch	3725.7	2764.6	3680.4	
ν_1 symmetric stretch	3632.5	2655.0	2705.1	
ν_2 bend	1596.9	1179.2	1405.4	
	Dimer H_2O		**Dimer D_2O**	
OH stretch	3714.4	ν_3	2756.6	OD stretch
	3625.6	ν_1	2650.0	
OH... stretch	3697.7	ν_3	2737.6	OD... stretch
	3547.5	ν_1	2599.1	
HOH bend	(1600.3?)	ν_2	(1181.5?)	DOD bend
HOH... bend	1618.1	ν_2	1193.3	DOD... bend

comparison with those of matrix-isolated HCN [17] which, with its single C - H bond provides a simpler spectral interpretation (Table II). Two of the H_2O dimer fundamentals (3714.4 and 3625.6 cm^{-1}) are only slightly displaced from the ν_3 and ν_1 frequencies of the monomer and the pair has essentially the same intensity ratio as the monomer ν_3/ν_1. This is suggestive that one unit of the dimer is an only slightly perturbed monomer. Of the remaining two frequencies in the stretching region, one (3697.7 cm^{-1}) lies between ν_3 and ν_1 monomer, while the other

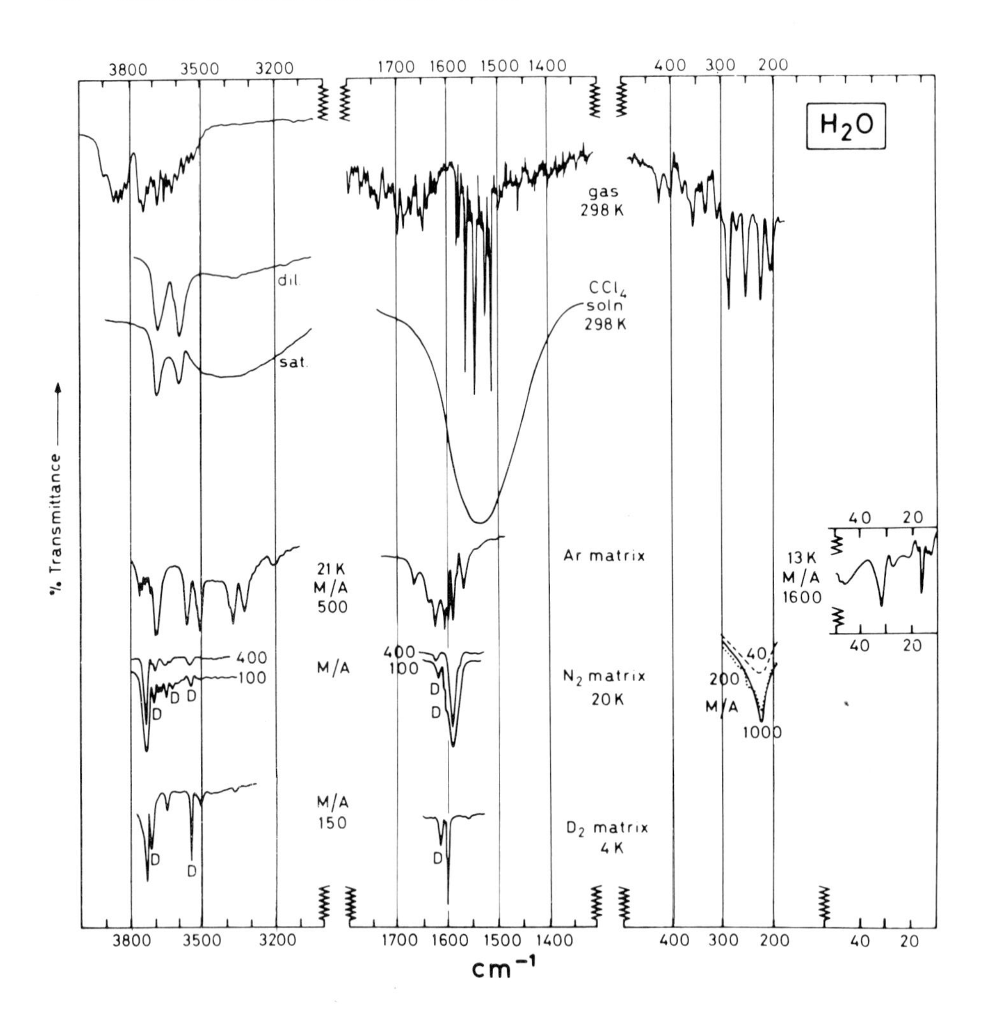

Fig. 4 Environmental effects of the infrared spectra of water from 4000 to 10 cm^{-1}. The matrix spectra are redrawn from the various literature sources referred to in the text.

TABLE II

Monomer and Dimer Absorptions in N_2 Matrices at 20 K

	Monomer	Dimer: Donor	Dimer: Acceptor	Ref.
	N≡C—H	N≡C—H · · · · N≡C—H		[17]
ν_3	3287.6	3204.9	3282.0	
ν_1	2097.3	2092.9	2110.9	
ν_2	745.6	799.0	757.6	
	H–O—H	H–O—H · · · · (H)O—H		[14]
ν_3	3725.7	3697.7	3714.4	
ν_1	3632.5	3547.5	3625.6	
ν_2	1596.9	1618.1	(1600.3?)	
	H–S—H	H–S—H · · · · (H)S—H		[18, 19]
ν_3	2632.6	2625.3	2631.1	
ν_1	2619.5	2580.3	2617.8	

(3547.5 cm^{-1}) is displaced considerably to a lower frequency. This is not the distribution of stretching fequencies expected for a structure with two equivalent "free" OH groups and two equivalent hydrogen-bonded ones which thus indicates an open structure with a single H-bond for the dimer, rather than a bifurcated or a cyclic structure. The fifth dimer band at 1618.1 cm^{-1} is assigned to the bending mode of the H-bonded unit and a weaker, and more questionable, peak at 1600.3 cm^{-1} to the bending mode of the "free" unit. This accounts for all of the dimer fundamentals, except for the lower frequency ones associated with the intermolecular bonding between the two water molecules. The results are summarised in Table II and are compared with those of matrix-isolated H_2S [18] also recently studied [19] along with H_2O in our laboratories. The absorption features of D_2O have an exact correspondence to those of H_2O (Table I); similar features in the stretching region are assigned, but expectedly with less confidence, for HDO. The absence of the OH stretch of DOH··· is interpreted as evidence for a stronger deuterium than hydrogen-bond. Force constant calculations are made for the dimer which are reasonable for an open structure with a moderately weak hydrogen-bond.

The N_2 experimental results of Tursi and Nixon have been confirmed by us [15] and extended to CO matrices at 20 K, and we feel their interpretation is extremely convincing. In CO which provides similar trapping sites there are some additional features which need further consideration, and the usual $\Delta\nu$ matrix shifts. However, the overall pattern is the same as in N_2 and we adopt the same model of an open-chain dimer.

The most recent matrix study of H_2O, reported [16] a few weeks prior to this Symposium utilises a deuterium support at 4K. The authors, after a careful concentration study over an M/A range of 83 to 935, are unable to find (Fig.4) more than three bands assignable to the dimer, from which they infer the presence of a cyclic dimer structure in the D_2 matrix.

The low frequency modes for water require more detailed study, the only report to date being that of Miyazawa [20] for H_2O and D_2O in Ar and N_2 at 20 K in the region 400 to 190 cm^{-1}. The monomer libration mode for H_2O is assigned to a broad band at 218 cm^{-1} in N_2 and a feature at 243 cm^{-1} to a dimer hydrogen-bond stretching vibration. A band at 265 cm^{-1} is tentatively attributed to the trimer vibration. No bands were observed in the 490 - 190 cm^{-1} region in Ar; the observations are thus in accord with the results of the near IR studies. These systems were however briefly investigated by Dr.B.Mann at Marburg during this Symposium week and extended down to _ca_ 70 cm^{-1} using the Polytec far-I.R. Interferometer and a Cryodyne closed cycle cryostat which were on exhibit, and this work has since been extended to the near I.R. and is still under active study.

The very far IR spectrum of H_2O and of D_2O isolated in Ar matrices at 13 K and at an M/A ratio _ca_ 1600 is shown in a paper by Cugley and Pullin [21] on isocyanic and iso-thiocyanic acids and which we include in Fig.4. In a preliminary analysis they identify the rotational transitions: 1(0,1) → 1(1,0) at 16.5 cm^{-1} for H_2O; 0(0,0) → 1(1,1) at 32.5 cm^{-1}, H_2O, 20.0 cm^{-1}, D_2O; 1(0,1) → 2(1,2) at 46 cm^{-1} (broad), H_2O, 27.5 cm^{-1}, D_2O; thus confirming the rotational motion in solid argon. They also have further studies in progress.

The results of Mann et al [22] indicate that the full story has yet to be unravelled. Firstly, although the broad outlines of their H_2O/N_2 spectra are similar to those of Miyazawa, the spectra obtained by Mann et al clearly show sub-structure features within the broad 218 cm^{-1} band. Secondly, in contrast to Miyazawa, they observe the same overall feature in D_2O/N_2 but with minor changes in the sub-structure. Similar results are also obtained for H_2O in Ar and for D_2O in Ar. Their results in the near I.R. are for H_2O/Ar and experimentally they confirm the results of Tursi and Nixon [14] . However, they adopt an interpretation based on the angle dependence of the hydrogen-bond, following that developed by Luck [23] in interpreting the earlier matrix data of water [14] and methanol [12].

He defines the angle β of H-bonds as the angle between the proton axis and the axis of the lone pair electron as shown below

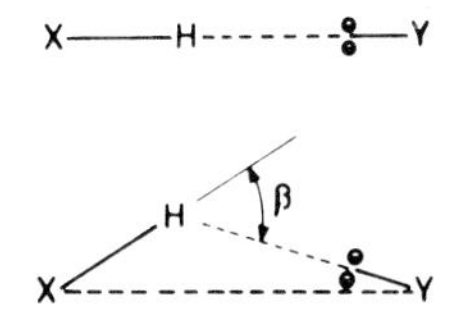

The angles are estimated by means of Stuart-Briegleb models for dimer, trimer, etc. species. It is accepted that the H-bond energy will be a maximum for a linear bond, ie for $\beta = 0$, thus for individual H-bonds in the series there will be an increase in energy from cyclic dimer, to cyclic trimer, ... to high multimer. Collectively of course the three H-bonds in a cyclic trimer, for example, may total in energy more than the two for a linear species. The $\Delta\nu = (\nu_{mon} - \nu_{n\text{-}mer})$ shifts are thus taken to parallel the H-bond energy, following the Badger-Bauer correlation. This is illustrated by plotting the ratio of the shifts $\Delta\nu/\Delta\nu_{max}$, where $\Delta\nu_{max} = (\nu_{mon} - \nu_{polymer})$, against the bond angle. Fig.5 depicts this correlation, updated to include the more recent $CH_3OH(D)$ data [6] and H_2O/D_2 data [16]. In attempting this interpretation for the latter data it has been assumed that the bands indicated by Strommen et al [16] as polymer can be assigned as: 3525 cm^{-1} (trimer), 3378 cm^{-1} (tetramer) with 3337 cm^{-1} (ν_{max}, polymer), whilst retaining their ν_3 monomer, 3737 cm^{-1}, and dimer, 3566 cm^{-1}, assignments. It is seen that there is a close correspondence with the H_2O/N_2 matrix data. The methanol curve is significantly higher and the angle-dependence interpretation would, for the dimer in particular with a $\Delta\nu/\Delta\nu_{max}$ value <u>ca</u> 0.34, indicate that a cyclic configuration would be less stable than the linear, which corroborates the far I.R. interpretation [6]. The fine structure in the vicinity of the monomer ν_3 peak is interpreted [23] (in the same manner as the free OH vibration of a linear bonded species) as being due to the free OH vibrations of the cyclic dimers, trimers, ..., slightly differently perturbed by the electron-donating effect on the oxygen atom of the H-bond interaction.

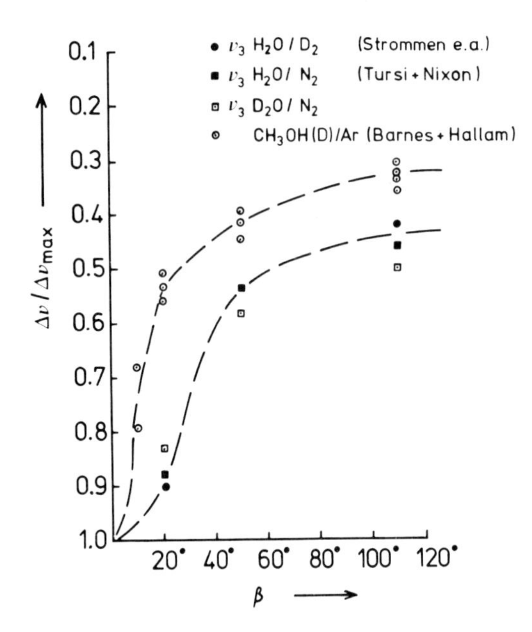

Fig.5

Plot of $\Delta\nu/\Delta\nu_{max} = f(\beta)$ of the H-bond $\nu(O - H \ldots)$ absorptions of water and methanol in low temperature matrices.

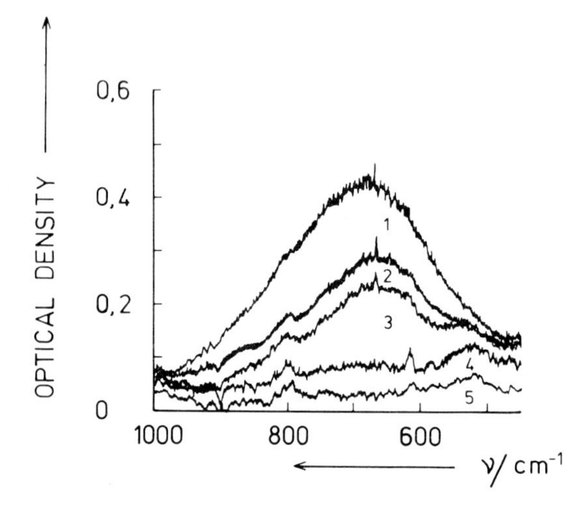

Fig.6

Librational absorption of H_2O in Ar matrices at 20 K: M/A ratios (1) 27; (2) 51; (3) 75; (4) 595; (5) 1448; [22].

Mann et al [22] also observe a new matrix feature at _ca_ 690 cm^{-1}, in a spectral region not examined in previous matrix studies. It corresponds with the broad absorption _ca_ 680 cm^{-1} (see figs. 1 and 2 Chap. III.1.) observed in the liquid. The general contour of the band in the liquid clearly indicates that it consists of several overlapping unresolved components. It shifts to lower frequencies and becomes broader with increasing temperature. They arise from librational or restricted rotational motions of H_2O molecules due to the restraints imposed by the hydrogen bonds. According to Walrafen [24] three librational components analogous to the rotations of isolated H_2O molecules are to be expected. The matrix spectra, although slightly interfered with by uncompensated atmosperic CO_2, clearly show (Fig.6) several components with the maximum of the overall contour moving to higher frequency with increasing concentration. At low concentrations the band clearly commences as a weak band _ca_ 550 cm^{-1} and presumably arises from dimer, possibly trimer, species. At higher concentrations the main feature at _ca_ 780 cm^{-1} emerges and can be assumed to arise from librational modes of higher multimers and possibly lower ones. This concentration behaviour in matrices thus parallels that of temperature in the pure liquid (note Raman bands for liquid water at 550 cm^{-1} and 720 - 740 cm^{-1} [24]). More detailed matrix studies of these features are required and indeed are planned, since the advantages of matrix-isolation spectroscopy are again evident.

III.4.5. CONCLUSIONS

The advantages of the cryogenic technique for investigating molecular association are amply borne out by these studies on water despite the fact that the results at this stage do not lead to an unequivocal interpretation. This may, however, be due the nature of the matrix technique. It should be stressed that the constraints imposed by the forced entrapment in the solid environment, even though a relatively inert one, does provide "unnatural" conditions. It is usually assumed, and there is much evidence for this [2], that the inert matrix spectrum of a substance is closely similar to that of its vapour-phase spectrum. It is certainly true that vapour-phase species will be trapped in the low-temperature matrix but there is now increasing evidence that other species may also be formed. The injection system through which the matrix gas/absorber vapour mixture is sprayed from _ca_ atmospheric pressure to low pressure, _ca_ 10^{-5} torr, may well be expected to cause aggregation as the molecules undergo expansion and thus internal cooling.

Also the rapid quenching process from ambient to cryogenic temperatures may itself lead to phenomena not encountered in the liquid state. In reviewing all of the published matrix studies on molecular association, even in cases where there is unequivocal evidence for a particular dimer configuration, it is usually this as the predominant species and there are indications of the limited presence of others.

Thus though one must make further use of the great advantages of the matrix technique one must be cautious in making extrapolations of the spectral interpretations to other phases. It should however be emphasised that the matrix spectra are the only ones to show clear evidence for the presence of several multimer species. It is therefore interesting to note that the theories of Ben Naim (see Chap. II.4.) and of Stillinger and Rahman [25] give indications of prefered sizes. In the case of Ben Naim, in particular, his energy distribution function plot (Fig. 18, Chap.II.4.) shows four distinct maxima.

REFERENCES

1 E.Whittle, D.A.Dows and G.C.Pimentel, J.Chem.Phys. 22, 1943 (1954)

2 "Vibrational Spectroscopy of Trapped Species - Infrared and Raman Studies of Matrix-Isolated Molecules, Radicals and Ions". Ed. H.E. Hallam, J.Wiley, London 1973

3 H.E.Hallam and G.F.Scrimshaw "Experimental techniques and properties of matrix materials", Chap. 2 in ref 2

4 H.E.Hallam, "Infrared studies of molecules trapped in low temperature matrices", Chap. 3 in ref. 2

5 H.E.Hallam p. 245 in "Spectroscopy, Proc. 3rd Inst. Petroleum Hydrocarbon Research Group Conference", Ed. M.J.Wells, Inst. of Petroleum, London 1962

6 A.J.Barnes and H.E.Hallam, Trans.Faraday Soc. 66, 1920 (1970)

7 E.Catalano and D.E.Milligan, J.Chem.Phys. 30, 45 (1959)

8 J.A.Glasel, J.Chem.Phys. 33, 252 (1960)

9 R.L.Redington and D.E.Milligan, J.Chem.Phys. 37, 2162 (1962); 39, 1276 (1963)

10 D.W.Robinson, J.Chem.Phys. 39, 3430 (1963)

11 H.P.Hopkins, jun., R.F.Curl, jun., and K.S.Pitzer, J.Chem.Phys. 48, 2959 (1968)

12 M.Van Thiel, E.D.Becker, and G.C.Pimentel, J.Chem.Phys. 27, 486 (1957)

13 K.B.Harvey and H.V.Shurvell, J.Mol.Spectroscopy, 25, 120 (1968)

14 A.J.Tursi and E.R.Nixon, J.Chem.Phys. 52, 1521 (1970)

15 M.L.Evans and H.E.Hallam, unpublished data (1973)

16 D.P.Strommen, D.M.Gruen, and R.L.McBeth, J.Chem.Phys. 58, 4028 (1973)

17 C.M.King and E.R.Nixon, J.Chem.Phys. 48, 1685 (1968)

18 A.J.Tursi and E.R.Nixon, J.Chem.Phys. 53, 518 (1970)

19 A.J.Barnes and J.D.R.Howells, J.Chem.Soc., Trans.Faraday II, 68, 729 (1972)

20 T.Miyazawa, Bull.Chem.Soc.Japan 34, 202 (1961)

21 J.A.Cugley and A.D.E.Pullin, Chem.Phys. Letters, 19, 203 (1973)

22 B.Mann, E.Schmidt, T.Neikes, and W.Luck, Abstracts of paper presented at 73rd meeting of Deutsche Bunsen-Gesellschaft für Physikalische Chemie, Kassel, May 1974

23 W.A.P.Luck, Chap.4 in Vol.2 "Water - a comprehensive treatise", Ed. Felix Franks, Plenum Press, New York, London 1973

24 G.E.Walrafen, Chap.5 in Vol.1 "Water - a comprehensive treatise" Ed. Felix Franks, Plenum Press, New York, London 1973

25 F.H.Stillinger and A.Rahman, J.Chem.Phys. 57, 1281 (1972)

IV RAMAN METHODS

IV.1. SPONTANEOUS AND STIMULATED RAMAN SPECTRA FROM WATER AND AQUEOUS SOLUTIONS

by

G. E. Walrafen
Bell Laboratories
Murray Hill, New Jersey 07974

ABSTRACT

Spontaneous Raman spectra in the OH and OD stretching regions from pure H_2O and D_2O, and stimulated Raman spectra in the OH and OD stretching regions from H_2O, D_2O, H_2O-D_2O mixtures, and aqueous solutions of $NaPF_6$, $NaSbF_6$, $NaClO_4$, and $NaBF_4$ have been obtained. The Raman results are consistent with a mixture model developed previously for water and aqueous solution structure in which hydrogen-bond breakage occurs with temperature rise, or with the addition of certain electrolytes. The stimulated Raman results, in particular, appear to rule-out continuum models of water structure.

IV.1.1. INTRODUCTION

Numerous Raman investigations of water and aqueous solution structure have now been conducted. Some of the spontaneous Raman spectral investigations have been reviewed in recent books.[1-4] Several reports of stimulated Raman investigations have also appeared.[4,5-8]

With regard to models of water structure, a long-standing controversy has existed among vibrational spectroscopists, namely, between advocates of mixture[9] or continuum[10] models. The mixture-model advocates have generally related distinct spectral features, for example, in spontaneous Raman contours, to structures differing in the extent of hydrogen bonding, whereas workers favoring continuum models have tended to consider water in terms of a continuous distribution of interactions presumed to be spectroscopically indistinguishable.

Since 1970, however, stimulated Raman data have been described[7,8] which indicate remarkably specific spectral detail, namely, mutually exclusive stimulation of components that in spontaneous scattering are normally overlapped. These stimulated effects tend to rule-out most, if not all, strict continuum models. Nevertheless, the mixture versus continuum controversy has recently been intensified by spontaneous Raman spectroscopists,[11-13] who have taken little note of the stimulated Raman data.

In the present work accurate spontaneous Raman data referring to the OH and OD stretching regions of pure H_2O and D_2O are presented and analyzed. Stimulated Raman data obtained from various aqueous solutions are also reported. The new Raman data provide evidence bearing directly upon the mixture--continuum arguments.

1.2 Spontaneous Raman Spectra from Pure H_2O and D_2O.

A. Experimental Procedures and Data.

The photoelectrically-recorded spontaneous laser-Raman spectra of this work were obtained with a Jarrell-Ash No. 25-102 double-monochromator, and a Carson Laboratories argon-ion laser. Detection was accomplished with an ITT FW-130 (S-20) photomultiplier tube, and a model 1120-1105 SSR photon counter. The focussed 4765-A radiation (280 mW) was passed once through a 10-cm Raman cell. Polarization measurements were made with a Polaroid sheet analyzer (HN 38), and with a compensated quartz wedge scrambler in front of the entrance slit. Scanning rates of 50 cm^{-1} min^{-1} were employed with a slit-width of 4 cm^{-1} and a time-constant of 1 sec.

Examinations of the response of the Jarrell-Ash instrument were made in conjunction with the ITT FW-130 tube. In the Raman shift region of 2000-4000 cm^{-1} from the 4765-A radiation, no significant corrections of the contour shapes were required. In addition, intensity measurements with incoherent unpolarized light in the region from $\Delta\bar{\nu}$ = 2800-3000 cm^{-1} indicated no observable differences for the $\perp$ and $||$ Polaroid positions.

Spontaneous Raman spectra obtained in the OH and OD stretching regions from pure H_2O and D_2O are shown in Figs. 1 and 2. The two upper sections of the figures refer to the Raman intensities, $I_\perp$ and $I_{||}$, obtained for the two Polaroid orientations and proportional to the quantities, $10\times3\ \beta^2$ and $45\alpha^2+4\beta^2$, indicated in the figures. The lower sections proportional to and labelled $45\alpha^2$ are discussed subsequently.

Depolarization ratios*, $\rho_\ell = I_\perp/I_{||}$, were obtained from the spectra of Figs. 1 and 2, using the base-lines indicated. The ratios are plotted versus $\Delta\bar{\nu}$ for H_2O and D_2O in Figs. 3 and 4. The dashed lines shown in the figures refer to subjective nonlinear interpolations, values from which are tabulated in Table I.

	$\rho_{\ell H_2O}$				$\rho_{\ell D_2O}$		
$\Delta\bar{\nu}$	Walrafen	Murphy & Bernstein	δ	$\Delta\bar{\nu}$	Walrafen	Murphy & Bernstein	δ
3000	0.075	0.077	+0.002	2200	0.081	0.078	-0.003
3050	0.063	0.061	-0.002	2250	0.063	0.059	-0.004
3100	0.055	0.056	+0.001	2300	0.048	0.044	-0.004
3150	0.054	0.052	-0.002	2350	0.041	0.040	-0.001
3200	0.063	0.061	-0.002	2400	0.055	0.059	+0.004
3250	0.082	0.083	-0.001	2450	0.091	0.103	+0.012
3300	0.112	0.112	0.000	2500	0.155	0.177	+0.022
3350	0.142	0.146	+0.004	2550	0.232	0.255	+0.023
3400	0.180	0.185	+0.005	2600	0.284	0.341	+0.057
3450	0.218	0.231	+0.013	2650	0.305	0.389	+0.084
3500	0.242	0.261	+0.019	2700	0.290	0.359	+0.069
3550	0.247	0.270	+0.023	2750	0.260	0.347	+0.087
3600	0.256	0.265	+0.020	2800	0.222	0.374	+0.152
3650	0.248	0.244	-0.004				
3700	0.224	0.270	+0.046				

Table I. Depolarization ratios, ρ_ℓ, compared to those from Murphy and Bernstein.[11]

*) See Appendix II.

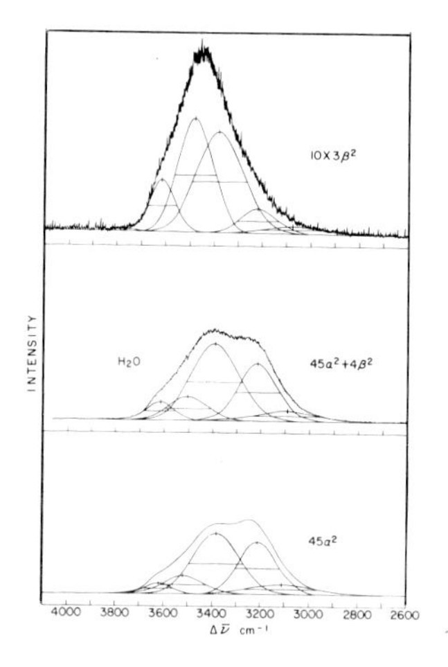

Fig. 1. Spontaneous Raman spectra in the OH stretching region from pure H_2O. The top and center figures refer to $I_\perp$ and to $I_{||}$ resp. The bottom figure is calculated. The intensities are proportional to the quantities indicated on the figure.

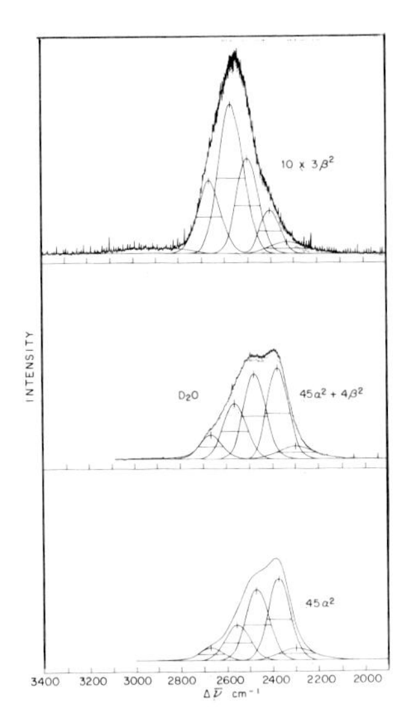

Fig. 2. Spontaneous Raman spectra in the OD stretching region from pure D_2O. See Fig. 1 caption for explanation.

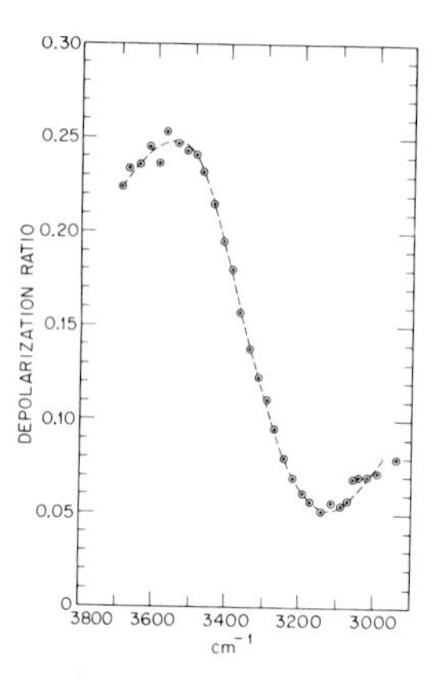

Fig. 3. Plot of depolarization ratio versus Raman shift in cm^{-1} obtained from Fig. 1 data.

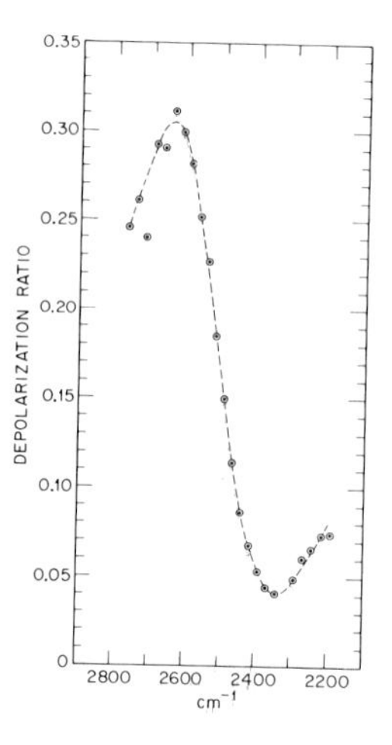

Fig. 4. Plot of depolarization ratio obtained from Fig. 2 data.

In Table I the present interpolated values are compared with values supplied by Murphy[14] resulting from the work of Murphy and Bernstein.[11] The agreements in the bracketed regions, where base-line assumptions contribute relatively less to the errors, vary from ~1-8% for H_2O, and ~2-20% for D_2O, with the 8% and 20% variations referring to the worst cases. This agreement, in addition to the agreement throughout the table, is considered very satisfactory in view of the very large corrections for instrumental response made by Murphy and Bernstein, and the fact that no corrections were required in this work.

B. Isotropic and Anisotropic Raman Spectra, and Computer Analysis.

Raman intensities, $I_\perp$ and $I_{||}$, obtained in $\perp$ and $||$ Polaroid orientations versus frequency shift in cm^{-1}, $\Delta\bar{\nu}$, can yield isotropic and anisotropic spectra, that is, spectral contributions proportional to α^2 and β^2. The quantities α and β are the isotropic and anisotropic polarizabilities, defined as $\alpha = 1/3(\alpha_{xx}+\alpha_{yy}+\alpha_{zz})$ and $\beta^2 = 1/2\{(\alpha_{xx}-\alpha_{yy})^2 + (\alpha_{yy}-\alpha_{zz})^2 + (\alpha_{zz}-\alpha_{xx})^2 + 6[(\alpha_{xy})^2+(\alpha_{yz})^2+(\alpha_{zx})^2]\}$ where all α_{ii} and α_{ij} terms refer to derivatives with respect to the normal coordinate, ξ_i.[15]

Consider that a vertical laser beam and its horizontal electric vector define a plane, and that a Polaroid sheet is placed parallel to that plane. When the Polaroid sheet does not transmit the laser light, it is in the $\perp$ position with respect to the electric vector. In this position a Raman spectrum proportional to $3\beta^2$ is obtained. However, when the Polaroid is rotated in its plane by 90°, it is in the $||$ position relative to the electric vector, and a spectrum proportional (essentially the same constant) to $45\,\alpha^2 + 4\,\beta^2$ results. A spectrum proportional to $45\,\alpha^2$ is thus not obtained directly, but it can be calculated. $I_\perp$ gives intensities proportional to $3\,\beta^2$, and $I_{||}-4/3\,I_\perp$ then yields the spectrum proportional to $45\,\alpha^2$.

The six spectra of Figs. 1 and 2 were analyzed into Gaussian components by means of a duPont 310 Curve Resolver, a ten-channel analog computer. Evidence for the use of Gaussian components has been detailed previously.[4,9] Five Gaussian components[9] were found to produce fits having residuals within the noise levels. The five components and their peak positions and half-widths are indicated in the figures. Values of $\Delta\bar{\nu}$, $\Delta\bar{\nu}_{\frac{1}{2}}$ and ρ_ℓ for the five Gaussian components are given in Table II. The ρ_ℓ values refer to $\perp$ and $||$ integrated component intensity ratios.

1.3. Stimulated Raman Spectra from H_2O, D_2O, H_2O-D_2O Mixtures and Aqueous Solutions of $NaPF_6$, $NaSbF_6$, $NaClO_4$, and $NaBF_4$.

A. Data and Detailed Experimental Procedures.

The stimulated Raman spectra obtained photographically in this work are reproduced from microdensitometer tracings in Figs. 5-7. Stimulated Raman data from other solutions were also obtained and are tabulated along with those from Figs. 5-7 in

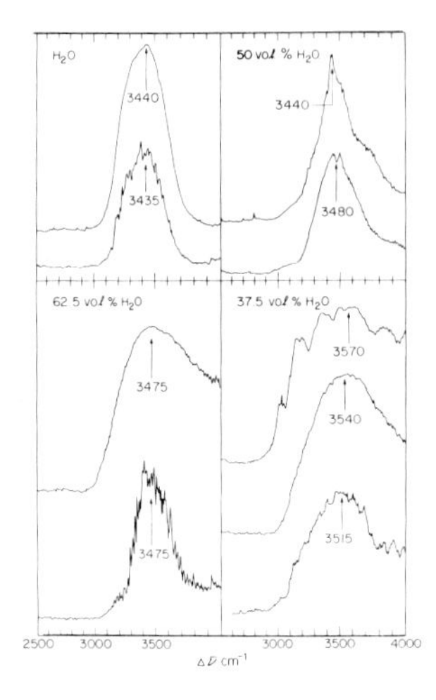

Fig. 5. Microdensitometer tracings corresponding to stimulated Raman spectra from H_2O-D_2O mixtures. Lower right panel, triplicate runs, all others duplicate.

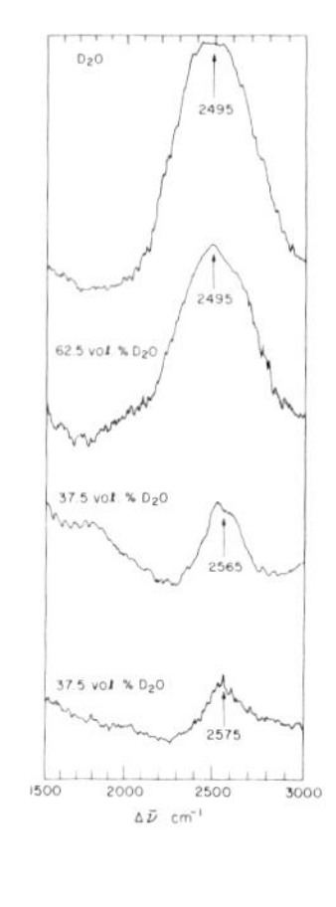

Fig. 6. Stimulated Raman tracings from D_2O-H_2O mixtures.

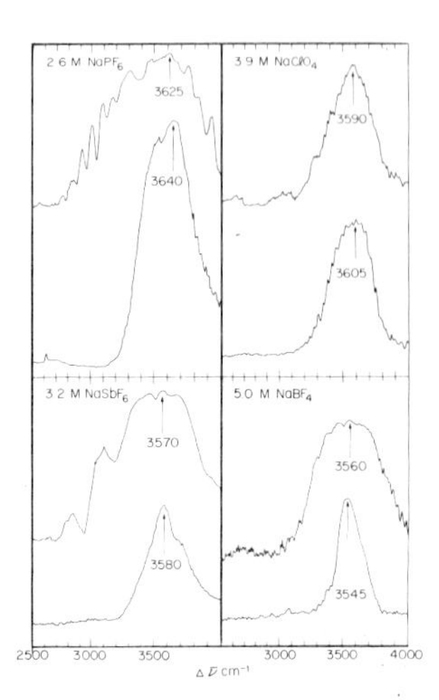

Fig. 7. Stimulated Raman tracings from electrolytes in 10 mole % D_2O in H_2O. All panels, duplicate runs.

		D_2O $\Delta\bar{\nu}$	$\Delta\bar{\nu}_{\frac{1}{2}}$	ρ_ℓ			H_2O $\Delta\nu$	$\Delta\nu_{\frac{1}{2}}$	ρ_ℓ
	a	2330	190	0.0_9		a	3075	270	0.0_7
1	b	2297	175		1	b	3090	275	
	c	2297	170			c	3110	300	
	a	2405	100	0.04_2		a	3218	175	0.03_7
2	b	2375	110		2	b	3214	185	
	c	2375	110			c	3212	185	
	a	2500	110	0.1_1		a	3380	228	0.1_3
3	b	2475	115		3	b	3393	230	
	c	2470	118			c	3383	228	
	a	2573	125	0.2_9		a	3480	175	0.4_2
4	b	2560	118		4	b	3505	190	
	c	2558	125			c	3520	193	
	a	2668	112	0.3_2		a	3618	125	0.2_7
5	b	2666	110		5	b	3616	125	
	c	2670	110			c	3620	120	

Table II. Gaussian component parameters from OH and OD stretching regions. a-depolarized scattering, $I_\perp$. b-polarized scattering, $I_{||}$. c-calculated, and proportional to $45\,\alpha^2$. $\Delta\bar{\nu}$-Raman shift in cm^{-1}. $\Delta\nu_{\frac{1}{2}}$-width at half-height. ρ_ℓ-depolarization ratio from component areas, $I_\perp/I_{||}$, for linearly polarized excitation.

Table III. Available spontaneous Raman intensity maxima values are included in the table for comparison.

Solution	Stimulated $\Delta\bar{\nu}_{MAX}$	Spontaneous $\Delta\bar{\nu}_{MAX}$
5.3 M $NaPF_6$ in 10 mole % D_2O in H_2O	$\nu(OH)$ 3630 cm^{-1}	$\nu(OH)$ 3640±5 cm^{-1} 3590±5 3450±10
4.0 M $NaPF_6$ in pure H_2O	3475	
2.6 M $NaPF_6$ in 10 mole % D_2O in H_2O	3625,3640	3630±5 3440±10
3.2 M $NaSbF_6$ in 10 mole % D_2O in H_2O	$\nu(OH)$ 3570, 3580	$\nu(OH)$ 3570±10 3445±10
4.0 M $NaClO_4$ in pure H_2O	$\nu(OH)$ 3470	$\nu(OH)$
3.9 M $NaClO_4$ in 10 mole % D_2O in H_2O	3590, 3605	3545±10 3455±10
5.0 M $NaBF_4$ in 10 mole % D_2O in H_2O	$\nu(OH)$ 3545, 3560	$\nu(OH)$ 3565±10 3435±10
4.0 M $NaBF_4$ in pure H_2O	3430	

Table III. Comparisons of stimulated and spontaneous Raman intensity maxima.

The stimulated Raman scattering system employed is shown schematically in Fig. 8, and it will be described here in detail. The physical processes involved in stimulated Raman scattering will be described in brief subsequently. Simplified descriptions of stimulated Raman scattering may be found in Refs. (4 and 8), see especially also Ref. (16). More specific discussions of stimulated Raman processes, with special reference to water, may be found in the chapter by Colles.

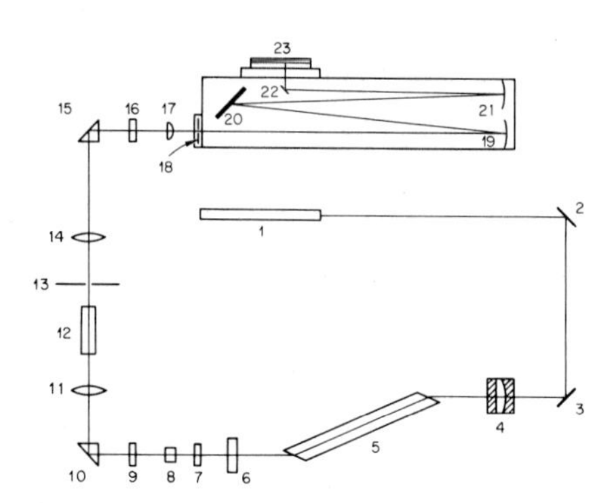

Fig. 8. Schematic illustration of stimulated Raman scattering system.

In this work a Spacerays neodymium-glass laser,(4-6) mode-locked and Q-switched by use of a bleachable dye, was employed in conjunction with a Jarrell-Ash No. 78-466 single-monochromator having a grating blazed at 5000 A and equipped for photographic detection, 18-23, Fig. 8. The Brewster-Brewster cut laser rod, 5, Owens-Illinois neodymium-glass, 4.5 in. × 3/8 in., was positioned between the high-reflectivity dye cell, 4, and a 55% reflectivity dielectric mirror having a 30' wedge, 6. The dye cell was composed of a 5 m high-reflectivity dielectric mirror and a window having a 6' wedge. The space between the window and the 5 m mirror, 0.125 in., was filled with a 5 volume % solution of Eastman Q switch solution 9860 in 1,2-dichloroethane. Radiation at 10600 A is produced by the laser, and the 10600-A radiation is typically composed of a train of ∿25 pulses separated by a round-trip time of ∿4 nsec. A pulse duration of 1-2 psec was reported in studies involving an identical laser system,[17] but a pulse duration of 5-6 psec is more generally obtained. The width of the 10600-A radiation from a neodymium-glass laser can vary from about 20-80 cm^{-1}, but large Raman conversions are considered to arise in the initial stages of the pulse train where the width is 20-30 cm^{-1}.[7] (In regard to this, stimulated Raman half-widths as low as about 17 cm^{-1} have been observed for the $\Delta\bar{\nu}$ = 990 cm^{-1} line from benzene in this work.) The laser mirrors were aligned interferometrically by use of a He-Ne laser, 1, and mirrors 2 and 3. Circular interference fringes were formed slightly below the He-Ne output mirror by back reflections from the laser mirrors at 4 and 6 - the path from 6 to 1 was about 2.2 m. A similar procedure involving 6328-A back reflections from the KDP (KH_2PO_4) crystal used for second harmonic conversion, 8, and the mirror, 6, was also initially employed, but the critical alignment to <1 mrad. of the axis perpendicular to the 10600-A laser beam and to the plane of Fig. 8 was made by maximizing the intensity of the picosecond pulses at 5300 A. The power at 10600 A was estimated as $\leq$1 GW, and that at 5300 A was $\leq$150 MW.[17] The 10600-A radiation passed through a Shott RG 10 filter, 7, and then through the KDP crystal. The 10600-A radiation remaining after frequency-doubling, was removed by a Corning CS 4-76 filter, 9, and the 5300-A radiation then passed through the prism, 10, and through the 19-cm lens, 11, where it was focussed beyond the front window of a 10-cm or 35-cm cell, 12. The peak laser power density at 5300 A was estimated to be ~20 GW/cm^2 or possibly above. (The dye concentration was about five-times greater than that employed previously, and focussing was accomplished with a 19-cm, rather than a 30-cm, lens.[7]) The stimulated Raman and 5300-A radiation was then passed (in some experiments) through a 1 mm pin hole, 13, which stopped radiation having a divergence >5 mrad. The radiation, made parallel by passage through the 19-cm lens, 1, was next passed through the prism, 15, into the filter (where radiation at 5300 A and above was removed), 16, then through the cylindrical lens, 17, and finally through the slit, 18, into the spectrograph. The cylindrical lens produced an image having approximately the same shape as the slit, and it caused the beam to diverge to the approximate size of the mirror, 19. A slit-width of 220 μ (6 cm^{-1}) was employed, and Kodak 103-F plates, 4x5 in., were used for photographic detection in addition to types 52(400 ASA) and 57 (3200 ASA) Polaroid. Exposed photographic plates were scanned photoelectrically with a Leeds and Northrup No. 6700 P-1 microdensitometer.

1.4 Interpretation.

A. Spontaneous Raman Data.

To shorten the discussion required in the interpretation of the spontaneous Raman data, it is expedient to refer the reader to Refs. (4,9,11, and 18).* In essence, there is now agreement that components contributing to the OH and OD stretching contours, irrespective of whether Gauss[9] or Gauss-Lorentz[11] shapes are invoked, can be divided into classes depending upon whether the component centers lie above or below experimentally observed isosbestic frequencies.[18,19] Specifically, there is agreement[9,11] that components near about 3616-3620 and 2666-2670 cm^{-1}, numbered 5 in Table II, refer to OH and OD bonds that are nonhydrogen-bonded. And, there is agreement that at least one of the components at about 3380-3393 or 3212-3218 cm^{-1} from H_2O and at about 2470-2500 or 2375-2405 cm^{-1} from D_2O, numbered 3 and 4 in Table II, refers to OH and OD units engaged in relatively strong hydrogen bonds, in which the O-H···O units do not deviate greatly from linearity. Some disagreements[11,20] exist on details, such as whether certain components should be assigned to Fermi resonance, rather than to intermolecular coupling between intramolecular vibrations, but the present interpretation will proceed from the component classification based on an isosbestic frequency or range.

From Figs. 1 and 2 and from Table II it is evident that component 5 is polarized, ρ_ℓ = 0.3. From the depolarization ratio, and from numerous studies, including temperature,[9,21] and electrolyte[22,23] dependences, component 5 is assigned to the nonhydrogen-bonded OH stretching of 3-bonded H_2O and D_2O molecules, <u>i.e.</u>, to molecules in which only one proton is hydrogen bonded, and in which the oxygen atom engages in two hydrogen bonds. The symmetry of the 3-bonded H_2O molecule must be C_s, according to Table II.

*) See chapter fundamental region, chapter III.2.3. and fig. III.2.8

Component 4 in Table II is also polarized for H_2O and D_2O, $\rho_\ell \approx 0.3$-0.4. Further, close inspection of the a, b, and c frequencies from H_2O (but not from D_2O) could suggest that sub-components are involved because of the frequency spread. Indeed, component 4 has been sub-divided into a polarized and a completely depolarized component by Murphy and Bernstein.[11] However, such a sub-division seems completely unwarranted in this work in view of the present experimental uncertainties, and at the present time component 4 is assigned either to an intermolecular coupling involving intramolecular vibrations, or to the envelope of the hydrogen-bonded valence vibration of 3-bonded C_s water with the highest-frequency valence vibration of 4-bonded water having either C_{2v} or C_s symmetry ($\sim C_{2v}$). Four-bonded water refers to H_2O molecules in which both protons are hydrogen-bonded, and in which the oxygen atom engages in two hydrogen bonds. A C_{2v} symmetry for 4-bonded water would imply that both proton interactions are exactly equivalent, whereas slight inequalities in these interactions ($\sim C_{2v}$) would lead in a strict sense, to C_s symmetry.

Component 3 in Table II is polarized, $\rho_\ell \approx 0.1$. Murphy and Bernstein assign component 3 to a Fermi resonance doublet which includes component 2. This assignment however, is not favored by this author for several reasons. The Fermi resonance assignment conflicts with high-pressure results,[20] with dilution results,[19] with results from the addition of structure-breaking electrolytes,[22,23] and with results from stimulated Raman scattering.[4,7,8] A more reasonable interpretation for component 3 would seem to involve the symmetric stretching of H_2O molecules that have C_{2v} symmetry, or the lowest frequency valence vibration, i.e., essentially the symmetric stretching, of ($\sim C_{2v}$) C_s molecules. Component 3 is thus thought to refer primarily to an intramolecular motion, that is, one that is not so strongly coupled to other molecules as in the case of component 2.

The Fermi resonance assignment is also questioned for component 2. A more favored assignment would involve the in-phase OH stretching motions of an aggregate consisting of a central H_2O molecule, and its nearest, or perhaps higher, neighbors. This assignment is in good agreement with the fact that the depolarization ratio of component 2 is the lowest of all components, $\rho_\ell \approx 0.04$, i.e. with the fact that component 2 has almost pure isotropic character. It is also in agreement with the high-pressure, dilution, electrolyte, and stimulated Raman results mentioned in regard to component 3. In particular, the abrupt intensification of component 2 relative to component 3 on formation of an extended lattice in ice VI at 28.8°C, is in excellent accord with the present assignment.[20] Component 2 is thus considered to have intramolecular and intermolecular character, and in this regard it is important to note that the frequency difference between components 3 and 2 is not far from the value of 170 cm^{-1} assigned to O-O stretching of O-H···O units.[25]

An exact assignment for the broad weak component 1 is not yet possible. Its intensity has been observed to increase relative to component 3 with F^- addition.[23] Thus, it probably also has intramolecular and intermolecular character, since F^- is thought to produce a special structuring in water.[23] In the Gauss-Lorentz treatment of Murphy and Bernstein, however, component 1 is included in the Lorentz tail of component 2. But, the Lorentz tails of the Murphy and Bernstein components are so extended, that it is difficult to understand, what, if any, physical meaning can be attached to the overlapping regions.

Finally, it should be mentioned that the present view is not that Fermi resonance is totally absent, but rather that other interactions give rise to the principal components that can be obtained from the present computer methods.

B. Stimulated Raman Data.

To shorten the discussion of the stimulated Raman data, the reader is referred to Refs. (4,7,8 and 16). Further, in the ensuing discussion only the positions of the stimulated intensity maxima shown in Figs. 5-7 and listed in Table III are considered significant. The spectra of Figs. 5-7 show considerable breadth, which in some cases is enhanced, as indicated. The breadth of stimulated Raman spectra can be greatly increased by frequency-sweeping in the laser and/or in the aqueous solutions when self-focussing occurs.[17] Some, if not a major part, of the presently observed breadth is thought to arise from self-focussing effects.

Briefly described, the stimulated Raman scattering process involves a two-photon interaction between a laser quantum $h\nu_L$ and Stokes quanta $nh\nu_S$ in which the material system is excited by an amount $h(\nu_L-\nu_S)$, and $(n+1)h\nu_S$ Stokes quanta are emitted. The stimulated intensity resulting when the laser beam travels a distance ℓ in a sample, $I_S(\ell)$, is related to the spontaneous intensity at $\ell = 0$, $I_S(0)$, as follows:

$$I_S(\ell) = I_S(0)e^{+g\,\ell\, I_o}, \qquad (1)$$

where I_o is the incident laser power density, and g is directly proportional to the total spontaneous differential scattering cross-section and inversely proportional to the spontaneous half-width. The total cross-section is related to the integrated component intensities, and the ratio of these integrated intensities to the component half-widths, yields a measure of the peak-heights. The true peak-heights, of course, are proportional to g.

Eqn. (1), however, refers to the case where the laser line half-width is very small compared to those of the spontaneous Raman lines. In other cases where the spontaneous Raman and laser line half-widths are equal, stimulation will be relatively efficient, compared to spontaneous Raman half-widths that are much larger than that of the laser line.

In previous work,[7,8] mutually exclusive stimulation of nonhydrogen-bonded and hydrogen-bonded OH stretching components from H_2O-D_2O mixtures was observed. The results shown in Figs. 5-7 constitute an extension of that work. From Fig. 5 intensity maxima are evident in the hydrogen-bonded OH stretching region from 3435-3480 cm^{-1} for concentrations to 50 vol. % H_2O, and in the nonhydrogen-bonded region, 3515-3570 cm^{-1}, for lower, 37.5 vol. % H_2O concentrations. (Stimulation in the nonhydrogen-bonded region at 3530-3550 cm^{-1} was also observed for a 30 vol. % H_2O mixture.) In the corresponding spontaneous Raman spectra, however, no intensity maxima are observed in the nonhydrogen-bonded region. For example, from 1 to 50 mole % H_2O, the spontaneous intensity maxima occur from 3435 cm^{-1} [26] to 3415 cm^{-1}.[4] Accordingly, the abrupt increase in the position of the stimulated intensity maximum from 50-37.5 vol. % H_2O, is in accord with the mutually exclusive OH stimulation observed previously at somewhat higher H_2O concentrations. (The cross-over point may depend upon power density.) And, the spectra of Fig. 5 constitute a violation of simple stimulated Raman scattering, Eqn. (1), in that the stimulated maxima should occur at positions corresponding to the maximum spontaneous peak heights.

Mutually exclusive stimulation of nonhydrogen-bonded and hydrogen-bonded OD components was not observed previously at the D_2O concentrations examined. Spectra of Fig. 6, however, indicate a fairly abrupt change in stimulation, analogous to Fig. 5. For pure D_2O and 62.5 vol. % D_2O, stimulated intensity maxima occur in the hydrogen-bonded OD stretching region near 2495 cm^{-1}, but the maximum shifts to 2565-2575 cm^{-1} when the concentration is lowered to 37.5 vol % D_2O. Again, the spontaneous spectra from 1 to 50 mole % D_2O show maxima near 2520 cm^{-1} and 2495 cm^{-1}.[4] Thus, the spectra of Fig. 5 constitute a second violation of Eqn. 1. More violations occur in results shown in Fig. 7 and Table III.

In regard to mixtures of H_2O and D_2O, it should be mentioned that an equimolar solution contains HDO, H_2O and D_2O in the approximate molar ratio of 2:1:1. However, the number of OD oscillators due to the presence of the HDO, is the same as the number from one-half that concentration of D_2O, and vice versa, for OH oscillators. Hence, it may not be very important to determine whether, for example, the OH stimulation arises from HDO or H_2O, particularly since the HDO components in the OH (and OD) regions occur at frequencies very close to those of Table II.

In Fig. 7 stimulated spectra from various electrolytes in 10 mole % D_2O in H_2O are presented. The intensity maxima occur in every case in the nonhydrogen-bonded OH stretching region, from 3545 cm^{-1} for $NaBF_4$, to a high value of 3640 cm^{-1} for $NaPF_6$. Spontaneous Raman data from the corresponding solutions are listed in Table III. Careful examinations of the original spontaneous Raman spectra,[22,23] suggest that the positions of the stimulated intensity maxima fail to correspond to the highest spontaneous peak heights in all cases. Let us examine this failure in more detail for Gaussian component data, Table IV.

Stimulated Frequency	Spontaneous Gaussian Component Peak Heights
3625-3640 cm^{-1} ($2.6\ M\ NaPF_6$)	<0.17 at 3640 cm^{-1} >0.25 at 3425 cm^{-1} ($5.0\ M\ NaPF_6$)
3570-3580 cm^{-1} ($3.2\ M\ NaSbF_6$)	0.02 at 3560 cm^{-1} 0.21 at 3425 cm^{-1} ($3.5\ M\ NaSbF_6$)
3590-3605 cm^{-1} ($3.9\ M\ NaClO_4$)	<0.09 at 3600 >0.23 at 3438 ($4.8\ M\ NaClO_4$)
3545-3560 cm^{-1} ($5.0\ M\ NaBF_4$)	0.04 at 3550 0.21 at 3435 ($5.0\ M\ NaBF_4$)

Table IV. Comparisons of stimulated Raman peak frequencies, with spontaneous Gaussian component peak heights and positions.

A quantity proportional to the spontaneous Gaussian component peak heights, namely, the ratios of the integrated Gaussian component percentages to the half-widths, are given in Table IV. There, the spontaneous components corresponding in frequency to the stimulated peaks, as well as the spontaneous components having the largest peak heights, are listed. In every case, stimulation occurs at a position that does not correspond to the maximum spontaneous component peak height. On the contrary, stimulation occurs at a position corresponding to one, or the other, of two narrow nonhydrogen-bonded components whose half-widths range from 60-80 cm^{-1}. Thus the simple theory is violated, but the observation that the spontaneous component half-widths are small, compared to the 200-225 cm^{-1} half-widths for the 3425-3438 cm^{-1} component suggests another relationship.

The narrowest pump pulses inferred in this work from stimulation of the 990 cm^{-1} vibration from benzene have half-widths of about 20 cm^{-1}. But some benzene spectra show widths 3-5 times larger. Thus, pump pulses of 20-60 cm^{-1} or perhaps even larger may have been employed. The spontaneous nonhydrogen-bonded OH components from the ternary electrolyte solutions range from 60-80 cm^{-1},[23] and from the binary solutions, about 110-125 cm^{-1} for OD or OH, Table II. Thus the spontaneous nonhydrogen-bonded widths are close to the pump pulse widths. The spontaneous hydrogen bonded half-widths, however, are near 110-118 cm^{-1} for OD, and 228-230 cm^{-1} for OH, component 3 of Table II, or about 200-225 cm^{-1} from Ref. (23). The nonhydrogen-bonded spontaneous OH components might therefore be expected to stimulate more readily than the hydrogen-bonded components. However, the half-widths of the major D_2O components, 2-5 are all roughly near 100 cm^{-1}, and they constitute a problem in terms of the observed mutually exclusive stimulation.

Even more striking results are evident from Table III, when data from the electrolytes in 10 mole % D_2O in H_2O are compared with results from the same or similar concentrations of electrolyte in pure water. The stimulated intensity maxima for 4 M $NaPF_6$, $NaClO_4$, and $NaBF_4$ are all found to occur in the hydrogen-bonded region from about 3475-3430 cm^{-1}. Thus the removal of only 10 mole % D_2O shifts the stimulation down to the hydrogen-bonded region, a shift as large as 175 cm^{-1}, in one case. Further, the stimulated and spontaneous intensity maxima correspond closely. A small amount of D_2O produces a large change in the stimulated frequency, under conditions where no corresponding changes are observed in spontaneous scattering. Simple stimulated theory is strongly violated once more.

When violations of simple stimulated Raman scattering theory occur, it should be noted that mixtures containing H_2O and D_2O are always involved (whether binary or ternary). Specifically even small amounts of D_2O can cause unexpected nonhydrogen-bonded OH stimulation. From this observation, and from the general approach used in Colles' chapter, an explanation for some, but not all, of the stimulated phenomena results.

In Colles' parametric coupling theory of mutually exclusive stimulation, three vibrations are stimulated simultaneously. These are here designated ω_{1e}, ω_{2a}, and ω_{3a}, where $\omega_{3a} = \omega_{2a} + \omega_{1e}$. Vibration ω_{3a} refers to the intramolecular spontaneous component 3 of Table II, and ω_{2a} refers to component 2. Vibration ω_{1e} refers to the intermolecular O-O stretching motion of O-H····O or O-D····O units that occurs near 170 cm^{-1}.[25] From Table II the equality $\omega_{3a} = \omega_{2a} + \omega_{1e}$ is almost perfectly obeyed for H_2O, and it may also be obeyed for parts of the D_2O bands because of large widths, but not very well for the band centers.

In previous work,[7] a stimulated Raman intensity maximum from H_2O near component 3, and a stimulated shoulder near component 2, were observed. In addition, stimulated intensity maxima at or near components 3 and 2 have been reported, with a shoulder at lower frequencies.[6] Such detailed features are not evident in Fig. 5 because they were probably obscured by frequency sweeping. Nevertheless, there is no serious doubt that components 3 and 2 stimulate simultaneously for pure H_2O.

When D_2O is added to H_2O, O-H····O interactions are certainly replaced by O-D····O interactions at all concentrations. However, the O-O stretching band observed in spontaneous scattering is not appreciably changed in frequency, intensity, or half-width,[27] but this is expected because the O-O stretching is a restricted translation of H_2O, D_2O, or HDO molecules whose masses are similar. (The individual H_2O,

D_2O, and HDO contributions are unresolved.) Nevertheless, there is evidence that the intensity of the spontaneous component 2 decreases with isotopic dilution.[19] Hence, from the previous assignment of component 2, it would appear that D_2O addition decreases the concentration of the O-H····O units, as expected. Also for the OD region from the D_2O present because of incompleteness in the reaction with H_2O or for the OD region from HDO,[18] the condition $\omega_{3a} = \omega_{2a} + \omega_{1e}$ is not met so well as for OH region of H_2O and HDO.[18] The parametric coupling leading to the simultaneous stimulation of ω_{1e}, ω_{2a}, and ω_{3a} is consequently thought to be impaired.

Of course, when simultaneous stimulation of ω_{1e}, ω_{2a} and ω_{3a} does not occur, stimulation of ω_{3a} alone might still be expected. But, the nonhydrogen-bonded half-widths are about 2 times smaller than the hydrogen-bonded half-widths for H_2O in H_2O-D_2O mixtures, and as much as 3 times smaller for $NaPF_6$ in the H_2O-D_2O solvent. Accordingly, stimulation of the narrow nonhydrogen-bonded components results when the parametric coupling is weakened.

As a specific example, let us consider the spontaneous Raman spectrum of 4 M $NaClO_4$ in 10 mole % D_2O in H_2O, see Fig. 2 of Ref. (22). Two spontaneous OH intensity maxima, one hydrogen-bonded, the other nonhydrogen-bonded, are present in the Raman spectrum with nearly equal apparent peak heights, but with dissimilar component peak heights, Table IV. Further, work with binary solutions of $NaClO_4$ in pure H_2O or in D_2O indicates no significant differences compared to the ternary spectra. Stimulation from 4 M $NaClO_4$ in H_2O occurs in the hydrogen-bonded region, at the position of the low-frequency intensity maximum. But when 10 mole % D_2O is added, stimulation occurs unexpectedly in the nonhydrogen-bonded region, at a frequency about 130 cm^{-1} higher.

In regard to $NaPF_6$, $NaSbF_6$, $NaClO_4$ and $NaBF_4$ it should be stated that these four electrolytes were chosen because they constitute the four strongest structure-breakers, that is, the strongest breakers of O-H····O or O-D····O bonds, known.[23] Thus, the parametric coupling interactions are weakened by the electrolytes, as well as by the D_2O addition. And this weakening may explain the observation that much less D_2O is required to change the region of stimulation for the ternary electrolyte solutions, than for the binary H_2O-D_2O solutions.

In summary, some principal observations and conclusions from the stimulated Raman work are as follows:

1. In aqueous mixtures, stimulation of two intramolecular OH stretching vibrations, ω_{3a} and ω_{2a}, and one intermolecular O-O stretching vibration of O-H····O units, ω_{1e}, is thought to occur simultaneously such that $\omega_{3a} = \omega_{2a} + \omega_{1e}$. In this parametric coupling process, the simultaneous stimulation of hydrogen-bonded spontaneous components is favored, over nonhydrogen-bonded components not involved in parametric coupling.

2. When sufficient D_2O is added the parametric coupling becomes less favorable because O-H····O bonds are replaced by O-D····O bonds. When structure-breaking electrolytes such as $NaPF_6$ are also added, O-H····O as well as O-D····O bonds are broken, and the parametric coupling is further weakened.

3. When sufficient weakening of the parametric coupling occurs, stimulation of narrow spontaneous OH stretching components results. The narrow components correspond to the nonhydrogen-bonded components. Addition of D_2O, or D_2O and structure-breaking electrolytes, shifts the stimulated frequency upward into the nonhydrogen-bonded region.

4. A mixed solvent, i.e., one containing H_2O and D_2O seems to be required to produce violations of ordinary stimulated Raman scattering theory. Structure-breaking electrolytes alone in unmixed solvents do not seem to yield unexpected results.

5. The stimulation of hydrogen-bonded and nonhydrogen-bonded OH and OD stretching components always appears to be mutually exclusive, regardless of whether simple stimulated Raman scattering theory is obeyed or not.

C. Mixture versus Continuum Model Controversy.

In the original version of the continuum model as proposed by Wall and Hornig,[10] water was considered to be engaged in a range of intermolecular interactions, which in terms of Raman spectra in the OD and OH regions of dilute HDO would yield symmetric, or very nearly symmetric, stretching contours. The Wall and Hornig picture was thought to be supported by their data, but the data were subsequently contested,[9] and there is now no serious doubt that the Raman contours are highly asymmetric and contain structure.[11-13] Recently, however, some spontaneous Raman studies have been reported in which attempts to revive,[13] or at least to modify,[12] the continuum ideas have been made.

The spontaneous Raman spectra from H_2O, D_2O, and HDO, of course, contain overlapping contributions from various sources. And, because of this it is not unexpected that controversy should arise over methods of contour decomposition,[9,11] or whether any decomposition is meaningful.[13] This controversy would almost certainly have been resolved in favor of contour decomposition, however, if attention has been paid to the remarkably well-separated hydrogen-bonded and nonhydrogen-bonded regions observed in the infrared overtone spectra by Luck.[28-30] But, the infrared overtone spectra and the stimulated Raman results, as well, were not emphasized in recent spontaneous Raman studies.[12,13]

From the present and previous stimulated Raman work, no exceptions to the observation of mutually exclusive stimulation of hydrogen-bonded and nonhydrogen-bonded OH and OD components has been found. Thus this fact, regardless of whether a detailed theory for it is presently available or not, is unequivocally in contradiction to the ordinary continuum model of water structure.[10] Great breadth of the spontaneous Raman components is certainly present, and this breadth must reflect a range of interactions in a given class, i.e., the presence of extensional and angular distortions, but there can now be no doubt that the spontaneous Raman contours contain component sub-structure, and that the spontaneous component positions correspond to stimulated Raman intensity maxima.[4,7,8] Accordingly, it would appear that a mixture model invoking broad classes of interactions referring to broken and unbroken hydrogen-bonds constitutes a beginning, from which future progress can be made consistent with the experimental Raman spectra, whereas continuum models seem to be ruled-out by the present stimulated Raman results.

REFERENCES

1. A. K. Covington and P. Jones, "Hydrogen Bonded Solvent Systems," (Taylor and Francis, London, 1968).
2. D. Eisenberg and W. Kauzmann, "The Structure and Properties of Water," (Oxford U.P., London, 1969).
3. R. A. Horne, "Water and Aqueous Solutions" (Wiley, New York, 1972).
4. F. Franks, "Water a Comprehensive Treatise," (Plenum, New York, 1972).
5. V. Parkash, M. K. Dheer, and T. S. Jaseja, Phys. Letters 29A, 220 (1969).
6. O. Rahn, M. Maier, and W. Kaiser, Opt. Commun. 1, 109 (1969).
7. M. J. Colles, G. E. Walrafen, and K. Wecht, Chem. Phys. Letters 4, 621 (1970).
8. G. E. Walrafen, Adv. Mol. Relax. Proc., 3, 43 (1972).
9. G. E. Walrafen, J. Chem. Phys. 48, 244 (1968).
10. T. T. Wall and D. F. Hornig, J. Chem. Phys. 43, 2079 (1965).
11. W. F. Murphy and H. J. Bernstein, J. Phys. Chem. 76, 1147 (1972).
12. K. Cunningham and P. A. Lyons, J. Chem. Phys. 59, 2132 (1973).
13. W. C. Mundy, L. Gutierrez, and F. H. Spedding, J. Chem. Phys. 59, 2173 (1973).
14. W. F. Murphy, Private communication.
15. G. Herzberg, "Infrared and Raman Spectra of Polyatomic Molecules," (Van Nostrand, New York, 1945), pp. 243-245.
16. N. Bloembergen, Am. J. Phys. 35, 989 (1967).
17. S. L. Shapiro, J. A. Giordmaine, and K. Wecht, Phys. Rev. Letters 19, 1093 (1967).
18. G. E. Walrafen and L. A. Blatz, J. Chem. Phys. 56, 4216 (1972).
19. G. E. Walrafen, J. Chem. Phys. 50, 567 (1969).
20. G. E. Walrafen, J. Solution, Chem. 2, 159 (1973).
21. G. E. Walrafen, J. Chem. Phys. 47, 114 (1967).
22. G. E. Walrafen, J. Chem. Phys. 52, 4176 (1970).

23. G. E. Walrafen, J. Chem. Phys. 55, 768 (1971).

24. The assignment of the OH and OD components at 3545 cm^{-1} and at 2610 cm^{-1}, Ref. (11), is questionable because hydrogen-bonded components are thought to occur at frequencies below the isosbestic frequencies of about 3460 cm^{-1} and 2570 cm^{-1}, resp., Refs. (22 and 9).

25. G. E. Walrafen, J. Chem. Phys. 44, 1546 (1966).

26. G. E. Walrafen, J. Chem. Phys. 50, 560 (1969).

27. G. E. Walrafen, unpublished work.

28. W. A. P. Luck, Ber. Bunsenges. Phys. Chem. 69, 626 (1965).

29. W. A. P. Luck and W. Ditter, Z. Naturforsch. 24b, 482 (1969)

30. W. A. P. Luck and W. Ditter, J. Phys. Chem. 74, 3687 (1970).

1A) G. Bolla, Nuovo Cimento 9, 920 (1932); 10, 101 (1933); 12, 243 (1935).

2A) L. A. Blatz and P. Waldstein, J. Phys. Chem. 72, 2614 (1968).

3A) L. A. Blatz, National Meeting, Society for Applied Spectroscopy, New Orleans, 1971.

4A) G. E. Walrafen, J. Chem. Phys. 44, 1546 (1966).

5A) Ref. 4, in preceding text, chapter by G. E. Walrafen in Vol. 1.

APPENDIX I

The Raman Band from Water at 60 cm^{-1}.

Considerable discussion at the Marburg conference was concerned with the presence of vibrations in water near 60 cm^{-1}. Reference was made to the large number of Raman observations in which a feature near $\Delta\bar{\nu} \approx$ 60 cm^{-1} was reported. For example, it was pointed-out that the 60 cm^{-1} feature was reported in the early 1930's in the excellent photographic Raman work of Bolla;(1A) then repeatedly verified by many other workers; and finally, established unequivocally in the excellent photoelectric Raman work of Blatz.(2A,3A) The 60 cm^{-1} Raman feature may be seen in Fig. 1A, in which low-frequency Raman spectra from the early work of Walrafen(4A) are shown for -5°C and 67°C. The 60 cm^{-1} feature is most clearly evident when the Raman intensitites are transferred to a horizontal baseline, lower curves in the two parts of the figure. (This feature is much clearer, however, in Blatz's spectra.(3A))

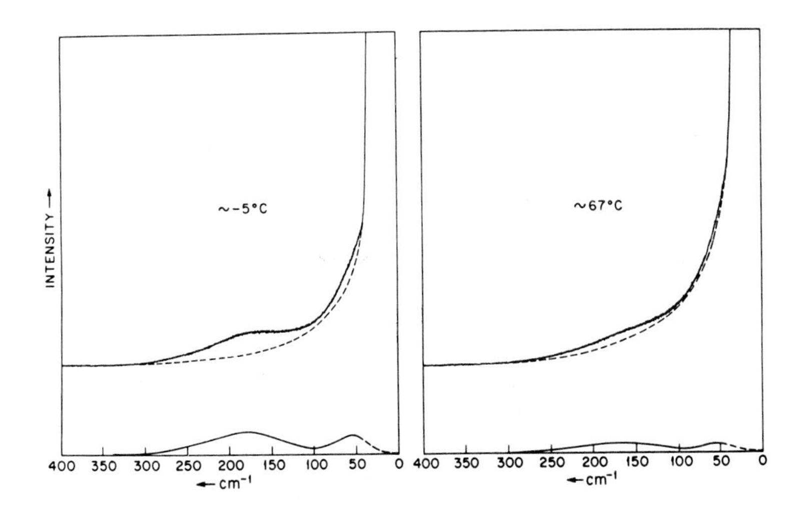

Fig. 1A. Photoelectric Raman tracings from water at ~-5 and ~67°C obtained in the low-frequency region, $\Delta\bar{\nu}$ < 400 cm^{-1}, with unpolarized 4358 Å mercury excitation. The baselines are indicated by curved dashed lines in the upper sections. The lower sections refer to the spectra transferred to horizontal baselines, i.e., to the differences between the photoelectric tracings and the curved baselines. The Raman intensities are quantitatively comparable at the two temperatures. The spectra were obtained with a Cary model-81 Raman spectrophotometer, using a slit width of 10 cm^{-1} and a time constant of ~0.1 sec.

The presence of a feature in the infrared spectra from water near 60 cm^{-1} was also debated at the conference, primarily between Williams and Walrafen. Specifically, the argument dealt with whether the 60 cm^{-1} infrared component was totally absent, Williams, or very weak and unresolved, Walrafen.

The feature near 170 cm^{-1} in Fig. 1A, is almost certainly related to O-O stretching of O-H····O units in water.(5A) The Raman feature near 60 cm^{-1} probably arises from bending of O-H....O units.(5A) Further, according to a 5-molecule treatment employing C_{2v} intermolecular symmetry,(5A) Raman and infrared activity would be expected at 170 cm^{-1}, as well as at 60 cm^{-1}. Accordingly, there is little reason to conclude that the 60 cm^{-1} absorption is forbidden on the basis of selection rules alone.

The infrared absorption at 60 cm^{-1} corresponding to the $\Delta\bar{\nu}$ = 60 cm^{-1} Raman feature, of course, could be very weak, but still allowed. From the Raman observations, the corresponding 60 cm^{-1} absorption would also be expected to be very broad.

A weak broad feature in the infrared spectrum near 60 cm^{-1} has been inferred from computer analysis,(5A) and this feature refers to asymmetry of the infrared component near 170 cm^{-1}, rather than to an absorption maximum of 60 cm^{-1}.

Accordingly, it would appear that there is some evidence for weak infrared activity at 60 cm^{-1} corresponding to the vibrations that give rise to the 60 cm^{-1} Raman band. The opposite view of Williams, namely, that no infrared contribution exists, seems somewhat more difficult to prove. Williams' view would be more convincing if the absorption coefficient were zero near 60 cm^{-1}, or if a pronounced minimum in the absorption coefficient were observed. However, these conditions are not met, at least in some reported infrared data.(5A)

A resolution of this infrared problem must probably await new data, beyond the several sets of data referred to by Williams at Marburg. At any rate, there now seems to be general agreement concerning the reality of the Raman feature at 60 cm^{-1} from water.

APPENDIX II

Simple Description of Raman Depolarization Ratios

In the preceding text, Raman depolarization ratios were used in conjunction with water, and the experimental procedures employed to obtain accurate depolarization ratios were described. However, a short, very elementary description seems to be in order.

Consider a vertical entrance slit in a Raman spectrometer, and consider that a vertical laser beam passes through, for example, a sample of CCl_4 in front of the slit. When the electric vector of the laser beam is perpendicular to the slit, the intensity of the strong line from CCl_4 at about $\Delta\bar{\nu} = 459\ cm^{-1}$ will be large. However, if the electric vector is rotated by 90 degrees so that it is parallel to the slit, that is, it lies in the plane formed by the slit and the beam, the intensity at 459 cm^{-1} will be very low.

From the observations described for the 459 cm^{-1} line of CCl_4, it can be stated that the line is polarized. This information, in turn, means that the molecular motion (of whatever type) is totally symmetric, <u>i.e.</u>, the atomic displacements are such that they correspond completely to the hypothetical static symmetry of the CCl_4 molecule.

If the ratio of the two intensities had been 6/7, on the other hand (or 3/4 as described for the more accurate method in the preceeding text), the line would be designated as depolarized. Depolarized Raman lines refer to displacements that do not correspond to all elements of the hypothetical static symmetry.

The depolarization ratios corresponding to Raman lines from liquids provide important and detailed information about the molecular vibrations. In many cases such information can provide useful structural information about the Raman active molecules as well, that is not obtainable from other methods, e.g., infrared absorption.

IV.2. STIMULATED RAMAN SCATTERING IN LIQUID WATER

by

M. J. Colles

Department of Physics, Heriot-Watt University,

Edinburgh.

ABSTRACT

A new stimulated process is proposed to explain the anomalous results obtained in stimulated scattering experiments in liquid H_2O and D_2O. The process envisaged is the parametric interaction of a strongly driven coherent vibrational wave at ω_i with two others at ω_j and ω_k where $\omega_i = \omega_j + \omega_k$. These waves are coupled through the mechanical anharmonicity of the hydrogen bonds which connect adjacent water molecules. It is shown that this driven interaction can explain the appearance of peaks in the scattered spectrum at shifts corresponding to the intra-molecular O-H stretching modes. In addition the interaction will lead to the appearance of stimulated librational scattering. The peaks in the spectra and librational scattering are not predicted on the basis of simple stimulated scattering theory. The theory supports a model of liquid water in which a division between hydrogen bonded and non hydrogen bonded molecules can be made. It is not consistent with a totally continuous distribution of intermolecular bonding.

IV.2.1. INTRODUCTION

Stimulated Raman scattering from liquid H_2O and D_2O has recently been reported by several groups. /1,2,3/ In all of the published spectra the principle Stokes scattering was generated with frequency shifts corresponding to the O-H stretching modes ($3500cm^{-1}$ for H_2O and $2500cm^{-1}$ for D_2O) . Our purpose here is to emphasize two anomalous results these spectra indicated and to suggest, as a possible explanation, a new form of driven coherent vibrational interaction. The results in question are (1) the appearance of two well defined stimulated peaks within the broad O-H stretch scattering contour and (2) the attainment of stimulated librational scattering. In what follows these two results will be discussed separately.

IV.4.2. STIMULATED SCATTERING FROM THE O-H STRETCHING MODES

It is well known that spontaneous Raman scattering from H_2O gives rise to a broad contour /4/. Although features such as points of inflexion are clearly present, separate components within the contour are not visually resolved. From the theory of the stimulated Raman effect /5/ the exponential gain factor at a frequency ω_{si} will be directly proportional to the cross section at ω_{vi} where $\omega_{vi} = \omega_L - \omega_{si}$ and ω_L is the incident laser frequency. This is true provided the laser linewidth is significantly less than the linewidth of the spontaneous Raman line and saturation is not occurring. Thus the gain profile will follow the exact shape of the scattering contour and should give rise to a stimulated output spectrum with a frequency profile an exponentially narrowed reproduction of this contour. Clearly this would result in a single peak rather than the two peaks obtained experimentally. /1,3/

Although the spontaneous scattering contour cannot be visually resolved, analog computer decomposition indicates the presence of four Gaussian components. These components correspond to the two classes or species of water making up the liquid at room temperatures, namely, hydrogen bonded (HB) and nonhydrogen bonded water (NHB). Each class will give rise to two basic normal modes. For a comprehensive discussion of this interpretation, see Walrafen /6/. A similar contour and decomposition into components can be obtained for D_2O.

Colles, Walrafen and Wecht /3/ (CWW) interpreted the peaks they obtained in the spectra of pure H_2O, D_2O, mixtures and solutions on the basis of these four components. Temperature changes and some solvents influence the relative contributions of the four components /7/. This assignment was considered reasonable in view of the close agreement in frequency shift between the stimulated peaks and the frequencies of the four components. In pure H_2O one of the HB components stimulated and in D_2O both appeared with varying relative intensity. Rahn, Maier and Kaiser (RMK) /2/ obtained two peaks at the two HB frequencies in H_2O. This assignment of the peaks, however, in no way explains their presence since the stimulated Raman gain is proportional to the total cross section and is insensitive as to how this total cross section is made up.

In order to explain this behaviour we wish to propose the following simple model of mechanical coupling between three hydrogen bonded H_2O molecules. This is part of the tetrahedrally coordinated five molecule unit introduced by Walrafen and shown in figure 1. Molecule 1 is assumed to be vibrating at $\sim 3250 cm^{-1}$ and molecules 2 and 3, both at $3425 cm^{-1}$. The atomic motions corresponding to these modes are taken, as a first approximation, as the motions of the atoms in the symmetric and asymmetric stretching modes of the isolated molecule. The anharmonicity of the hydrogen bond (this weak bond is highly anharmonic) will give rise to a coupling between these two intramolecular modes and a third intermolecular translational mode. This mode corresponds to the lattice type vibration of molecule 1 against molecules 2 and 3, has a frequency $\approx 170 cm^{-1}$, and has been identified in spontaneous Raman /6/ and infra-red absorption studies /8/.

For simplicity, consider the single anharmonic hydrogen bond between molecules 1 and 2. The cubic anharmonicity will provide a term in the potential energy $\sigma(y_2-y_1)^3$ where y_2 and y_1 are the vibrational displacements of the hydrogen atom of molecule 2 and the oxygen atom of molecule 1, respectively and σ is the anharmonic coefficient. In terms of the normal vibrational coordinates y_2 and y_1 may be written as:

$$y_2 = R_i + \tfrac{1}{2}\beta R_k \quad (1a)$$

$$y_1 = \beta\left[\frac{2m_h}{2m_h+m_O} R_j - \tfrac{1}{2} R_k\right] \quad (1b)$$

where the suffixes i, j, k refer respectively to the two intramolecular modes at $3425 cm^{-1}$ and $3250 cm^{-1}$ and the translational mode at $170 cm^{-1}$. β is a factor depending on the relative orientation of the two molecules. Introducing the normalized coordinates $Q = \rho^{1/2}R$ with ρ the reduced mass density, the resonant term in the potential energy will be:

$$- 6\sigma\,\beta^2 \frac{2m_h}{2m_h+m_O} \frac{1}{(\rho_i\rho_j\rho_k)^{1/2}} Q_iQ_jQ_k \quad (2)$$

The molecule attached to the second hydrogen bond will provide an equal energy of interaction when the vibrations Q_{i2} and Q_{i3} are in phase as will be the case for a coherently driven wave. The potential energy will be expressed as $\xi Q_iQ_jQ_k$ with ξ twice the multiplier of $Q_iQ_jQ_k$ in Eq.(2).

The equations of motion for the three vibrational waves and the Stokes fields may be obtained from the Lagrangian density and Maxwell's equations in the form: /9/

$$\ddot{Q}_i + \omega_{io}^2 Q_i + \Gamma_i\dot{Q}_i = N(\partial\alpha/\partial Q)E_LE_1 + N\xi Q_jQ_k \quad (3)$$

$$\nabla \times (\nabla \times E_1) - \frac{\omega_1^2 n_1^2}{c^2} E_1 = \frac{4\pi\omega_1^2}{c^2} N(\partial\alpha/\partial Q)E_LQ_i \quad (4)$$

with similar equations for Q_j, Q_k and where E_1 and E_2 refer to the Stokes fields at

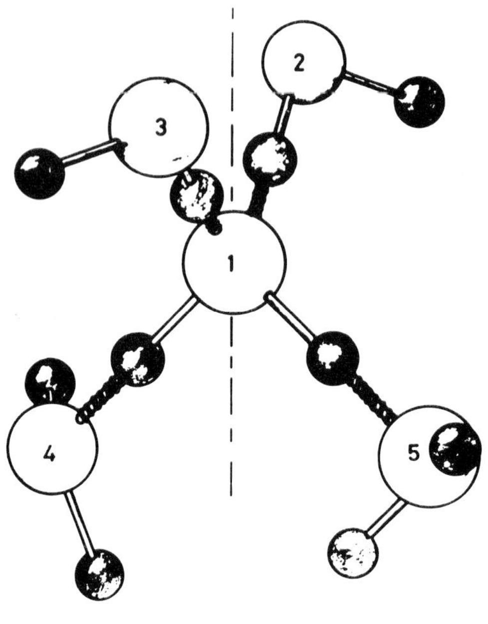

FIGURE 1

Tetrahedrally coordinated 5 molecule unit. The hydrogen bonds are indicated as spirals.

shifts corresponding to Q_i and Q_j respectively. In Eq.(3) the damping term has been added phenomenologically, ω_{io} is the resonant frequency of the ith mode, N is the number of molecules per cc and α is their polarisability.

Take the fields and waves in the form: *

$$E = \tfrac{1}{2}\, U(z)e^{-i(\omega t-kz)} + cc \tag{5}$$

$$Q = \tfrac{1}{2}\, q(z)e^{\,i(\omega t-kz)} + cc$$

making the usual assumptions of slowly varying amplitudes, highly damped vibrational waves and phase-matched processes these equations (3) and (4) lead directly to the equations for the growth of the field and wave amplitudes:

$$\frac{\partial U_1}{\partial z} = i\cdot \frac{2\pi\omega_1^2}{c^2k_1} N(\partial\alpha/\partial Q)_1 \; U_L q_i^* \tag{6}$$

$$\frac{\partial U_2}{\partial z} = i \frac{2\pi\omega_2^2}{c^2k_2} N(\partial\alpha/\partial Q)_2 \; U_L q_j^* \tag{7}$$

$$q_i = \frac{N(\partial\alpha/\partial Q)_1}{2\omega_i\Gamma_i A_i} U_L U_1^* + \frac{N\xi}{2\omega_i\Gamma_i A_i} q_j q_k \tag{8}$$

$$q_j = \frac{N(\partial\alpha/\partial Q)_2}{2\omega_j\Gamma_j A_j} U_L U_2^* + \frac{N\xi}{2\omega_j\Gamma_j A_j} q_i q_k^* \tag{9}$$

$$q_k = \frac{N\xi}{2\omega_k\Gamma_k A_k} q_i q_j^* \tag{10}$$

with $A_i = 2\left[\frac{\omega_i - \omega_{io}}{\Gamma_i} - i\right]$ and similar definitions for A_j, A_k.

In Eq.(10) the electromagnetic contribution to the driving term has been assumed negligible compared with the vibrational contribution. Equations (8), (9) and (10) may be solved simply to give:

$$q_i\left[1 - \frac{N^2\xi^2}{4\omega_i\omega_k\Gamma_i\Gamma_k}\frac{1}{A_i A_k^*} q_j q_j^*\right] = \frac{N(\partial\alpha/\partial Q)_1}{2\omega_i\Gamma_i A_i} U_L U_1^* \tag{11}$$

$$q_j\left[1 - \frac{N^2\xi^2}{4\omega_j\omega_k\Gamma_j\Gamma_k}\frac{1}{A_j A_k} q_i q_i^*\right] = \frac{N(\partial\alpha/\partial Q)_2}{2\omega_j\Gamma_j A_j} U_L U_2^* \tag{12}$$

The growth equations for the fields become:

$$\frac{1}{U_1}\frac{\partial U_1}{\partial z} = \frac{i\pi\omega_1^2 N^2(\partial\alpha/\partial Q)_1^2 U_L U_L^*}{c^2 k_1\omega_i\Gamma_i A_i^*}\left[\frac{1}{1 - \frac{N^2\xi^2 q_j q_j^*}{4\omega_i\Gamma_i\omega_k\Gamma_k A_i A_k^*}}\right] \tag{13}$$

$$\frac{1}{U_2}\frac{\partial U_2}{\partial z} = \frac{i\pi\omega_2^2 N^2(\partial\alpha/\partial Q)_2^2 U_L U_L^*}{c^2 k_2\omega_j\Gamma_j A_j^*}\left[\frac{1}{1 - \frac{N^2\xi^2 q_i q_i^*}{4\omega_j\Gamma_j\omega_k\,{}_k A_j A_k^*}}\right] \tag{14}$$

These equations are in the form:

$$\frac{1}{U_1}\frac{\partial U_1}{\partial z} = \gamma_1\left[\frac{1}{1-F}\right] ; \frac{1}{U_2}\frac{\partial U_2}{\partial z} = \gamma_2\left[\frac{1}{1-G}\right] \tag{15}$$

where γ_1 and γ_2 are the exponential gain coefficients which would have been calculated in the absence of any vibrational coupling. These equations are not simply solved since F and G are functions of U_2 and U_1 respectively - that is the equations are still coupled. However, in the small gain approximation up to the point where U_2 becomes comparable with U_1 they have the approximate solution $U = U_0 e^{\gamma' z}$ with:

$$\gamma_1' = \frac{\gamma_1}{1-F} ; \quad \gamma_2' = \frac{\gamma_2}{1-G} \tag{16}$$

*) cc : complex conjugate

When the three modes are resonant, as in the case of this model of the liquid water interaction, $A_i = A_j = A_k = -i$ and $F < 0$, $G > 0$. Thus $\gamma_1' < \gamma_1$ and $\gamma_2' > \gamma_2$ and an equalizing of the two gains will take place.

Again in the small gain approximation the parameter G may be written as:

$$G = \frac{N^2\xi^2}{4\omega_j\omega_k\Gamma_j\Gamma_k} \cdot \frac{N^2(\partial\alpha/\partial Q)^2}{4\omega_i^2\Gamma_i^2} \cdot U_L U_L^* U_1 U_1^* \qquad (17)$$

In terms of the gain parameter γ_1, this becomes:

$$G = \frac{N^2\xi^2\gamma_1 I_1}{2\omega_1 \; (\omega_i\omega_j\omega_k) \; (\Gamma_i\Gamma_j\Gamma_k)} \qquad (18)$$

with I_1 the intensity in ergs sec^{-1} cm^{-2}.

IV.2.3. STIMULATED LIBRATIONAL SCATTERING

In addition to the two stimulated scattering peaks in the O-H stretch region in H_2O, RMK also reported the observation of stimulated scattering from an intermolecular librational mode. Due to the large disparity in scattering cross sections between the librational and vibrational modes, the direct stimulation of the librational mode is extremely unlikely. The ratio of the two cross sections /11/ is $\frac{\partial\sigma}{\partial\Omega}_{Lib} \simeq 3 \times 10^{-2} \frac{\partial\sigma}{\partial\Omega}_{Vib}$ and will give the theoretical ratio of the stimulated exponential gain factors. This compares with an observed ratio (RMK) of 0.75. If the librational gain is indeed so small compared to the vibrational gain then one would not expect to see stimulated librational scattering under any circumstances since severe pump depletion due to vibrational scattering would occur first. An additional reason for believing that librational scattering does not occur in a simple fashion is that the H-O-H bending vibration at $\simeq 1650 cm^{-1}$ failed to stimulate despite its higher spontaneous cross section /11/.

The appearance of librational scattering can be explained as the consequence of a driven vibrational interaction similar to the mechanism giving rise to the two peaks in the O-H stretching region. In this case the interaction is between the central molecule and the two other molecules of the tetrahedrally coordinated system shown in figure 1. A parametric process is envisaged in which the two outer molecules, vibrating at $\sim 3425 cm^{-1}$ couple via the anharmonic hydrogen bond to the central molecule which is both librating and vibrating. The frequency of vibration of the central molecule will correspond to the difference frequency between $3425 cm^{-1}$ and the frequency of libration $\simeq 475 cm^{-1}$. This difference frequency wave will cause additional stimulated scattering at a shift of $\simeq 2950 cm^{-1}$. Evidence for this scattering is apparent in the spectrum published by RMK.

Figure 2 illustrates molecules 1 and 4 with their atoms in vibrationally displaced positions. As before it is assumed that the atomic motions correspond to those of the

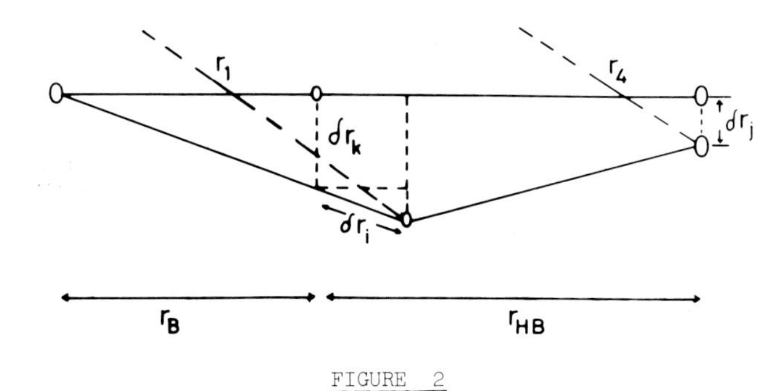

FIGURE 2

Librational and vibrational mode

symmetric and asymmetric stretching modes of the isolated molecule. The contribution to the potential energy from the anharmonic part of the hydrogen bond stretching constant will be given by $\sigma|(\bar{r}_1-\bar{r}_4)|$ where $\bar{r}_1$, $\bar{r}_4$ are the general positional coordinates shown in Fig 2. With the vibrational displacements $\delta r_{i,j,k}$ as shown $(\bar{r}_1-\bar{r}_4)$ can be expressed as:

$$|(\bar{r}_1-\bar{r}_4)|^3 = \left[r_{HB}^2 + \left\{\delta r_k - \delta r_j + \frac{\delta r_i \delta r_k}{r_B}\right\}^2\right]^{3/2} \tag{19}$$

$$\approx 3\frac{r_{HB}}{r_B}\delta r_i\ \delta r_j\ \delta r_k + \text{nonresonant terms in } \delta r^2$$

Replacing the δr with normalized coordinates, Q, the resonant term in the potential energy becomes:

$$(\text{PE})\ \text{resonant term} = 3\sigma\beta' \frac{r_{HB}}{r_B} \times \frac{2m_h}{2m_h+m_o} \times \frac{1}{(\rho_i\rho_j\rho_k)^{1/2}} \times Q_iQ_jQ_k \tag{20}$$

Where, in Eqs. (19) and (20), r_B and r_{HB} are the equilibrium bond and hydrogen bond lengths, and β' is a geometrical factor dependent on the relative orientation of these two molecules. With the coherent addition of a similar contribution from molecule 5 the potential energy may be expressed in the same form as before, namely $\eta\ Q_iQ_jQ_k$ with η twice the multiplier of $Q_iQ_jQ_k$ in Eq. (20). With the potential energy in this form the equations of motion will be given by Eqs. (6)-(10), on reinterpretation of the suffixes 1, 2, 3 as fields scattered from the wave at 3425cm^{-1}(i), the librational wave, (j) and the difference frequency wave (k) respectively and with the addition of a third field equation for the amplitude $_3$. The solutions in the small gain approximation will be given by $U = U_0e^{\gamma' z}$ with $\gamma_2' = \frac{\gamma_2}{1-H^*}$, $\gamma_3' = \frac{\gamma_3}{1-H}$ where

$$H = \frac{N^2\eta^2\gamma_1 I_1}{2\omega_1\ (\omega_i\omega_j\omega_k)\ (\Gamma_i\Gamma_j\Gamma_k)A_jA_k^*} \tag{21}$$

The differences between this interaction and the one considered earlier are solely connected with the parameter $A_jA_k^*$. This parameter did not appear in the expression for the gain enhancement in the O-H stretching region because a resonant interaction had been assumed. Since we are looking for a large enhancement of the librational gain, H must approach a real value of 1. This will only be possible if the parameter $A_jA_k^*$ is real. There are two conditions which will satisfy $A_jA_k^*$ real, namely both parametrically generated waves on resonance or both off resonance by equal amounts relative to their linewidths. In the case of the librational interaction we have to consider the second of these conditions. The specific model envisaged is that the librational mode of the central molecule and its vibrational mode correspond to the libration out of the plane of the molecule ($\simeq 770\text{cm}^{-1}$) and the 3250cm^{-1} symmetric stretching mode both off resonance by $\simeq 300\text{cm}^{-1}$. The parameter $A_jA_k^*$ then becomes $\simeq 10$.

In order to evaluate the coupling parameters G and H a knowledge of the anharmonic coefficient σ is required. This is not known with any accuracy, however, an estimate based on the effect of the anharmonicity on the width Raman lines in hydrogen bonded materials /10/ gives $\sigma = 5.4 \times 10^{12}$ dynes cm^{-2} molecule $^{-1}$. If one assumes that all the molecules have identical resonant frequencies and that the broad linewidths ($\sim 200\text{cm}^{-1}$) of the components are due entirely to lifetime effects then the following parameters can be evaluated. $\rho_i \simeq \rho_j \simeq 0.1$ grams cm^{-3}; $\rho_k \simeq 0.5$ grams cm^{-3}, $N \simeq 3.3 \times 10^{22}\text{cm}^{-3}$ and $\Gamma_i \simeq \Gamma_j \simeq \Gamma_k \simeq 10^{13}\text{sec}^{-1}$. In fact this situation is very unlikely. What is more probable is that the broad linewidth is due, at least in part, to a range of molecular environments. For the purposes of providing an estimate we take the 'natural' linewidth of a single molecular environment as 20cm^{-1} consistent with the linewidth used for the evaluation of the anharmonic coefficient. This leads to $\Gamma_i' \simeq \Gamma_j' \simeq \Gamma_k' \simeq 10^{12}\ \text{sec}^{-1}$ and to modified values of the reduced mass densities and molecular densities, $\rho \simeq \rho/_{10}$, $N \simeq N/_{10}$. These numbers together with $\omega_1 = 2 \times 10^{15}$, $\omega_i = 6 \times 10^{14}$, $\omega_j = 5.4 \times 10^{14}$, $\omega_k = 0.3 \times 10^{14}$ (all in sec^{-1}) lead to a value for the coupling parameter:

$$G \simeq 4.5 \times 10^{-22}\ \gamma_1 I_1$$

With the additional values $r_{HB} = 1.95\text{Å}$, $r_B = 0.95\text{Å}$ and with $\beta' = 0.866$, $\rho_k = 0.1$ grams cm^{-3}, $\omega_j = 5 \times 10^{14}$, $\omega_k = 0.9 \times 10^{14}\ \text{sec}^{-1}$ the coupling coefficient for the librational interaction may be evaluated as $H \simeq 4 \times 10^{-22}\ \gamma_1 I_1$ or $H \simeq G$.

The experimental conditions obtained by CWW provide $\gamma_1 \simeq 20\text{cm}^{-1}$ and $I_1 \simeq 20$ GW cm^{-2} and lead to $G \simeq H \simeq 2 \times 10^{-3}$ or a 0.2% modification in the gain coefficient. It should be noted that, since $G, H \propto \gamma_1 I_1$ they are both proportional to $^1/_{d^4}$ with d the beam diameter and are therefore very sensitive to self-focussing effects. If one assumes a reduction in beam diameter of around 5 due to self-focussing (a quite conservative figure) then $G \simeq H \simeq 1$, $\gamma'_{j,k} \gg \gamma_{j,k}$ and large enhancements of the librational gain can take place. With the experimental conditions of RMK rather more self-focussing would be required but would indeed be expected due to the longer pulse lengths used in their experiments.

IV.2.4. CONCLUSIONS

Based on an extremely simple model of the anharmonic coupling between adjacent water molecules it has been shown that, with the power densities obtained in stimulated scattering experiments, a driven, coherent vibrational interaction can take place. This interaction can give rise to an enhanced stimulated scattering from regions where one would normally predict little or no scattering. This model, although the calculations must be regarded in order of magnitude terms only, is fully capable of explaining the appearance of peaks within the O-H stretching region and the appearance of stimulated librational scattering, two anomalies hitherto unexplained.

It is possible that a related coupling mechanism may affect the behaviour of the NHB components. The coupling in this case would be provided by dipole-dipole interactions of neighbouring molecules. The coherent coupling introduced between the components of either the HB or the NHB class could then account for the exclusive stimulation of either of these classes but not both as was observed experimentally /3/. It should be noted that the model also predicts a zero coupling to the bending mode and is therefore consistent with the experimental observation that stimulated scattering from this region does not occur.

Attempts to explain the stimulated Raman scattering results on the basis of a truly continuous distribution of molecular environments over the whole width of the vibrational scattering contour have proved unsuccessful. It would seem, therefore, that the stimulated scattering data together with the concept of a driven parametric interaction are consistent with a model of water structure based on the tetrahedrally coordinated unit and the broad division between hydrogen bonded molecules and non hydrogen bonded molecules. Within each of these classes a range of local environments is to be expected and will give rise to the large widths of the scattering profiles attributable to each component.

REFERENCES

1 O. Rahn, M. Maier and W. Kaiser; Opt. Commun. 1, 109, (1969).

2 V. Parkash, M.K. Dheer and T.S. Jaseja; Phys. Lett. 29A, 220, (1969)

3 M.J. Colles, G.E. Walrafen and K.W. Wecht; Chem. Phys. Lett. 4, 621, (1970)

4 G.E. Walrafen; J. Chem. Phys. 47, 114, (1967)

5 N. Bloembergen; Amer. J. Phys. 35, 989, (1967)

6 G.E. Walrafen; J. Chem. Phys. 40, 3249, (1964)
J. Chem. Phys. 44, 1546, (1966)

7 G.E. Walrafen; Chapter 5 in 'water, a comprehensive treatise', Vol. 1 pp 151-214 Edited F. Franks, Plenum Press 1972.

8 D.A. Draegert, N.W.B. Stone, B. Curnette and D. Williams; J. Opt. Soc. Amer. 56, 64, (1966)

9 Y.R. Shen and N. Bloembergen; Phys. Rev. 137, A1787 (1965)

10 E.R. Lippincott and R. Schroeder, J. Chem. Phys. 23, 1099, (1955).

11 G.E. Walrafen; Private Communication.

IV.3. RAMAN SPECTRAL STUDIES OF ION-ION AND ION-SOLVENT INTERACTIONS

D.E. Irish
University of Waterloo, Department of Chemistry,
Waterloo, Ontario, Canada N2L 3G1.

ABSTRACT

Raman spectroscopy is a powerful technique for the study of water and aqueous solutions. Raman lines characteristic of ion-pairs, complex ions and solvated ions provide direct evidence for the existence of these species in solution and information concerning their structure and properties. This paper describes data for three cations, Mg^{2+}, Be^{2+} and Zn^{2+}, and one anion, NO_3^-. Comparisons are provided between spectra of aqueous solutions of $Mg(NO_3)_2$ and three crystalline hydrates. Both intermolecular and intramolecular vibrations are discussed. Facts relating to the doublet structure of the $\nu_3(E')$ band of nitrate ion in water are reviewed and a tentative model for the aquated nitrate ion is advanced.

IV.3.1. INTRODUCTION

Much of the effort in our laboratory has been directed toward the study of the ion-pairs or ion-aggregates which form in aqueous, nonaqueous and mixed solvents. These ion-pairs have been detected by their distinctive Raman and infrared spectra. In particular the anions nitrate and nitrite generate spectra when they are bonded to metal cations which are quite different from the spectra of the dissociated solvated anion. It has been possible in many cases to measure the integrated Raman line intensity of free and bound forms of the anions, and to compute stability constants for the processes which occur in those solutions (1,2). In effect the anion provides a probe and alterations in its spectrum provide insight into the events which are taking place and the structure in its immediate surroundings (3). Changes in the half-widths of Raman lines provide information about the lifetime of the species generating the line and both collisional processes (4) and ultra-fast proton transfer processes in acid media (5) have been studied by this technique.

In this paper attention is to be focused on the water in aqueous electrolyte solutions; in particular we seek information about the ion-water interactions. According to the model proposed by Frank and Wen (6) a relatively small specific number of water molecules exist in the primary solvation sheath of a cation. This aquated cation is separated from bulk water by a secondary solvation sphere in which water molecules must compromise between orientations which satisfy the cation and orientations which satisfy bulk water. For small cations of high charge density the molecules in this region may also be strongly oriented by the electric field of the cation. For larger cations, water molecules in this region are considered to be more disordered than in bulk water. Anion solvation is markedly different. Two models are readily imagined. Some electronegative anions will form hydrogen bonds with water molecules. In other cases the anion may occupy a cavity in the water structure; hydrogen bonding then occurs between water molecules giving rise to a clathrate-like cage around the anion. Data from Raman spectral studies are presented and correlated with these models.

IV.3.2. CATION HYDRATION

Magnesium(II) Ion.

Magnesium nitrate is a strong electrolyte in water at 25°C. Raman spectra show no features characteristic of $Mg^{2+}NO_3^-$ contact ion-pairs up to saturation (7). When the water content is lowered below six moles water per mole of salt at a suitable higher temperature spectral changes are observed which clearly indicate that a significant fraction of the nitrate ions directly encounter Mg^{2+}. The absence of such features in more dilute solutions supports the view that we are dealing with a system containing aquated Mg^{2+} cations and nitrate ions.

Raman lines characteristic of the cation-water moiety are observed. We detected two lines at 240 and 357 cm^{-1}; the latter is polarized (6). Such lines have been observed and reported before and the substantial contributions of DaSilveira, Mathieu, Lafont, Marques and Ananthanarayanan are acknowledged (8-13). The number of water molecules in the primary solvation sphere of an ion in solution has never been measured, to our knowledge, by Raman spectroscopy. Spectra of solutions are generally compared with spectra of crystalline hydrates for which the structure has been determined by X-ray diffraction. Based on such a comparison, the lines reported above are attributed to a $Mg(H_2O)_6{}^{2+}$ species with octahedral symmetry. A third line at 315 cm^{-1} reported by Silveira, Marques and Marques (12) was not observed. The crystal $Mg(NO_3)_2 \cdot 6H_2O$ generates lines at 362, 312, and 222 cm^{-1} which show appropriate shifts to 344, 300, and 208 cm^{-1} when H_2O is replaced by D_2O (14). The assignments are reasonable and in keeping with the hydration number of six inferred from n.m.r. studies (15). The intensity of the 357 cm^{-1} line, assigned to the symmetric stretch of the aquated cation, is directly proportional to concentration, up to saturation (Fig. 1). This result supports the conclusion that Mg^{2+} is not associated with

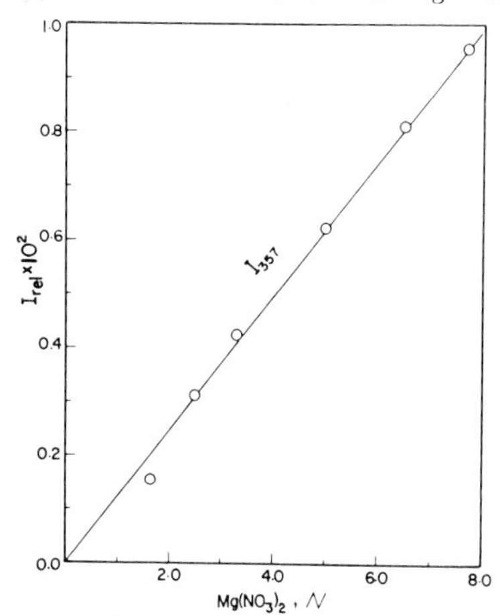

Fig. 1. The relative integrated intensity of the 357 cm^{-1} band of $Mg(H_2O)_6{}^{2+}$ plotted against stoichiometric magnesium nitrate concentration.

nitrate ion in this system. The existence of the Raman line also implies that covalent forces exist between H_2O and Mg^{2+}; an ion-dipole interaction would not give rise to Raman intensity (16). Raman lines of aquated alkali metal cations are not observed, presumably for this reason.

Let us now consider the intramolecular vibrations of water - the stretching and bending vibrations. It has been suggested that cations primarily affect the ν_2 bending mode of water and anions influence the stretching modes ν_1 and ν_3. The broad Raman contours in the ν_1 and ν_3 region are illustrated in Fig. 2. As the con-

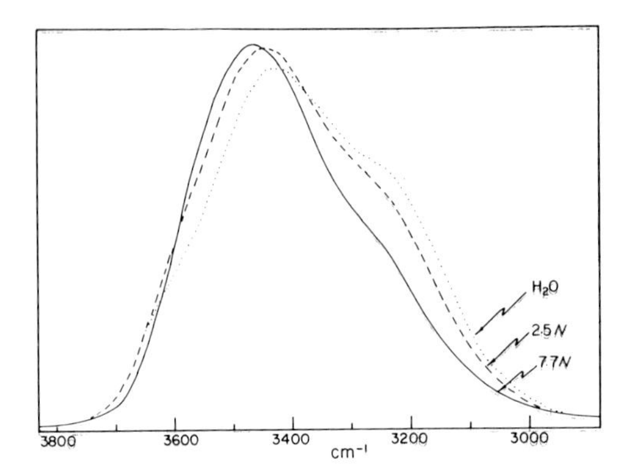

Fig. 2. Raman contours in the O-H stretching region for pure H_2O, 2.5 N and 7.7 N aqueous solutions of magnesium nitrate.

centration of $Mg(NO_3)_2$ increases the intensity in the 3200 cm^{-1} region decreases and the maximum at ~ 3430 cm^{-1} shifts high. The changes are consistent with disruption of hydrogen bonded structure in water and formation of bonds between water and ions (17). These changes are not entirely to be attributed to the anions however. Let us compare these spectra with those of variously hydrated crystals of $Mg(NO_3)_2$.

Consistent with solubility studies of Sieverts and Petzold (18), Chang obtained Raman spectra of mixtures of polycrystalline hydrates of $Mg(NO_3)_2$ which clearly revealed the existence of the tetra-, and dihydrates (14). The tetrahydrate is metastable and its existence had been in doubt (19). The spectra suggest that four equatorial sites around Mg^{2+} in the tetrahydrate are occupied by four water molecules and the two axial sites by nitrate ions, oriented in monodentate fashion. In the dihydrate approximately octahedral coordination is maintained by having two nitrate ions oriented in bidentate fashion and two water molecules trans to each other on the remaining sites (14).

In Fig. 3 the 3000-3800 cm^{-1} Raman spectral regions of the three hydrates are compared with the spectrum of a 3.9 M aqueous solution. It is significant that the O-H stretch of water bound to a cation occurs at 3515 cm^{-1} for the dihydrate. Even for the hexahydrate the major intensity is in the region 3405-3500 cm^{-1}. Thus water molecules which are rather rigidly bound to the cation can be responsible for the changes observed in the solution spectra of Fig. 2. It is also noteworthy that the ν_3 antisymmetric stretch of water generates very little Raman intensity. In Fig. 3 a very weak 3550 cm^{-1} feature can be seen for the dihydrate. On the other hand, the infrared intensity is substantial, rendering i.r. spectra more difficult to interpret (Fig. 4).

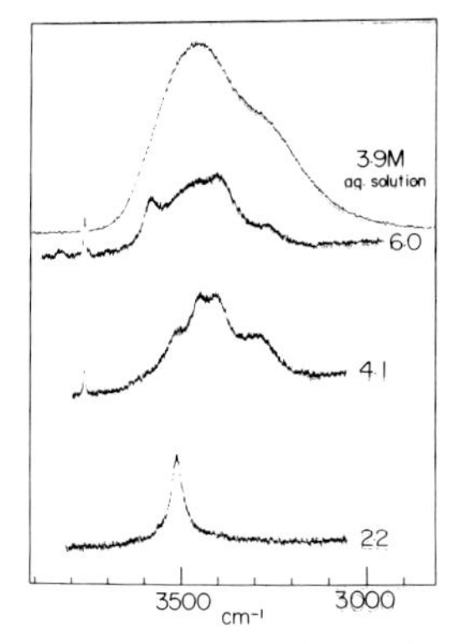

Fig. 3. Raman spectra, excited by the 488 nm line of an argon ion laser, for solid hydrated magnesium nitrates and the 3.9 M aqueous solution. The molar ratios of water to magnesium nitrate are shown beside the contours.

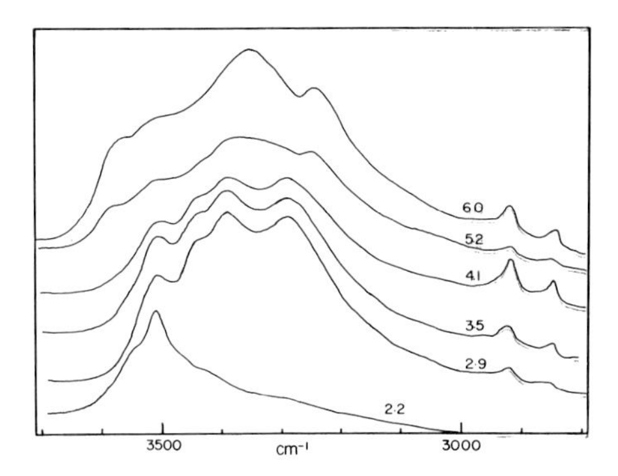

Fig. 4. Multiple internal reflectance infrared spectra in the O-H stretching region for hydrated magnesium nitrate solids. The molar ratios of water to magnesium nitrate are shown beside the contours.

The infrared active deformation modes of water, ν_2, are illustrated in Fig. 5 for the same series of polycrystalline hydrate mixtures. The X-ray study of $Mg(NO_3)_2 \cdot 6H_2O$ indicates that there are two significantly different Mg-O distances (2.053 and 2.061 to 2.063 Å) and two O-Mg-O angles (91.3 and 90.7 to 90.4°). Thus four water molecules experience an environment which differs slightly from the other two. Consistent with this, the deformation mode is a doublet (1632 and 1647 cm^{-1}). These shift to 1195 and 1210 cm^{-1} on deuteration. For the tetrahydrate three bands (1632, 1650 and 1683 cm^{-1}) suggest three distinct sites for water, although the breadth of the 1683 cm^{-1}

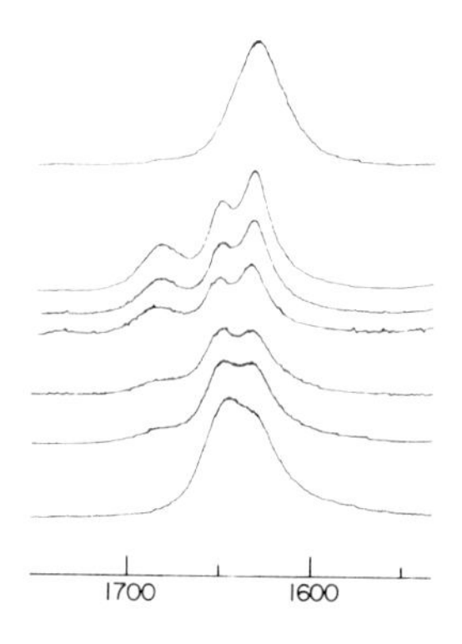

Fig. 5. Multiple internal reflectance infrared spectra of the water deformation modes for $Mg(NO_3)_2 \cdot xH_2O$ polycrystalline mixtures. KRS-5 plates cut at 45° were employed. The molar ratios of water to magnesium nitrate are, from top to bottom, 2.2, 2.9, 3.5, 4.1, 4.4, 5.2 and 6.0.

line may indicate that it encompasses two components. For the dihydrate a single broad 1632 cm^{-1} line is observed and therefore the Mg-O distances are probably equivalent. Hydrogen bonding is much less extensive in the dihydrate (cf. ν_1 region). The water molecules can be considered isolated, discrete units in the inorganic matrix, oriented by the cations. Let us compare these data with results for aqueous solutions. The infrared-active, water bending mode occurs at 1636 cm^{-1} for dilute solutions. It broadens, shifts to higher frequency (1645 cm^{-1}) as the salt concentration increases, and becomes asymmetric indicating the presence of several water environments. It is inferred that a line at 1636 cm^{-1} from bulk water is replaced by a line with greater molar intensity at 1650 cm^{-1} from the water of $Mg(H_2O)_6{}^{2+}$. The molar intensity increase is attributed to polarization of H_2O by Mg^{2+}. Although the lines are more diffuse due to rapid exchange in the liquid phase, the structure of the concentrated solution can be considered to approach that of the hexahydrate (7). The breadth of the 357 cm^{-1} line is probably also linked to the lability of the water and the variation in cation-water distances.

Beryllium(II) Ion.

The radius of Be^{2+} is only 0.31 Å and it thus possesses a relatively high charge-to-radius ratio. Consequently it is the most heavily hydrated of all the bivalent ions. A hydration number of four water molecules in the primary solvation sheath is consistent with n.m.r. data (15, 20, 21). Chang (22) observed four Raman lines which can reasonably be assigned to $Be(H_2O)_4^{2+}$ with T_d symmetry: $\nu_1(A_1)$ 535 cm^{-1}; $\nu_2(E)$ 82 cm^{-1}; $\nu_3(F_2)$ 880 cm^{-1}; $\nu_4(F_2)$ 355 cm^{-1}. The 880 cm^{-1} line is intense in the infrared spectrum. Raman spectra in the region 270-600 cm^{-1} are illustrated in Fig. 6.

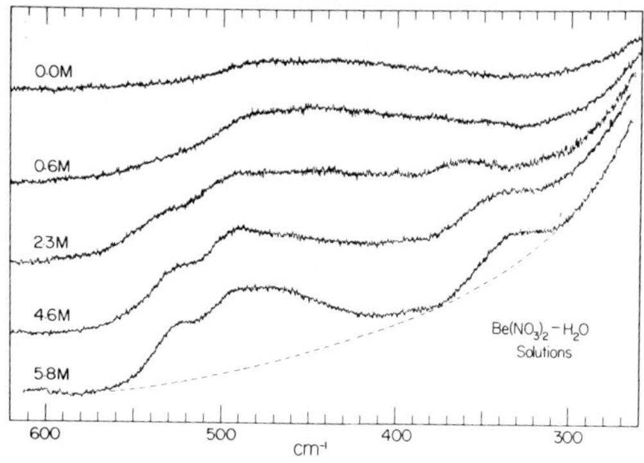

Fig. 6. Raman spectra of aqueous solutions of beryllium nitrate in the 250-600 cm^{-1} region.

Hydrolysis products contribute bands to these spectra. Lines at 535 and 355 cm^{-1}, characteristic of the aquospecies, increase in intensity and sharpen on addition of acid (Fig. 7), a 490 cm^{-1} band disappears, and a pair of bands at 466 and 420 cm^{-1}.

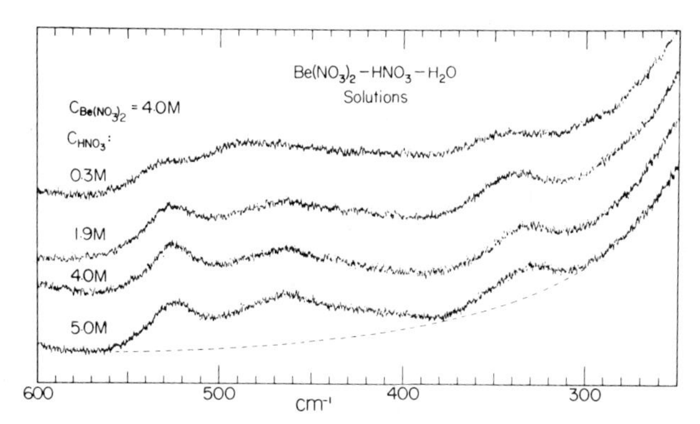

Fig. 7. Raman spectra of aqueous solutions of beryllium nitrate containing nitric acid in the 250-600 cm^{-1} region.

are resolved. Thus the 490 cm^{-1} band is attributed to a hydrolysis product. Addition of acetonitrile also causes the 535 and 355 cm^{-1} lines to sharpen. Raman lines at 420 cm^{-1} (depolarized) and 466 cm^{-1} (polarized) are tentatively assigned to the structured water immediately beyond the first coordination sphere of Be^{2+} ion. The interpretation advanced here differs from that proposed by Marques et al. (23).

Zinc(II) Ion.

Zinc nitrate is also a strong electrolyte in water. The low degree of association is indicative of the stability of the peripheral hydration sphere of Zn^{2+}. This stability is apparent from a comparison of the stability constants, K_1, for the process:

$$Zn^{2+} + NO_3^- \rightleftarrows ZnNO_3^+.$$

In anhydrous methanol with dielectric constant 33.6 K_1 is 2.6 mol^{-1} ℓ (24). In a water-dioxane medium of dielectric constant 33 K_1 is 0.07 mol^{-1} ℓ or about forty times smaller (25). Thus bulk dielectric constant is not the predominant property in determining the degree of association in the $Zn(NO_3)_2$/dioxane/water system. We ascribe the low degree of association of zinc nitrate in the mixed solvents to the large affinity of the zinc ion for water. X-ray diffraction (26) and n.m.r. data (27) suggest that zinc ion has a well-defined, octahedrally-coordinated hydration shell in aqueous solutions. Raman lines at ∿390 (polarized), 335 and 230 cm^{-1} are attributed to $Zn(H_2O)_6^{2+}$.

IV.3.3. ANION HYDRATION

Nitrate Ion.

The nitrate ion has six normal modes of vibration which span the representation $A_1'(R) + A_2''(ir) + 2E'(R, ir)$. In 1968 Irish and Davis (28) drew attention to the fact that in aqueous solutions the $\nu_3(E')$ mode (ca. 1380 cm^{-1}) occurs as a broad doublet; for dilute solutions the two components are centered at about 1348 cm^{-1} and 1404 cm^{-1} in the Raman effect. It has been illustrated several times (e.g. references 2, 28 and 29). The doublet is present even for dilute solutions of alkali metal nitrates where no cation-nitrate ion binding is present to perturb the spectrum of solvated nitrate ion. An unequivocal interpretation of this observation is not possible but consideration of the following facts suggest a tentative model (30).

A Fermi resonance interaction between ν_3 and $2\nu_4$ is excluded from isotope enrichment studies (31). The doublet shifts down (1408 and 1350 to 1375 and 1323 cm^{-1}) when ^{14}N is replaced by ^{15}N but no change in shape of the contour is observed. Because ν_4 remains constant at 720 cm^{-1} such a shift should decouple modes in Fermi resonance and result in a marked change in the intensity ratio. Also, in certain solvents such as chloroform and in certain crystals the line is a singlet.

The positions of Raman and infrared components are not identical, as would be expected for a site splitting or lifting of the degeneracy of an E' mode. Thus in dilute solution we observe:

Raman	1404	1348	$\Delta\nu$ = 56 cm^{-1}
Infrared	1395	1347	$\Delta\nu$ = 48 cm^{-1}

The cations which form the "ion atmosphere" (beyond the primary solvation sphere) have an effect on the doublet, however. The separation of the two components is 56 cm^{-1} for dilute nitrate electrolytes of 1:1 charge type, 63 cm^{-1} for 2:1 and 74 cm^{-1} for 3:1 charge type.

Spectra are independent of high frequency, low intensity electric fields (29). This observation is consistent with the view that the splitting of ν_3 is not due to anisotropy of the coulombic field imposed by the environment; such asymmetry might be expected to collapse under the influence of a high frequency electric field according to the model of the Debye-Falkenhagen effect (32). (This argument is not considered strong because relatively concentrated solutions were studied and alternative explanations of the lack of sensitivity to the field are possible.) Similarly magnetic fields of 12,100 gauss had no effect on the spectra (29).

There is definite evidence that chloroform hydrogen-bonds to nitrate ion. A 2984 cm^{-1} hydrogen-bonded C-H stretch occurs beside the 3019 cm^{-1} non-hydrogen-bonded C-H stretch, when tetraphenylarsonium nitrate is dissolved in chloroform (33). (Spectra are illustrated in reference 34.) The ν_3 region is a singlet when subjected to this perturbation, although it is broad. Thus hydrogen bonding does not seem to be sufficient to lift the degeneracy of ν_3. Carbonate ion has the same D_{3h} symmetry as nitrate ion and in both aqueous solutions and molten salt ν_3 also appears as a doublet, suggesting a common origin (35).

Chisler (36) has provided an interesting comparison between crystal, melt and solution spectra. The single, sharp 1386 cm^{-1} line of $NaNO_3$ crystal broadens as the temperature rises above 275°C and shows doublet structure as the crystal passes from the ordered calcite structure to a disordered phase through a second order phase transition. Similarities between the shape of the doublet observed for this solid phase, the melt and aqueous solutions were pointed out. This provides a clue, linking the observation with orientational freedom.

Possibly the nitrate ion exists in a cavity created by hydrogen-bonded water, or created by positively charged cations in the molten salt. The cavity geometry is such that it lifts the degeneracy of the E' mode. The magnitude of the perturbation is dependent on the charge type of cations in the surroundings; this in turn affects the separation of the line components. In such a cavity the nitrate ion is free to librate. For an oblate top the coupling between the antisymmetric stretch and the librational modes can be quite large. This factor, superimposed on the perturbation arising from the environment, may rationalize the lack of coincidence between infrared and Raman spectra, since different selection rules apply to the librational motions in these two forms of spectroscopy. Calculations to clarify this point are currently in progress (37). Obviously this model is incomplete and based only on circumstantial evidence. Additional empirical facts and theoretical analyses are needed to ascertain its validity. In this regard the light-scattering data of T.A. Litovitz are significant (38). This model of disordered, librating and non-correlated nitrate ions in the fluid state is seen as an alternative to the explanation of the ν_3 split in terms of transverse and longitudinal optic modes - an explanation which would imply significant structure and correlation between particles in a solution or melt (39).

ACKNOWLEDGMENTS

It gives me great pleasure to acknowledge the many contributions and the stimulation of my graduate students. Particularly, with reference to the content of this chapter, I mention Dr. M.H. Brooker, Dr. T.G. Chang, Dr. A.R. Davis, Dr. D.J. Lockwood (P.D.F.) and Dr. Y-K. Sze. I am also grateful to the National Research Council of Canada for support for this work.

REFERENCES

1. D.L. Nelson and D.E. Irish, J.C.S. Faraday I, 69, 156 (1973).

2. J.D. Riddell, D.J. Lockwood, and D.E. Irish, Can. J. Chem., 50, 2951 (1972).

3. D.E. Irish, in Ionic Interactions - From Dilute Solutions to Fused Salts. Ed. S. Petrucci, Academic Press, New York, 1971. Vol. II, Chapter 9, pp. 187-258.

4. D.E. Irish and M.H. Brooker, Trans. Faraday Soc., 67, 1916 (1971).

5. H. Chen and D.E. Irish, J. Phys. Chem., 75, 2672 (1971).

6. H.S. Frank and W-Y. Wen, Discussions Faraday Soc., 24, 133 (1957).

7. T.G. Chang and D.E. Irish, J. Phys. Chem., 77, 52 (1973).

8. A. DaSilveira, Compt. Rend. 197, 1033 (1933).

9. A. DaSilveira and E. Bauer, Compt. Rend. 195, 416 (1932).

10. J.P. Mathieu, Compt. Rend. 231, 896 (1950).

11. R. Lafont, Ann. Phys. (Paris), [13], 4, 905 (1959).

12. A. DaSilveira, M.A. Marques, and N.M. Marques, Compt. Rend. 252, 3983 (1961); Mol. Phys. 9, 271 (1965).

13. V. Ananthanarayanan, J. Chem. Phys. 52, 3844 (1970) and references therein.

14. T.G. Chang and D.E. Irish, Can. J. Chem., 51, 118 (1973).

15. A. Fratiello, R.E. Lee, V.M. Nishida, and R.E. Schuster, J. Chem. Phys., 48, 3705 (1968).

16. J.H.B. George, J.A. Rolfe, and L.A. Woodward, Trans. Faraday Soc., 49, 375 (1953).

17. G.E. Walrafen, J. Chem. Phys., 36, 1035 (1962).

18. A. Sieverts and W. Petzold, Z. anorg. allgem. Chem., 205, 113 (1932).

19. W.W. Ewing, J.D. Brandner, C.B. Slichter, and W.K. Griesinger, J. Am. Chem. Soc., 55, 4822 (1933).

20. R.E. Connick and D.N. Fiat, J. Chem. Phys., 39, 1349 (1963).

21. M. Alei, Jr., and J.A. Jackson, J. Chem. Phys., 41, 3402 (1964).

22. T.G. Chang, Ph.D. Thesis, University of Waterloo, Waterloo, Ontario, Canada, 1972.

23. M.A. Marques, B. Oksengorn and B. Vodar, in Advances in Raman Spectroscopy, Vol. 1, Proceedings of the Third International Conference on Raman Spectroscopy, Reims, France, 1972. Heyden and Son Ltd., London 1973, p. 585.

24. S.A. Al-Baldawi, M.H. Brooker, T.E. Gough, and D.E. Irish, Can. J. Chem., 48, 1202 (1970).

25. Y-K. Sze, Ph.D. Thesis, University of Waterloo, Waterloo, Ontario, Canada, 1973.

26. I.M. Shapovalov and I.V. Radchenko, J. Struct. Chem., 12, 705 (1971). (Zh. Strukt. Khim., 12, 769 (1971)).

27. A. Fratiello, V. Kubo, S. Peak, B. Sanchez, and R.E. Schuster, Inorg. Chem., 10, 2552 (1971).

28. D.E. Irish and A.R. Davis, Can. J. Chem., 46, 943 (1968).

29. D.J. Lockwood and D.E. Irish, Chem. Phys. Letters, in press.

30. M.H. Brooker, T.G. Chang, A.R. Davis, D.J. Lockwood and D.E. Irish, In preparation.

31. M.H. Brooker, private communication.

32. J.R. Williams and H. Falkenhagen, Chem. Revs., 6, 317 (1929).

33. A.R. Davis, J.W. Macklin, and R.A. Plane, J. Chem. Phys., 50, 1478 (1969).

34. D.E. Irish, in Physical Chemistry of Organic Solvent Systems, Eds. A.K. Covington and T. Dickinson, Plenum Press, London, 1973.

35. J.B. Bates, M.H. Brooker, A.S. Quist, and G.E. Boyd, J. Phys. Chem., 76, 1565 (1972).

36. E.V. Chisler, Fiz. Tverd. Tela, 11, 1272 (1969); Soviet Physics, Solid State, 11, 1032 (1969).

37. D.J. Lockwood, private communication.

38. T.A. Litovitz, this conference.

39. J.P. Devlin, D.W. James, and R. Frech, J. Chem. Phys., 53, 4394 (1970).

V SCATTERING METHODS

V.1. X-RAY AND NEUTRON DIFFRACTION FROM WATER AND AQUEOUS SOLUTIONS*

A. H. Narten
Chemistry Division, Oak Ridge National Laboratory
Oak Ridge, Tennessee 37830, U.S.A.

ABSTRACT

The x-ray and neutron diffraction methods are complementary, and it is valuable that suitably chosen aqueous systems be studied by both methods. X-ray diffraction data yield information about the positional correlation while neutron diffraction data yield information about the orientational correlation between pairs of water molecules and between water molecules and ions. In water and aqueous solutions of LiCl and HCl orientational correlations are found to be of much shorter range than positional correlations between molecule and ion pairs, but the existence of well-oriented hydration shells about Li^+ and Cl^- seems firmly established.

Only the Ben-Naim/Stillinger model of water, tested in computer experiments, agrees with the diffraction as well as other data on liquid water. It is not clear whether the predictions of the "primitive" model frequently used to discuss dilute electrolyte solutions can be tested with diffraction data.

V.1.1. INTRODUCTION

The structure of liquid water is for many purposes sufficiently described by probability functions of position and orientation, and the diffraction pattern of water contains information about such correlations. The main purpose of diffraction experiments is to provide a sensitive test of models and, eventually, a molecular theory of liquid water.

The structure of aqueous solutions may be described in a similar manner. However, the correlation functions predicted by theoretical models are realistic only for very dilute solutions, for which the diffraction method may not be sufficiently sensitive for testing. Hence, the correlation functions obtainable from diffraction experiments at intermediate and high ionic concentration represent only the beginning of a description of the structure of aqueous solutions. The broader problem is to infer from the data something of the typical arrangement of ions and water molecules, insofar as this is non-random.

The x-ray and neutron diffraction methods are complementary, and it is valuable that suitably chosen aqueous systems be studied by both methods and over wide ranges of concentration. Such extensive studies reduce greatly the non-uniqueness of interpretation usually associated with the diffraction data from liquids.

V.1.2. DIFFRACTION PATTERN AND STRUCTURE

The mutual configuration of pairs of water molecules is given by a separation vector $\underset{\sim}{r}$ and three Euler angles $\underset{\sim}{\Omega} \equiv \alpha,\beta,\gamma$, and the molecular pair distribution function may be designated $g(\underset{\sim}{r},\underset{\sim}{\Omega})$. This function cannot in general be obtained from the one-dimensional diffraction data of a macroscopically isotropic fluid.

The molecular pair distribution function, as well as other molecular properties defined in pair configuration space, can be written /1-5/ as a series expansion in the complete set of orthonormal polynomials in the orientation angles, the coefficients in such a series being functions of the intermolecular separation distance only. The use of such an expansion yields equations for the scattering from fluids of nonspherical molecules in which the leading terms are functions of the spherically symmetric contributions to $g(\underset{\sim}{r},\underset{\sim}{\Omega})$, and the higher terms involve orientational correlations in the fluid. For molecules that are not too asymmetric, in the sense that their scattering density can be represented by a rapidly converging multipole expansion, only the leading terms need to be retained, and the resulting expression for the observable intensity (per molecule) of coherently scattered radiations is /6/:

$$I(s) = \langle F^2 \rangle + \langle F \rangle^2 \int_0^\infty 4\pi r^2 \rho h_{OO}(r) \frac{\sin(sr)}{sr} dr \qquad (1)$$

The quantity $\langle F \rangle^2$ describes the average scattering from a molecule of random orientation with respect to any other molecule taken as the origin, i.e., the spherical part of the total scattering from one molecule, denoted by $\langle F^2 \rangle$. The molecular scattering amplitude F is a function of the scattering variable $s = (4\pi/\lambda)\sin\theta$, with λ the wavelength and 2θ the scattering angle. The quantity $h_{OO}(r) \equiv g_{OO}(r) - 1$ is the orientationally averaged correlation function for molecular centers, describing deviations from the bulk density ρ around any origin molecule.

The molecular x-ray scattering amplitudes for water have been calculated /2,6/, and the quantities $\langle F_x^2 \rangle$ and $\langle F_x \rangle^2$ were found to differ by less than ~1%, an amount comparable to the accuracy of the calculations as well as the overall precision presently achievable in x-ray diffraction experiments. We conclude that orientational correlation between molecules in water cannot be detected with x-rays, and that Eq. (1) can be used to yield molecular structure and correlation functions from x-ray diffraction data.

In contrast to the x-ray case, a multipole expansion of the neutron scattering density of a water molecule is poorly converging, and the leading spherical terms in Eq. (1) are inadequate for an analysis of the neutron scattering from liquid water. Although orientational correlations between water molecules are "seen" by neutrons, this information cannot be directly extracted from the neutron scattering measurements. Assumptions about the molecular structure and orientation may of course be introduced in the initial stage in the analysis of neutron diffraction data, and this method has been explored by Powles /7/.

An alternate approach is to describe the structure of liquid water (D_2O) in terms of the three atom pair correlation functions $h_{OO}(r)$, $h_{OD}(r)$, and $h_{DD}(r)$. The coherently scattered intensity (per molecule) is then related to the atom pair correlation functions $h_{\alpha\beta}(r)$ descriptive of interactions between atoms of type α,β by the expression

$$I(s) = \sum_{\alpha=1}^{n} f_\alpha^2 + \sum_{\alpha=1}^{n} \sum_{\beta\neq\alpha=1}^{n} f_\alpha f_\beta (\rho_\alpha \rho_\beta)^{\frac{1}{2}} \int_0^\infty 4\pi r^2 dr h_{\alpha\beta}(r) \frac{\sin(sr)}{sr} , \qquad (2)$$

with f_α the atomic scattering amplitude, ρ_α the bulk density of atoms of type α, and summation over the atoms in the water molecule.

Eq. (2) is also suitable for the analysis of diffraction data from aqueous solutions. In this case, I(s) would be the coherently scattered intensity normalized to a stoichiometric unit representative of the solution, and summation would be over all atoms and ions in this structural unit.

V.1.3. REDUCTION OF DIFFRACTION DATA

A. Experimental Corrections. In conventional diffraction experiments a detector is used to count x-rays or thermal neutrons at a preset scattering angle 2θ. The problem is to relate the measured intensity to the liquid structure function, defined below, which in turn is closely related to the Fourier transform of the atom pair correlation functions descriptive of liquid structure.

The measured intensity of scattered radiation must be corrected for background, absorption, double scattering, and other effects which are highly dependent on the scattering geometry. For an exhaustive discussion of these corrections the reader is referred to the literature /8,9/.

B. Dynamic Corrections. The relationship between the corrected intensity and the liquid structure function is straightforward only in the static approximation, which assumes that the scattering atoms are rigidly bound, so that all exchange of energy between radiation and sample can be neglected. In a liquid the atoms may be considered as "bound" only if their mass M is very large compared to the mass m of the scattered photons or particles, and if the energy E_o of incident radiation is much larger or smaller than the energy transfer in scattering processes. If these conditions are not met, departures from the static approximation must be considered, and lead to corrections of the order m/M to the effective cross section obtained from the scattering experiment. These dynamic corrections have been discussed in detail elsewhere /10,11/; we here describe only the method used in the reduction of our neutron diffraction data from water /12/ and aqueous solutions /13,14/.

The effective cross section derived from experiment may be written

$$I^*(s) = \sum_{\alpha=1}^{n} f_\alpha^{*2} + \sum_{\alpha=1}^{n} \sum_{\beta\neq\alpha=1}^{n} f_\alpha f_\beta \overline{h}_{\alpha\beta}(s) + \sum_{\alpha=1}^{n} f_{i,\alpha}^{*2} , \quad (3)$$

with f_α^{*2} and $f_{i,\alpha}^{*2}$ the effective coherent and incoherent scattering cross sections, f_α the static coherent scattering amplitude of atom α, and $\bar{h}_{\alpha\beta}(s)$ the liquid structure function defined by the relation

$$\bar{h}_{\alpha\beta}(s) = (\rho_\alpha\rho_\beta)^{\frac{1}{2}} \int_0^\infty 4\pi r^2 dr h_{\alpha\beta}(r) \frac{\sin(sr)}{sr} . \quad (4)$$

A basic assumption in (3) is that the static approximation is valid for the structurally sensitive terms proportional to $\bar{h}_{\alpha\beta}(s)$. The effective "self" terms $\sum_{\alpha=1}^{n} f_\alpha^{*2}$ and $\sum_{\alpha=1}^{n} f_{i,\alpha}^{*2}$ are related to the corresponding static quantities through the approximate expressions /10/

$$\sum_{\alpha=1}^{n} f_\alpha^{*2} = \sum_{\alpha=1}^{n} f_\alpha^2 \,\mathcal{E}(\lambda_0)[1 + A_\alpha - B_\alpha (s/s_0)^2]$$

$$\sum_{\alpha=1}^{n} f_{i,\alpha}^{*2} = \sum_{\alpha=1}^{n} f_{i,\alpha}^2 \,\mathcal{E}(\lambda_0)[1 + A_\alpha - B_\alpha (s/s_0)^2] , \quad (5)$$

with $\mathcal{E}(\lambda_0)$ the detector efficiency for the incident wavelength λ_0, $s_0 = 2\pi/\lambda_0$, and $s = 2s_0 \sin\theta$. The constant $A_\alpha = (m/2M_\alpha)(k_B T/E_0)$ depends on the mass M_α of atom α, the mass m and energy E_0 of the incident neutrons, the Boltzmann constant k_B and the absolute temperature T. In addition, the constant $B_\alpha = (m/M_\alpha)[C_1 + C_3 (k_B T/E_0)]$ depends on the detector efficiency represented by the coefficients C_1 and C_3, defined /10/ to be positive for most detectors /15/.

For x-rays ($\lambda_0 = 0.71$ Å, $E_0 = 1.7 \times 10^4$ eV) the dynamic corrections are negligible, and the effective quantities in (3) can be replaced by the static values. For neutrons ($\lambda_0 = 1.1$ Å, $E_0 = 0.07$ eV) the incident energy is comparable to the energy transfer in molecular librations, and the dynamic corrections are significant. For water and other liquids containing nuclides of mass comparable to that of the neutron these corrections become uncomfortably large and must be considered approximations /11/.

C. Refinement of the Liquid Structure Function. We define an average liquid structure function by the expression

$$H(s) \equiv \sum_{\alpha=1}^{n} \sum_{\beta\neq\alpha=1}^{n} W_{\alpha\beta}\, \bar{h}_{\alpha\beta}(s) \quad (6)$$

as a weighted sum of the partial structure functions $\bar{h}_{\alpha\beta}(s)$ descriptive of atom pair interactions. The weight functions

$$W_{\alpha\beta} = f_\alpha f_\beta \Big/ \left[\sum_{\alpha=1}^{n} f_\alpha\right]^2 \quad (6a)$$

depend only on the static coherent scattering amplitudes of the constituent atoms. The function H(s) can always be constructed from the measured cross section (3) by subtraction of the effective "self" terms for the coherent and incoherent scattering from independent atoms, and division by the quantity $\left[\sum_{\alpha=1}^{n} f_{\alpha}\right]^2$.

The partial liquid structure functions $\bar{h}_{\alpha\beta}(s)$ oscillate around the asymptotic value $\bar{h}_{\alpha\beta}(\infty) = 0$, with frequencies determined by the mean separation $r_{\alpha\beta}$ between atom pairs in the liquid, insofar as these distance distributions are not uniform ($0.5\ Å \lesssim r_{\alpha\beta} \lesssim 7\ Å$). The functions H(s) accessible from diffraction data always contain additional, error, components, which can be eliminated if their frequency is sufficiently different from that due to the liquid structure.

Random errors arising from counting statistics correspond to a white frequency spectrum and can be minimized by standard smoothing procedures. The high angle part ($s \gtrsim 10\ Å^{-1}$) of the structure functions derived from our data has been smoothed in such a manner /16/.

The main result of systematic errors is a smoothly varying, low-frequency "background." Consequently, a tolerable separation of this background can be accomplished by standard techniques used for many years in the reduction of electron diffraction data /17/. One such procedure is to fit by least-squares low-order polynomials or low-frequency trigonometric functions to the curves derived from experiment. An equivalent method, adopted by us /6,12/, is based on repeated Fourier transformations. In either case, the elimination of Fourier components $c_i \sin(sr_i)$, with $r_i < 0.5\ Å$, yields structure functions which meet the available criteria /18,19/ for the overall accuracy of diffraction data from liquids.

V.1.4. RESULTS AND DISCUSSION

There is a limit to the amount of information which can be obtained by diffraction methods. The liquid structure "seen" by x-rays or neutrons is an average of any actual instantaneous configuration over a volume large compared to atomic dimensions. Since the instantaneous environments of different but chemically equivalent atoms may well be quite different from one another, it must not be concluded that the liquid structure function uniquely describes the environment of all atoms of a given type.

A further source of difficulty in interpreting the average

structure function accessible from a single diffraction experiment lies in its composite nature. For polyatomic liquids with ν different types of atoms the function H(s) cannot be uniquely resolved into all components $\bar{h}_{\alpha\beta}(s)$. For this purpose $\nu(\nu - 1)/2$ experiments involving sufficiently different values for the atomic scattering amplitudes are necessary, and this condition has proved to be very difficult to realize for water and aqueous solutions. In view of these difficulties, we have to consider models for the structure of these liquids. In order to be realistic, structural models must have only a small number of adjustable parameters (such as interatomic distances), and these parameters must be physically reasonable. We further require that a set of model parameters derived from x-ray diffraction must also describe the neutron diffraction pattern and, for solutions, must do so over a wide concentration range. These requirements are necessary but not sufficient for the model to be tenable. The ultimate criterion must be its usefulness in predicting a wide range of properties.

In the following discussion we will frequently refer to the average pair distribution functions $G_X(r)$ and $G_N(r)$ derived from the x-ray and neutron structure functions $H_X(s)$ and $H_N(s)$ by Fourier transformation according to

$$G(r) \equiv 1 + (2\pi^2\rho_o r)^{-1} \int_o^M sH(s)\sin(sr)ds \ , \tag{7}$$

with M the maximum value of s accessible in diffraction experiments. We emphasize that the direct interpretation of these distribution functions can be dangerous because they contain features attributable /8/ to the cut-off of the integral (7) at finite values of the variable s. All quantitative information presented here has been derived from the structure functions H(s), the distribution functions G(r) being used only to illustrate these results.

A. Liquid Water. Orientational correlation between neighboring molecules must be expected to play an important role, but it is scarcely detectable /8/ with x-rays, because the electron density is very nearly spherical /2/. Hence, x-ray scattering data can be deconvoluted to yield structure and correlation functions descriptive of molecular centers (oxygen atoms). The structure functions $\bar{h}_{OO}(s)$ for liquid water (Fig. 1) below 100°C rise monotonically through a pronounced double maximum centered around 2.5 Å^{-1} and oscillate with decreasing amplitude around the asymptotic value $\bar{h}_{OO}(\infty) = 0$. The oscillations beyond ~4 Å^{-1} in the functions $\bar{h}_{OO}(s)$ are almost completely determined by interactions between nearest-neighbor molecules. The frequency, amplitude, and damping of these oscillations correspond to the position,

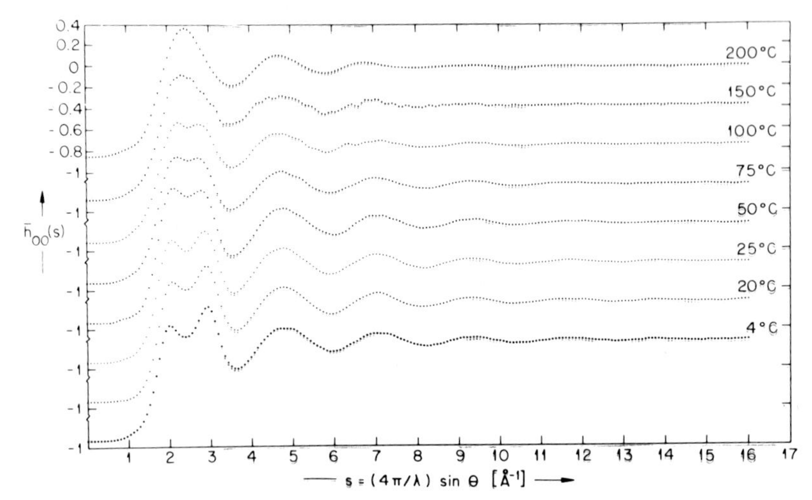

Figure 1. Structure functions descriptive of molecular centers (oxygen atoms) for H_2O /6/.

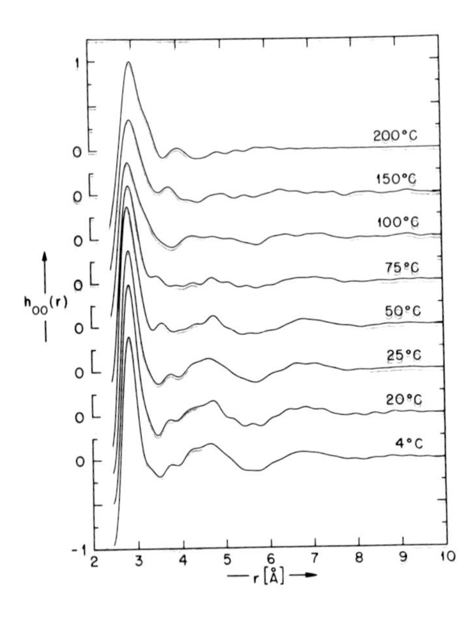

Figure 2. Correlation functions descriptive of molecular centers (oxygen atoms) for H_2O /6/.

area, and width of the first major peak in the correlation functions $h_{OO}(r)$, shown in Fig. 2. The position of this peak shifts gradually from 2.84 Å at 4°C to 2.94 Å at 200°C, its width increasing progressively. The area under the first maximum in the functions $h_{OO}(r)$ corresponds to an average coordination number slightly larger than four, independent of temperature, indicating predominantly tetrahedral coordination. The distance ratio for the maxima in the correlation functions which correspond to first and second neighbor O···O interactions indicates that at low temperatures the average deviation from ideal tetrahedral coordination in the liquid must be quite small. The maxima and minima between ~3.5 Å and ~8 Å in the correlation functions disappear gradually with increasing temperature. This behavior of the functions $h_{OO}(r)$ corresponds in the functions $\bar{h}_{OO}(s)$ to an equally gradual merger of the double maximum around 2.5 $Å^{-1}$ into a single peak. The limiting value $\bar{h}_{OO}(0)$ at zero scattering angle is equal to the zeroth moment of the correlation functions $h_{OO}(r)$, which is in turn related to bulk thermodynamic properties. These small-angle features imply that at large distances $r \gtrsim 8$ Å the correlation functions $h_{OO}(r)$ in Fig. 2 show only the small deviations from the asymptotic value $h_{OO}(\infty) = 0$ caused by random equilibrium fluctuations in the particle density.

With the structure and correlation functions for oxygen atom pair interactions thus known from x-ray diffraction, two additional scattering experiments are necessary for the unique determination of the corresponding quantities involving hydrogen atoms. The beautiful electron diffraction experiment on very thin films of liquid water, reported at this conference /20/, is unfortunately not suitable for this purpose. High-energy electrons are predominantly scattered by the atomic nuclei. The electron scattering amplitudes are hence proportional to the atomic number at all but inaccessibly small scattering angles, and the ratio of the atomic scattering amplitudes for oxygen and hydrogen for high-energy electrons is insufficiently different from that for x-rays. The structure and correlation functions derived from electron diffraction /20/ are in excellent agreement with the corresponding quantities derived from x-ray scattering. This fact lends additional support to the validity of our assumption that the electron density distribution in a water molecule is "seen" by x-rays as spherically symmetric.

The neutron scattering amplitudes of hydrogen, deuterium, and oxygen, as well as the ratios f_H/f_O and f_D/f_O, differ sharply from the corresponding quantities for x-ray scattering. However, in practice it has thus far been feasible to perform only a single type of neutron diffraction experiment with sufficient precision, namely with D_2O /12,21/. The average structure function accessible from neutron data on D_2O may be written

$$H_N(s) = H_1(s) + 0.09\,\overline{h}_{OO}(s) + 0.42\,\overline{h}_{OD}(s) + 0.49\,\overline{h}_{DD}(s), \qquad (8)$$

with $H_1(s)$ the intramolecular contribution to $H_N(s)$, and the values for the weights of the intermolecular terms (Eq. (6a)) being obtained from the neutron scattering amplitudes /12/ of oxygen and deuterium.

The structure function $H_N(s)$ derived from our data, shown in Fig. 3, differs significantly from that obtained by Page and Powles /21/ at values of the variable $s \gtrsim 2$ Å^{-1}. Since this region of the curve $H_N(s)$ is dominated /12/ by the intramolecular term $H_1(s)$ in Eq. (8), the main result of this discrepancy is a broadening of the first major peak in the correlation function $G_N(r)$ derived /22/ from the data presented by Page and Powles /21/. Hence, the discrepancy between the two data sets should not significantly affect the interpretation of the intermolecular terms in (8), descriptive of water structure.

In contrast to the x-ray case, the neutron diffraction pattern of water (D_2O) is almost completely determined by interactions involving deuterium atoms (Eq. (8)), oxygen atom pair interactions contributing only 9% of the scattering. This means that the neutron data contain predominantly information about the orientational correlation between pairs of water molecules. The structure and correlation functions derived from neutron diffraction can be uniquely resolved into contributions from oxygen atom pair interactions alone and an almost evenly weighted sum of contributions from OD and DD interactions (see Fig. 2 of ref. 12). Unique separation of the OD and DD interactions is not possible until a second neutron experiment involving a sample of different isotopic composition has been carried out with high precision. We have therefore used a simple model to separate the contributions from OD and DD interactions.

The model is based on the function $\overline{h}_{OO}(s)$ derived from x-ray diffraction, the functions $\overline{h}_{OD}(s)$ and $\overline{h}_{DD}(s)$ being calculated from the OD and DD interactions corresponding to orientational correlation of first neighbors only, as illustrated in Fig. 7a. All but one of the interatomic distances necessary for the description of this model are determined by a tetrahedral constraint and the intramolecular OD and DD distances. Structure and correlation functions calculated for this model are shown in Fig. 3 along with the curves derived from experiment, and the agreement is satisfactory. Fig. 4 shows the atom pair

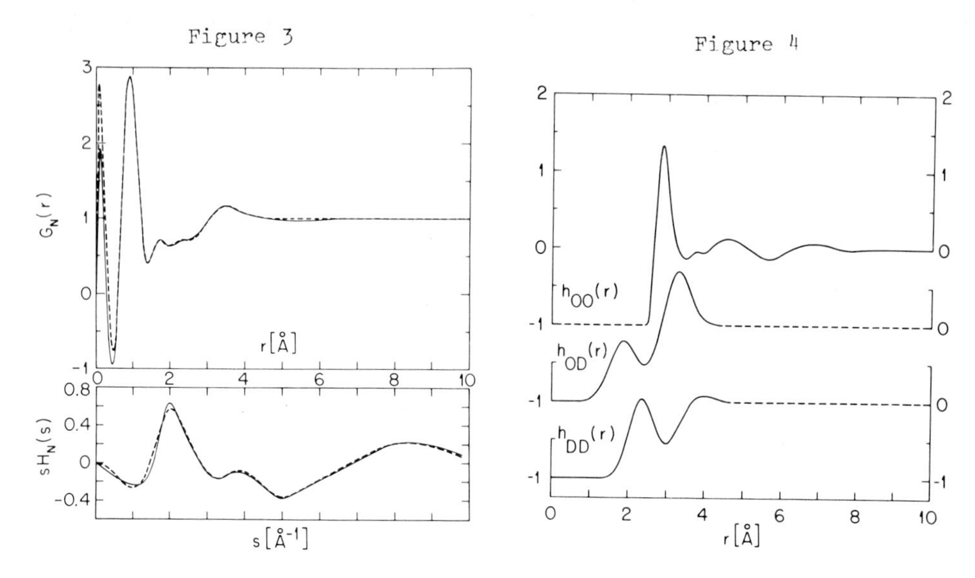

Figure 3. Weighted structure functions (bottom) and distribution functions (top) for D_2O at 25°C. Solid curves are derived from neutron data. Dashed curves were calculated from the near-neighbor model shown in Fig. 7a /12/.

Figure 4. Atom pair correlation functions for liquid water at 25°C /12/.

correlation functions for liquid water.

The function $g_{OO}(r)$, derived from the x-ray data, has already been described. The function $g_{OD}(r)$ shows two broad maxima at 1.92 Å and 3.28 Å, with equal rms variations of 0.33 Å. The function $g_{DD}(r)$ shows maxima at 2.40 Å and 3.58 Å, with rms variations of 0.33 Å and 0.44 Å, respectively. The very large rms variations in the mean intermolecular OD and DD distances indicate that water molecules exist in local and instantaneous environments which are grossly distorted from the average configuration shown in Fig. 7a.

The structure and correlation functions derived in this section are sufficiently accurate to provide a sensitive test for future work on a molecular theory for liquid water. In the concluding section of this chapter we will compare our data with results obtained from other methods.

B. Aqueous Solutions. Ionic solutions were among the first liquids to which x-ray diffraction was applied /23/, and a number of systems have been examined over the years. The neutron diffraction method has only

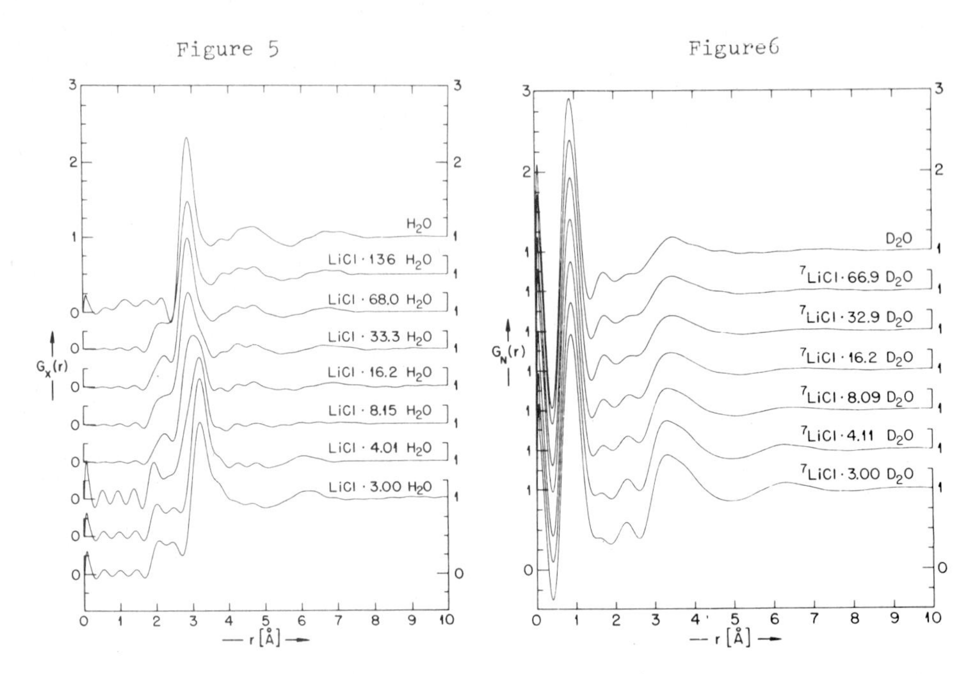

Figure 5. Distribution functions for $LiCl/H_2O$ solutions derived from x-ray diffraction /13/.

Figure 6. Distribution functions for $^7LiCl/D_2O$ solutions derived from neutron diffraction /13/.

recently been applied to an aqueous solution /13/. We here summarize the main conclusions of an extensive study of aqueous LiCl /13/, and we describe results obtained but not as yet analyzed for aqueous HCl /14/. Both systems were studied with x-rays and neutrons over a wide concentration range, to saturation.

Average pair distribution functions $G_X(r)$ derived from x-ray diffraction data on the system $LiCl/H_2O$ are shown in Fig. 5. The corresponding functions $G_N(r)$ for the system $^7LiCl/D_2O$ are shown in Fig. 6. The scattering amplitudes ensure that interactions involving oxygen dominate the x-ray curves while those involving D dominate the neutron curves. With the results for pure water as a point of departure, the variation of $G_X(r)$ and $G_N(r)$ with concentration yields directly some valuable information. The dominant peak at 0.94 Å in $G_N(r)$ is ascribed to the intramolecular OD interaction and the next peak at 1.7 Å to a combination of intramolecular DD at 1.5 Å and first intermolecular OD at 1.9 Å. With increasing concentration, the second peak in $G_N(r)$ shifts from 1.7 Å to 1.6 Å and becomes progressively less prominent; this is understood as the disappearance of the hydrogen bonded OD interaction as water molecules adjacent to each other become scarce, leaving only the intramolecular DD. In $G_X(r)$ for the most concentrated solution (Fig. 5), the peak at 3.2 Å is ascribed predominantly to Cl-O interactions. If this interaction is a hydrogen bond Cl···H-O, a Cl-D peak should appear in $G_N(r)$ at 2.3 Å; this is in fact found with prominence increasing with LiCl concentration. A hydrated Li^+ ion should have /24/ a distance to coordinated water of about 2.0 Å. Such a peak is found in $G_X(r)$, although, because of the low x-ray scattering power of Li^+, it is not well isolated. The negative scattering amplitude of ^7Li for neutrons should result in a minimum in $G_N(r)$, which is in fact found.

The preceding interpretation of the pair distribution functions in terms of nearest neighbor atom pair interactions suggests the existence of well-oriented hydration shells about Cl^- and Li^+ in the concentrated solutions. Further analysis of the x-ray and neutron structure functions, described in detail elsewhere /13/, yielded the following results:

The coordination of the Cl^- ion is octahedral on the average, with a deuterium atom near each connecting line. A hydrogen bonded Cl-D distance of 2.27 Å and the longer Cl-D, OD, and DD distances implied by the octahedral geometry of Fig. 7c gave the best fit to the data. The coordination around Li^+ appears to be tetrahedral on the average. A single LiD distance at 2.71 Å and the longer OD and DD distances characteristic of the tetrahedral geometry of Fig. 7b gave the best fit. The more dilute solutions were assumed to consist of independent configurations of $Cl^-(H_2O)_6$ and $Li^+(OH_2)_4$. Structure parameters for the hydrated ions were those found for the most concentrated solution, with small but significant adjustment of the Cl-O, LiO, and the shortest Cl-D and LiD distances /13/. For mol ratios H_2O/LiCl greater than 8, part of the remaining water was assumed to occur in a tetrahedrally linked environment (Fig. 7a), described by the properly weighted structure functions for pure water. Structure functions calculated from the parameters of the model agree well with the curves derived from x-ray and neutron diffraction (see Fig. 4a and 4b of ref. 13).

The system HCl/H_2O and DCl/D_2O has also been studied with both x-rays and neutrons over a wide concentration range /14/. Analysis of these data is still underway, and we will here only present the average pair distribution functions derived from experiment. The functions $G_X(r)$ for the system HCl/H_2O are shown in Fig. 8. With increasing HCl concentration the peak characteristic of nearest neighbor OO pairs

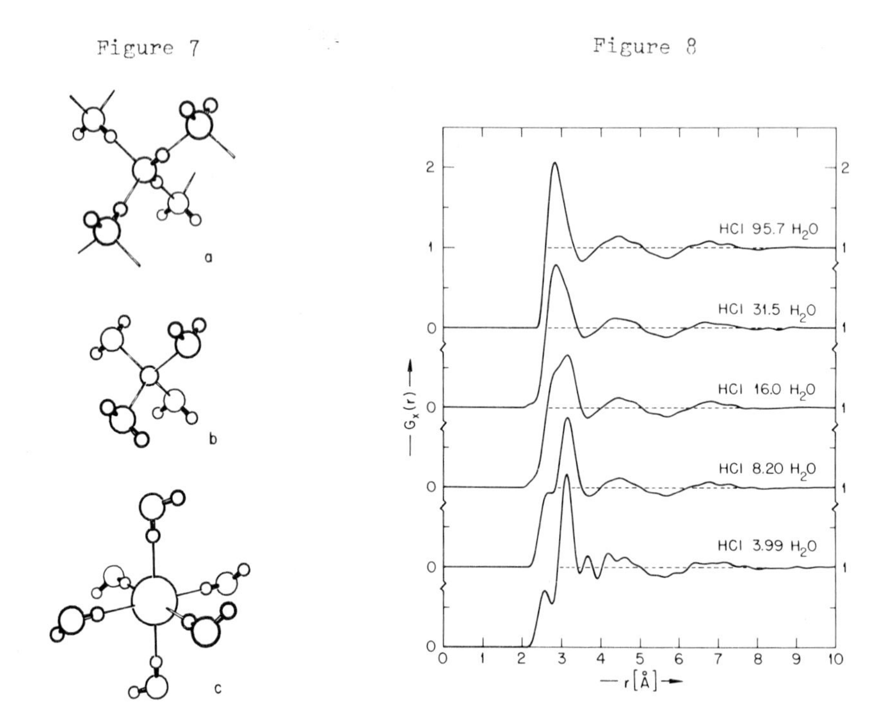

Figure 7. (a) Average coordination of D_2O molecules in water, (b) average coordination of D_2O molecules around Li^+ ions, (c) average coordination of D_2O molecules around Cl^{-1} ions.

Figure 8. Distribution functions for HCl/H_2O solutions derived from x-ray diffraction.

shifts from 2.85 Å in pure water to about 2.5 Å. Since these short OO distances are not observed in the LiCl solutions, we attribute them to the presence of excess protons in the HCl solutions. This result and our conclusion agree with results derived in a previous high-resolution x-ray study /25/. The short OO distances are barely /26/ or not at all /27,28/ resolved in x-ray diffraction studies of relatively low resolution. The maximum at 3.2 Å in the concentrated HCl solutions is ascribed to near neighbor ClO interactions. This feature in $G_X(r)$ is consistent with six-fold coordination of water molecules around Cl^-, as was found for the LiCl solutions.

The functions $G_N(r)$ for the system DCl/D_2O are shown in Fig. 9. These curves show better resolution in the small distance region, because the neutron data on the DCl solutions extend /14/ to values of the scattering variable $s \leqslant 16$, as do the x-ray data, while the neutron

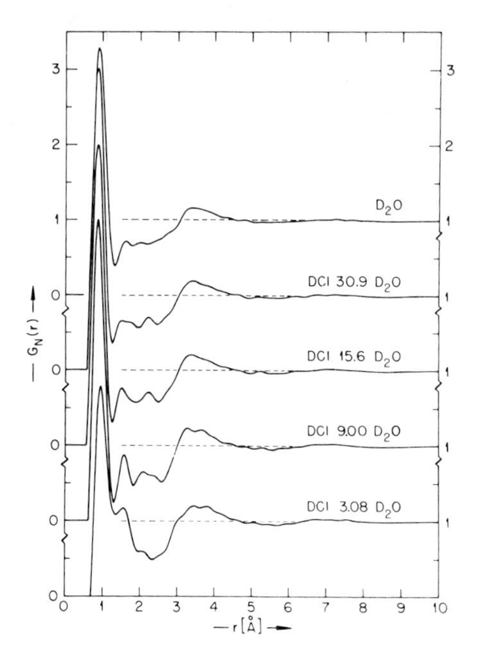

Figure 9. Distribution functions for DCl/D_2O solutions derived from neutron diffraction.

data on LiCl solutions extend /13/ only to values of $s \leq 10$. A striking feature of the functions $G_N(r)$ is the increasing prominence of the second peak at 1.6 Å with increasing HCl concentration. This feature is closely correlated with the appearance of short OO distances in the functions $G_X(r)$ shown in Fig. 8; it is understood as the increasing strength of short D···OD hydrogen bonds in the concentrated acid solutions. The maximum at 2.2 Å in the functions $G_N(r)$ of Fig. 9, ascribed to Cl···DO hydrogen bonds, shows a more complex concentration dependence than found for the LiCl solutions, and no detailed interpretation will be attempted here. Our preliminary interpretation of the average pair distribution functions for hydrochloric acid solutions suggests the existence of strongly bonded oxygen pairs, with a deuteron near each connecting line.

V.1.5. SUMMARY AND CONCLUSIONS

It has already been mentioned that, for liquids, the amount of information that can be extracted from diffraction data alone is inherently limited and, in practice, is often not unique. The non-uniqueness usually associated with diffraction data from liquids can be greatly reduced by combined x-ray and neutron diffraction studies, which are always feasible. The feasibility of the isotopic substitution method in neutron diffraction studies of aqueous systems has yet to be demonstrated in practice and is, in any case, restricted to relatively few nuclides which are not of the greatest interest to solution chemists. In the meantine, we have to consider models for the structure of water and aqueous solutions.

By definition, models do not provide rigorous descriptions. A model that achieves near-perfect agreement with the measured x-ray and neutron scattering is not very helpful if it does not predict thermodynamic, kinetic, and spectroscopic properties, and vice versa. A realistic and useful model must try for a balanced degree of precision for a wide range of properties.

A. Liquid Water. We have defined the structure of liquid water in terms of the atom pair correlation functions $h_{OO}(r)$, $h_{OD}(r)$, and $h_{DD}(r)$. Only the structure and correlation functions descriptive of oxygen atom pair interactions has so far been uniquely derived from x-ray diffraction. The uniqueness of this derivation has been possible because the electron distribution in an isolated water molecule is "seen" by x-rays as spherically symmetrical. The validity of this assumption, questioned by some /21,29/, has now received additional support from an electron diffraction experiment /20/ on liquid water. The dominant peak in $h_{OO}(r)$ (Fig. 4) corresponds to the near neighbor OO interaction of a tetrahedral network, with additional broad interactions at distances somewhat larger than the first-neighbor distances. At distances $r \gtrsim 8$ Å the functions $h_{OO}(r)$ show only the small deviations from the asymptotic value $g_{OO}(\infty) = 0$ caused by random equilibrium fluctuations in the particle density. In this respect, liquid water is no different from simple monatomic liquids. The entire curve has been well fitted on the basis of a model /8/ based on the ice I structure with some additional molecules. However, we have stressed that this agreement, while necessary, is not sufficient proof of the reality of the ice-I model for liquid water, and this model has not been useful in the prediction of other properties. In a similar manner, other simple models /30/ designed to describe specific properties of water have been unable to describe the results derived from diffraction data /31/. The only model which reproduces the over-all properties of water reasonably well is that proposed by Ben-Naim and Stillinger /32/ (BNS model) and tested by Stillinger and Rahman in computer experiments by the molecular dynamics method. The BNS model is being continuously refined, and each refinement has resulted in structure and correlation functions agreeing more closely /29/ with those derived from x-ray diffraction.

We have concluded that the neutron diffraction pattern of D_2O is almost completely described by the orientational correlation between nearest neighbor molecules, second and higher neighbors being "seen" as randomly oriented by any origin molecule. Most simple models /30/ of liquid water assume near-neighbor configurations similar to the ones shown in Fig. 7a; the various models differ in the assumed arrangement of oxygen atoms from second and third neighbors around an origin molecule, the relative orientation of these pairs being not specified or assumed random. Hence, the neutron data cannot be used to test any of these simple models.

Unlike the correlation function $h_{OO}(r)$, the functions $h_{OD}(r)$ and $h_{DD}(r)$ shown in Fig. 4 are model dependent. Comparison of the curves in Fig. 4 with the correlation functions obtained from the BNS model at slightly different nominal temperatures /29/ shows good qualitative agreement. The two sets of curves have maxima and minima at almost exactly the same radial distances, the maxima being, however, much sharper in the curve derived from the molecular dynamics study. We have concluded /12/ that Rahman and Stillinger's simplifying assumption of rigid water molecules is one reason for this discrepancy. The same conclusion was reached by Popkie, Kistenmacher, and Clementi /33/. These authors computed the Hartree-Fock (HF) energy of the water dimer for a large number of configurations. The computed curves were then used to derive a simple analytical expression reproducing the HF potential energy surface. This analytical HF potential was compared with the BNS potential, and used in a Monte Carlo simulation of liquid water. The correlation functions computed for the HF model /33/ are in very good agreement with the curves derived from the diffraction data.

B. Aqueous Solutions. The only completed study of an ionic solution with both x-rays and neutrons concerns LiCl. The existence of well-oriented hydration shells about Li^+ and Cl^- seems firmly established by this study. As indicated in Fig. 7b, Li^+ appears to engage a lone pair of O, producing a tetrahedral environment about O of one Li^+, two D and, presumably, a hydrogen bond to a neighboring D_2O or, at high LiCl concentration, to a neighboring Cl^-. Chloride ions appear to coordinate water by hydrogen bonding as an acceptor atom.

Surrounding these hydrated species, in sufficiently dilute solutions, one visualizes layers of water of more or less random configuration and, beyond that, water molecules in a tetrahedral configuration like that in pure water. The average configurations shown in Fig. 7 should not be taken as separate species in the thermodynamic sense. They exist in environments that are distorted from the average, and these distortions are implied by sizeable /13/ mean-square variations in interatomic distance.

The advantage of complementary data from x-ray and neutron diffraction is clearly demonstrated, as is also the utility of data covering a wide range of concentration.

The diffraction method is most sensitive to solute-solute and solute-water interactions at high ionic concentration. For LiCl these interactions are barely detectable at concentrations of ~1 M. It is not clear whether the predictions of the "primitive" model /34/ frequently used to discuss dilute electrolyte solutions can be tested with diffraction data. This model neglects solute-water interactions which become significant at ~1 M and, for LiCl, dominate the diffraction pattern at high concentration. We hope to discuss this problem in more detail in the future.

REFERENCES

*Research sponsored by the U. S. Atomic Energy Commission under contract with the Union Carbide Corporation.

1 W. A. Steele and R. Pecora, J. Chem. Phys. 42 (1965) 1863.

2 L. Blum, J. Computat. Phys. 7 (1971) 592.

3 L. Blum and A. J. Torruella, J. Chem. Phys. 56 (1972) 303.

4 L. Blum, J. Chem. Phys. 57 (1972) 1862.

5 L. Blum, J. Chem. Phys. 58 (1973) 3295.

6 A. H. Narten and H. A. Levy, J. Chem. Phys. 55 (1971) 2263.

7 J. G. Powles, "The Structure of Water by Slow Neutron Scattering," p. , this volume.

8 A. H. Narten and H. A. Levy, "Liquid Water: Scattering of X-rays," in "Water a Comprehensive Treatise," F. Franks, Editor, Plenum Press, New York, 1972, Vol. 1, p. 311.

9 A. H. Narten, J. Chem. Phys. 56 (1972) 1185.

10 J. L. Yarnell, M. J. Katz, R. G. Wenzel, and S. H. Koenig, Phys. Rev. A7 (1973) 2130.

11 J. G. Powles, Adv. Physics 22 (1973) 1.

12 A. H. Narten, J. Chem. Phys. 56 (1972) 5681.

13 A. H. Narten, F. Vaslow, and H. A. Levy, J. Chem. Phys. 58 (1973) 5017.

14 R. Triolo and A. H. Narten, "Diffraction Pattern and Structure of Hydrochloric Acid Solutions," to be submitted to J. Chem. Phys.

15 For the neutron detectors used in the experiments at Oak Ridge the efficiency was found to be of the form $\mathcal{E}(\lambda) = 1 - \exp[-a\lambda_o/\lambda]$. Typical values of the constant a, determined experimentally, were of the order 1.4.

16 C. Lanczos, "Applied Analysis," Prentice Hall, Englewood Cliffs, N. J., 1956, p. 336.

17 M. I. Davies, "Electron Diffraction in Gases," Marcel Dekker Inc., New York, 1971, p. 146.

18 A. Rahman, J. Chem. Phys. 42 (1965) 3540.

19 R. D. Mountain, J. Chem. Phys. 57 (1972) 4346.

20 E. Kalman, S. Lengyel, L. Haklik, and A. Eke, "The Use of Electron Diffraction in Studies of the Structure of Liquid Water," p. this volume.

21 D. I. Page and J. G. Powles, Mol. Physics 21 (1971) 901.

22 A. H. Narten, unpublished work. The first maximum in the $G_N(r)$ curve for D_2O derived from our data /12/ is quantitatively described by a mean distance $r_{OD} = 0.934$ Å, with rms variation $\langle \Delta r_{OD}^2 \rangle^{\frac{1}{2}} = 0.138$ A. The values obtained from our analysis of the data of Page and Powles /21/ are $r_{OD} = 0.934$ Å, and $\langle \Delta r_{OD}^2 \rangle^{\frac{1}{2}} = 0.195$ Å.

23 J. A. Prins, Z. Phys. 56 (1929) 617.

24 L. Pauling, "The Nature of the Chemical Bond," Cornell U.P., Ithaca, N. Y., 1945, p. 346.

25 S. C. Lee and R. Kaplow, Science 169 (1970) 477.

26 D. L. Wertz, J. Sol. Chem. 1 (1972) 489.

27 D. S. Terekhova, J. Struct. Chem. 11 (1970) 483; translated from Zh. Struct. Khim. 11 (1970) 530.

28 G. Licheri, G. Piccaluga, and G. Pinna, Chem. Phys. Letters, 12 (1971) 425.

29 F. H. Stillinger and A. Rahman, "Improved Simulation of Liquid Water by Molecular Dynamics," J. Chem. Phys., in press. I am grateful to Dr. Stillinger for making the manuscript available to me prior to publication.

30 H. S. Frank, "Structural Models," in "Water A Comprehensive Treatise," F. Franks, Editor, Plenum Press, New York, 1972, Vol. 1, p. 515.

31 A. H. Narten and H. A. Levy, Science 165 (1969) 447.

32 A. Ben-Naim and F. H. Stillinger, "Aspects of the Statistical Mechanical Theory of Water," in "Structure and Transport Processes in Water and Aqueous Solutions," R. A. Horne, Editor, Wiley-Interscience, New York, 1972, p. 295.

33 H. Popkie, H. Kistenmacher, and E. Clementi, J. Chem. Phys. 59 (1973) 1325. I am grateful to Dr. Popkie for making the manuscript available to me prior to publication.

34 H. L. Friedman, "Solvation and Solute-Solute Interactions in Aqueous Solutions," p. , this volume.

V.2. ELECTRON DIFFRACTION STUDIES OF LIQUID WATER

E. Kálmán, S. Lengyel, G. Pálinkás, L. Haklik and A. Eke
Central Research Institute for Chemistry of the
Hungarian Academy of Sciences, Budapest, Hungary

ABSTRACT

A new electron diffraction device has been developed for producing stable liquid layers of controlled temperature. A theoretical review of electron scattering of liquids is given. Diffraction patterns of liquid water were taken at $4^{o}C$. Intensity distributions were determined by a microdensitometer. The electron diffraction pair correlation function is discussed in comparison with the X-ray diffraction data.

Formerly: Centre for Studies on Chemical Structures of the
Hungarian Academy of Sciences

V.2.1. INTRODUCTION

After several research on liquid water and aqueous electrolyte solutions /1/ and X-ray diffraction /2/ we commenced methods on liquid water structure by diffraction of electron beams in transmission.

Special characteristics, described below, of the diffraction of electrons in comparison to X-rays gave us the idea as how to overcome the experimental difficulties of electron diffraction of condensed phases with higher vapour pressure. Such difficulties are the production of thin liquids layers and maintenance of high vacuum despite the evaporation of the liquid.

In the interaction between radiation and matter the main difference between X-rays and the electron beam is that the former are diffracted by the electron clouds in the molecules of the condensed phase, while the latter are diffracted by the electric potential depending on the space configuration of the nuclei and the density distribution of the electrons, as well. Atomic scattering factors of X-rays increase with a significantly higher power of the atomic number than do the electrons. Consequently, electron diffraction allows the determination of the positions of light atoms in the presence of heavy stoms. Therefore, in the case of electron diffraction of water, information on hydrogen positions can be expected.

The energy of the electron beam generally used in diffraction experiments is at least ten times less than the wavelength of the X-rays applied to structural studies. Thus the accessible range of the scattering variables is much greater for electrons.

The ratio of the intensity of the coherently scattered radiation to that of incident radiation is approximately six orders of magnitude higher for electrons than for X-rays. Thus an exposure time of a few seconds is sufficient in the case of electrons, while several hours are necessary for X-rays.

V.2.2. THEORY OF ELECTRON SCATTERING OF LIQUIDS

A. Elastic scattering of fast electrons by molecular liquid. The correlation radii of intra- and intermolecular interactions in liquids and thus the characteristic lengths of the structure are about in the same region (1 - 10 Å). If the energy of scattered electron beams is high enough (∿ 50 keV) and so their wavelengths are sufficiently low (∿ 0,05 Å) the scattered electrons "can see" both structures and so the diffraction patterns give information on those.

There are two approaches to the theory of electron scattering on molecular liquids: in the first the molecules, in the second the independent atoms are considered as scattering centres. Determination of the scattering factors of molecules is generally a difficult task. In some cases, however, the molecular scattering amplitude can be determined by considering the molecule as a "quasi atom" in an approximation of a one centre molecular wave function. To such molecules belongs the water molecule among some other ten-electron-systems. /3/ In this case, however, the potential of the molecule is of spherical symmetry, which excludes the investigation of molecule-molecule orientation correlation /4/. In the second approach on the other hand, all atomic positions can be observed, but charge and potential distribution differences between free atoms and bonded atoms in the molecule are neglected. In the first approximation the diffraction equation can be solved, i.e. the molecule-molecule spatial correlation function G_M can be determined by Fourier transformation. In the case of free atom approach direct inverse Fourier transformation is impossible and the correlation functions can be obtained only by further approximations or fitting the model intensities to the experimental ones.

The general theory of electron scattering by atoms has been treated in well-known monographs /5-7/ and the scattering theory of isolated molecules and related problems of gas electron diffraction investigations has also been widely discussed /8-10/. In the following we wish to emphasise aspects related directly to liquid structure determinations. The theory of scattering will be treated in the same way as neutron and X-ray diffraction /11,12/.

According to the first Born approximation the elastic intensity of an isolated molecule at an angle θ in a distance R from the molecule is /8/:

$$I^M(\theta) = I\left|F^M\right|^2 \quad (1)$$

Here $I^M(\theta)d\Omega$ is the number of electrons per unit time elastically scattered into a given solid angle $d\Omega$, around θ and

$$I = \left[\frac{I_o}{R^2 \frac{2\pi m e^2}{h^2}^2}\right]$$

I_o is the intensity of the incident monoenergetic electron beam. The following equation gives the dependence of electron scattering amplitude of the molecule $F^M(\vec{s})$ on $\vec{s}$.
The vector $\vec{s}$ is the difference between the wave number vectors of primary and scattered electrons:

$$\vec{s} = \vec{k}_1 - \vec{k}_1 \qquad |\vec{s}| = 4\pi/_\lambda \sin \ /_2$$

$$F^M(\vec{s}) = \int \Phi_0^* (\vec{x})\ V(\vec{r})\ \Phi_0 (\vec{x})\ \exp\{i\vec{s}\vec{r}\}\ d\vec{x}\ d\vec{r} \quad (2)$$

Here $\vec{x}$ stands for the ensemble of coordinates of nuclei and electrons composing the molecule, Φ_0 is the wave function of the initial state of the molecule, $\vec{r}$ is the vector coordinate of the scattering point and $V(\vec{r})$ denotes the full electrostatic potential of the molecule at point $\vec{r}$.

In the approximation of independent atomic scattering centers according to Debye-Ehrenfest theory, the intensity may be described as

$$I^M_{D.E.} = I\left|\sum_{i=1}^{n} f_i\ (\vec{s})\ \exp\ \{i\vec{s}\vec{r}_i\}\right|^2 \quad (3)$$

where $\vec{r}$ is the radius vector of the i.-th atom from an arbitrary center and $f_i(s)$ is the electron scattering amplitude of the i.-th atom, which, in the first Born-approximation is as follows:

$$f_i(\vec{s}) = \int \Psi_0^*\ V_a\ \Psi_0 \exp\ \{i\vec{s}\vec{r}\}\ d\vec{r}_i\ d\vec{r} \quad (4)$$

Here $V_a(r)$ is the electrostatic potential of the atom in point $\vec{r}$, is the wave function of the atom in the initial state, a_o is the Bohr radius, and $\vec{r}_i$ denotes the ensemble of coordinates of the electrons in the i.-th atom. The relation between the atomic scattering amplitudes of electrons and X-rays, form-factor, in the first Born approximation is /9/:

$$f_i(s) = \frac{Z_i - f_i^x(s)}{s^2} \qquad (5)$$

where Z is the atomic number.

B. Molecule-molecule spatial correlation.

Let the scattering volume consist of N identical molecules and let us denote the radius vector of the μ-th molecule by $\vec{r}_\mu$, and the amplitude of electrons scattered on one molecule by F^M. According to the Debye-Ehrenfest theory, the relative elastic intensity is as follows:

$$\frac{I(s)}{I} = \left| \sum_{\mu}^{N} F_\mu^M(s) \exp\{i\vec{s}\vec{r}_\mu\} \right|^2 = \sum_\mu \sum_\nu F_\mu^M F_\nu^M \exp\{i\vec{s}(\vec{r}_\mu - \vec{r}_\nu)\} \qquad (6)$$

The summation with respect to μ and ν extends over all molecules in the system. Since measurements required much longer time than the relaxation period of molecular motions, the measured relative intensity I^T is the time average of eq.6. This corresponds to the average on the canonical ensemble.

In order to obtain this average, the pair correlation function

$$G_M = \frac{1}{N} \sum_{\mu,\nu} \left\langle \delta(\vec{r} + \vec{r}_\mu - \vec{r}_\nu) \right\rangle = \delta(\vec{r}) + \rho_o g_M(\vec{r}) \qquad (7)$$

of Van Hove /13/ will be used.

Here the average <> has been performed over the classical canonical ensemble.

The expression has the following physical interpretation; the term $G_M(\vec{r})\, d\vec{r}$ is the average number of particles in the elementary volume $d\vec{r}$ around $\vec{r}$, if there is one particle at $\vec{r}$ = o. In other words $G_M(\vec{r})$ is the conditional probability density of finding a particle at $\vec{r}$ knowing that there was a particle at the origin.

$\rho_o = \frac{N}{V}$ is the average molecular density and

$$\lim_{r \to r_c} g_M(\vec{r}) = 1$$

where r_c is the correlation radius of the liquid structure. The $g_M(\vec{r})$ function denotes the correlation between the positions of two molecules. By applying the definition of the $G_M(\vec{r})$ function and assuming that each pair orientation occurs with the same probability, we have for the average of eq.6.:

$$I^T = \left\langle \frac{I(s)}{I} \right\rangle = N\left\langle \left|F^M\right|^2 \right\rangle + N\left\langle \left|F^M\right| \right\rangle^2 \int_0^\infty 4\pi r^2 \rho_o g_M(r) \frac{\sin sr}{sr} dr \tag{8}$$

Here $4\pi r^2 \rho_o g(r)$ is the number of molecules in a spherical shell of radius r and thickness dr around a given molecule. The first term expresses the structure of the molecule while the second term of eq.8. characterizes the structure of the liquid.

By evaluating the first term of eq.8. by means of the Debye-Ehrenfest expression and forming the average with the $G_A(r)$ correlation function of the atoms of the molecule, the scattered intensity I^T will be composed of an atomic intensity component I^A, and components characteristic for the molecular structure $I^{M'}$ and the liquid structure I^L, respectively:

$$I^T = I^A + I^{M'} + I^L \; ; \qquad I^A = N \sum_{i=1}^{n} \left|f_i\right|^2$$

$$I^{M'} = N \sum_{i<j}\sum f_i f_j \int_0^\infty 4\pi r^2 G_A(r) \frac{\sin sr}{sr} dr$$

$$I^L = N \left\langle \left|F^M\right| \right\rangle^2 \int_0^\infty 4\pi r^2 \rho_o g_M(r) \frac{\sin sr}{sr} dr \tag{9}$$

In addition the total intensity also contains background scattering inelastic scattering $S(s)$, multiple scattering $\Delta(s)$, and extraneous scattering E, which is independent of the structure.

The pair correlation function $g(r)$ which characterizes the structure of the liquid, can be derived from I^L by Fourier transformation. Since the integral in I^L diverges it must previously be made convergent. This can be done by substituting $\rho_o g_M(r) - \rho_o$ for $\rho_o g_M(r)$ in the integrand /12/ and finally obtain

$$g_M(r) = 1 + \frac{1}{2\pi^2 \rho_o r} \int_0^\infty s\, M(s) \sin sr\, ds \tag{10}$$

where

$$M(s) = \frac{I_t^E(s) - I^B(s)}{I^M(s)} \tag{11}$$

Here I_t^E is the total observed scattered intensity. I^B comprises all components not depending on the liquid structure, i.e. atomic, molecular, inelastic, extraneous, multiple scattering and the correction for the above mentioned substitution.

V.2.3. THE DIFFRACTION EXPERIMENT

There are only few reports of studies of liquid structure by high energy electron diffraction. Twenty years after his first attempt /14/ Maxwell /15/ obtained a diffraction pattern from a 50 keV electron beam penetrating thin films of non-volatile liquids e.g. diffusion pump oil, glycerol, silicon oils.

Roth /16-19/ used a pressurized chamber, through the apertures of which, a thin layer of liquid could be penetrated by the electron beam. So, in addition to non-volatile liquids he could study water, also. The diffraction pattern was photographed with a normal camera and evaluated visually. Following Maxwell, Roth used Bragg's equation to calculate intermolecular distances characteristic for the liquid. His diffraction patterns for water were of intermolecular origin for he was unable to obtain any diffraction rings from water vapour. Four years later, however, Bartell and Shibata /20/ obtained information on water vapour using the sector microphotometer technique.

Roth encountered the following difficulties:

(a) the electrons had to travel a long distance through a high vapour pressure region, since the height of the diffraction chamber could not be made shorter.

(b) the temperature of the liquid film could neither be regulated nor be measured.

The production of a thin water film in high vacuum (10^{-4} - 10^{-5} Torr) is, of course, a difficult task. Utilizing the above mentioned experiences of Roth /19/ and those of other authors /21-23/ on the electron microscopy of hydratised biological materials, we developed a device - a chamber under pressure-for producing a stable thin liquid film for the diffraction of penetrating electron beams.

In the chamber near-equilibrium vapour pressure was maintained by thermostating a large volume of bulk water. In order to minimize gas scattering, the thickness of water vapour crossed by the electron beam had to be kept as small as possible a condition fairly well fullfilled in our chamber /24/, (Fig.1.).

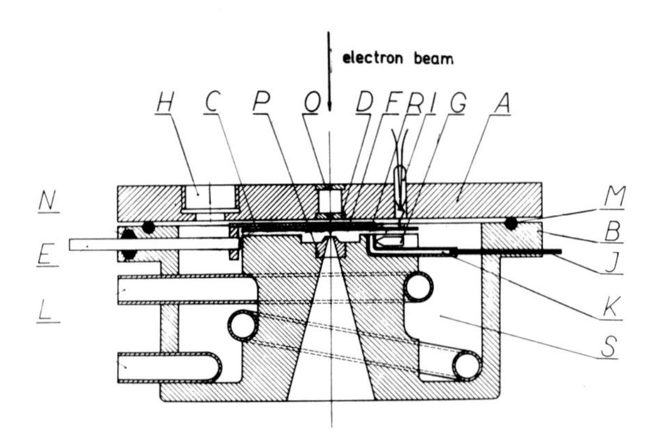

Fig.1. Scheme of the chamber under pressure.

Upper part (A) with entrance aperture 0.07 mm Ø; H: connection to the vacuum measuring device; I: temperature controlling thermistor; C: copper foil for the liquid sample (thickness 0.02 mm, opening diameter 2 mm).
Lower part (B) with the exit aperture 0.2 mm Ø. L: thermostate; S: volume container for the bulk liquid. The liquid film is produced by introducing water through the tube J-K into the hole R and then smearing it over the copper foil by the wiper mechanism E.

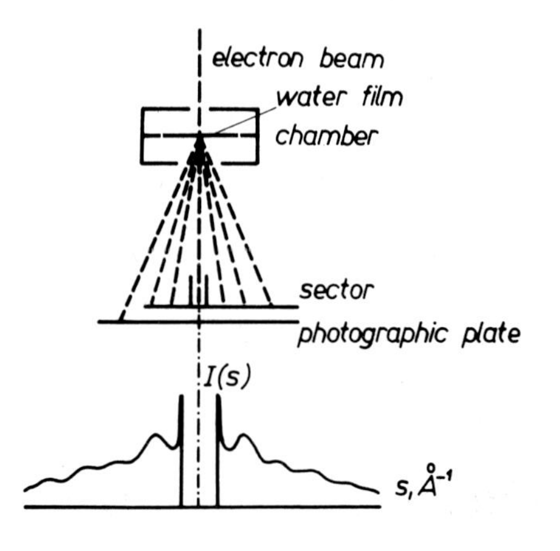

Fig. 2.

System of liquid electron diffraction

Fig.2. shows the scheme of liquid electron diffraction. The diffraction pattern is photographed on a photoplate (Agfa-Gevaert Scientia 23 150) using a so called s^2 - rotating sector, /25/ which serves to counterbalance the very steep decrease of the electron scattering function with growing s.

During the experiments the following observations were made on the fluorescent screen:

1. With no liquid film present, the sharp focus of the primary beam can be seen.

2. At the beginning of the experiment the water film is too thick, the screen is almost dark, because all electrons are absorbed. In a few seconds the primary beam gets visible and the whole screen gradually becomes clearer.

3. After a few additional seconds diffuse concentric rings appear and the photoplate can be exposed. Exposure time is about 3 seconds. At this moment, the thickness of the water layer amounts to a few hundred Angströms. The lifetime of such a liquid layer is a few seconds after which it bursts.

Our electron-optical device is not suited for the study of a thicker film since the maximal accelerating voltage is 75 keV.

With this method, the diffraction pattern water could be examined between 0° and 50^oC.

V.2.4. OBTAINING STRUCTURE FUNCTION AND PAIR CORRELATION FUNCTION OF LIQUID WATER FROM ELECTRON DIFFRACTION EXPERIMENT

The electron diffraction pattern of water at 4° was taken with a medium camera distance (of 42 cm).

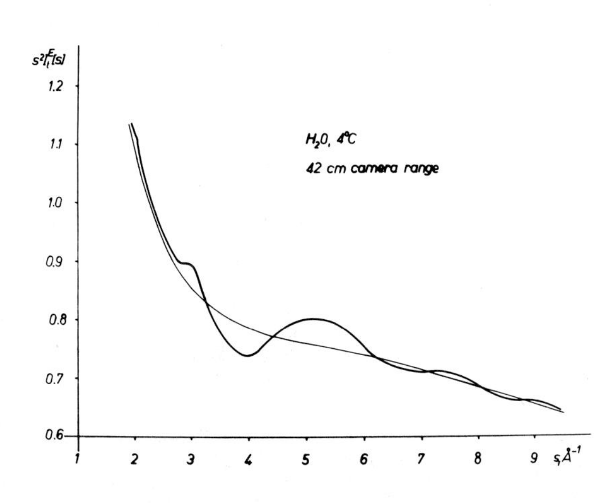

Fig.3.

Total experimental intensity, I_t^E, (the background drawn in), for liquid water, at 4^oC.

By the use of a microdensitometer the density distribution in the diffraction were converted into intensity distributions and these plotted against s (see Fig.3.). From this averaged experimental intensity $I_t^E(s)$, in parallel measurements, the modified structure-function $M^E(s)$

$$M^E(s) = \frac{I_t^E(s) - I^{B'}(s)}{I^{B'}(s)} \tag{13}$$

was calculated (Fig.4.). This differs from M(s) defined in eq.11. in the denominator and in that the background I^B in eq.13. does not contain I^M of eq.9. because the background in the averaged experimental intensity curve was drawn, to some extent arbitrarily, as a smooth curve. Variations of the background curve affect mainly that part of the pair correlation function which correponds to small values of r. These variations may lead to changes in the shapes of the peaks of the pair correlation function but do not move the positions of their maxima.

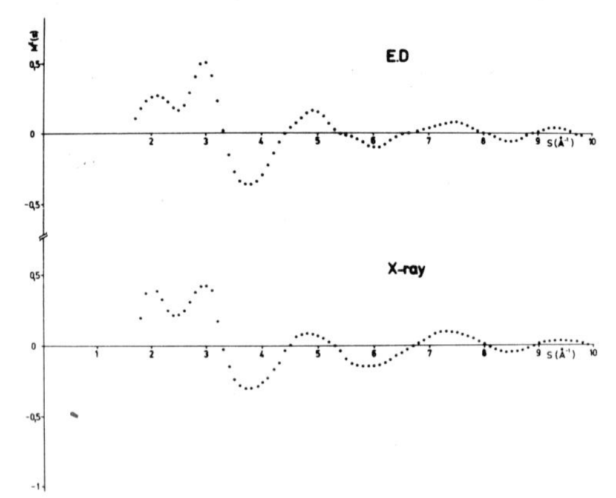

Fig.4. Experimental structure-function, $M^E(s)$, for water at 4^oC, in comparison to X-ray data.

Finally, from the modified structure function M^E the experimental pair correlation function g(r), (eq.14.) was calculated. In contrast to g_M in eq.10., g in eq.14. contains also information on the structure within the molecule (O-H bond distance, and H...H distance).

$$g(r) = 1 + \frac{1}{2\pi^2 r\rho_0} \int_{s_{min}}^{s_{max}} s\, M^E(s) \exp(-as^2) \sin sr\, ds \tag{14}$$

Experimental conditions always ensure that contributions to the integral in the ranges 0 to s_{min} and s_{max} to infinity are neglected. For the latter a correction is made by using the damping factor $\exp(-as^2)$ which makes the neglected integral irrelevant. Extension of the s interval particulary in the upper range increases resolution of the peaks of the pair correlation function mainly for small r values. Thus, by this extension we may get more detailed information about the structure of the molecule.

In order to show the applicability of electron diffraction to the study of liquid water structure, the pair correlation function g(r),

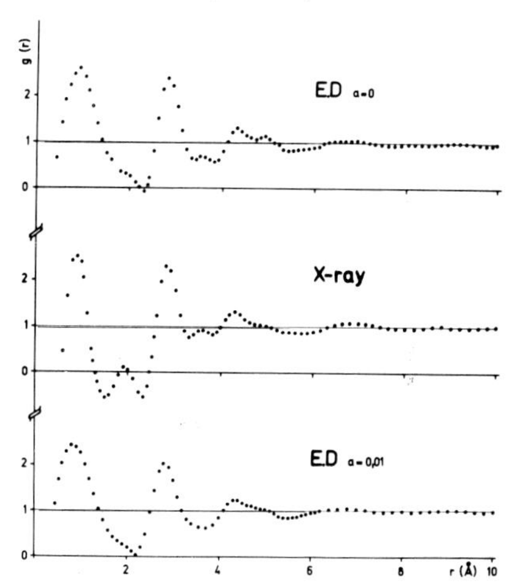

Fig.5. Experimental pair correlation function, g(r), for liquid water (with and without damping), at 4°C, in comparison to X-ray data.

Fig.5., was calculated from the $M^E(s)$ curve Fig.4. obtained on water at 4°C in the given s range, and compared with the corresponding function derived from X-ray diffraction /2/. In the calculations (performed on a CDC 3300 computer) the integral in eq.14. has been replaced by Simpson rule summation (s = 0,1). The overall pattern of our electron diffraction pair correlation function (Fig.5.) can be discussed according to the interpretation of Narten /27-28/ as follows:

a) The first maximum, at 0.95 Å is ascribed to the intramolecular O-H bond distance.

b) The large maximum at 2.84 Å is due to the distance between oxygen atoms of neighbouring water molecules.

c) On the other hand, the three maxima 4.3 Å; 4.9 Å and 6.85 Å correspond to second and more distant neighbour interactions.

When using a damping factor, the pair correlation function was obtained without any significant change in the positions of the maxima. Detailed interpretation of the pair correlation function however, will follow more thorough data reduction and structure analysis which is in progress.

We thank many colleagues especially Dr.I.Hargittai for useful discussions.

REFERENCES

1 S.Lengyel, Far.Soc.Disc. 24, 223 (1957); Acta Chim.Hung. 37, 87, 319 (1963)

2 F.Hajdu, S.Lengyel and G.Pálinkás: International Symposium "Structure of Water and Aqueous Solutions" Marburg (GFR) 1973

3 C.Tavard, Cah. de Phys. 20, 1 (1966)

4 D.I.Page and J.G.Powles, Mol.Phys. 21, 901 (1971)

5 N.F.Mott and H.S.W.Massey, The Theory of Atomic Collision, Oxford University Press, Oxford (1965)

6 L.Schiff, Quantum mechanics, McGraw-Hill Book Co, New York, 1955

7 Z.G.Pinsker, Electron diffraction, Butterworths Scientific Publications, London 1953

8 T.Iijima, R.A.Bonham and T.Ando, J.Phys.Chem. 67, 1472 (1963)

9 C.Tavard, Cah. de Phys. 17, 165 (1963)

10 S.H.Bauer, Physical Chemistry an Advanced Treatise, Vol.IV. Ch.14, Academic Press, New York (1971)

11 P.A.Egelstaff, An Introduction to the Liquid State. Academic Press, London and New York, 1967

12 H.H.Paalman and C.J.Pings, Rev.Mod.Phys. 35, 389 (1963)

13 L.van Hove, Phys.Rev. 110, 999 (1954)

14 L.R.Maxwell, Phys.Rev. 44, 73 (1933)

15 C.W.Lufcy, F.S.Palubiskas and L.R.Maxwell, J.Chem.Phys. 19, 217 (1951)

16 G.Roth, Dissertation, Freie Universität Berlin, 1961

17 G.Roth, Z.Naturforschung, 17a, 1022 (1962)

18 G.Roth, Z.Naturforschung, 18a, 516 (1963)

19 G.Roth, Z.Naturforschung, 18a, 520 (1963)

20 S.Shibata and L.S.Bartell, J.Chem.Phys. 42, 1147 (1965)

21 D.F.Parson and R.C.Moretz, Septieme Congres International de Micorscopie Electronique, Grenoble, 497, 1970

22 H.G.Heide, J.Cell. Biol. 13, 147 (1962)

23 I.G.Stoyanova and G.A.Mikhalovski, Biofizika, USSR 4, 1483 (1959)

24 E.Kálmán, S.Lengyel, L.Haklik and A.Eke, Z.Phys.Chem., in press

25 Ch.Finbak, Avh.Norsk Vidensk-Akad. Oslo, M.-N.Kl 13, 2 (1937)

26 W.Witt, Z.Naturforschung 19a, 1363 (1964)

27 A.H.Narten, "X-ray Diffraction Data on Liquid Water in the Temperature Range 4°C - 200°C, ORNL-4578, 1970

28 A.H.Narten, H.A.Levy, J.Chem.Phys. 55, 2263 (1971)

VI DIELECTRIC METHODS

VI.1. THE DIELECTRIC PROPERTIES OF WATER

J. B. Hasted

Birkbeck College
(University of London)

Malet Street, London, WC1E 7HX England

ABSTRACT

The static field and time-dependent dielectric properties of liquid water are discussed in terms of the static and time-dependent dipole correlation functions, which can be calculated by computer dynamics and using more approximate models; in particular, the statistics of hydrogen bond-breakage are used in these calculations.

Recently published data at submillimetre wavelengths are analysed by subtraction of the contributions from the principal (microwave) dielectric relaxation and from infrared absorptions. A second relaxation process is proposed, which is interpreted as arising from water molecules with no more than one hydrogen bond. The proportions of these molecules are thereby estimated and their temperature variation compared with bond-breaking statistics. With all bond energies equal the agreement is poor, but with the inclusion of cooperation, using scaled molecular orbital energies of Del Bene and Pople, a great improvement is found.

It is possible in 1973 to make use of the advances in dielectric theory, in computerized numerical analysis, in microwave and submillimetre-wave spectroscopy and in data handling, in order to throw light on the problems of water structure, by understanding its electrical polarization.

Although the water molecule possesses a range of electrical multipole moments, both permanent and induced, dielectric theory is still only able to interpret the time-dependent dielectric properties of the liquid in terms of the permanent H_2O dipole moment $\mu = 1.84 \times 10^{-18}$ esu and the scalar polarizability $\alpha = 1.444 \times 10^{-24}$ cm^3. Even within these limitations there is considerable difficulty about the effects of inter-dipole correlation upon macroscopic dielectric properties. One current formulation /1/ of the problem of how the time-dependent correlation function

$$\gamma(t) = \frac{\langle \underline{\mu}(t) \cdot \underline{\mu}(o) \rangle}{\langle \underline{\mu}(o) \cdot \underline{\mu}(o) \rangle} , \qquad (1)$$

for response of dipole to a rotational displacement at time t = o, is related to the frequency-dependent complex dielectric constant $\hat{\varepsilon}(\omega) = \varepsilon' - j\varepsilon''$, is as follows:

$$\frac{(\hat{\varepsilon} - \varepsilon_\infty)(2\hat{\varepsilon} + \varepsilon_\infty)\varepsilon_s}{(\varepsilon_s - \varepsilon_\infty)\hat{\varepsilon}(2\varepsilon_s - \varepsilon_\infty)} = \int_o^\infty - \frac{d\gamma}{dt} \exp(i\omega t)\, dt \qquad (2)$$

In principle it is possible to calculate $\gamma(t)$ from the experimental $\varepsilon(\omega)$, but owing to mathematical difficulties and certain gaps in the data, this has not yet been achieved for water. The converse process, of calculating $\varepsilon(\omega)$ from a $\gamma(t)$ derived from theory, is somewhat easier. The computer-dynamical calculations of Rahman and Stillinger /2/ for liquid water have in fact enabled a function $\gamma(t)$ to be calculated at T = 34.4°C. Its form is shown in Figure 1, and the $\varepsilon''(\varepsilon')$ function derived from it is compared with experiment in Figure 2. Although the detail of this comparison is inexact, the fact that the (microscopic) principal relaxation time is predicted to be 5.6×10^{-12} s, in comparison with the best experimental value /3/ at this temperature of 6.7×10^{-12} s, is a remarkable achievement.

The calculations also yield the static dipole correlation coefficient g; by definition $g\mu$ is the total dipole moment of the

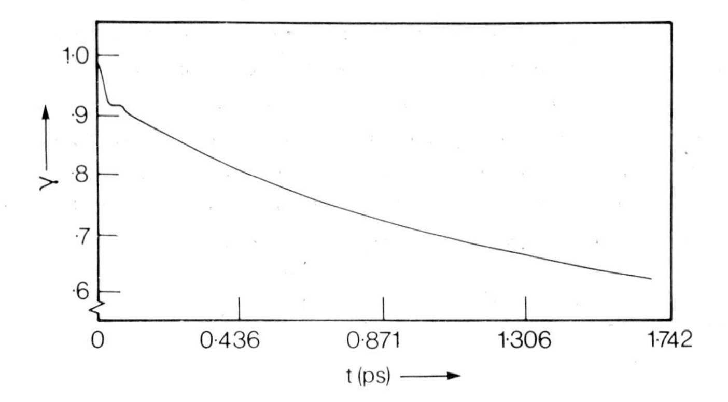

Fig.1 Time-dependent dipole correlation function for water at 34.3°C calculated by computer dynamics /2/.

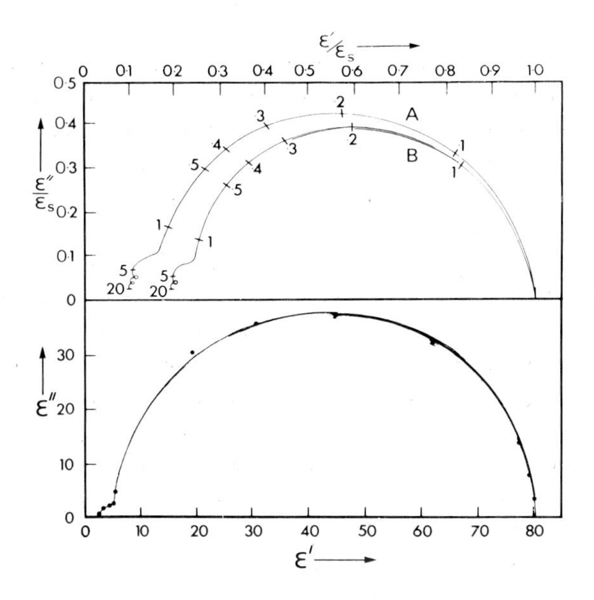

Fig.2 Comparison of experimental ε" (ε') data (lower curve) at 25°C with that calculated for water at 34.3°C (upper curve) by computer dynamics, using two different ratios $\varepsilon_S/\varepsilon_\infty$.

molecules in a sphere of the liquid surrounding an H_2O dipole of moment μ and fixed orientation. It is thus a structural parameter which measures the instantaneous correlation between neighbouring dipoles. The calculation of g for a polymeric configuration is purely a problem of vector analysis. According to the Kirkwood-Fröhlich theory /4,5/ the static field dielectric constant ε_s is related to the correlation parameter as follows:

$$\frac{(\varepsilon_s - n^2)\,(2\varepsilon_s + n^2)}{\varepsilon_s\,(n^2 + 2)^2} = \frac{4\pi\, N_o}{9\,kT\,V}\, g\,\mu^2 \qquad (3)$$

where n is the refractive index, V the molar volume, N_o the Avogadro constant, k the Boltzmann constant, μ the free molecule dipole moment.

The temperature variation of static dielectric constant is such that the correlation parameter may be regarded as taking the values: 2.87 at 0°C, 2.56 at 100°C, 2.24 at 200°C, 1.64 at 300°C. The computer-dynamically calculated value is g = 2.96 at 34.3°C, but the temperature dependence is not yet available.

It may be that the value of g is not a particularly sensitive measure of water structure, but it does give a general indication that the hydrogen bonding (or similar form of correlation between molecules) is considerable, and falls monotonically with increasing temperature. A number of previous attempts to calculate g(T), using various models of water, have appeared. Calculated and experimental variations $\varepsilon_s(T)$ are illustrated in Figure 3. There is no very marked difference in the g(T) calculated on bond-bending /6/ and bond-breaking /7/ models, but bond-breaking statistics allows some comparisons between different structural proposals to be made. In bond-breaking statistics the liquid is assumed to be composed of a mixture of species i, j, which are inter-converted by the breakage and formation of hydrogen bonds, the energies ΔE_{ij} of which determine the temperature variation of the species populations n_i, and the populations of bond types m_{ij} which exist between different species. The appropriate equations are

$$i + j \longrightarrow (i \pm 1) + (j \pm 1) + \Delta E_{ij\pm} \qquad (4)$$

$$\sum_i n_i = 1 \qquad (5)$$

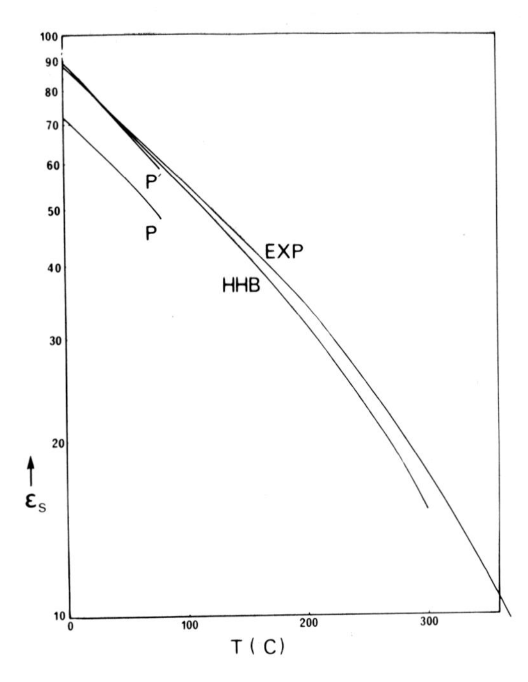

Fig.3 Measured (exp) and calculated temperature variation of static dielectric constant ε_S of water, P, Bond-bending calculations of Pople /6/; P', calculations of Pople with modified dipole moment; HHB, bond-breaking calculations of Haggis et.al. /7/.

$$i\, n_i = m_{ii} + \sum_{i,j} m_{ij} \tag{6}$$

$$m_{ij} \exp\left(- \frac{\Delta E_{ij\pm}}{kT}\right) = n_{i\pm 1}\, n_{j\pm}\, r \tag{7}$$

When the matrix solution of a large number of simultaneous equations of the type of equation 7 is carried out, it is found that the unknown factor r is unity. The energies ΔE_{ij} must be shown to be consistent with the temperature variation of latent heat of evaporation of the liquid, using two further equations

$$\sum_{i,j} m_{ij} = 2(1 - p) \tag{8}$$

and

$$p = 1 - \left(\frac{L - W}{2 \langle \Delta E_{ij} \rangle} \right) \tag{9}$$

where L(T) latent heat of evaporation, W(~~α~~ T) the van der Waals energy and p is the proportion of bonds broken, compared with a completely four-bonded ice-like system. A reasonable, positive and fairly temperature-independent value of W must be found.

Three models have been so treated:

i) Broken networks with all bond energies equal /7/. The $\varepsilon_s(T)$, as shown in Figure 3, is satisfactory, but the temperature variation of infinite frequency dielectric constant $\varepsilon_\infty(T)$ is unsatisfactory.

ii) Ring and chain species as considered by Del Bene and Pople /8/ without bonding between them. It has been shown /21/ that for consistency with the thermal properties the MO energies calculated for these species must be scaled down by a large factor. The species include chains of up to 5 members, and planar rings of between 6 and 3 members, both symmetrical about the plane of the ring, and unsymmetrical, i.e. with all the free hydrogen atoms on one side of the ring. Clearly the symmetrical planar rings are almost non-polar, and could not occur frequently in such a polar liquid as water. When the scaling factor for these species is raised to $\simeq 9$ and that for the remaining species held at $\simeq 6$, not only is thermodynamic consistency maintained, but the calculated $\bar{g}(T) = \sum_i n_i g_i$ are quite reasonable (2.93 at 0°C, 2.22 at 100°C, 1.67 at 200°C, 1.42 at 300°C); again, however, the temperature variation of infinite frequency dielectric constant is unsatisfactory.

iii) Broken networks with cooperative hydrogen bond energies. Following the original proposal by Frank that a hydrogen bond is strengthened by the formation of a neighbouring bond, this cooperation effect was clearly demonstrated in the calculations of Del Bene and Pople. Unfortunately these calculations have not yet been extended into network polymers, but the calculated chain energies can be used as a guide to the cooperation effect, in a broken network statistical calculation. As before, scaling down is necessary for thermodynamic consistency, but if this is done only by a factor of 3 (e.g. ΔE_{11} = 2.03 kcal/mole, ΔE_{21} = 2.87 kcal/mole, etc), then an interesting result is achieved. The temperature variation of infinite frequency dielectric constant $\varepsilon_\infty(T)$ becomes reasonable. The static dielectric constant is not reproduced so

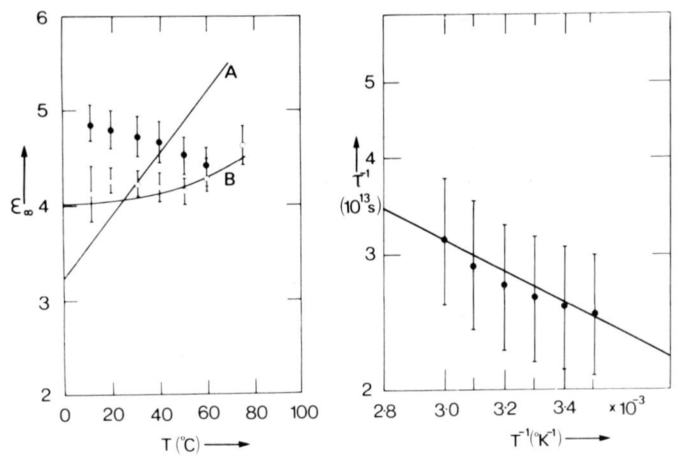

Fig.4 a) Measured and calculated temperature variation of high frequency limiting dielectric constant ε_∞ of water, open circles, derived from principal relaxation data. Closed circles, derived from second relaxation data. A, calculated assuming all hydrogen bands of equal energy. B, calculated assuming cooperation between bond energies.

b) Activation enthalpy plot for the second relaxation in water.

faithfully, but on the other hand it is more sensitive to bond-bending effects, and is a less stringent test of bond-breaking statistics. The variation ε_∞ (T) so calculated is compared with the experimental variation, and with that calculated with all bond energies equal, in Fig. 4. The improvement may be taken as supporting the cooperation effect. The species populations n_i , i = 0 to 4, are represented at T = 0° , 50° , 100°C in Figure 5, being compared with previous calculations /7/ and with other proposals /9,10,11/. The wide variation between different calculated values may of course be taken as a criticism of bond-breaking statistics.

We return now to the time-dependent dielectric properties of the liquid, which are illustrated in Figure 6. The principal dielectric relaxation has now been subjected to a regression analysis /12/ by means of which it has been shown that it is more than 99% certain that there is a small spread of relaxation times (or possibly two or more closely spaced times), as represented by the Cole-Cole equation for the complex dielectric constant

$$\hat{\varepsilon} = \varepsilon_\infty + \frac{\varepsilon_s - \varepsilon_\infty}{1 + (j\omega\tau)^{1-\alpha}} \qquad (10)$$

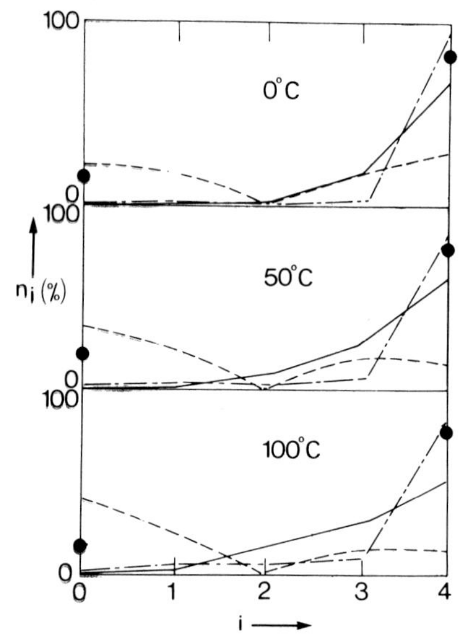

Fig.5 Comparison of bond species populations calculated according to different mixture theories. ——— Dielectric properties, assuming all hydrogen bonds of equal energy. — – — – Dielectric properties, assuming cooperation between bond energies. — — — — Thermodynamic properties /9/
● x-ray diffraction model /10/.

The principal relaxation times τ , spread parameters α, and infinite frequency dielectric constants ε_∞ are tabulated over a range of temperature in Table 1. Data at 25°C are illustrated in Figure 6.

Table 1. H_2O relaxation parameters derived from the Cole-Cole equation

T(°C)	ε_∞	$\tau(10^{-11}s)$	α
0	4.46 ± 0.17	1.79	0.014
10	4.10 ± 0.15	1.26	0.014
20	4.23 ± 0.16	0.93	0.013
30	4.20 ± 0.16	0.72	0.012
40	4.16 ± 0.15	0.58	0.009
50	4.13 ± 0.15	0.48	0.013
60	4.21 ± 0.16	0.39	0.011
75	4.49 ± 0.17	0.32	-

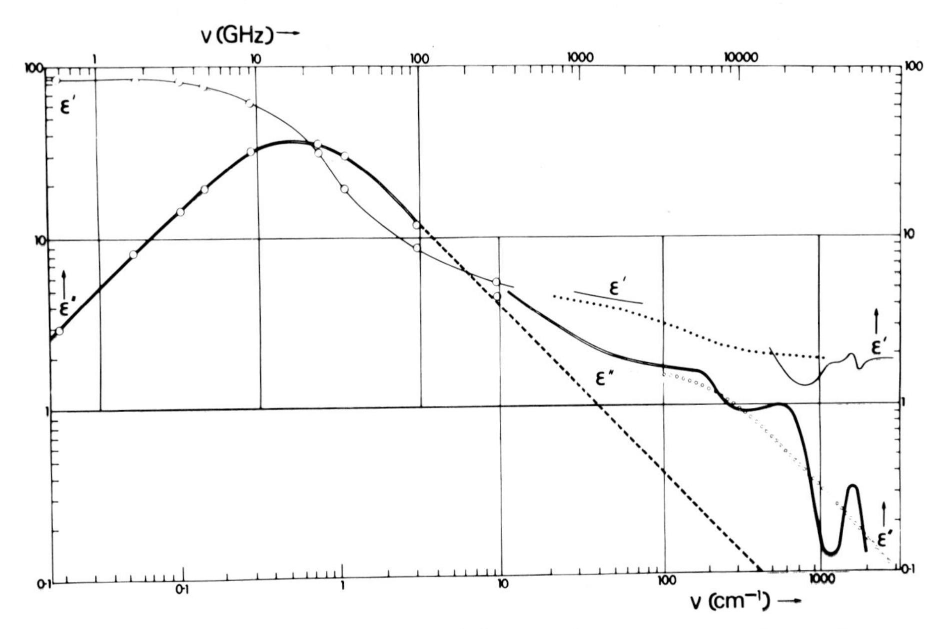

Fig.6 Frequency variation of ε' (single line and open circles) and ε" (double line and open circles) for water at 25°C. Broken double line represents the contribution of the principal relaxation time to ε". Dotted lines (open circles for ε", closed for ε') represent a first attempt /3/ to fit the second relaxation process.

The temperature variation of τ is consistent with an activation enthalpy of about 4.5 kcal/mole which is approximately the energy of an (isolated) hydrogen bond between two water molecules; but there is some curvature in the semilogarithmic plot. The reorientation process, involving as it does the breakage of one hydrogen bond, might well involve the rotation of those molecules which make two and only two bonds to neighbours.

It is apparent that the infinite frequency dielectric constant ε_∞ is considerably larger than the square of the infrared refractive index $n^2 = 1.8$. Its temperature variation is shown in Figure 4. It will be recalled that the corresponding value of ε_∞ for ice I at 0°C is 3.2. There is some process unique to the liquid which causes this enhancement. Recent investigations of refractive index and absorption coefficient in the waveband 20 - 100 cm^{-1} /13/, as well as earlier measurements of absorption coefficient /14/ show an apparent broad absorption band and monotonic decrease of dielectric constant, similar in form to a dielectric relaxation.

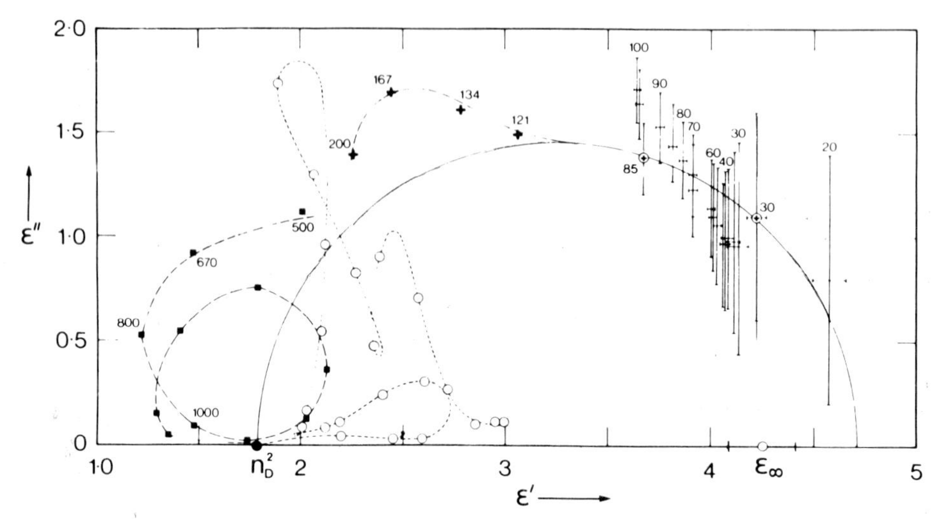

Fig.7 ε"(ε') (full line) representation of the second relaxation, with curlecues, of water at 25°C. Circles and data with error bars, from Zafar et.al. /13/, corrected for contribution from principal relaxation process. Crosses, Cartwright and Errera /15/. Squares, Querry et.al. /21/. $\bar{\nu}$ (cm^{-1}) marked against data points. Broken line and open circles ε"(ε') for ice at 0°C /20/.

It was decided /14/ to analyse these data by subtracting the contribution from the tail of the principal relaxation and displaying the ε"(ε') on a Cole-Cole diagram, which is shown in Figure 7. The infrared absorption bands at 165 and 700 cm^{-1} show up as curlecues superposed upon a semicircular arc. In the 200 cm^{-1} region the only refractive index data are those due to Hawley Cartwright /15/, which appear to have stood the test of time remarkably well. If it is accepted that the fall of refractive index can be analyzed in terms of a Debye equation, then the usual graphical analysis of chords yield a relaxation time of 0.53×10^{-13} s at T = 25°C. The temperature variation, illustrated in Figure 4, shows that the activation enthalpy of the process is 1.0 ± 0.5 kcal/mole. However the dielectric constant corresponding to the low frequency side of this process ("static" dielectric constant, or extreme righthand value of ε' in Figure 7) is not precisely equal, as it should be in this analysis, to the 'infinite frequency' dielectric constant of the principal relaxation. In Figure 4 both values of this quantity are shown as a function of temperature, and it is apparent that although the error bars are large, they are insufficient to cover the discrepancy.

There are two possible interpretations of the discrepancy. One is that the principal relaxation behaves according to the Fatuzzo-Mason equation /16/, which displays a bulge on the high frequency side of the Cole-Cole diagram. If analyzed in terms of a Cole-Cole equation, the principal relaxation will yield an artifically low value of ε_∞. An analysis in terms of the Fatuzzo-Mason equation is planned.

Although this interpretation was preferred in the first discussion of this relaxation /13/, it is significant that a discrepancy of this type is also to be expected when hindered rotation is taken into account /17/. A resonance hindered rotation of the dipole gives rise to a minimum in the real part of the dielectric constant; such behaviour has been observed in non-hydroxylic liquids. If this process also contributes to the dielectric constant, in a region between the two relaxations, then we would expect the principal relaxation ε_∞ to appear low and the second relaxation ε_s to appear high when analyzed without taking account of hindered rotation.

The ε_∞ discrepancy should not be allowed to confuse the issue by obscuring the reality of the second relaxation. It is worthwhile to consider how this relaxation might arise, and the obvious interpretation, consistent with the low enthalpy of activation and with the absence of the relaxation in ice, is that it arises from molecules in the liquid which are free to rotate without 'breakage' of a hydrogen bond. The dielectric behaviour of ice is shown as a broken line in Figure 7, and it is clear that no appreciable relaxation is displayed. On the other hand, the submillimetre-band absorption is greatly enhanced in a dilute solution of water in carbon tetrachloride /18/, under which conditions a high proportion of unbonded molecules would be expected.

The second relaxation process, therefore, is produced by those molecules in water which possess less than two hydrogen bonds. The ratio $\varepsilon_\infty/\varepsilon_s$ will indicate the magnitude of the proportion of these molecules in the liquid, i.e. $(n_o + n_1)/\sum_i n_i$ (.045 at 0°C, .05 at 25°C, .075 at 60°C).

The ε_∞ value calculated from the principal relaxation should be preferred in the calculation, since hindered rotation may be contributing to the second relaxation. It is this procedure which has been used to test the bond breaking statistical calculations described above. It is remarkable how slowly this proportion increases with temperature. It is of some importance to extend measurements to higher temperatures, especially over 100°C, where the proportion of free molecules is believed to be quite large. The extension of water refractive index measurement to the region 100-500 cm^{-1}, as well as 0.2 - 20 cm^{-1} is also necessary.

Some unpublished submillimetre band measurements on NaCl solutions have been made by Zafar et al /20/. The value of ε_{∞} is enhanced, which is consistent with the well-known lowering of principal relaxation time /19,7/; the effect of the hydration sheath is on the whole to increase the proportion of broken bonds.

REFERENCES

1 D.D. Klug, D.E. Kranbuehl and W.E. Vaughan, J. Chem. Phys. 50, 3904 (1969)

2 A. Rahman and F.H. Stillinger, J. Chem. Phys. 55, 3336 (1971)

3 J.B. Hasted, in WATER, A Comprehensive Treatise (edit. by Felix Franks) Vol 1, chap. 7. Plenum Press:New York (1972)

4 J.G. Kirkwood, J. Chem. Phys. 4, 592 (1936)

5 H. Fröhlich, Theory of Dielectrics, Oxford University Press, 2nd ed., (1958)

6 J.A. Pople, Proc. Roy. Soc. A 221, 498 (1954)

7 G.H. Haggis, J.B. Hasted and T.J. Buchanan, J. Chem. Phys. 20, 1452 (1952)

8 J. Del Bene and J.A. Pople, J. Chem. Phys. 52, 4858 (1970)

9 G. Nemethy and H.A. Scheraga, J. Chem. Phys. 36, 3382 (1962); J. Chem. Phys, 41, 680 (1964)

10 A.H. Narten, M.D. Danford and H.A. Levy, Disc. Faraday Soc. 43, 97 (1967)

11 K. Buijs and G.R. Choppin, J. Chem. Phys. 39, 2035 (1963); J. Chem. Phys. 40, 3120 (1964); V. Vand and W.A.Senior, J. Chem. Phys. 43, 1878 (1965)

12 P.R. Mason, J.B. Hasted and L. Moore, In course of publication

13 M.S. Zafar, J.B. Hasted and J. Chamberlain, Nature Physical Science, 243, 106 (1973); J. Chamberlain, M.S. Zafar and J.B. Hasted, Nature Physical Science, 243, 117 (1973)

14 J.E. Chamberlain, G.W. Chantry, H.A. Gebbie, N.W.B. Stone, T.B. Taylor and G. Wyllie, Nature, 210, 790 (1966); D.A. Draegert, N.W.B. Stone, B. Curnette, and D.Williams, J. Opt. Soc. Amer. 56, 64 (1966); V.M.Zolotarev, B.A. Mikhailov, L.I. Alperovich, and S.I. Popov, Optics and Spectroscopy, 27, 430 (1969)

15 C.H. Cartwright and I. Errera, Proc. Roy. Soc. A,154, 138 (1936); Phys. Rev. 49, 470 (1936)

16 E. Fatuzzo and P.R. Mason, Proc. Phys. Soc. 90, 741 (1967)

17 Nora E Hill, Chem. Phys. Letters, 2, 5 (1968); J. Phys. A. 2, 398 (1969)

18 G W.F. Pardoe and H.A. Gebbie, Symposium on Submillimetre Waves, Polytech. Inst. Brooklyn, p 643 (1970)

19 J.B. Hasted, D.M. Ritson and C.H. Collie, J. Chem. Phys. 16, 1 (1948)

20 J.B. Hasted, "Aqueous Dielectrics", ed. A.D. Buckingham, Chapman and Hall, London. To be published November 1973.

21 M.R Querry, B. Curnutte and D. Williams, J. Opt. Soc. Amer. 59, 1299 (1969)

VI.2. DIELECTRIC RELAXATION OF WATER IN AQUEOUS SOLUTIONS

R. Pottel, K. Giese and U. Kaatze

Drittes Physikalisches Institut der Universität
Göttingen, Germany

Abstract

Thermal molecular motion give rise to a fluctuating dielectric polarization. Dielectric relaxation measurements yield the time auto-correlation function of these thermally induced fluctuations. The relation of this function to correlation functions of the molecular electric dipole moments is considered. Different types of electric dipole moment fluctuations are mentioned. In view of aqueous solutions approximations and models are discussed necessary to derive molecular electric dipole correlation times, especially reorientation times, from dielectric relaxation measurements.

Some experimental results are presented in order to show the influence of different types of solute-solvent interaction on the water molecular motion in aqueous solutions. The motion of the hydration molecules in the cases of "positive", "negative", hydrophobic and hydrogen bond hydration is briefly discussed.

VI.2.1. Introduction to the Relation between Molecular Motion and Dielectric Relaxation

A. Any Pure Dipolar Liquid

At first we consider the molecules in any pure dipolar liquid. The arbitrary molecule "i" has a total electric dipole moment $\underline{\mu}_i$ [+)], which consists of two parts. One is of permanent amount characteristic to the isolated molecule. The other part is due to the interaction with other molecules in the liquid. Due to thermal motion the orientation of the molecules and the interaction between them are varying. The spontaneous fluctuation of both the direction and the amount of the total molecular electric dipole moment causes $\underline{\mu}_i(t)$ to be a more or less random function of time t.

The sum over the total electric dipole moments of all N molecules within a macroscopic spherical volume V embedded in the same infinitely extended medium, divided by V, yields the dielectric polarization of the liquid

$$\underline{P}(t) = \frac{1}{V}\sum_{i=1}^{N} \underline{\mu}_i(t) \tag{1}$$

As we assume the liquid to be in thermal equilibrium and any macroscopic electric field to be absent, $\underline{P}(t)$ is the spontaneous thermal fluctuation of the dielectric polarization with the ensemble average being

$$\langle \underline{P}(t) \rangle = 0 \tag{2}$$

The information about molecular motions as contained in $\underline{P}(t)$ can be obtained by measuring the time auto-correlation function

$$\Psi(t) = \langle \underline{P}(t)\cdot\underline{P}(0) \rangle \tag{3}$$

of the polarization. Accessible to dielectric relaxation measurements, which are treated in this paper, are the initial value $\Psi(0)$ and the course in time of $\Psi(t)$ for times larger than about 1 psec. With respect to this time scale polarization fluctuations occuring much faster are considered to happen instantaneously.

After inserting the sum (1) into eqn. (3), and separating the products of the dipole moments of equal from that of different molecules, the relation between $\Psi(t)$ and the molecular correlation functions is found to be

+) The bar below a symbol denotes a vector

$$\Psi(t) = \frac{N}{V^2} < \underline{\mu}(t) \cdot \underline{\mu}(0) > + \frac{N}{V^2} < \underline{\mu}(t) \cdot \sum_j{}' \underline{\mu}_j(0) > \tag{4}$$

The first term, the dipole moment auto-correlation will be abbreviated by

$$A(t) = \frac{N}{V^2} < \underline{\mu}(t) \cdot \underline{\mu}(0) > \tag{4a}$$

It describes the average motion of a single molecule. The second term, the dipole moment cross-correlation will be denoted by

$$C(t) = \frac{N}{V^2} < \underline{\mu}(t) \sum_j{}' \underline{\mu}_j(0) > \tag{4b}$$

With respect to further discussion it is useful to distinguish between that part of C(t) due to short range specific interaction, $C^{(s)}$, and that part due to long range electric dipole-dipole interaction $C^{(d)}$.

To separate the measured polarization correlation function into contributions due to the various molecular processes is, in general, a task of great difficulty. It is aproximately possible if not all of the following influences on $\Psi(t)$ are equally important:

Fluctuations of the amount and the direction of $\underline{\mu}_i(t)$;

specific short range interaction part and electric dipole interaction part of C(t).

To estimate the relative importance of these influences additional experimental facts and approximative theory based upon models are required.

B. Pure Liquid Water

The permanent amount of the electric dipole moment of an isolated water molecule is known. But the variation of the molecular electric dipole moment and the cross-correlation between the directions of neighboring dipoles due to short range specific molecular interaction (e.g. H-bonds) in the liquid is not or insufficiently known. So one has to change this interaction experimentally by changing the surroundings of a water molecule. This has been done by adding appropriately choosen solutes: monovalent ions and neutral solute particles without own electric dipole moments and with sizes slightly larger than that of the water molecules, which means solutes with weak solute solvent interaction.

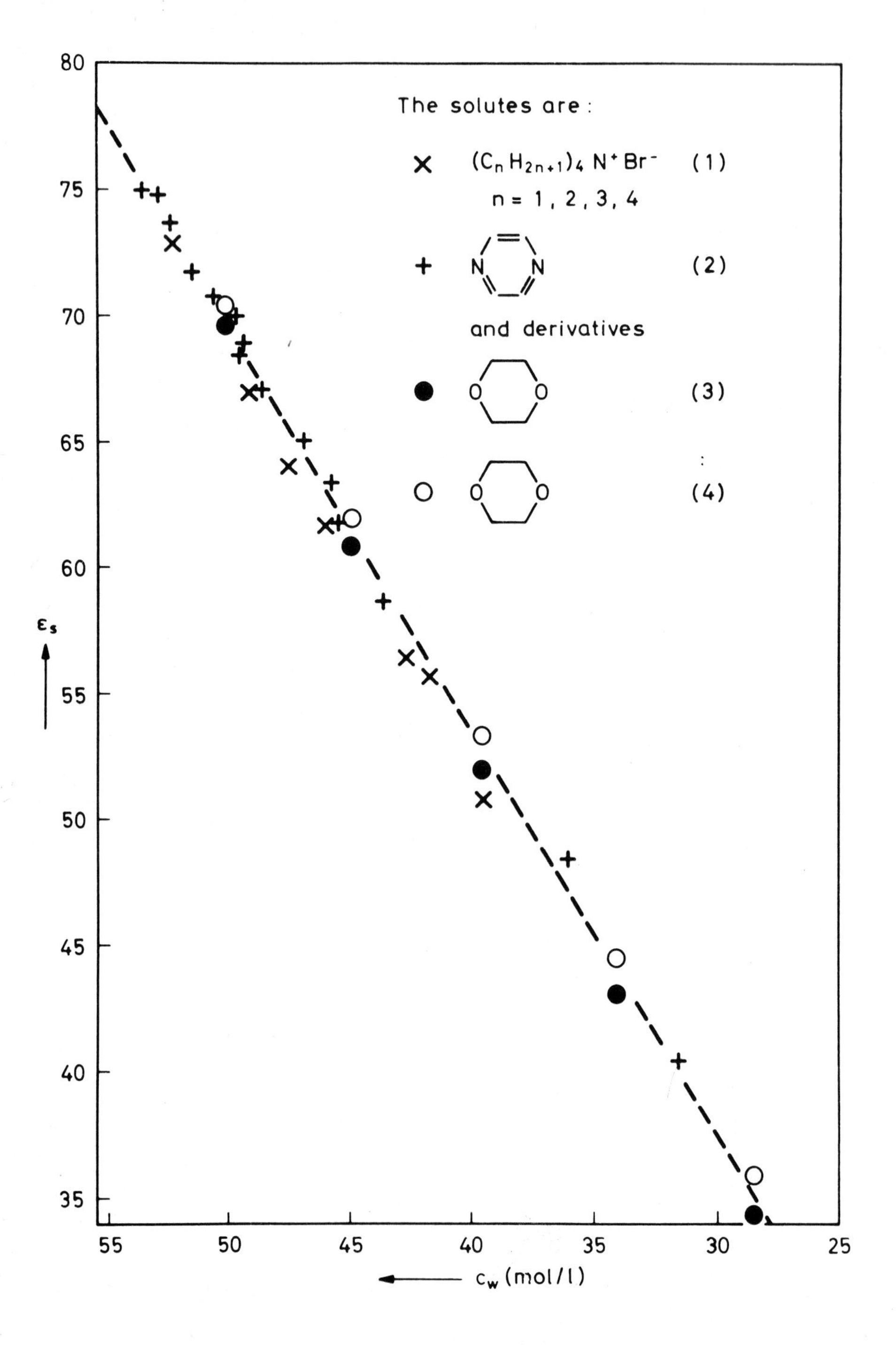

Figure 1 : The static permittivity ε_s of aqueous solutions versus the water concentration c_w for different types of solutes.

From experimental values of the static permittivity ε_s of numerous solutions of such solutes, Fig. 1, the following statement can be made ($\varepsilon_s \propto \Psi(0)$): The initial value of the measured correlation function $\Psi(0) = \langle \underline{P}^2 \rangle$, the mean square polarization fluctuation, is roughly independent of the solute type, that means roughly independent of slight variations of the short range specific interaction between the water molecules. This statement leads to the two conclusions:

a) The fluctuations of the amount μ of the molecular electric dipole moment due to variations in the short range specific molecular interaction cannot be dominant. So the dipole moment auto-correlation

$$A(t) = \frac{N}{V^2} \langle \mu(0)\mu(t)\cos\theta(t) \rangle \approx \frac{N}{V^2} \mu^2 \langle \cos\theta(t) \rangle \tag{5}$$

is predominantly determined by the auto-correlation function $\langle \cos\theta(t) \rangle$ of the molecular orientational motion with $\theta(t)$ being the angle through which the permanent molecular dipole moment $\underline{\mu}(t)$ rotates during the time t. The correlation time τ of A(t), defined by

$$\tau = \int_0^\infty (A(t)/A(0))\,dt \approx \int_0^\infty \langle \cos\theta(t) \rangle\,dt \tag{6}$$

then approximately is the reorientation time of the water molecule.

b) That part of the dipole moment cross-correlation being due to the short range specific interaction between the water molecules, $C^{(s)}$, can only be of small amount:

$$|C^{(s)}(t)| \ll |C(t)| \tag{7}$$

There is still needed some knowledge about that part of the dipole moment cross-correlation being due to the electric dipole-dipole interaction between the water molecules, $C^{(d)}$. As it cannot be determined correctly, the Onsager model of a dipolar liquid is used to obtain an estimate. In this model the arbitrary molecule under consideration is embedded in a continuous dielectric medium with the macroscopic properties of the liquid characterized by $\Psi(t)$ /5/. Here $C^{(d)}$ is decaying much more quickly than A(t) so that at times t larger than about $\tau/6$

$$|C^{(d)}(t)| \ll A(t) \tag{8}$$

holds.

On the basis of the approximations given by eqns. (5), (7), and (8) the relation between the measured correlation function of the polarization fluctuations and the molecular behaviour is

$$\Psi(t) = \langle \underline{P}(t)\, \underline{P}(0) \rangle \approx \frac{N}{V^2}\, \mu^2 \langle \cos \Theta(t) \rangle \tag{9}$$

as far as times t larger than about one sixth of the reorientation time are considered. The range of our measurements lies above that limit. They show a nearly exponential decay of the correlation function according to

$$\Psi(t) \approx \hat{\Psi}(t) = \hat{\Psi}(0)\, e^{-t/\tau} \tag{10}$$

where $\hat{\Psi}(0)$ is a parameter as obtained by extrapolation of the measured $\Psi(t)$ -values to t = 0. At 25°C one has τ = 8.3 psec and $\hat{\Psi}(0) = 0.94\Psi(0)$.

VI.2.2. Dielectric Relaxation and Molecular Motion in Aqueous Solutions

A. A Model of the Solutions

In the scope of this treatment we will restrict the discussion to solute particles without permanent dipole moment. If electrolytes are considered, the concentration of ion pairs is assumed to be negligible.

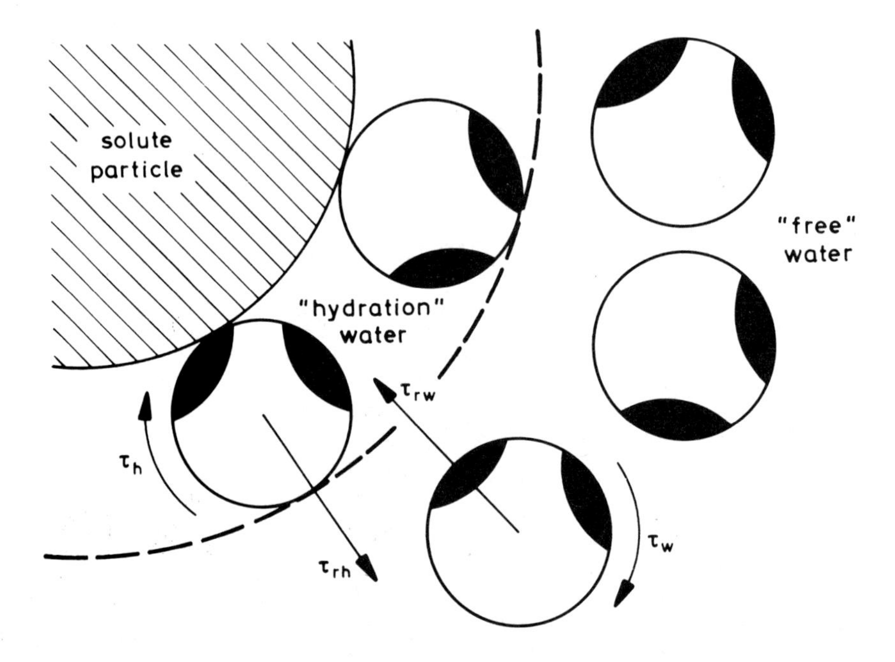

Figure 2 : The model of the solutions.

For the interpretation of the measured correlation function $\Psi(t)$ a model of the solution is required. It is introduced as shown in Fig. 2. The water molecules in the solution are subdivided in at least two classes, depending upon the number of different solutes. Those molecules around one kind of solute particles form one class, the reorientational motion of which is influenced by the solute. There is some uncertainty in deliminating this class. Apart from those cases, where the interpretation of the experimental results requires an extension it is assumed that only the adjacent water molecules are influenced by the solute particles. To these "hydration" water molecules (index h) are attributed the uniform reorientation time τ_h and the mean residence time τ_{rh}. All the water molecules between the hydrated solute particles are denoted as "free" water (index w). The corresponding uniform reorientation time τ_w is assumed to be that of pure water. τ_{rw} is the mean residence time of the water molecules in the free water.

Within this model, with the previously made approximations, and with the restriction to times larger than about 1 psec, the polarization correlation function is obtained as a sum of exponentials. Correspondingly the measured correlation function of a solution is analyzed in this manner.

B. Weak Solute Solvent Interaction

For a large number of solutes the decrease of $\Psi(0)$ or, corresponding, the decrease of static permittivity with increasing solute concentration only depends on the dilution of water by the solute particles (see Fig. 1). We denote this behaviour to be due to weak solute solvent interaction, which is characteristic not only to non-polar solutes but also to the larger mono-valent ions.

For the model of solutions described above the polarization correlation function in the case of one kind of solute particles is /6/:

$$\hat{\Psi}(t) = \hat{\Psi}(0)\left\{\frac{q-q_2}{q_1-q_2}e^{-q_1 t} - \frac{q-q_1}{q_1-q_2}e^{-q_2 t}\right\} \tag{11}$$

with

$$q = \frac{1}{N}\left(\frac{N_w}{\tau_w} + \frac{N_h}{\tau_h}\right) \tag{12}$$

and

$$q_{1,2} = \frac{1}{2}\left\{\frac{1}{\tau_w} + \frac{1}{\tau_{rw}} + \frac{1}{\tau_h} + \frac{1}{\tau_{rh}} \pm \left[\left(\frac{1}{\tau_w} + \frac{1}{\tau_{rw}} - \frac{1}{\tau_h} - \frac{1}{\tau_{rh}}\right)^2 + \frac{4}{\tau_{rw}\tau_{rh}}\right]^{1/2}\right\} \tag{13}$$

where N_w and N_h with $N_w+N_h=N$ are the numbers of water molecules in the free water and the hydration water, respectively. In the limit of very slow exchange of water molecules between the two subliquids, that is $\tau_{rw}, \tau_{rh} >> \tau_w, \tau_h$, eqn. (11) is reduced to

$$\hat{\Psi}(t) = \hat{\Psi}(0)\left\{\frac{N_w}{N}e^{-t/\tau_w} + \frac{N_h}{N}e^{-t/\tau_h}\right\} \tag{14}$$

An analysis of the correlation function then directly yields the reorientation times τ_w and τ_h and the relative numbers of water molecules in the two subliquids.

In the limit of fast exchange, that is $\tau_{rh} << \tau_w, \tau_h$, the time dependence of the correlation function is described by one exponential only according to

$$\hat{\Psi}(t) = \hat{\Psi}(0)e^{-qt} \tag{15}$$

with q given by eqn. (12).

In this limiting case an analysis of the measured correlation function does not permit the evaluation of all the parameters of interest. Additional assumptions or experimental facts are neccessary to reduce the number of parameters. In most cases we therefore introduce the number of hydration water molecules from geometrical considerations.

C. Strong Solute Solvent Interaction

In aqueous solutions of some kinds of small ions the dielectric saturation of the hydration water causes a depression of the static permittivity, which exceeds that due to the dilution effect by a considerable amount. Induced by the strong ionic electrical field, the hydration water dipoles show a radial preferential orientation. Correspondingly there is a strong cross-correlation between the dipoles of the n_{kh} water molecules which form the hydration layer of the arbitrary k-th ion, $\langle n_{kh}\rangle = n_h$. The amount of the correlation function of the dipole moments of the hydration molecules

$$< \mu_{ki}(0) \sum_{j=1}^{n_{kh}} \mu_{kj}(0) > \equiv G\mu^2 \qquad (16)$$

characterizes the degree of dielectric saturation of the hydration water ($0 \leq G \leq 1$). The case of weak solvent solute interaction is described by G = 1. With increasing degree of dielectric saturation G = 0 is approached. The polarization correlation function in this special case (G = 0) is given by /6/:

$$\hat{\Psi}(t) = \hat{\Psi}(0) \left\{ \frac{\frac{1}{\tau_w} - p_2}{p_1 - p_2} e^{-p_1 t} - \frac{\frac{1}{\tau_w} - p_1}{p_1 - p_2} e^{-p_2 t} \right\} \qquad (17)$$

with

$$p_{1,2} = \frac{1}{2} \left\{ \frac{1}{\tau_w} + \frac{1}{\tau_{rw}} + \frac{1}{\tau_{rh}} \pm \left[\left(\frac{1}{\tau_w} + \frac{1}{\tau_{rw}} - \frac{1}{\tau_{rh}} \right)^2 + \frac{4}{\tau_{rw}\tau_{rh}} \right]^{1/2} \right\} \qquad (18)$$

As opposed to the weak interaction formula (11), eqn. (17) in the slow exchange limit reduces to one exponential only

$$\hat{\Psi}(t) = \hat{\Psi}(0) e^{-t/\tau_w} \qquad (19)$$

VI.2.3. Description of Dielectric Relaxation in the Frequency Domain

Until recently only frequency (f) domain measurements of the complex permittivity $\varepsilon(f) = \varepsilon'(f) - j\varepsilon''(f)$ were applicable to the investigation of fast dielectric relaxation processes as given by the reorientational motion of the water molecule. Due to improvements in instrumentation, time domain techniques are available now in a range corresponding to an upper frequency limit of about 15 GHz /7/. The evaluation of the experimental data obtained by these fast time domain spectroscopy methods, however, requires mathematical operations which have to be performed in the frequency domain. Even here, therefore, the information is primarily preserved in terms of the complex permittivity.

The fluctuation-dissipation theorem

$$\varepsilon''(f)/f = \pi(\varepsilon_s - 1) S(f) \qquad (20)$$

yields the relation between the loss-number function $\varepsilon''(f)$ and the Fourier transform of the normalized polarization correlation function

$$S(f) = \int_{-\infty}^{+\infty} \frac{\Psi(t)}{\Psi(0)} e^{-i2\pi ft} dt \quad , \quad \Psi(t) = \Psi(-t) \tag{21}$$

The spectrum S(f) contains contributions of both the fastly decaying part of the correlation function ($t < 1$ psec) and the part due to the reorientational motion of the water molecules. In a general formulation the latter may be written as

$$\hat{\Psi}(t) = \hat{\Psi}(0) \int_0^{\infty} g(\tau) e^{-t/\tau} d\tau \tag{22}$$

by introducing a relaxation time distribution function $g(\tau)$ with

$$\int_0^{\infty} g(\tau) d\tau = 1 \tag{23}$$

Measurements in the microwave frequency range up to about 100 GHz show that the contribution of the fastly decaying part of the correlation function to S(f) is of negligible amount. With ε_∞ defined by

$$\varepsilon_\infty - 1 = (\varepsilon_s - 1)(\Psi(0) - \hat{\Psi}(0))/\Psi(0) \tag{24}$$

eqn. (20) therefore reduces to

$$\varepsilon''(f)/f = 2\pi(\varepsilon_s - \varepsilon_\infty) \int_0^{\infty} \frac{g(\tau)\tau}{1+(\omega\tau)^2} d\tau \tag{25}$$

If a mean relaxation time $\bar{\tau}$ is defined by

$$\bar{\tau} = \int_0^{\infty} g(\tau)\tau \, d\tau \tag{26}$$

the low frequency value of $\varepsilon''(f)/f$ yields $2\pi(\varepsilon_s - \varepsilon_\infty)\bar{\tau}$. If microwave measurements are performed in a fairly extensive frequency range, $g(\tau)$ can be obtained from $\varepsilon'(f)$ or $\varepsilon''(f)$ by integral transformations /8/. In most cases, however, especially if the permittivity curves $\varepsilon(f)$ of aqueous solutions show only small deviations from that of pure water, the measurements essentially yield the values of the dielectric increment $\varepsilon_s - \varepsilon_\infty$ and of the mean relaxation time $\bar{\tau}$.

In the case of ionic solutions a contribution to the complex permittivity due to the ionic conductivity σ has to be taken into account. Usually this contribution is regarded by correcting the experimental loss-number function $\varepsilon''_{exp}(f)$ according to

$$\varepsilon''(f) = \varepsilon''_{exp}(f) - 2\sigma(0)/f \tag{27}$$

where $\sigma(0)$ is the dc conductivity. The application of eqn. (27) has been criticized as the conductivity itself is a frequency dependent quantity /9/. This fact is neglected, if eqn. (27) is substituted into eqn. (25) and the relaxation time distribution function $g(\tau)$ is assigned to the reorientational motion of the water molecules. Indeed the Born-Fuoss-Boyd-Zwanzig theory /10/ yields a frequency dependence of the conductivity as given by

$$\sigma(f) = \sigma(0)\left[1 + b\left(\frac{a_o}{a}\right)^4 \frac{i\omega\tau^*}{1+i\omega\tau^*}\right] \tag{28}$$

where a is the radius of the ion and a_o a radius characteristic to the solvent (= 1,26 Å for water at 25°C). The relaxation time τ^* depends on the reorientation time τ of the solvent molecules by

$$\tau^* = \tau\frac{\varepsilon_\infty}{\varepsilon_s}\left[1 + b\left(\frac{a_o}{a}\right)^4\right]^{-1} \ll \tau \tag{29}$$

The parameter b ($1 \leq b \leq 2$) describes the hydrodynamic boundary condition introduced into the theory.

It may be shown by the use of eqn. (28) that, if eqn. (27) is applied, the corrected permittivity function $\varepsilon = \varepsilon' - j\varepsilon''$ still contains a conductivity term according to:

$$\varepsilon_\sigma(f) = \frac{4\pi\sigma(0)\tau^* \cdot b \cdot \left(\frac{a_o}{a}\right)^4}{1+i\omega\tau^*} \tag{30}$$

A numerical estimation shows that this contribution is too small to be detected by microwave measurements below about 100 GHz.

The conductivity of ionic solutions prevents an accurate determination of the low frequency value $\varepsilon''(f)/f$ and by this a direct evaluation of $\bar{\tau}$. It has been found that in these cases another mean dielectric relaxation time $\tau_M = 1/2\pi f_M$, defined by the frequency f_M at the maximum of the loss-number function, is a characteristic time, obtainable with large accuracy /11/. In general the relation between $\bar{\tau}$ and τ_M is rather complicated. A simple recomputation is possible if one works within the model described in the previous section.

VI.2. 4. Experimental Results

Extensive experimental studies have been performed on aqueous solutions of mono-valent ions. Preponderantly 1 molar solutions were investigated to obtain significant deviations from the pure water data. The characteristic quantities derived from permittivity measurements, that are the mean relaxation times τ_M or $\bar{\tau}$ and the dielectric increment $\varepsilon_s - \varepsilon_\infty$, up to about this concentration show a linear dependence on solute concentration. This finding justifies to apply the formulas of section 2 in the limit of vanishing solute concentration. In this limit for solutions containing more than one kind of solute particles, for example cations and anions, the total variations of τ_M or $\bar{\tau}$ and $\varepsilon_s - \varepsilon_\infty$ are obtained as a linear superposition of contributions due to the different solutes /11,14/.

As it has been shown in section 2 the course in time of the polarization correlation function not only depends on the reorientation times but also on the mean residence times of the water molecules in the different sub-liquids. The hydration water exchange is of importance especially in the case of strong solute solvent interaction. Then apart from a rotation of the hydrated ion as a whole, the only possible reorientational motion of a hydration water dipole aligned parallel to the radial electric field of the ion is a proton around proton rotation. No information about this motion can be obtained from dielectric measurements. Therefore, if the hydration water shows dielectric saturation, only the exchange rate influences the value τ_M.

In solutions containing particles of the weak interaction type the hydration water exchange is of lower importance. A numerical estimation shows /6/ that, if the reorientation time ratio τ_h/τ_w is not very different from unity, $0.4 \lesssim \tau_h/\tau_w \lesssim 2.5$, the exchange influences the value of τ_M by a negligible amount.

The Fig. 3 shows the static permittivities of 1 molar solutions of the alkali chlorides, of NH_4Cl and HCl at 25°C versus the water concentration c_W. The curve is taken from Fig. 1. It holds for weak solute solvent interaction. All the experimental points lie below this curve, which means that besides the dilution of water by the ions the dielectric saturation of the hydration water causes an additional depression of the static permittivity. Within the limits of experimental uncertainty this additional depression can be completely attributed to the cations: even if the Cl^- anion is substituted by the smaller F^- anion almost the same additional decrements are obtained /14/.

From Fig. 3 an estimation of the degree of dielectric saturation of the cationic hydration water can be obtained. The horizontal distances between the experimental points and the straight line yield the number of moles H_2O per mole of solute which seem not to contribute to the orientational polarization. In the description of the

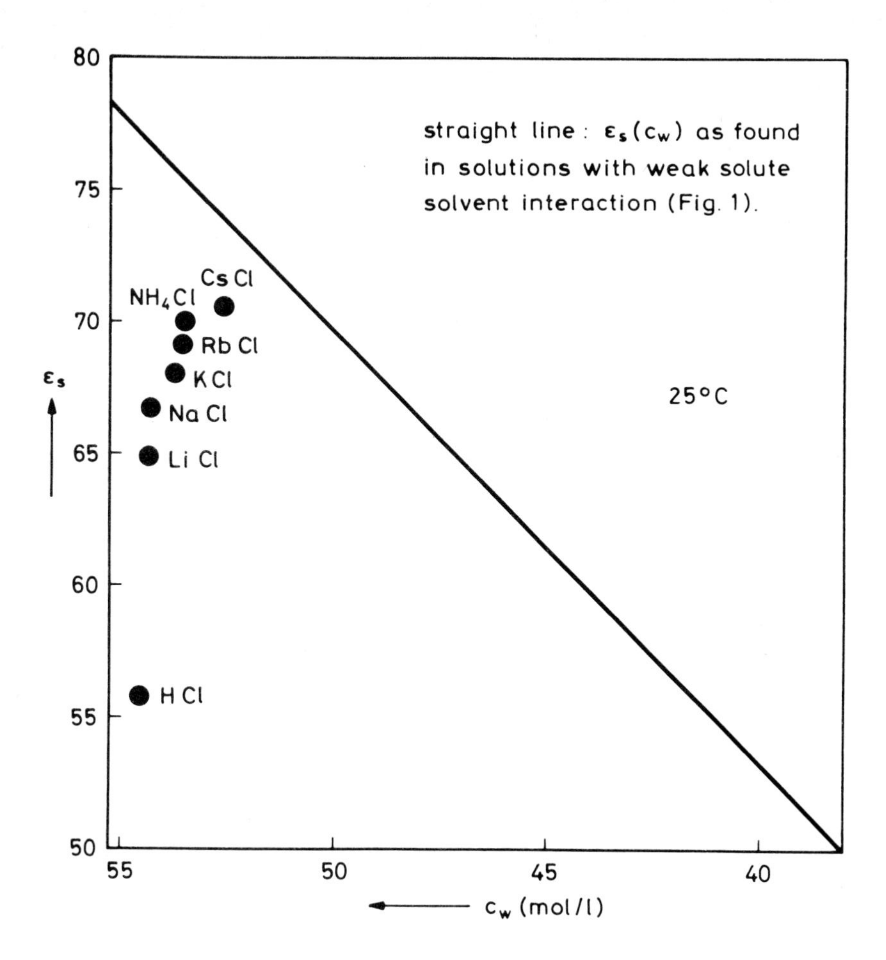

Figure 3 : The static permittivities of 1 molar solutions of some chlorides vs. the water concentration c_w.

dielectric saturation by eqn. (16) we attribute the permittivity depression to a mean radial preferential dipole orientation of all the hydration water molecules. In the case of the larger cations Cs^+, NH_4^+ and Rb^+ this preferential orientation is only weakly marked ($G \geq 0.6$). With decreasing radius of the cations the alignment of the water dipoles parallel to the ionic electric field becomes stronger. In the case of the smallest alkali ion, Li^+, the experimental data show the complete dielectric saturation of the primary hydration layer. The analysis of the mean dielectric relaxation time τ_M in this case yields a mean residence time of the hydration water molecules of $\tau_{rh} \approx 3\tau_w$.

The extreme saturation effect caused by the proton can not be induced by the Coulomb field only. The hydronium ion $(H_3O)^+$ (primary hydration of the proton) has a diameter comparable to that of the potassium ion which shows only an incomplete saturation of its primary hydration layer. Therefore one has to assume the saturation effect mainly to be caused by the formation of a strongly marked hydrogen bonded structure around the hydronium ion. It is restricted not only to the hydration complex $(H_9O_4)^+$ but comprises at least one further layer of water molecules (tertiary hydration).

Two processes contribute to the hydration water exchange in the case of the proton: the exchange of individual water molecules of the outer hydration layer and the diffusion of the total hydrogen bonded hydration structure accompanied with the displacement and change of the central proton. According to an estimation of Eigen /16/ the time necessary for a displacement of the centre of the proton hydration complex by the length of a hydrogen bond is about 3 psec. The mean time a water molecule belongs to the proton hydration complex is larger because several of these elementary steps are necessary for an inner hydration water molecule to become a free water molecule. The interpretation of the shift of the mean dielectric relaxation time due to the proton requires to assume $\tau_{rh} \approx \tau_w$.

Similar to the proton even the hydroxyl ion causes a depression of the static permittivity by the formation of a strongly hydrogen bonded structure in the hydration water. Compared to the depression due to the proton it is less extended. The number of water molecules per hydroxyl ion which seem not to contribute to the orientational polarization is about 4 /15/.

The Fig. 4 gives a survey of reorientation times as found for the hydration water molecules in solutions of mono-valent ions. The ratio τ_h/τ_w of the reorientation times of the hydration and free water molecules is plotted versus the ratio r/r_w of solute and solvent particle radii. The data for the larger particles and the small F^- anion (●) have been obtained from dielectric relaxation measurements in agreement with n.m.r. relaxation rate measurements. In those cases where the dielectric saturation of the hydration water prevents information about the reorientational motion to be obtained from dielectric measurements, n.m.r. data (○) of Hertz et al. /12,13/ are taken to complete the diagram.

As far as the ion radius is smaller than about twice the radius of the water molecule, for both cations and anions the diagram shows a

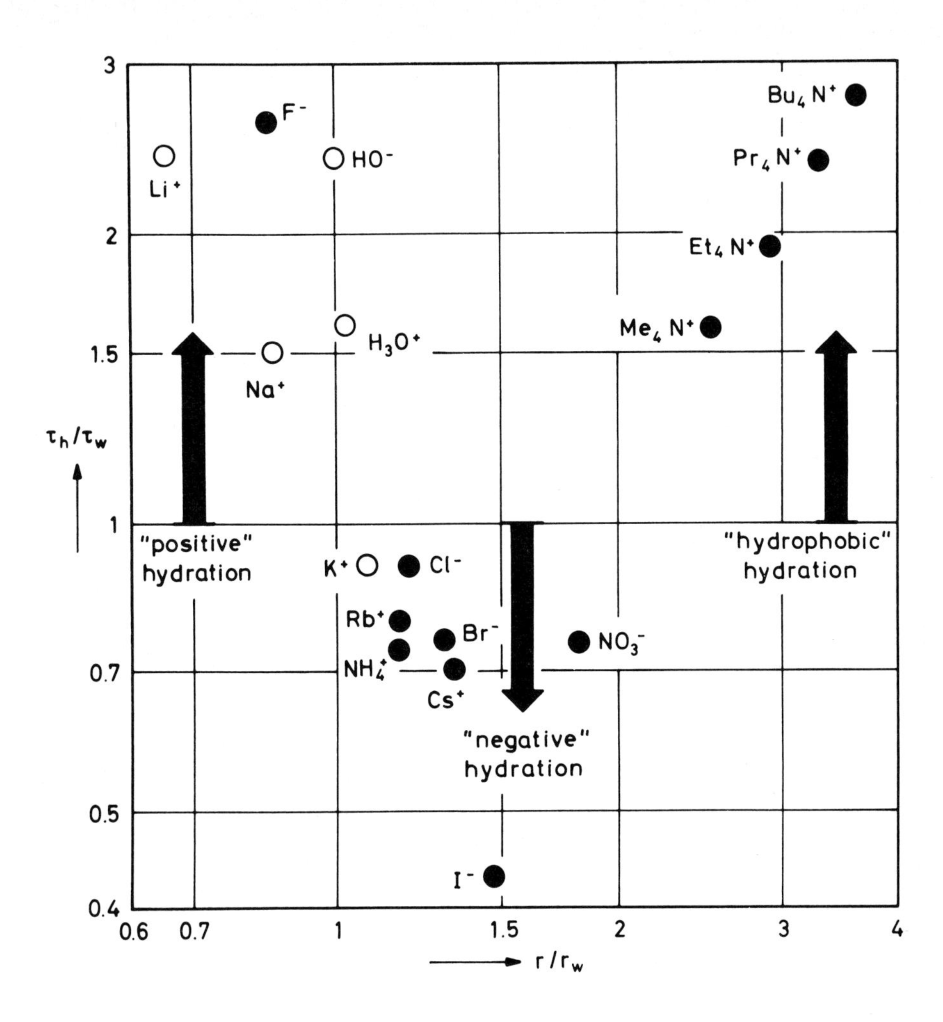

Figure 4 : The reorientation time ratio τ_h/τ_w vs. the ratio r/r_w for some ions at 25°C. (r is the ionic radius, r_w =1.38 Å).

o denotes ions with strong solute solvent interaction. The τ_h/τ_w - values are n.m.r. data from Hertz et al. (12), (13).

● denotes ions with weak solute solvent interaction. The τ_h/τ_w - values are dielectric data from ref. (1), (14) and (15).

rather uniform decrease of τ_h/τ_w with increasing ion size from $\tau_h/\tau_w>1$, "positive" hydration, towards $\tau_h/\tau_w<1$, "negative" hydration. Apart from the F^- anion the positively hydrated ions are those which cause a strong decrease of the static permittivity. In the case of negative hydration the reorientational motion of the hydration water molecules is faster than in pure water. Here the structure breaking effect of the ions becomes dominant.

For r/r_w-values larger than about 2.5 the reorientation time ratio τ_h/τ_w is found to be larger than one. An example is given in Fig. 4 by the tetra-alkyl ammonium ions. With increasing radius the electric field strength at the surface of the ions decreases. Therefore negative hydration is reduced. However, this cannot explain the very large τ_h/τ_w-values obtained for the hydration water of the tetra-alkyl ammonium ions. These high values are thought to be due to the special solvent-solvent interaction around inert solute molecules causing the so called hydrophobic hydration or hydration of the second kind /17/. Hydrophobic hydration has been found for a lot of solutes - ions and neutral organic molecules - containing inert groups /2,18,19/. But never reorientation time ratios τ_h/τ_w larger than about 3 were obtained. From the τ_h/τ_w values of the tetra-alkyl ammonium ions (Fig. 4) one could suppose that there is a strong dependence on the radius of the solute, and that the hydrophobic hydration is enlarged with increasing size of the solute particle. To prove this assumption, solutions of cationic surfactants have been investigated /19/. The solutes were mono-n-alkyl ammonium ions, which spontaneously form micelles in water. The radius of the micelles was about 7 r_w to 15 r_w depending on the alkyl chain length. In these solutions no reorientation times τ_h larger than about 3 τ_w were found.

As mentioned above the solute concentration must amount to about 1 mol/l in order to get distinct results from dielectric relaxation measurements. These large concentrations can be reached only if the solute particles contain hydrophilic parts, i.e. electrically charged, polar, or hydrogen bonding groups. All experimental results up to now show that the tendency to the hydrophobic hydration is counteracted by the hydrophilic groups. A study of the influence of some hydrogen bonding groups on the hydrophobic interaction yields the following sequence of groups /18/:

$$H_2N-, -COO^- < -OH < -COOH < \rangle CO$$

In this sequence these groups more and more prohibit the hydrophobic hydration of neighboring inert groups.

The number of water molecules influenced by an organic solute particle turns out to be at least the number of nearest neighbors. There are indications /2,18,19/ that some more molecules than the nearest neighbors may change their reorientation time with respect to pure water.

References

1 R. Pottel and O. Lossen, Ber. Bunsenges. physik. Chem. 71 (1967) 135

2 R. Pottel and U. Kaatze, Ber. Bunsenges. physik. Chem. 73 (1969) 437

3 G. Akerlöf and A. O. Short, J. Am. Chem. Soc. 58 (1936) 1241

4 F. E. Critchfield, J. Am. Chem. Soc. 75 (1953) 1991

5 T. Nee and R. Zwanzig, J. Chem. Phys. 52 (1970) 6353

6 K. Giese, Ber. Bunsenges. physik. Chem. 76 (1972) 495

7 A. Suggett, in "Dielectric and Related Molecular Processes", M. Davies (Ed.), The Chemical Society, London, 1972, Ch. 4

8 K. Giese, Advan. Mol. Relaxation Processes 5 (1973) 363

9 See the discussion cited in the paper of R. Pottel, J. Chim. Phys. (Numero Special) 1969 (Octobre) 115

10 R. Zwanzig, J. Chem. Phys. 52 (1970) 3625

11 R. Pottel, in "Water, A Comprehensive Treatise", F. Franks (Ed.), Vol. III, Plenum, New York, N.Y., 1973, Ch. 8

12 L. Endom, H. G. Hertz, B. Thül and M. Zeidler, Ber. Bunsenges. physik. Chem. 71 (1967) 1008

13 G. Engel and H. G. Hertz, Ber. Bunsenges. physik. Chem. 72 (1968) 808

14 K. Giese, U. Kaatze and R. Pottel, J. Phys. Chem. 74 (1970) 3718

15 U. Kaatze, Ber. Bunsenges. physik. Chem. 75 (1973) 447

16 M. Eigen, Angewandte Chem. 75 (1963) 489

17 H. G. Hertz, Ber. Bunsenges. physik. Chem. 68 (1964) 907

18 R. Pottel and D. Adolph, to be published

19 U. Kaatze, C. H. Limberg and R. Pottel, to be published

VI.3. STRUCTURAL INFORMATION FROM DIELECTRIC PROPERTIES OF ICE

C. Jaccard
Institut de Physique de l'Université, Neuchâtel, Switzerland

ABSTRACT

Temperature dependence and magnitude of the static permittivity indicate disorder of the proton configuration in ice I_h, III, V, VI and VII, and order in ice II, VII and IX, with a molecular dipole moment of 2.6 Debye in ice I_h. Frequency dependence (a Debye relaxation) and DC conductivity suggest the presence of complementary defects behaving as charged particles : proton excess or deficiency on a bond (D and L Bjerrum defects) or in a molecule (H_3O^+ and OH^- ions). Their motion affects the order of the proton arrangement depending on type and charge. Doping with HF shows that in pure ice D and L defects are responsible for polarization and relaxation, and ions for DC conductivity. Pair formation enthalpy is about 16 kcal/mol for both types. The motion of D and L is thermally activated (enthalpy of 5.5 kcal/mol, mobility of 10^{-4} cm^2/Vs at -10^oC) and their concentration is about 2.10^{-5} mol/l, but for ions, which are in much smaller concentration of about 10^{-10} mol/l, proton transfer is probably due to quantum mechanical tunnelling coupled with lattice vibrations (mobility near 10^{-3} cm^2/Vs). Activation volumes, obtained from pressure dependence, amount to +3 cm^3/mol for D and L and -3.5 cm^3/mol for ions in pure ice.

VI.3.1. INTRODUCTION

Dielectric properties express the response of a system to an electric field, response which is characterized by a variable macroscopic polarization. The phenomena occuring at frequencies higher than 100 GHz and due to electronic excitations, molecular and lattice vibrations, pertain rather to optical, infrared and Raman properties, and their only contribution considered in this paper is the so-called "high frequency permittivity", which makes a quasi constant background of $\varepsilon_\infty = 3.2$ for all the processes related with configurational changes, i.e. changes in the relative molecular positions. Whereas these changes, responsible for dielectric relaxation, occur within 10^{-11} s^{-1} in water, they are much slower in ice by five orders of magnitude at least, thereby providing a clear separation between deformation and rearrangement processes [1].

Apart from this marked difference in the time scale, ice is evidently characterized by a crystalline lattice with well defined molecular locations, obtained from X-ray and neutron diffraction measurements [2]. Each oxygen atom is tetrahedrally coordinated to four neighbours by hydrogen bonds. In the ordinary low pressure phase I_h, the tetrahedra are regular and the oxygen atoms form planes of pluckered hexagons which are piled up with successive rotations by 180° in a sequence ABAB..., leaving open hexagonal channels along the c-axis (Fig. 1).

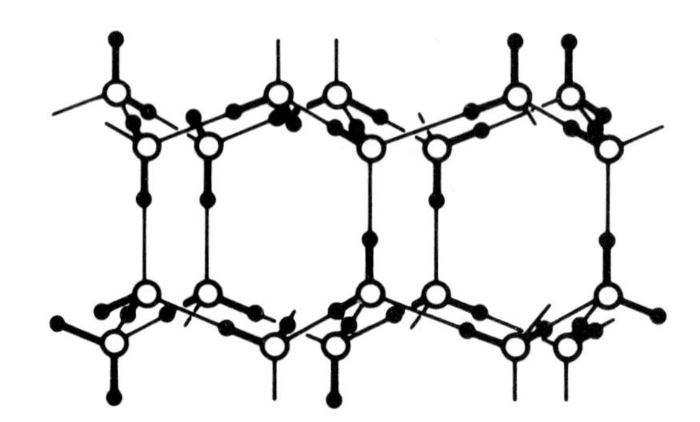

Fig. 1 : Lattice of ice I_h seen perpendicularly to the hexagonal c-axis with a disordered proton configuration complying to Bernal-Fowler rules

The intermolecular distance r_{O-O} is 2.76 Å. In the low pressure, cubic phase I_c, which has been observed only below -70°C, the planes are piled up without rotation in a sequence ABCABC..., giving a diamond-like lattice. In high pressure forms, the coordination is maintained, but the tetrahedra are not always regular, and the highest pressures phases are made up of two interpenetrating sublattices (e.g. ice VII and VIII : two I_c sublattices) [3]. The position of the protons (or of deuterons in heavy ice) is obtained from neutron diffraction and it is consistent with Bernal-Fowler rules : a single proton between two coordinated oxygen atoms (H-bonding) and two protons near each oxygen atom (H_2O molecules), at a distance r_{O-H} = 1.0 Å in I_h. These rules are not very restrictive and allow for ordered and for disordered protonic configurations. Infrared and Raman spectroscopy, and thermal analysis, give valuable information on the short range structure and on the order problem [4].

As it will be shown in the following section, the static value of the dielectric permittivity is intimately connected with the protonic order, and can be related quantitatively with the molecular dipole moment in the ice. But the frequency dependence reflects changes in the protonic configuration occuring only either by intramolecular proton jumps (corresponding to a "rotation" of the molecules) or by intermolecular jumps. This requires however the presence of structural defects, which are so diluted that they cannot be examined by the methods cited above. Only transport properties, such as dielectric permittivity, electrical conductivity and thermoelectricity [5] are sensitive to their interplay. These defects will constitute the main object of the following section and an independent physicochemical method of determining some of their parameters presented during this Symposium will be reported in the last section.

Before discussing the dielectric properties, it has to be stressed that their reliable interpretation is bound to several difficulties of purely experimental nature, which are mainly responsible for the large scatter between the results obtained in the different laboratories during the last two decades. In the first place, crystal quality is a decisive parameter. Many measurements made on polycrystals are useless for quantitative analysis, and even if monocrystals seem to be easy to produce in large volumes, great care is needed to control their purity and physical perfection, influencing very sensitively certain electrical properties; the problem of their ageing has not yet been solved. Another delicate question, implicitely raised by the dielectric method, pertains to the electrodes. If they do not ensure a perfect transfer of the electron to the proton current, they give rise to a space charge which can hide completely the intrinsic bulk properties. Finally, the surface of the ice has been discovered to be quite anomalous with respect to the bulk, and this has to be obviated by using systematically guarded electrodes on suitably shaped samples. Owing to these difficulties, older results which were thought to be quite reliable, have to be reconsidered with scepticism, as far as they have not been overthrown by newer measurements performed under better conditions.

VI.3.2. STATIC DIELECTRIC PERMITTIVITY

Elementary thermodynamics relates entropy and permittivity of a system with the electric field according to the equation

$$S(T,E) = S_o(T) + \frac{\varepsilon_o \partial \varepsilon_s}{2\partial T} E^2$$

Therefore, since the electric field creates some order in a disordered system by, say, aligning some dipoles, the entropy of such a system decreases when the field is applied, and the temperature derivative of its permittivity must be negative. If the system is ordered, the field introduces some disorder, and the sign is reversed. The temperature dependence of the permittivity gives a direct qualitative information on the order of the system. Moreover, there are much more possible configurations with a given net dipole moment in a disordered system than in an ordered one, so that the permittivity should be higher in the former than in the latter. Since this parameter is about 100 in ice I_h, with a negative temperature derivative ($\varepsilon_s \simeq$ const./T) it suggests that the protonic arrangement is completely disordered in this phase, an assumption verified in all the other experiments.

Fröhlich's theory [6] gives a quantitative value to the static permittivity :

$$\varepsilon_s - 1 = 3\varepsilon_s(2\varepsilon_s+1)^{-1} \langle M^2 \rangle / 3\varepsilon_o VkT$$

$\vec{M}$ is the spontaneous electrical moment at temperature T of a macroscopic sphere of volume V imbedded in its own material, and $\langle M^2 \rangle$ is the average of its square. It can be split into a first component due to the configurational changes and a second one due to deformation of the lattice :

$$\vec{M} = \vec{M}_c + \vec{M}_d$$

Since the time scales of these processes are well separated by many orders of magnitude, the cross-correlation vanishes

$$\langle \vec{M}_c \cdot \vec{M}_d \rangle = 0$$

If the configuration is kept "frozen", as it is the case in experiments at high frequency, an analogous formula gives the h.f. permittivity

$$\varepsilon_\infty - 1 = 3\varepsilon_\infty(2\varepsilon_\infty+1)^{-1} \langle M_d^2 \rangle \,/\, 3\varepsilon_o VkT$$

Substracting this equation from the first one gives the configurational contribution

$$\varepsilon_s - \varepsilon_\infty = 3\varepsilon_s(2\varepsilon_s+\varepsilon_\infty^{-1})^{-1} \langle M_c^2 \rangle \,/\, 3\varepsilon_o VkT$$

If the sphere is made of N identical dipoles of moment μ_c, the average can be expressed with the help of Kirkwood's correlation factor g, which is determined by the lattice structure and the short range interactions, i.e. Bernal and Fowler rules :

$$\langle M_c^2 \rangle = N.g.\mu_c^2$$

For the I_h lattice with protonic disorder, the value of the correlation factor has been shown to lie very near 3 [7] and inserting in the last equation the experimental values $\varepsilon_s - \varepsilon_\infty = 100$, $T = 263^oK$, $N/V = 3.1\times10^{22}$ cm^{-3} yields for the molecular dipole moment in ice the value of 8.3×10^{-30} Asm or 2.5 Debye. It is larger than in the vapour (1.87 Debye) but the difference can be explained by the reaction of the environment. In a simple model, the molecule is assumed to be a sphere of radius a and of polarizability $\alpha = 4\pi a^3 \varepsilon_o(\varepsilon_\infty-1)/(\varepsilon_\infty+2)$, with a moment μ_v in vacuum. If brought into material of permittivity ε_∞, the reaction field R of the polarized medium increases the moment to $\mu_c = \mu_v + \alpha R$, where $R = 2(\mu_c/4\pi a^3\varepsilon_o)(\varepsilon_\infty-1)/(2\varepsilon_\infty+1)$. The ratio of both moments is then $\mu_c/\mu_v = (\varepsilon_\infty+2)(2\varepsilon_\infty+1)/9\varepsilon_\infty = 1.33$, just the ratio of both experimental values. Although this agreement is fortuitous due to the oversimplification of the molecular model, it nevertheless supports the initial assumptions.

These considerations have been extended to the high pressure modifications on which several dielectric measurements have been performed [8], with the following result :
I_h, I_c, III, (IV), V, VI and VII are disordered (some of them show a partial order at lower temperatures) and II, VIII and IX are ordered, the best fit with neutron diffraction being attained with an antiferroelectric structure (no net moment in the elementary cell). The drastic change occuring as a function of temperature at a phase separation is exemplified by Fig. 2 (from Whalley et al. [9]) : in the transition from ice III to ice IX the configurational component $\varepsilon_s - \varepsilon_\infty$ of the permittivity vanishes completely.

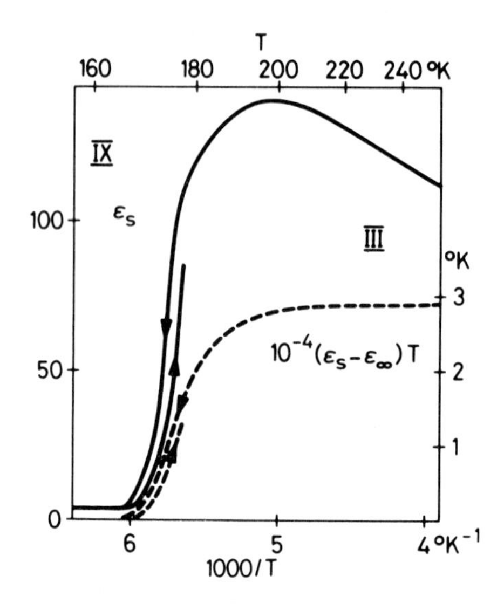

Fig. 2 : Effect of the order-disorder transition IX-III in the static permittivity at 2.3 kbar (after Whalley et al. [9])

VI.3.3. DYNAMICAL PROPERTIES AND LATTICE DEFECTS

The model and the theory given thus far concern only equilibrium states and have to be completed in order to explain the processes by which a protonic configuration evolves into another one if the applied field is changed. They manifest themselves macroscopically by the frequency dependence of the permittivity. In the frequency domain considered here, the response of ice is a superposition of pure relaxation processes, indicating a very strong coupling of any protonic motion with lattice vibrational modes. Analyzing very precise measurements in a large frequency domain, von Hippel and his collaborators [10] have discovered up to eight different relaxation processes, with characteristic times ranging from a microsecond to ten seconds. The fast ones, with a relaxation strength (i.e. a contribution to the static permittivity) smaller or equal to unity, are attributed to unknown bound defects. The slow ones have large relaxation strengths, between hundred and thousand, and they are certainly due to electrode effects, space charge and extended structural faults subject to ageing. We shall restrict the discussion to the only well investigated relaxation mechanism, which is intrinsic and characteristic of the bulk substance and therefore always present.

At -10°C the relaxation strength amounts to 100 (nearly proportional to 1/T) and the relaxation time is 50 μs, with an activation enthalpy of 13.6 kcal/mol. In heavy ice, this time is increased by 40%, with the same activation enthalpy, indicating a many phonon process with an harmonic oscillation of the protons responsible for the frequency factor of the transition probability. The frequency dependence and the observed enthalpy preclude for configurational changes any cooperative rearrangement of the protons along paths extending from one electrode to the other (a closed path does not alter the polarization)[11]. The only as yet satisfactory mechanism requires local violations of Bernal-Fowler rules, constituting the defects postulated by Bjerrum [12]. As a possible consequence of thermal fluctuations, a proton can perform an intramolecular jump on a neighbouring bond (equivalent to a rotation of the molecule) as in Fig. 3, with a finite probability that the motion extends step by

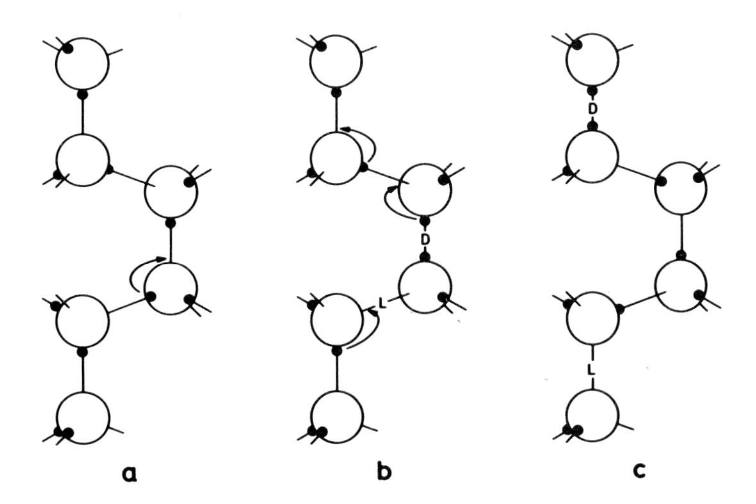

Fig. 3 : Formation and separation of a pair of Bjerrum defects D and L by intramolecular proton jumps (from ref. 1b)

step, leaving finally somewhere in the lattice a region with a proton deficiency, represented as an empty bond (L defect) and somewhere else a region with a proton excess, represented as a doubly occupied bond (D defect). This symbolism is evidently a gross oversimplification : the defects are rather localized elementary excitations of the lattice, extending over several molecules and involving significant distortion of the lattice and of the electron distribution. Their concentration is determined by the balance between dissociation and recombination and they respond to an electric field according to their effective charge ($e_{D,L} = \pm e_B$) by adding to their fast brownian motion a drift parallel to the field producing more and more order in the protonic arrangement. In the case of a static field, this drift stops when the gain in electrical energy is compensated by the decrease of configurational entropy, leaving a strong lattice polarization proportional to the field. If the field is turned off, the polarization vanishes as a result of the random motion, with a single relaxation time.

Since the only effect of this process is to turn the molecules, it cannot give rise to the DC conductivity observed even in very pure and perfect crystals. It is necessary to postulate intermolecular protons jumps. These are provided by complementary defects, the ions H_3O^+ and OH^-, which are created and move by stepwise proton shifting along the bonds (Fig. 4) and constitute also localized elementary

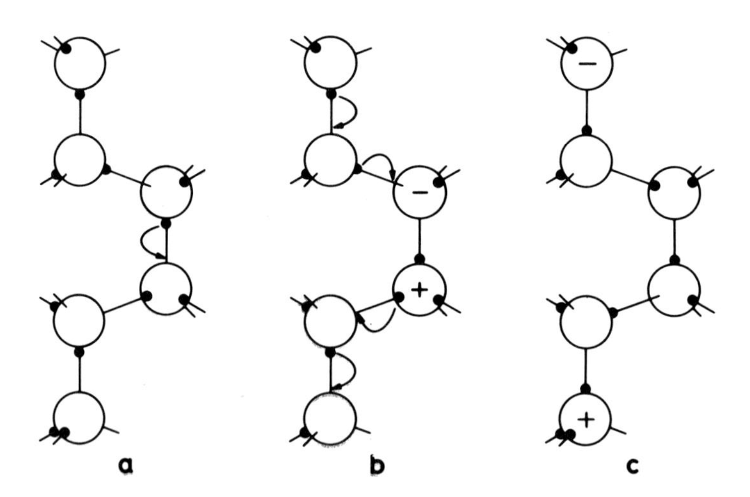

Fig. 4 : Formation of a pair of ions H_3O^+ and OH^- by intermolecular protons jumps (from ref. 1b)

excitations of the lattice with effective charges $e_{\pm} = \pm e_I$. However their structure is more apparent than for the bond defects, since they exist in the free state too. In a DC process, protons are transferred stepwise alternatively by Bjerrum and by ionic defects from one electrode to the other, both mechanisms acting in series and requiring $e_I + e_B = e$ = protonic charge, whereas in a high frequency process ($\omega >> 1/\tau_{relax}$), each species oscillates for itself on a very short path and both mechanisms act in parallel.

A decisive correlation between Bjerrum and ionic defects is mediated by the protonic configuration. As it is shown in Fig. 5, the order and thereby the polarization

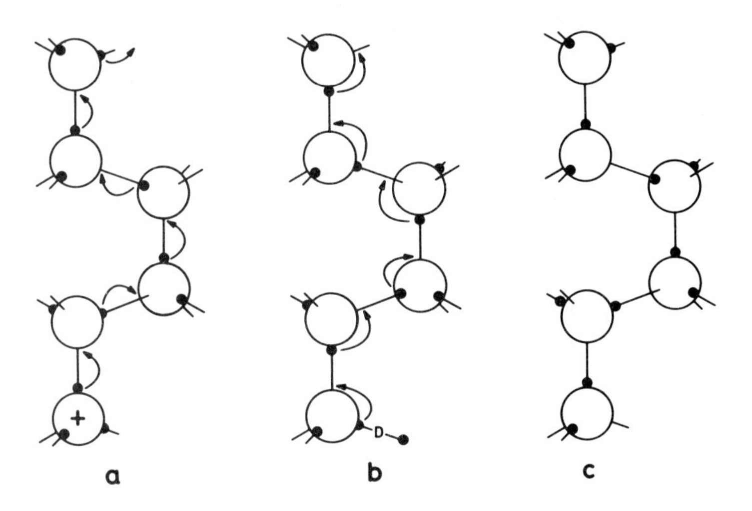

Fig. 5 : Restauration of the initial proton configuration by a Bjerrum defect along a path used by an ion (from ref. 1b)

produced along a chain by the passage of a defect of a given type is destroyed by the passage of a defect of the other type. In the theoretical treatment [13] this introduces a configuration vector representing a trace density [14] which is a weighted average of the integrated defect currents :

$$\vec{\Omega} = \int (\vec{j}_+ - \vec{j}_- - \vec{j}_D + \vec{j}_L)\, dt$$

Its square is proportional to the configurational entropy per unit volume :

$$Ts = \phi\, \Omega^2$$

If correlations due to short range interactions are neglected, the value of the coefficient is $(16/\sqrt{3})kTr_{00}$. The calculation of conductivity and permittivity yields the following relation, in agreement with the experiments :

$$[\sigma(\omega) - \sigma_o] / i\omega\varepsilon_o = \varepsilon(\omega) - \varepsilon_\infty = (\varepsilon_s - \varepsilon_\infty)/(1 + i\omega\tau)$$

The parameters can be expressed by the specific conductivities σ_I and σ_B :

$$\sigma_\infty = \sigma_I + \sigma_B \qquad e^2/\sigma_o = e_I^2/\sigma_I + e_B^2/\sigma_B$$

$$1/\tau = \phi\ (\sigma_I/e_I^2 + \sigma_B/e_B^2)$$

$$\varepsilon_o\ (\varepsilon_s - \varepsilon_\infty) = e_P^2/\phi \qquad e_P = (\sigma_I/e_I - \sigma_B/e_B)/(\sigma_I/e_I^2 + \sigma_B/e_B^2)$$

If majority and minority character are defined by the ratio of the specific conductivities, the formulae above show that DC conductivity is determined by minority carriers, high frequency conductivity and static permittivity by majority carriers. The critical intermediate case of matching between both types of defects ($\sigma_I/e_I = \sigma_B/e_B$), characterized by the disappearance of the configurational polarization ($\varepsilon_s = \varepsilon_\infty$) will be treated in the following section.

Bjerrum defects predominate in pure ice, and the polarization charge e_P is equal to $-e_B$. The polarization being produced by "rotation" of the molecules, there is a definite relationship between e_B and the molecular dipole moment μ_c : $e_B r_{OO} = \mu_c\sqrt{3}$. Consistence with the theory of the preceding section requires to divide the factor ϕ by Kirkwood's correlation factor g, in order to take into account the short range interactions originally neglected. This yields the value 0.32 for the ratio e_B/e, smaller than previous estimates ranging around 0.5.

VI.3.4. EFFECT OF CHEMICAL IMPURITIES AND OF PRESSURE

Chemical impurities modify all the dielectric parameters, certain of them, such as the static conductivity, being quite sensitive to very small electrolyte concentrations. A typical effect is to enhance space charge, thereby obscuring the interesting relaxation processes and making a quantitative evaluation of the experiments very difficult. It is therefore not surprising that the situation is far more confused for solutions than for the pure substance. A specific problem, not encountered in the study of liquid solutions, results from the segregation of foreign species occuring during crystal growth. Even if the overall concentration can be determined in the melt water after the experiments, the location in the lattice is still problematic : substitutional, interstitial or aggregated at extended lattice defects ? Many experiments have been performed with various substances, but general lack of agreement does not yield a clear picture, except for the simplest cases to which the following discussion will be restricted. (Fig. 6.)

Hydrofluoric acid has the most significant effect, as well in magnitude as theoretically. It is believed to replace water molecules in the lattice, an assumption justified by proton channeling experiments showing that chlorine from HCl is

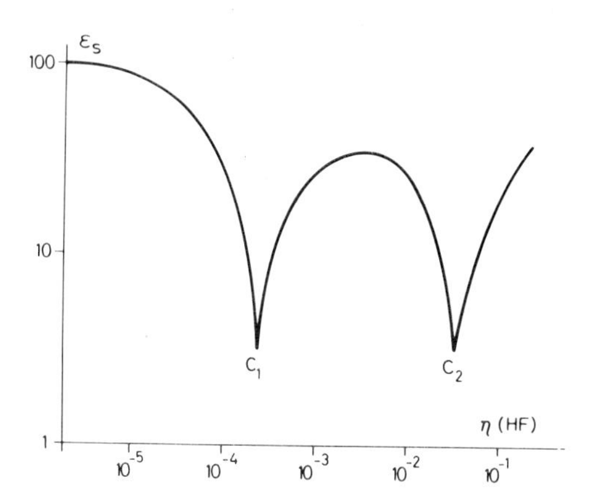

Fig. 6 Effect of HF on the static permittivity at $-10^{\circ}C$. At the critical concentrations C_1 and C_2, the partial conductivities are matched : $\sigma_I/e_I = \sigma_B/e_B$ (after Steinemann [16])

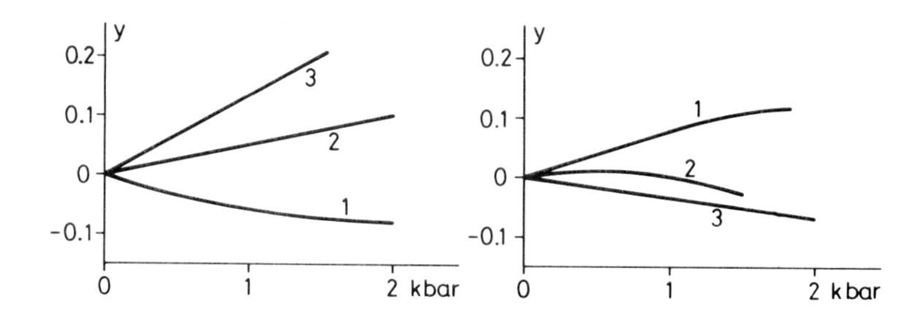

Fig. 7 Effect of pressure on the conductivities at $-25^{\circ}C$. Left side : $y = \log \sigma_\infty(p)/\sigma_\infty(0)$; right side : $y = \log \sigma_0(p)/\sigma_0(0)$. 1) pure; 2) $C_{HF} = 5\cdot10^{-6}$ mol/ℓ ; 3) $C_{HF} = 3\cdot10^{-5}$ mol/ℓ (after Taubenberger et al. [1a])

substitutional up to high concentration [15]. Each HF molecule introduces in the lattice a proton deficiency, which can diffuse almost freely from the acid molecule. At high acid concentration, the number of L defects is therefore equal to the number of HF, and D defects disappear. Moreover, the acid molecules dissociate weakly into negative fluorine ions and positive hydronium ions, the concentration of which tends to become proportional to the square root of the acid concentration, with the OH^- ions disappearing too. Inclusion of hydrofluoric acid modifies then the ratio of the specific conductivities σ_I/σ_B by increasing the concentrations in different ways. This is revealed drastically by the dependence of the static permittivity on HF content [16] : the configurational contribution $\varepsilon_S(C_{HF})-\varepsilon_\infty$ shows two pronounced minima of the order of unity near acid concentrations of $c_1 = 10^{-4}$ and $c_2 = 10^{-2}$ mol/l (at -10^oC). They are readily explained by the theory of the precedent section and the assumption made for the acid dissociation : for $c > c_2$, and $c < c_1$ Bjerrum defects are majority carriers, whereas ions predominate between c_1 and c_2, the critical concentrations being determined by the shift of the majority character from one defect type to the other. This experiment supports the picture that polarization in pure ice is due to molecular reorientation.

At -10^oC the relaxation time decreases for acid concentrations higher than about 2.10^{-5} mol/l. This should correspond to the concentration of Bjerrum defects in pure ice; their mobility should then range about 10^{-4} cm^2/Vs to account for the observed high frequency conductivity. This agrees with experiments performed in a temperature and concentration range in which this last parameter is proportional to the acid concentration [13], where it has an activation energy of 5.5 kcal/mol. This quantity represents the activation enthalpy for the mobility of the Bjerrum defects E^m_{DL}. Comparing with the activation enthalpy for the relaxation time E_τ, the dissociation enthalpy for a pair of Bjerrum defects E^d_{DL} amounts then to 16±1 kcal/mol ($E_\tau = E^m_{DL} + \frac{1}{2} E^d_{DL}$). The value of the static conductivity and its temperature dependence have given rise to much controversy because of the experimental difficulties mentioned in the Introduction. Its value at -10^oC is likely to be around 10^{-10} Ω^{-1} cm^{-1} with an activation energy of 8 kcal/mol. It is sensitive to HF in concentration higher than 10^{-7} mol/l, (the activation energy becomes 7.5 kcal/mol) but doping cannot be used to determine the carrier mobility, since the dissociation constant is not known. Basing on measurements of the saturation current, Eigen et al. [17] obtained a mobility of 10^{-1} cm^2/Vs, suggesting a fast quantum mechanical proton transfer for the reaction $H_2OH^+ - OH_2 \rightarrow H_2O - HOH_2^+$ (this is also supported by the fact that the static conductivity in heavy ice is two orders of magnitude lower than in light ice). However, recent measurements by von Hippel et al. [10] invalidated Eigen's results and the mobility value at high temperature is probably not far from its value at 125^oK of 10^{-3} cm^2/Vs, measured directly with electrodes injecting protons in short bursts by Eckener et al. [18]. This implies the low value for the ionic concentration in pure ice of about 10^{-10} mol/l (at -10°C). On the other hand, Eckener's results still confirm the hypothesis of a proton transfer which does not depend classically on temperature according to an Arrhenius law. The activation enthalpies for DC conductivity are then due only to the dissociation process, giving 16 and 15 kcal/mol for the ionic dissociation enthalpy of H_2O and of HF, respectively.

Doping with ammonia produces qualitatively complementary effects : NH_3 brings D defects in the lattice and it dissociates weakly into ammonium and hydroxyl ions. This impurity has not been studied as extensively as HF, and the case is more complicated [19]. It seems that the D defects do not diffuse freely from their parent molecule but need some activation energy near 6 kcal/mol. More complex impurity systems, such as ammonium fluoride, have been investigated too [20] but the lack of precise knowledge of the dissociation constants renders a quantitative description uneasy. Increasing the number of chemical species multiplies the reactions and the physico-chemical constants and generally it hides the elementary processes which are looked for.

Another way of influencing the permittivity, allowing an easier control than chemical doping but requiring a precise measuring technique, is to vary the pressure. Static permittivity is related with density ρ according to the equation $d \ln \varepsilon_s / d \ln\rho = 1.2 \pm 0.3$ [21], which shows that no significant change of the molecular moment is involved : the increase in ε_s is accounted for essentially by the higher density. Dynamical parameters such as the conductivities and the relaxation frequency depend on the pressure* according to an exponential law $e^{-pV_a/kT}$ introducing an activation volume. It represents the static volume change due to the defect formation plus the transient volume change occuring during its stepwise motion. In pure ice [22] the activation volume for the high frequency conductivity is near 3 cm^3/mol and is positive, in agreement with the picture of Bjerrum defects producing a dilatation of the lattice. For the static conductivity it is slightly larger (3.5 cm^3/mol) but negative and it can be accounted for by electrostriction. In HF doped ice, the activation volume for both conductivities change their sign (according to the change of the majority character revealed by a minimum of the static permittivity) and their magnitude, due to the interaction with the acid. A determination of the defect concentration is possible in principle, but further measurements are needed for a detailed interpretation.

* See fig. 7.

VI.3.5. SEGREGATION OF HYDROGEN FLUORIDE AND LATTICE DEFECTS

A physicochemical method for testing the models of the preceding sections has been presented by J. Bilgram at the Symposium. Although it does not involve any electrical measurement, it is briefly reviewed here since it confirms by quite independent means the results obtained from dielectric properties [23] and it explains the basic mechanism of desalination by freezing.

When ice grows from a dilute hydrofluoric acid solution, it rejects most of the fluorine into the melt, a small fraction between 10^{-2} and 10^{-3} being incorporated in the lattice. If the fluorine concentration measured in the solid (c_S) is compared with the known activity of the acid in the solution ($a_L = \gamma_L c_L$) in a log-log plot, several concentration ranges can be distinguished : for $c_S > 3\times10^{-4}$ mole/l and for $2\times10^{-5} > c_S > 10^{-6}$ mole /l the concentration in the melt is proportional to the activity in the liquid ($c_S \sim a_L$), whereas it grows with its $2/3^d$ power in the medium range ($c_S \sim a_L^{2/3}$). This applies to quasi-equilibrium growing conditions with forced convection maintaining an homogeneous concentration in the liquid and preventing any thermodielectric effect, the chemical potentials being equal in both phases. In the liquid, HF dissociates into F^- and OH_3^+ and its chemical potential is $\mu_L = \mu_L^o + 2RT \ln a_L$. In the solid, we have the same ionic dissociation, but the fractional power in the medium concentration range indicates that the acid may introduce into the lattice non chemical defects of structural nature identified as Bjerrum L defects. In pure ice, these and the ions exist already in concentrations c_B and c_I. The HF chemical potential in the solid is therefore $\mu_S = \mu_S^o + RT \ln (c_B + \gamma_B c_S) + RT \ln (c_I + \gamma_I c_S) + RT \ln (\gamma_I c_S)$.

In the high concentration range, the extrinsic defects predominate ($c_S > c_I, c_B$; $\gamma_B = 1$) the electrolyte is weak and it dissociates according to the mass action law with $\gamma_I c_S = (k_{HF} c_S)^{\frac{1}{2}}$. The equation $\mu_L = \mu_S$ gives then the linear relationship $c_S = a_L (k_{HF})^{-\frac{1}{2}} \exp[(\mu_L^o - \mu_S^o)/RT]$ between c_S and a_L. In the medium range, the electrolyte is strong and dissociates completely ($\gamma_I = 1$). The same equation yields the correct dependence $a_L^2 = c_S^3 \exp[(\mu_S^o - \mu_L^o)/RT]$. From the experimental data on segregation, the free enthalpy difference amounts to 13 kcal/mole, representing the formation of an ionic pair and of a L defect in the ice.

In the intrinsic low concentration range, the only relation consistent with the observations are $c_I < c_S < c_B$ ($\gamma_B = 0$, $\gamma_I = 1$) and the equation reduces to $a_L = c_S c_B^{\frac{1}{2}} \exp[(\mu_S^o - \mu_L^o)/2RT]$, again a linear relationship between a_L and c_S, with a value of 2×10^{-5} mole/l for c_B. At very low concentration, the thermodynamical properties of

HF are independent of any structural defect concentration and $a_L = c_S \exp[\mu^* - \mu^o)/2RT]$. This enthalpy difference corresponds to the formation of an ionic pair and amounts, according to segregation experiments, to 5 kcal/mole, about the activation enthalpy for DC conductivity in HF doped ice. From these results, the formation enthalpy for a DL pair in pure ice is 16 kcal/mole, in agreement too with dielectric measurements.

The large segregation coefficient of HF is due therefore to the introduction into the lattice of ionic and Bjerrum defects which require some energy. On the other hand ammonium fluoride produces a much lower concentration of ions and of Bjerrum defects and it has a segregation coefficient smaller by one or two orders of magnitude.

VI.3.6. CONCLUSION

Basic information on the lattice structure being given by X-ray and neutron diffraction, the interpretation of permittivity and conductivity measurements by standard theories applied to simple models gives much insight on short and long range interactions, localized excited states and their microscopical parameters. Confirmation is obtained by spectroscopy (IR and Raman) and from the analysis of thermal behaviour and of segregation of certain impurities during crystal growth. Experimental difficulties require sensitive and refined methods of measurement to obtain reliable results. The static dielectric permittivity reflects the order of the proton arrangement and, with Fröhlich's theory, gives the magnitude of the molecular dipole moment. The frequency dependence of permittivity and of conductivity is accounted for by localized excitations, the Bjerrum and ionic defects. Statistical treatment connects the macroscopic quantities with their specific parameters, and their role un pure ice is elucidated by doping with hydrofluoric acid. Measurements as a function of temperature determine the enthalpies for formation and motion of the defects, whereas pressure dependence gives the activation volumes.

REFERENCES

1 Recent advances and reviews of the subject, with the references to previous original papers, can be found in the following books :

1a E. Whalley, S.J. Jones and L.W. Gold, eds.: Physics and Chemistry of Ice. The Royal Society of Canada, Ottawa 1973

1b R.A. Horne, ed.: Water and Aqueous Solutions. J. Wiley, New York 1972

1c N.H. Fletcher: The Chemical Physics of Ice. Cambridge University Press, Cambridge 1970

1d N. Riehl, B. Bullemer and H. Engelhardt: Physics of Ice. Plenum Press, New York 1969

1e D. Eisenberg and W. Kauzmann: The Structure and Properties of Water, Clarendon Press, Oxford 1969

2 J.S. Chamberlain, F.H. Moore and N.H. Fletcher, ref. 1a, p. 283

3 Ref. 1c, Ch. 3

4 E. Whalley, ref. 1d, p. 19

5 C. Jaccard, ref. 1d, p. 348

6 H. Fröhlich; Theory of Dielectrics, Clarendon Press, Oxford 1950, p. 36

7 W. Gobush and C.A.J. Hoeve, J. Chem. Phys. 57, 3416 (1972)
F.H. Stillinger and M.A. Cotter, J. Chem. Phys. 58, 2532 (1973)
J.F. Nagle, ref. 1a, p. 175

8 G.P. Johari and E. Whalley, ref. 1a, p. 278

9 E. Whalley, J.B.R. Heath and D.W. Davidson, J. Chem. Phys. 48, 2362 (1968)

10 A. von Hippel, R. Mykolajewycz, A.H. Runck and W.B. Westphal, J. Chem Phys. 57, 2560 (1972)
A. von Hippel, D.B. Knoll and W.B. Westphal, J. Chem. Phys. 54, 134 (1971)

11 H. Gränicher, Z. Kristallographie 110, 432 (1958)

12 N. Bjerrum, Kgl. Danske Videnskab, Selskab. Mat.-fys. Medd. 27, 56 (1951)

13 C. Jaccard, Helv. Phys. Acta 32, 89 (1959); Phys. Kondens. Materie 3, 99 (1964)
L. Onsager and M. Dupuis, Rendiconti Scuola Intern. Fis. Enrico Fermi (X. Corso) 234 (1960) and in B. Pesce, ed. : Electrolytes, Pergamon Press, New York 1962, p. 27

14 R.V. Babcock and R.L. Longini, J. Chem. Phys. 56, 344 (1972)

15 H. Huber, C. Jaccard and M. Roulet, ref. 1a, p. 137

16 A. Steinemann, Helv. Phys. Acta 30, 553 (1957)

17 M. Eigen, L. De Mayer and H.C. Spatz, Ber. Bunsen Ges. 68, 19 (1964)

18 U. Eckener, D. Helmreich and H. Engelhardt, ref. 1a, p. 242

19 L. Levi, O. Milman and E. Suraski, Trans. Faraday Soc. 59, 2064 (1963)

20 L. Levi and L. Lubart, Phys. kondens. Materie 7, 368 (1968)

21 R.K. Chan, D.W. Davidson and E. Whalley, J. Chem. Phys. 43, 2376 (1965)

22 R. Taubenberg, M. Hubmann and H. Gränicher, ref. 1a, p. 194

23 J. Bilgram, thesis (ETH Zürich), to be published.

VII NMR Methods

VII 1. NMR STUDIES OF PURE WATER

Jay A. Glasel
Department of Biochemistry
University of Commecticut Health Center
Farmington, Connecticut 06032, U.S.A.

ABSTRACT

Experimental results are available for: nuclear magnetic resonance relaxation of ^{1}H, ^{2}H and ^{17}O and nuclear magnetic resonance chemical shifts of ^{1}H and ^{17}O in water. Theories pertaining to the microscopic interpretations of these macroscopic results are well developed for relaxation, much less so for chemical shifts. However, the interpretations while indicating that water is different than other liquids are not detailed enough at the present time to connect with specific structural models.

VII.1.1. INTRODUCTION

All of the measurable NMR quantities are ensemble and time averaged. That is, classical equations suffice to explain the behavior of nuclear magnetism in systems such as bulk water. Since water "structure", if it exists, is a short lived microscopic phenomenon considerable effort must be expended to connect macroscopic with microscopic behavior.

The sum total of NMR results on pure water can be expressed very briefly as follows. Water is an associated liquid whose constituent molecules behave in a microdynamical sense as if they were Stokes-Einstein particles distributed in a continuous, featureless medium. This follows from a very straightforward analysis of experimental data. In fact the analysis is so direct, and so simple, that no worker in the field believes that the results are anything but fortuitous and therefore the past years have seen many attempts to develop complex theories to produce the same results as a simple one thus violating the principle of Occam's Razor.

The nuclei whose NMR signals are of interest here are ^{1}H, ^{2}H and ^{17}O. The measurable parameters are: the chemical shifts of ^{1}H and ^{17}O in the various forms of water, and the nuclear magnetic spin-lattice relaxation times (T_1) of ^{1}H, ^{2}H and ^{17}O in liquid water.[a]

---

[a] The spin spin relaxation times T_2 for ^{1}H and ^{2}H in liquid water of normal isotopic composition are equal to the appropriate T_1's. However, for ^{17}O this is not true and the difference is pH dependent /19/.

VII.1.2. THE CHEMICAL SHIFT

The nuclear magnetic resonance frequency for a given nucleus and magnetic field is not a strictly constant quantity. Rather it depends on the molecule and environment within the molecule in which the nucleus resides. This effect is due to the electron cloud surrounding the nuclei, and the electronic structure in turn is determined by the type of chemical binding and other chemical influences, the slight variation in frequency is denoted as chemical shift. Assume that the magnetic field in the laboratory is H_o, then the resonance frequency ω_o is given as

$$\omega_o = \gamma H_o (1 - \delta)$$

γ is the gyromagnetic ratio , σ is the shielding parameter depending on the electronic environment. With respect to a reference substance (with shielding σ_r) we quantitatively define $\delta = \sigma - \sigma_r$ as the chemical shift, δ is of the order of 10^{-6}, i.e., p.p.m., for protons. More details regarding the nature of the chemical shift for a nucleus is given in any number of texts /1a, 1b, 1c/. The results of measurements for ^1H_2O /2/ and $H_2{}^{17}O$ /3/ are that the nuclei become <u>deshielded</u> as the temperature decreases in liquid water (see Figures 1 and 2). This deshielding effect must be due to association and it amounts to about half of the total range of shielding which has been observed for protons in all chemical compounds. One finds similar magnitude of deshielding for the OH proton of methanol and ethanol when transferring the molecule from the gas to the liquid. A complete theory of chemical shifts is not available at the present time. However, the results qualitatively indicate that deshielding represents a decrease in the <u>average</u> electron density around the nucleus in question. In putting down the world "average" the problem of connecting macroscopic experiments with microscopic interpretations has been entered into. <u>A priori</u> one might expect that if there are some water molecules in a

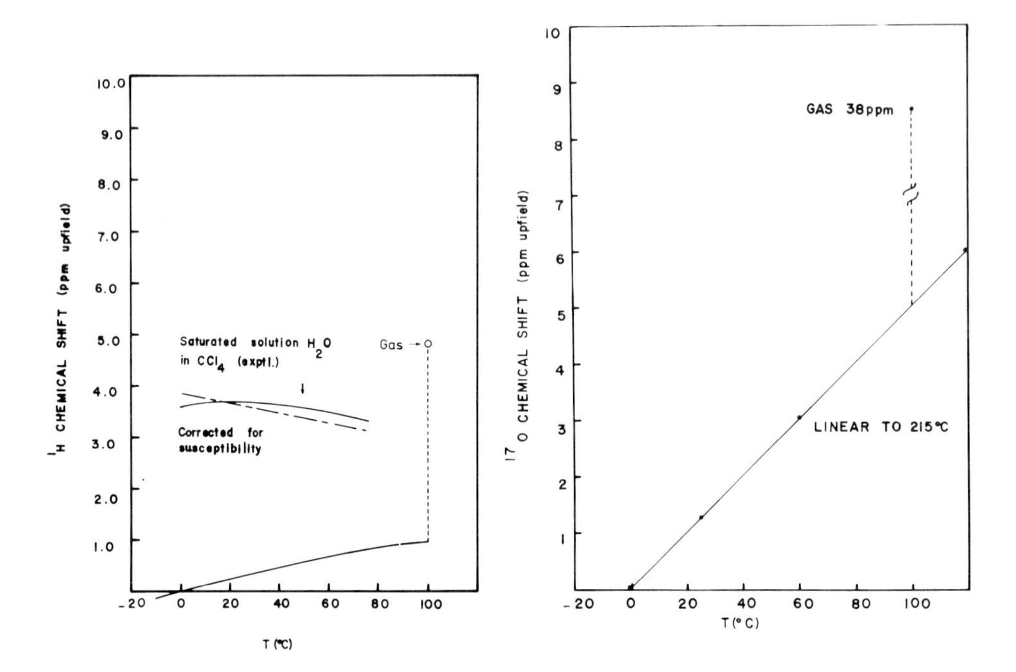

Figure 1. ^{1}H-chemical shift of water versus temperature.

Figure 2. ^{17}O-chemical shift of water versus temperature.

structured environment and some in an unstructured one the nuclei within the molecules would experience two different chemical shifts and therefore two different signals might be observed. In fact, if there are two species whose resonant frequencies are ω_A and ω_B and whose lifetime is τ before species interchange (i.e., corresponding to the lifetime of a water molecule in a structured environment) then the signal observed will be a single line when /1/,

$$1/\tau \approx 2\pi (\omega_A - \omega_B) \qquad (1)$$

The maximum range of all known chemical shifts for protons is something like 3000 Hz. Hence, if there is exchange between environments faster than 0.5 msec there will exist a single NMR line whose shift will be the weighted average of the instantaneous amounts of species A and B. Of one knew ω_A for "pure structured water" and ω_B for "pure unstructured water" and that the mixture was simple than the relative proportions as a function of temperature could be found easily. However, neither is known experimentally (e.g. The chemical shift of ice is not known) nor can be predicted theoretically. The simple fact is that the observed shifts for water are very temperature dependent in contrast to those of, for example, liquid methane. Thus there is strong temperature dependent intermolecular pertubation of molecular electronic structure in liquid water and this is perhaps one definition of an associated liquid.

Various theoretical calculations of the chemical shift of protons in an isolated water molecule relative to the fairly accurate theoretical value for the H_2 molecule /4/ have been made /5/; more or less successfully. As discussed by Aleksandrov /6/ extension of these to predictions of chemical shifts in bulk water can only be very crude approximations since the magnitude of the gas-liquid ($\sim 5\ 10^{-6}$) is the result of several effects whose shifts have positive and negative signs but whose magnitudes are similar. The conclusion is that the magnitude of the chemical shift upon H-bond formation can be explained by juggling factors of approximately equal magnitude /6/. The major factors are probably: an increase in the polarity of the O-H bond and an increase in bond length (both of which result in deshielding) balanced against an increase in electron density at the acceptor proton upon bridge formation.

There has been essentially no theoretical work done on ^{17}O chemical shifts. One can predict, however, that the paramagnetic terms /7/ will be of great importance. At the present time chemical shift measurements in hydrogen bonded systems can't be used to prove theoretical models of aqueous solution structure.

VII.1.3. SPIN-LATTICE RELAXATION TIMES

Let the magnetic moment per cm^3 due to the nuclear magnets in thermal equilibrium be M_o. M_o is called the nuclear magnetization in thermal equilibrium. M_o is a vector whose direction coincides with the magnetic field H_o, which is the z-axis. Now assume that at a given time the nuclear magnetization has any other value M_o in particular, assume that the z component has an instantaneous value $M_z \neq M_o$. In this event a change with time of M_z which is given by the equation

$$\frac{dM_z}{dt} = -\frac{1}{T_1}(M_z - M_o)$$

is observed.

The integrated form of this equation is

$$M_z = M_o + (M_z(0) - M_o)e^{-t/T_1}$$

where $M_z(0)$ is the z component at time t = 0. The time constant T_1 is denoted as the spin-lattice relaxation time or longitudinal nuclear magnetic relaxation time. There is another time constant T_2, for the "movement" of the transversal component of the magnetization, however, T_2 need not be considered when questions of the structure of water are concerned. Notice that T_1 is a "macroscopic" time: for the water protons T_1 is of the orders of seconds (see Fig. 3), for the deuterons T_1 is about one order of magnitude shorter (0.41 sec at 25 °C), and $T_1 = 6.7 \cdot 10^{-3}$ sec at 25 °C for ^{17}O. In Fig.3 the temperature dependence of T_1 of the protons in the range from 0 to 100 °C is shown.

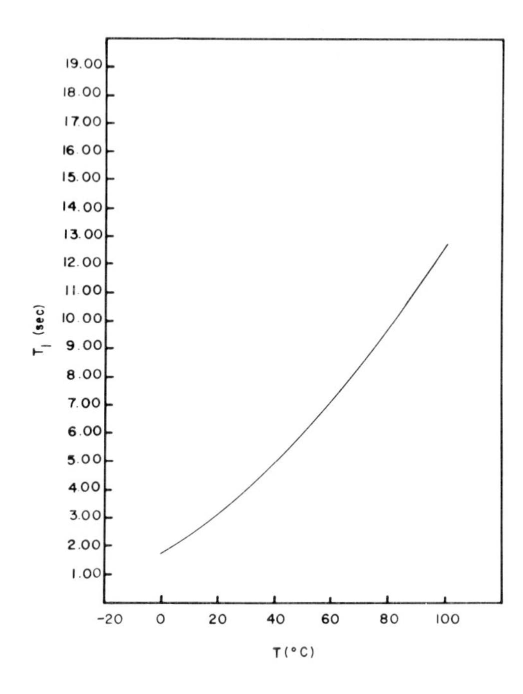

Figure 3. Spin-lattice relaxation time T_1 of water versus temperature.

The theory of nuclear relaxation which is of most current use stems from Kubo's work /8/. Observed macroscopic relaxation times are related to the Fourier transform integral of the appropriate angular autocorrelation function. The latter may be described as follows. Choose an arbitrary set of angular coordinates for a molecule. As the molecule undergoes collisions with others its angular position varies in some fashion, perhaps randomly. However, if one examines its position at some initial time, pauses and then examines again; a position very close to the initial one is expected provided the pause is very short. As the pause gets longer other angular positions become more probable. After a long time the molecule's position is no longer correlated with its initial position. The angular autocorrelation function describes the approach of the system to randomization. The persistence of correlation in molecular position betrays the existence of order. Information about the Brownian process is tied to the existence of this correlation and as the correlation decays so does the information. An underlying assumption is that the "structure" of a liquid is related to the degree of order pertaining to the autocorrelation function.

In the frictional limit where there are many molecular collisions per unit time, and doubtless applicable to low temperature liquid water, Steele /9/ has shown how to derive theoretical values for the appropriate angular autocorrelation functions for spherical and symmetric top molecules. It has not been done for other shapes including the asymmetric top H_2O molecule, and hence all treatments assume the water mole to be spherical. For the spherical case the angular autocorrelation function $G_{(2)}(t)$ is:

$$G_{(2)}(t) = e^{-\frac{6KT}{\xi}t} \qquad (2)$$

where ξ is the frictional coefficient. The Fourier Transform integral in its simplified form is /9/,

$$J(\omega) = 2\int_0^\infty \cos \omega t \; G_{(2)}(t)dt \qquad (3)$$

It should be clear intuitively that in a liquid correlation does not persist for very long. Hence, despite the limits on the integral essentially all the information is contained within the limits zero to a few pico-seconds. The factor,

$$\frac{\xi}{6KT} \qquad (4)$$

has the dimensions of time and is often called the rotational correlation time. At 25°C for water one finds from a variety of physical measurements that $\tau_c \approx 2.5 \cdot 10^{-12}$ sec. We may introduce another time, the reorientation time τ_r of the water molecule. This is the time after which a vector fixed in the molecule has changed its direction by roughly 180° due to rotational diffusion. In the frictional limit just mentioned the theory tells us that we should have $\tau_r = 3\ \tau_c$.

The factor 3 stems from the particular angular function which is implied in $G_2(t)$. If the value of ξ is,

$$\xi = 8\pi\eta a^3 \tag{5}$$

where η is the macroscopic viscosity of the continuous medium and a is the molecular "radius" then the rotational motion is defined as "classical". It is a fact that for many liquids satisfactory agreement with experiment is found only if a phenomeno-logical parameter - the microviscosity /10/ - replaces η . Its value averages $\frac{\eta}{12}$ /11/. However, water appears fully "classical" in both the value of ξ and its temperature dependence /12/. This must presumably be rationalized as accident. In order to do so one would have to show that the magnitude and functional dependence of $G_{(2)}(t)$ is such that the transform integral over all time is the same as the classical value. Studies of the frequency dependence of the dielectric constant show an apparant single macroscopic relaxation time τ_d (proportional to $G_{(1)}(t)$ for the molecular electric dipole moment vector). It is not clear whether or not this implies a single microscopic relaxation time or a distribution of times. However, if it is a single relaxation time than, remarkably, frictional behavior is observed for the entire time range in question. In water, at 25 ^{o}C $\tau_d = 8.2\cdot10^{-12}$ sec. Unfortunately, it is not entirely clear yet how to derive a reorientation time τ_r from these measurements. Whether $G_{(2)}(t)$ follows the same behavior is not known. However, it may be possible to study the problem experimentally since there is a theoretical connection between Raman line shapes and the angular autocorrelation functions /13/. However, it has been only recently that this has been done for anything so complicated as a symmetric top molecule /14/.

An alternate explanation for water might be that observed rotational correlation times are insensitive to the "structures" of interest. One must remember that the observed relaxation times are again averages.

If there is so-called fast exchange between two species which have different relaxation times the observed relaxation rate is:

$$\frac{1}{T_{1\ obs}} = X_1 \frac{1}{T_{1,1}} + X_2 \frac{1}{T_{1,2}} \qquad (6)$$

X_1 is the mole fraction of species 1 and $T_{1,1}$ is its relaxation time. Let species 1 be the unstructured water and 2 the structural water. This equation holds when the lifetime of the longest lived species is shorter than the shortest relaxation time of either species and, of course, assumes a 2-species model. Assume that the T_1 for the structural species is about 10^{-2} sec. This estimate is taken from our knowledge of the reorientation time of the hydration sphere of strongly hydrated ions, i.e., of definitely structured entities. On the other hand, T_1 for the unstructures species should be $T_1 > 10$ sec, which corresponds to the relaxation time of a liquid with non-associated molecules and of the same size as water. Then we see immediately from eq. (6), that X_2, the mole fraction of the water molecules in the structured form comes out as $X_2 \leqslant 3 \cdot 10^{-3}$. This is at variance with many but not all such estimates. Equation (6) implies fast exchange as compared with T_1, however slow exchange as compared with the correlation times τ_{c1}, τ_{c2}. If we have also fast exchange as compared with the correlation times eq. (6) has to be replaced by the formula:

$$(T_1)_{obs} = X_1 T_{1,1} + X_2 T_{1,2} \qquad (6a)$$

Now, with $T_1 \approx 1/5\ T_{1,1}$ and $T_{1,2} \approx 0$ we obtain $X_1 \approx 0.2$ which again is in accord with some but not all estimates. If this is true then whatever the structured state of water is, the residence time of a water molecule in such a state must be very short, probably $\tau \approx 10^{-12}$ sec. Were this argument correct it can be seen from eq. (6a) that due to the fast fluctuations in the environment of a given H_2O molecule, T_1 is indeed rather insensitive to the "structure" present in water.

The relations between observed relaxation rates and the transform integrals for liquid water fall into two catagories. For ^{2}H and ^{17}O the relations are,

$$(1/T_1)_{j,k} = E_{j,k} \int_0^\infty 2\cos \omega t \; G_{(2)_{j,k}}(t)dt \qquad (7)$$

where j refers to the nucleus and k to the species of molecule. For these two nuclei the $E_{j,k}$ are called the quadrupole coupling constants. The value of these "constants" depends somewhat upon molecular electronic structure. Since the latter is not known the coupling factors must be assumed to be something of a sliding parameter although attempts have been made to derive their values in the liquid /15, 16/. For proton relaxation in H_2O (as opposed to $^{2}H_2O$ doped with ^{1}H) the observed relaxation rate is,

$$(1/T_1)_{1H,k} = E_{^1H,k} \int_0^\infty 2\cos \omega t \; G_{(2)_{^1H,k}}(t)dt + f(D_{^1H}) \qquad (8)$$

where $E_{^1H,k}$ is called the dipolar coupling constant and is an accurately known value derived from nuclear constants and molecular geometry. The factor $f(D_{^1H})$ indicates that the observed relaxation rate is dependent upon intermolecular translational diffusion. This dependence may in principle be used to determine the self-diffusion coefficient of water. But such a method is very crude. It should be mentioned however, that the self-diffusion coefficient can be measured by special NMR pulse techniques with an appropriate instrument. The values found and their temperature dependence are again "classical" /17/. It is doubtful. however, that self diffusion coefficients are at all sensitive to the physical model of liquid structure /18/.

Since the possible conclusions were emphasized at the beginning of this discussion it remains to state where future advances may be made. From the experimental standpoint one can point out that ^{17}O chemical shift measurements have not been done to a very high degree of accuracy nor do proton shifts in ice nor as a function of pressure exist. Furthermore, it would be very interesting to do a parallel study of all the parameters mentioned above with liquid hydrogen sulfide where presumably liquid "structure" is absent or insignificant. From the theoretical standpoint it should be possible to predict rotational correlation functions for asymmetric tops and to compare these with experimental values from Raman spectra. Theoretical prediction of ^{17}O shifts are needed as well as predictions of quadrupole coupling factors for structured units.

REFERENCES

1a J. A. Pople, W. G. Schneider and H. J. Bernstein: High-resolution Nuclear Magnetic Resonance, McGraw-Hill, Inc., New York, 1959

1b C. P. Slichter: Principles of Magnetic Resonance, Harper and Row, New York, 1963

1c A. Covington and A. D. McLachlan: Introduction to Magnetic Resonance, Harper and Row, New York, 1967

2 J. C. Hindman, J. Chem. Phys. 44, 4582 (1966)

3 Z. Luz and G. Yagil, J. Phys. Chem. 70, 554 (1966)

4 W. G. Schneider, H. J. Bernstein and J. A. Pople, J. Chem. Phys. 28, 601 (1958)

5 C. W. Kern and M. Karplus: in Water, A Comprehensive Treatise, Vol. I, ed., F. Franks, Plenum Press, New York, 1972

6 I. Aleksandrov: The Theory of Nuclear Magnetic Resonance, Academic Press, New York, 1966

7 J. D. Memory: Quantum Theory of Magnetic Resonance Parameters, McGraw-Hill, New York, 1968

8 R. Kubo: In Fluctuation, Relaxation and Resonance in Magnetic Systems, ed., D. ter Haar, Oliver and Boyd, London, 1962

9 W. A. Steele, J. Chem. Phys. 38, 2404, 2411, 2418 (1963)

10 A. Gierer and K. Wirtz, Z. Naturforsch. Ba 532 (1953)

11 J. A. Glasel, J. Am. Chem. Soc. 91, 4569 (1969)

12 J. A. Glasel: In Water, A Comprehensive Treatise, Vol. I, ed., F. Franks, Plenum Press, New York, 1972

13 R. G. Gordon, Adv. Magn. Resonance 3, 1 (1967)

14 E. F. Johnson and R. S. Drago, J. Am. Chem. Soc. 95, 1391 (1973)

15 J. A. Glasel, Proc. Nat. Acad. Sci. (USA) 58, 27 (1967)

16 J. G. Powles, M. Rhodes and J. H. Strange, Mol. Phys. 11, 515 (1966)

17 R. Hausser, G. Maier and F. Noack, Z. Naturforsch. 21a, 1410 (1966)

18 P. K. Sharma and S. K. Joshi, Phys. Rev. 132, 1431 (1963)

19 S. W. Rabideau and H. G. Hecht, J. Chem. Phys. 47, 544 (1967)

VII.2. N.M.R. METHODS, AQUEOUS ELECTROLYTE SOLUTIONS

H. G. Hertz

Institut für Physikalische Chemie und Elektrochemie der Universität Karlsruhe,Germany

ABSTRACT

If we make a microscopic snapshot of a liquid we will find that the instantaneous picture is that of a heap of molecules. Now imagine that we make several such snapshots. We will get a number of pictures of heaps which, as regards the instantaneous details, at first sight seem to be entirely different. Thus we might come to the result that everything can occur in such a heap, i.e. in a liquid. How then can one describe the structure of the liquid? On closer inspection of our series of pictures showing the moelcular heap we will find that certain intermolecular arrangements occur more freouently than others. A marked preference of relative molecular arrangements will be the more pronounced the more modest we are regarding the number of molecules involved in an arrangement of interest. The simplest object of such a study is just a pair of molecules, a pair selected at random, i.e. formed of any two molecules being anywhere in our heap. We shall find that certain pair configurations - given by the vector $\vec{r}_o$ connecting the two centres of mass and the relative orientation Ω of the pair partners - occur very often, other pair configurations can be found only very rarely. Thus, we may define a probability to find a given pair configuration as being the fraction of observations in which - on selecting a pair at random - we found the desired pair configuration to the total number of observations. This probability will be a particular function $p(\vec{r}_o,\Omega)d\vec{r}_o d\Omega$ of the configuration $\vec{r}_o \cdot \Omega$, when the two pair partners are close neighbours, but for large separations we have complete randomness, $p(\vec{r}_o,\Omega)$ = $(8\pi^2 V)^{-1}$, this being the typical property of a liquid (V = volume of the liquid).

For the present article we have to keep in mind that we are concerned with electrolyte solutions, i.e. with mixtures. In mixtures we may be interested in several types of pairs of particles when pursuing a structural investigation. Firstly there is the solvent which may be described by a solvent-solvent molecular distribution function. In the mixture, the (mean) probability to find a solvent molecule with given $\vec{r}_0$ and Ω relative to another one will in general differ from that in the neet liquid. Next the distribution of the solute molecules relative to each other may be of interest, and lastly we have molecular distribution functions which regard the pair solute-solvent. Let A be the solvent molecule, and B the solute molecule, then the pair distribution functions which can be investigated are

A - A

B - B

2(A - B).

The factor two in the A-B case arises because in favourable cases we can expect to have two independent sources of information regarding the unlike particle distribution function: The relaxing nucleus may be either on A or on B.

But even with the knowledge of the pair distribution function we are at the lowest level in the hierarchy of structural description by molecular distribution functions. The local hydration phenomena will be adequately described by three particle or even higher distribution functions.

In the present contribution we shall use the convention that a set of molecular pair distribution functions $\rho(\vec{r}_0,\Omega)$ describes the structure of the electrolyte solution. Again, $\rho(\vec{r}_0,\Omega)$ is the probability to find a given particle at $\vec{r}_0$ with orientation Ω relative to another one. For our purpose the pair distribution functions are understood to be simple model distribution functions which are characterized by a small number of parameters. It will be shown in which way nuclear magnetic relaxation rates of the water nuclei or of the ion nuclei can be utilized to determine one or more parameters in the corresponding distribution function.

The basic principle underlying the possibility to study the pair distribution function by nmr relaxation techniques is as follows: Consider two particles in the liquid which approach towards one another, denote them as particle No. I and particle No.II, respectively. Assume that one of the atomic nuclei of particle II possesses a magnetic moment. This nuclear magnet produces a magnetic field, the field strength is proportional to $1/r_0^3$ where r_0 is the distance of the nucleus of particle II from any selected point in the liquid. Now let this reference point be a nucleus of particle No.I. Then, if this nucleus as well has a magnetic moment, the torque exerted on this nucleus by the magnetic field of the nucleus on particle II is proportional to $1/r_0^3$, the same holds true for the magnetic interaction energy between these two nuclei. This magnetic interaction energy together with the molecular motion in the liquid causes a flipping process of the nuclear magnet on particle No.I (and of course vice versa), and the sum of the flipping processess for all nuclear magnets in the liquid represents the change with time of the nuclear magnetisation if its instantaneous value is different from the equilibrium value. The change of the nuclear magnetization in such a situation is called the nuclear magnetic relaxation and the time constant describing this "motion" of the magnetization is the nuclear magnetic relaxation time. Now it is plausible that the change in time of the magnetization is the faster the stronger the local magnetic field at nucleus I due to the nucleus on particle II. Thus the study of nuclear magnetic relaxation times permits a determination of the relative approach of the two pair partners I and II, or expressed in another way, it gives the probability that such an approach occurs and this in turn, as has been described above, is our convention to describe the structure of the liquid.

Chemical shift studies will also be described briefly in those cases where they contribute to our knowledge of the pair distribution function.

VII.2.1. INTRODUCTION

This article will mainly be concerned with nuclear magnetic relaxation time measurements. These relaxation times (denoted as T_1, T_2), which describe the approach of the nuclear magnetization towards thermal equilibrium, may be used as a fairly efficient tool for the investigation of the structure of electrolyte solutions. The study of the other nuclear magnetic resonance parameter, the chemical shift, yields less information pertinent to our subject and thus shift data will only be discussed very briefly. The fundamental physical significance of the nuclear magnetic relaxation time and the chemical shift has been explained in the preceding chapter. The third type of nmr parameter, the spin-spin coupling constant, shall not be treated at all, only a very small number of investigations has been devoted to the behaviour of coupling constants in electrolyte solutions.

The central quantity for the facts to be discussed here is the time correlation function $g(t)$ of the function $Y^{(2)}(\theta,\phi)/r^3$, where $Y^{(2)}$ is the second order spherical harmonic and $\vec{r} = \{\theta, \phi, r\}$ is the vector connecting two nuclei which interact as (nuclear) magnetic point dipoles (relaxation by magnetic dipole-dipole interaction):

$$g(t) = \mathcal{N} \int \frac{Y^{(2)*}(\theta_0,\phi_0)}{r_0^{\,3}} \frac{Y^{(2)}(\theta,\phi)}{r^3}\, p(\vec{r}_0)\, P(\vec{r}_0,\vec{r},t)\, d\vec{r}_0\, d\vec{r} \tag{1}$$

The special functional form of $Y^{(2)}(\theta,\phi)$ is not of importance here, it may suffice to state that the $Y^{(2)}$'s are certain products of $\sin\theta$, $\cos\theta$, $\sin\phi$, ...

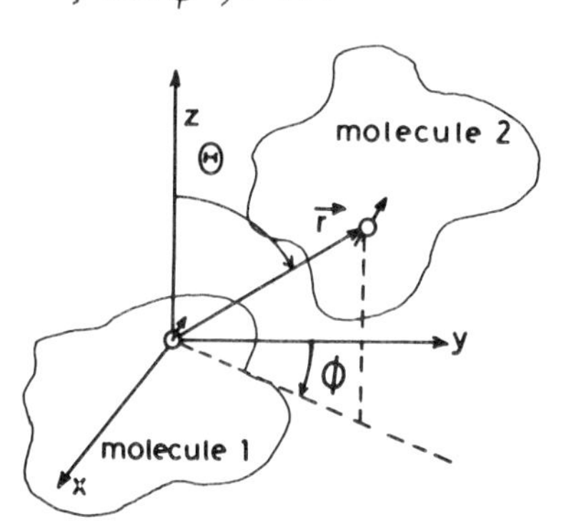

Fig.1 The vector $\vec{r}$ connects two nuclei in two molecules. If the time dependence of $\vec{r}$ is not important $\vec{r}$ may also be denoted as $\vec{r}_0$.

Fig.1 shows the vector $\vec{r}$. Now the two molecules carrying the nuclei perform molecular motions in the liquid, thus $\vec{r}$ will change with time, $\vec{r} = \vec{r}(t)$. At time t = 0 we have $\vec{r} = \vec{r}_0$. $\mathcal{N}$ is the number of interacting nuclei in the system, the time dependent quantity $P(\vec{r}_0, \vec{r}, t)$, the propagator, gives the probability that the nuclear separation has moved from $\vec{r}_0$ to $\vec{r}$ after a time t. $p(\vec{r}_0)$ is the pair distribution function with respect to the two interacting nuclei, i.e. $p(\vec{r}_0)$ gives the probability density to find a given interaction partner at $\vec{r}_0$ relative to the reference nucleus ("the relaxing nucleus"). We notice that $p(\vec{r}_0)$ coincides with the "molecular distribution function" only if the nuclei are at the centres of mass of the molecules or ions. Generally this is not so, then $p(\vec{r}_0)$ is given by an integral of $p(\vec{r}_s, \Omega)$ over $\vec{r}_s$ and Ω , where the molecular distribution function $p(\vec{r}_s, \Omega)$ represents the probability density to find a molecule with the centre of mass at $\vec{r}_s$ and having the orientation Ω , both quantities taken relative to the molecule containing the reference nucleus /1/.

We shall use the convention that <u>$p(r_s, \Omega)$ describes the structure of the electrolyte solution</u>. However, since in general the interrelation between $p(\vec{r}_s, \Omega)$ and $p(\vec{r}_0)$ is complicated, simplifying, we consider the structure of the electrolyte as to be given by $p(\vec{r}_0)$ and we may say: <u>$p(\vec{r}_0)$ is the quantity we are interested in.</u>

Some comments are in order to explain the physical background behind eq.(1). As the vector connecting two interacting nuclei varies with time, the magnetic interaction energy $\mathcal{H}$ fluctuates as well. This is shown schematically in Fig.2. It may be seen that after a time τ_c correlations between successive values of the interaction energy are lost. If we take an ensemble average we see that correlations in interaction decay as a smooth curve which is depicted in Fig.3. Eq.(1) represents such a curve for $g(t)$. For the solutions of interest here the time after which $g(t)$ has decayed is of the order 10^{-11} sec.

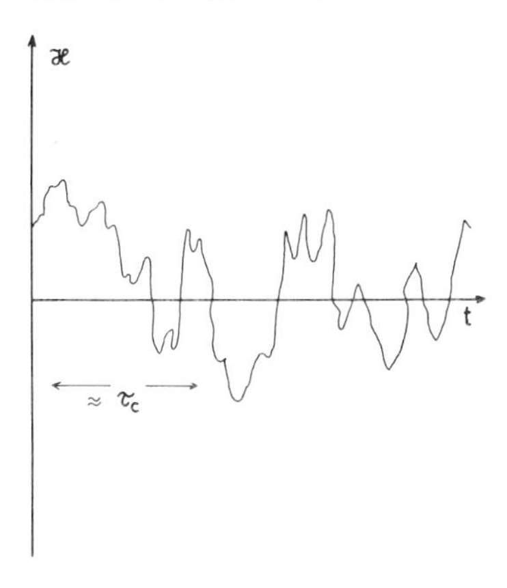

Fig.2 Fluctuations of the magnetic interaction energy with time

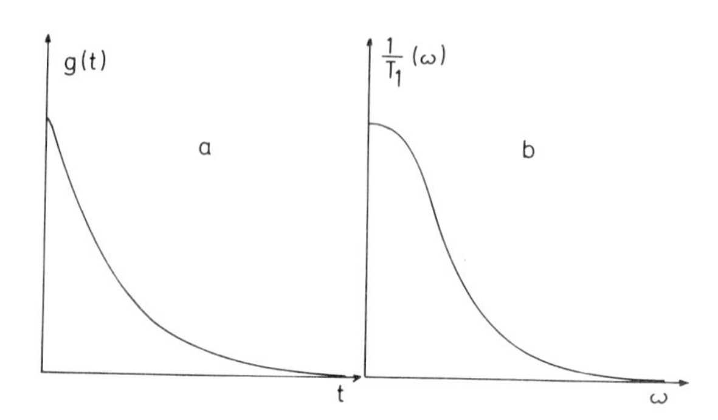

Fig.3 a) Time correlation function g(t) in the "time language".
b) Time correlation function in the "frequency language" which is (roughly) proportional to $1/T_1(\omega)$

As a next step we need the relation between $g(t)$ and the nuclear magnetic relaxation times. It is sufficient here to write /2-4/

$$\frac{1}{T_1} \sim \int_{-\infty}^{+\infty} e^{-i\omega t} g(t)\,dt \quad . \tag{2}$$

$1/T_1$ is denoted as (longitudinal) relaxation rate, for $1/T_2$ a similar relation holds, only some factors of proportionality differ. However, we shall not discuss quantitative expressions for $1/T_2$. The integral in eq.(2) defines the Fourier transformation, it may be understood as a translation from the "time language" to the "frequency language". According to eq.(2), in general, $1/T_1$ is frequency dependend, a rough and qualitative sketch of the resulting $1/T_1(\omega)$ is given in Fig.3b.

We now see of what general type our program is. The measured quantity is the relaxation rate $1/T_1$ on the left-hand side of eq.(2). The quantity we wish to know is the pair distribution function $p(\vec{r}_0)$ on the right-hand side of eq.(1).

For the chemical shift δ we may write

$$\begin{aligned} \delta = {} & w_0 \int \delta(\vec{R}_1, \vec{R}_2, \ldots)\, p(\vec{R}_1, \vec{R}_2, \ldots)\, d\vec{R}_1 d\vec{R}_2 \ldots \\ & + w_1^+ \int \delta(\vec{R}_+, R_1, \ldots)\, p(\vec{R}_+, \vec{R}_1, \ldots)\, d\vec{R}_+ d\vec{R}_1 \ldots \\ & + w_1^- \int \delta(\vec{R}_-, R_1, \ldots)\, p(\vec{R}_-, R_1, \ldots)\, d\vec{R}_- d\vec{R}_1 \ldots + \ldots \end{aligned} \tag{3}$$

which only holds for the sitation of fast exchange (see chapterVII.1). $\vec{R}_i = \{\vec{r}_i, \Omega_i\}$ is the generalized vector describing the position and orientation of the i-th neighbour molecule of the reference molecule, $p(\vec{R}_1, \vec{R}_2, \ldots)$ is the probability density that a neighbour configuration $\vec{R}_1, \vec{R}_2, \ldots$ occurs and $\delta(\vec{R}_1, \vec{R}_2, \ldots)$ is the chemical shift of the water protons in this configuration. w_0 gives the probability that the H_2O molecule is surrounded only by water molecules, $w_1^{\pm}$ is the probability that one of the nearest neighbours is a cation or anion respectively,

R_+, R_- is the vector connecting the water molecule with the cation and anion, respectively. In the event that we have a water exchange rate slow compared with the difference of chemical shifts in different sites, two separated lines for the hydration and bulk water, respectively, are observed. Then the integrated signal intensity of hydration water is:

$$J_{hydr.} = \frac{n_h^+ c^*}{55,5} J_{bulk} \left(1 - \frac{n_h^+ c^*}{55,5}\right)^{-1} \tag{4}$$

with

$$n_h^+ = \mathcal{N} \int_c^{r_+^*} p(r_+^o)\, dr_+^o$$

where r_+^* is the distance of the first minimum in the water-ion distribution function $p(r_+^o)$. J_{bulk} is the integrated signal intensity of the bulk water including the anion hydration spheres, n_h^+ is the hydration number of the cation, c^* is the cation concentration in molality scale.

Next we give some introductory remarks regarding the pair distribution function. A typical schematic representation of a radial distribution function is shown in Fig.4. In general $p(\vec{r}_o)$ is given as

$$p(\vec{r}_o) = N e^{-E(\vec{r}_o, T)/kT}, \tag{5}$$

N is a normalization factor, $E(\vec{r}_o, t)$ is the potential of mean force /5/. An analogous expression holds for $p(\vec{r}_o, \Omega)$ where $E(\vec{r}_o, \Omega, T)$

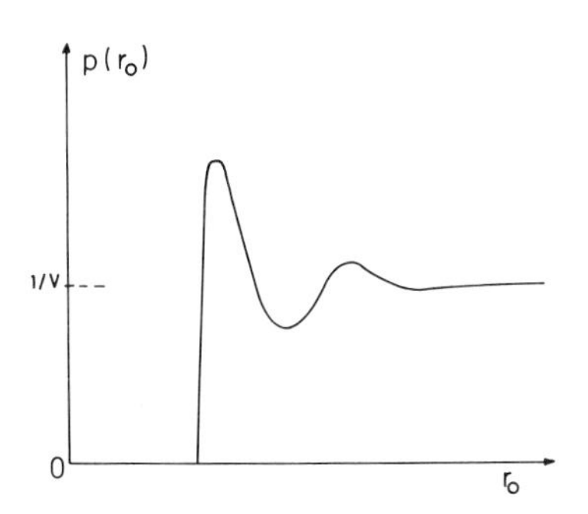

Fig.4 Schematic representation of the radial distribution function $p(r_o)$

is the potential of mean force and torque. For the maximum occurence probability $p(r_o)_{max}$ we may write:

$$p(r_o)_{max} = N e^{-E_{max}(T)/kT} \tag{6}$$

Now all information regarding $p(\vec{r}_0)$ is reduced to one parameter only. Since in many cases only one relaxation rate can be measured, this approximation is suitable to relate one experimental quantity, T_1, to one structural parameter $E_{max}(T)$. We may denote $E_{max}(T)$ as maximum effective potential depth or often we shall simply say maximum potential depth. But it should be kept in mind that actually a potential of mean (generalized) forces is concerned, i.e. a potential between two particles in the presence of all other molecules and averaged over all their configurations.

There are three types of pair distribution functions in our electrolyte solutions: The water-water, the water-ion, and the ion-ion distribution functions.

Summarizing, in the present chapter we have to discuss the following problem: What information regarding the various $p(\vec{r}_0)$'s in eletrolyte solutions can be obtained from nuclear magnetic resonance measurements.

VII.2.2. THE WATER-WATER DISTRIBUTION FUNCTION

No direct nmr studies of the water-water pair distribution function have been undertaken yet, although such a study should be possible in principle. Before we pass to the ion-water distribution function we wish to present an evaluation of nmr data where the water-water and the ion-water distribution functions are involved. This point regards relaxation time measurements and chemical shift measurements of the solvent water in diamagnetic electrolyte solutions.

First we describe relaxation time measurements. Consider the proton-proton vector in the water molecule (see Fig.5). Now clearly $p(\vec{r}_0)$ in eq.(1) is a δ-function, $p(r_0) = \delta(r_0 - b)$ where b is the intramolecular proton-proton separation. As a consequence of this, $P(\vec{r}_0, \vec{r}, t)$ degenerates to a rotational propagator /4,6/ since only rotational motion is concerned which leaves r_0 constant. Let us characterize the rotational motion by the rotational correlation time τ_c. Now $1/T_1$ is called an "intramolecular relaxation rate", $(1/T_1)_{intra}$, and it may be shown that /2/

$$\left(\frac{1}{T_1}\right)_{intra} \sim \tau_c \,.$$

In pure water at 25°C $\tau_c = \tau_c^0 = 2.5 \cdot 10^{-12}$ sec /6a/. In Fig. 6 some examples are given, the ordinate represents $1/T_1$, the total relaxation rate, but $(1/T_1)_{intra} \sim 1/T_1$ /7/, thus, apart from a constant factor, the mean τ_c as a function of the concentration is given. Then, from the mean τ_c, local $\tau_c^{\pm}$ in the hydration spheres of the ca-

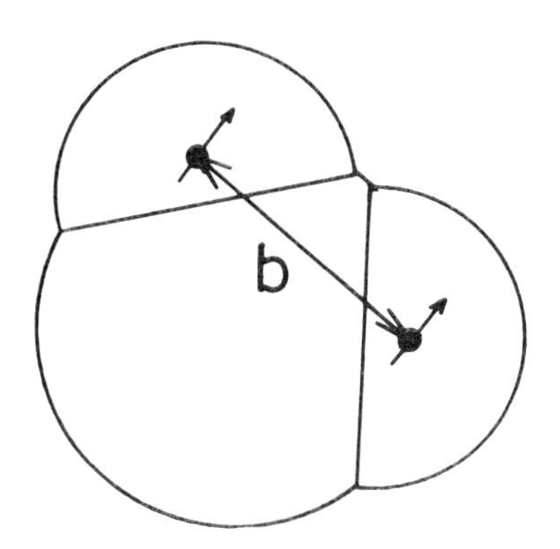

Fig.5 Intramolecular proton-proton vector in the water molecule

tion and anion, respectively, may be extracted. These local $\tau_c^{\pm}$ vary from $\approx 1 \cdot 10^{-12}$ sec to $\approx 12 \cdot 10^{-12}$ sec as we go from I^- to Mg^{2+}. We define the quantity

$$B^{\pm\prime} = \frac{n_h^{\pm}}{55.5}\left(\frac{\tau_c^{\pm}}{\tau_c^{0}} - 1\right)$$

which represents an analogue to the Jones-Dole viscosity B coefficient. In Fig. 7 we show various $B^{\pm}$ values for ions of different types. It will

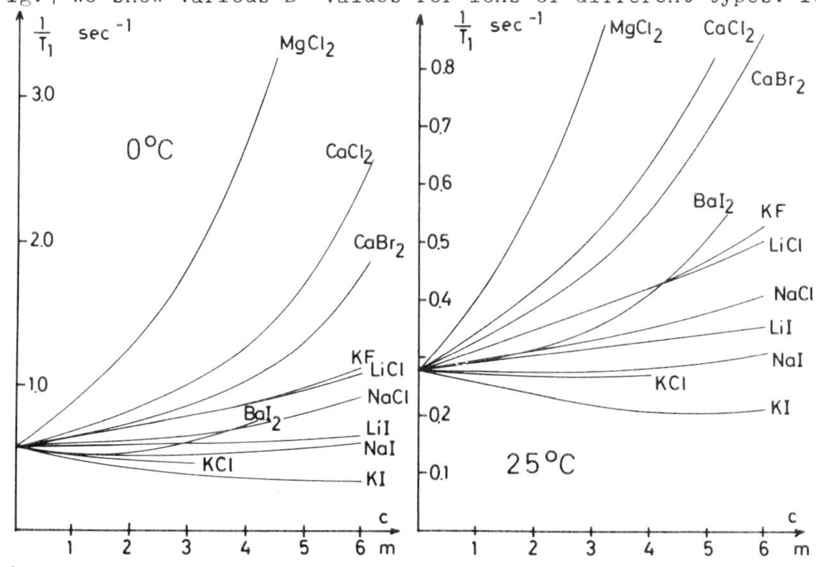

Fig.6 Proton relaxation rates in aqueous electrolyte solutions as a function of concentration m (moles/kg H_2O) /7/

be seen that an appreciable number of ions has a $B^{\pm}$ value < 0, and that among these ions we find some anions with charge two /8/.

So far the $\tau_c^{\pm}$ give a dynamical information, we wish to correlate the reorientation times with structural facts. In a qualitative way this may be done as follows /9-11/. We consider the Debye equation

$$\tau_c = \frac{4\pi \left(\tilde{a}(|E_{max}|)\right)^3 \eta^*}{3kT} \tag{7}$$

$\tilde{a}$ is the effective radius of the water molecule which is understood as a function of E_{max}, E_{max} being given according to eq.(6). If $|E_{max}|$ is large, then the mean kinetic energy is unsufficient to compete with the potential energy, aggregates of several water molecules will be long-lived as compared with the appropriate time scale implied in eq.(7) /9-11/. This makes $\tilde{a}$ large and consequently τ_c will be comparatively long. On the other hand, for low $|E_{max}|$ the kinetic energy strongly tends to destroy aggregates, thus $\tilde{a}$ and thereby τ_c becomes small. The quantity η^* which may be denoted as "microviscosity" is defined by the relation

$$\tau_{co} = \frac{4\pi \tilde{a}_o^3 \eta^*}{3kT},$$

where $\tilde{a}_o$ is the radius of the water molecule in the absence of attractive potentials $\lesssim kT$ (that is, $\tilde{a}_o$ is the normal molecular radius). τ_{co} is the correlation time observed for such a molecule e.g. dissolved in an inert solvent. In water $\tilde{a}(E_{max})$ is relatively large, i.e. τ_c is much larger than expected for a molecule of mass 18. Likewise, the self-diffusion coefficient of water, $D = 2.3 \cdot 10^{-5}$ cm^2/sec (at 25°C) is much smaller than it should be for a normal liquid, i.e. $D \approx 4 \cdot 10^{-5}$ cm^2/sec. Again, this may be accounted for by the translational analogue of eq.(7), the Stokes-Einstein relation. If one follows up these considerations more quantitatively one finds $|E_{max}| \gtrsim 2.3$ kcal for water at 25°C /10/ . In electrolyte solutions E_{max} in the hydration sphere also involves a contribution from the ion-water distribution function. We give some numerical results for $|E_{max}|$ in the first hydration sphere: I^-: $\gtrsim 1.3$, Br^-: $\gtrsim 1.9$, Cl^-: $\gtrsim 2.2$, F^-: $\gtrsim 4.5$, Na^+: $\gtrsim 4.5$, Li^+: $\gtrsim 4.5$ kcal/mole /10/. It will be seen that for some ions $|E_{max}| > |E_{max}(\text{water})|$. Since here the potential depth is larger than for pure water, i.e. the distribution function according to eq.(6) has a steeper maximum than in water, we denote these ions as structure forming ions. On the other hand, for I^- and other ions $|E_{max}| < |E_{max}(\text{water})|$ i.e. $p(\vec{r}_o, \Omega)$ is flatter than in pure water. Thus it is natural to denote ions of this type as structure breaking ions. These considerations give the basis for the classification: structure forming - structure breaking. It is clear that this denotation is still not very sharp, in fact, $E(\vec{r}, \Omega)$ may be defined by a set of parameters and it is by no means sure that all these parameters vary in the same direction as we go from water to the ionic hydration sphere.

Similar results as shown in Fig.6 have also been obtained by aid of the deuteron relaxation rate in D_2O /6,12,13/ and the ^{17}O relaxation in $H_2{}^{17}O$ solutions /14,15/. Here the relaxation mechanism is the nuclear quadrupole interaction and the intramolecular quantity whose reorientational motion is studied is the electric field gradient tensor /2/. A systematic comparison between the proton and deuteron relaxation rates, however, have not been performed yet /6,15/. It is likely , that some information about anisotropic reorientational motion may be obtained from such a study.

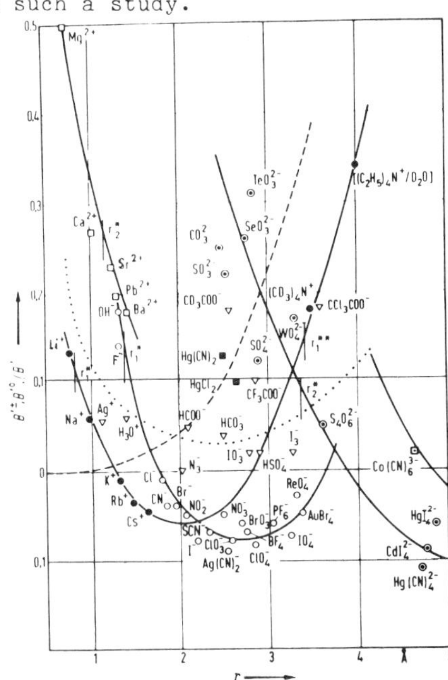

Fig. 7 $B^{\pm}$ coefficients as a function of ionic radius, after /18/

We turn to proton chemical shift measurements in diamagnetic electolyte solutions /16-21/. Fig.8 gives a typical example for a number of 1-1 and 2-1 electrolytes. In eq.(3) the first term on the right-hand side gives a positive contribution to δ as ions are added to water, the same is true for the contributions of the water neighbours in the hydration sphere, i.e. regarding the second and third terms of eq.(3). Now the water-water configurations no longer are those usually denoted as hydrogen bonded configurations and this causes a positive shift. For weakly hydrated - or negatively hydrated ions (= structure breaking ions) the contribution from the ion-water interaction is small, thus, the total shift is positive relative to pure water. On the other hand, if the ion-water interaction is stronger the ion produces a negative shift which then will be stronger than the previously mentioned contribution, the net effect will be a downfield shift. It is clear from eq.(3) that an evaluation which transgresses qualitative arguments is difficult to perform. The chemical shifts of the ^{17}O resonance follow an entirely different pattern /14,22/. Any connection of the shift with hydrogen bonding is not apparent here, i.e. a detailed understanding of the behaviour

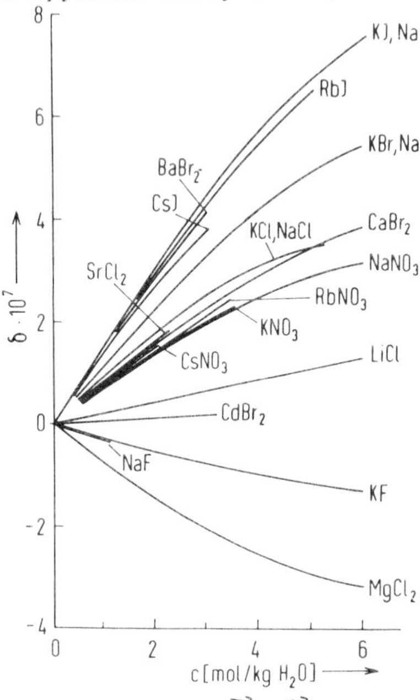

Fig.8 Proton chemical shifts in aqueous electrolyte solutions /17/, after /62/

of the coefficients $\delta(\vec{R}_+, \vec{R}_1, \ldots)$, $\delta(\vec{R}_-, \vec{R}_1, \ldots)$ in eq.(3) is completely lacking. It should be pointed out that the general pattern of the behaviour of the ^{17}O shifts is similar to that of the ionic nuclear magnetic resonances in the electrolyte solutions /23/.

VII.2.3. WATER-ION DISTRIBUTION

Here we have two types of experiments: The relaxing nucleus is the water proton (III.1) and the relaxing nucleus is the ion nucleus (III.2).

A. The Relaxing Nucleus is the Water Proton (or Other Water Nucleus)

a) Solutions of paramagnetic ions. Here the interaction partner of the relaxing nucleus is the electron spin. The resonance frequency of the electron spin is by three orders of magnitude greater than that of the nuclear magnets. Thus, we may easily work in a range where $\tau_c \omega \approx 1$, which means that the frequency dependence of $(1/T_1)$ and $(1/T_2)$ can be studied. In hexa-aquo ions like Mn^{2+}, Cu^{2+}, Fe^{3+}, Cr^{3+}, Gd^{3+} $\rho(\vec{r}_0)$ has a sharp maximum, that is, the potential has a deep minimum. In this situation $|\vec{r}_0|$ does not vary during a time of the order of τ_c^*, the rotational correlation time of the vector connecting the interacting spins. This means that again $P(\vec{r}_0, \vec{r}, t)$ in eq.(1) degenerates to the rotational propagator. In this event, the detailed computation shows that $g(t)$ is an exponential function. The Fourier transform of an exponential is a Lorentzian curve, thus $1/T_1(\omega)$ is of Lorentzian form. In other words, if we find that $1/T_1(\omega)$ is a Lorentzian, then we know that E_{max} is deep, i.e. $|E_{max}| \gtrsim 4.5$ kcal /4,6,24-26/. On the other hand, if $p(\vec{r}_0)$ is essentially repulsive, then $1/T_1(\omega)$ has less intensity in the low frequency region and has more intensity in the high frequency wings as compared with the Lorentzian frequency distribution /27-30/. A typical example for this situation is the complex anion $Cr(CN)_6^{3-}$ /31/. In the intermediate situation the primary quantity obtained from a measurement of the frequency dependence of $1/T_1$ is the product $p(\vec{r}_0)\cdot P(\vec{r}_0,\vec{r}_1 t)$. It is not quite easy to separate the structural contribution $p(\vec{r}_0)$ from the motional part $P(\vec{r}_0,\vec{r}_1 t)$ /32/. Of course, both of them are closely interconnected. With the ions Ni^{2+}, Co^{2+} and Fe^{2+} a new effect appears. Here the electron spin relaxation time T_s is very short compared with the rotational or translational diffusion times. Thus T_s acts as the correlation time. It seems as if these circumstances allow a direct determination of $p(\vec{r}_0)$ from the experimental $1/T_1$ without any consideration of the propagator $P(\vec{r}_0,\vec{r}_1 t)$ /33/.

We saw that $P(\vec{r}_0,\vec{r}_1 t)$ is substantially of rotational nature if $|E_{max}| \gtrsim 4.5$ kcal/mole. In other words $P(\vec{r}_0, r, t)$ has almost no radial or translational component. The small translational contribution involved in $P(\vec{r}_0,\vec{r}_1 t)$ is usually denoted as exchange of water molecules in and out of the hydration sphere, whose rate is characterized by the residence time of the water molecule in the hydration sphere. These residence times are of the order of 10^{-10} to 10^{-5} sec (and longer) and may be determined by T_2 measurements of the proton and ^{17}O resonance. The relaxation mechanism is the scalar or contact interaction between nuclear and electron spin. Details cannot be given here /4,6,25,34-37/.

b) Diamagnetic electrolyte solutions. In diamagnetic electrolyte solutions our experimental relaxation rates are again zero frequency quantities. Thus, here much less information is available, we measure only one parameter. We may now rewrite eq.(1) as /1/:

$$g(t) = {}_{c}V e^{-t/\tau_c^*} \int \frac{p(r_0)\, r_0^2\, dr_0}{r_0^6} \tag{8}$$

Here the time correlation function has been approximated to be exponential. Eq.(8) has the advantage that it contains the pair distribution function in a form which can be easily treated as mathematics is concerned. Now we introduce the model distribution function (see Fig.9)

$$p(r_0) = \frac{p_0}{(\bar{r} - r_0)^n} \quad \text{for} \quad a < r_0 < b \tag{9}$$

and $p(r_0) = 1/V$ for $r_0 \geq b$.

V is the volume of the solution, a is the closest distance of approach, thus it gives the repulsive contribution to the effective pair potential. n is a measure of the steepness of the distribution, that is, the steepness of the water ion potential. b is about the radius of the second coordination sphere of the ion.

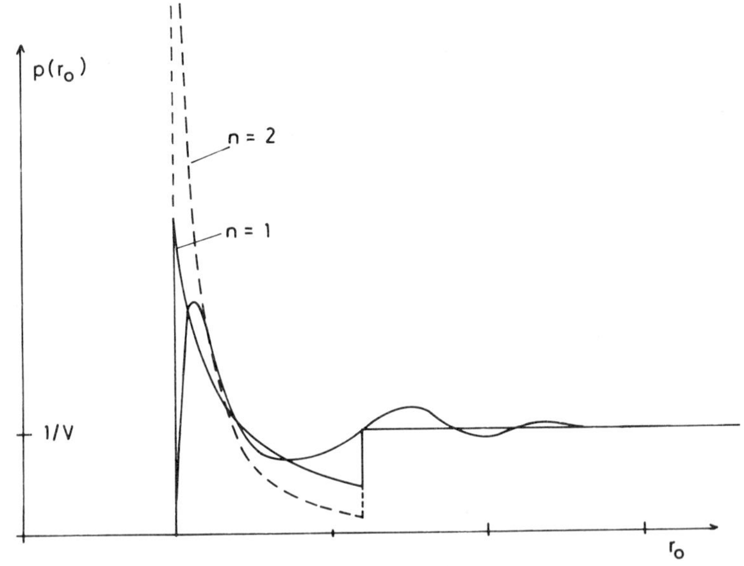

Fig.9 Model distribution function $p(r_0)$

With eq.(9) and (2) we obtain (ω = 0)

$$\left(\frac{1}{T_1}\right)_{inter} = \frac{4}{3} K \frac{n_h}{a^6} \tau_c^* g(n,k,l) + \delta^* \tag{10}$$

$$\delta^* = \frac{16}{27} K \frac{\mathcal{N}}{Vb\bar{D}} , \qquad b = la ,$$

$$K = \gamma_I^2 \gamma_S^2 \hbar^2 S(S+1) .$$

With the estimate

$$\tau_c^* \approx \frac{a^2}{3\bar{D}}$$

we have

$$\left(\frac{1}{T_1}\right)_{inter} = \frac{4}{9} K \frac{n_h}{a^4 \bar{D}} g(n,k,l) + \delta^* \tag{10a}$$

$\bar{D}$ is the mean self-diffusion coefficient of the two interaction partners. The function g(n,k,l) is given elsewhere /1/, g(n,k,l) has the properties: $g(n,k,l) \rightarrow 1$ as n large, $n > 4$ say, $g(n,k,l) \gtrsim 0.2$ for $n = 1$ and $l \approx 2$.

It may be seen from eq.(10a) that n, the steepness parameter of the potential can be determined from the experimental $(1/T_1)_{inter}$ when suitable numbers are introduced for the geometrical parameters a and b.

The following pair distribution functions have been investigated: $O^1H_2 - {}^7Li^+$, $O^1H_2 - {}^{19}F^-$, $O^1H_2 - {}^{27}Al^{3+}$, $O^1H_2 - {}^{81}Br^-$, and $O^1H_2 - {}^{127}I^-$. The experimental procedure is as follows: In a solution of the salt in question in D_2O which contains a small amount of HDO the proton relaxation rate is measured. Since the proton-proton contribution is very weak, the proton-ion nucleus contribution may be observed. For Al^{3+} and Li^+ $g(n,k,l) \approx 1$ was found i.e. putting $g = 1$ in eq.(10) the experimental correlation time was $\tau_c^* = (1.2 \pm 0.3)\cdot 10^{-11}$ sec for Li^+ and $\tau_c^* = (5.3 \pm 1.3)\cdot 10^{-11}$ sec for Al^{3+} /38/ which correspond to values to be expected from solutions of paramagnetic ions. For F^- /39,40/ $g(n,k,l)$ may be slightly less than 1, i.e. the potential is not quite as deep. But for ${}^{127}I^-$ $g(n,k,l) \approx 3$ was found which is physically meaningless within the framework of the theory given. The experiment regarding the water-iodide distribution function is of particular interest, since I^- is a structure breaking ion. There has been some controversy in the literature whether the more fluid zone connected with structure breaking ions extends right to the ion surface or whether the first hydration layer has a rigid structure in all cases /41-44/. Unfortunately, as we have seen, this question cannot be answered yet, evidently another relaxation mechanism, which is still unknown, is effective at the I^- ion. However, a deep localized potential well may be excluded because for $O^1H_2 - {}^{81}Br^-$ $g(n,k,l)$ is found to be smaller than for I^- inspite of the fact that Br^- is a smaller ion where the electrostatic potential depth should be greater.

It may be mentioned that in KI-glycerol mixtures it was possible to show directly that the rotational correlation time of the vector connecting a glycerol proton with the I^- ion is shorter than that of a vector connecting two glycerol protons /45/. KI - and similar salts - produce a structure breaking effect in glycerol similar to that in water /8/ and thus here we are indeed able to show that the regions of increased fluidity extends up to the surface of the I^- ion.

We must now describe a number of experiments which regard a three body distribution function, namely the water-water-ion distribution function. In Fig. 10 a pair of water molecules is shown which are both neighbours of an ion. We ask: What is the potential depth when we rotate one water molecule (a) relative to the other one (b) when the axis of rotation is the radial direction connecting the centre of the ion with the oxygen of molecule (a). If the corresponding potential depth is $\gtrsim$ 4.5 kcal/mole then the hydration sphere would behave as a rigid body during the time of reorientation. In a rigid body which performs isotropic reorientational motion all vectors have the same reorientation time. Thus, the rotational correlation time of the ion-proton vector and that

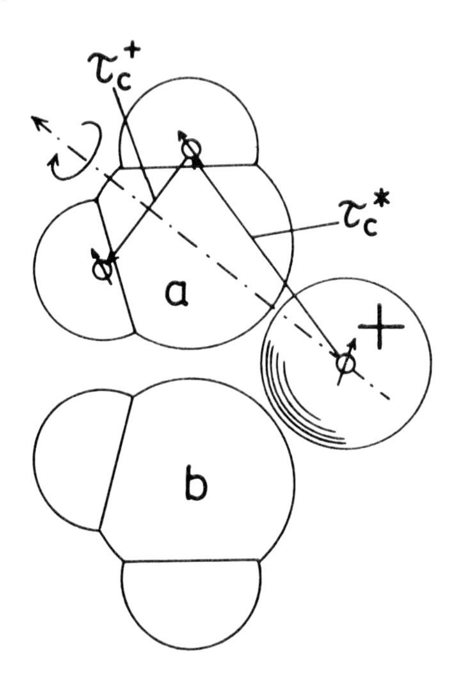

Fig.10 τ_c^* is the correlation time for the ion-proton vector and τ_c^+ is the correlation time for the proton-proton vector

of the proton-proton vector in the same hydration sphere was measured. The results are given in table 1 /38,40/.

Table 1

	τ_c^+(proton-proton) 10^{-12} sec	τ_c^*(proton-ion) 10^{-12} sec	τ_c^*(ion-proton) 10^{-12} sec
Al^{3+} /38/	44 ± 6	53 ± 13	-
Li^+	4.5 /38/	13 ± 3 /39/	12± 3 /38/
F^-	5.1± 0.3 /40/	9.4± 2 /39/	7 /40/

We see that the potential depth with regard to rotation about the dipole axis is indeed $\gtrsim$ 4.5 kcal/mole for Al^{3+}, but for Li^+ the hydration sphere does not behave like a rigid body in the time scale given by the reorientation time of the hydration complex. In the hydration sphere of F^-, though the method is not very sensitive with respect to rotations about the O-H-F axis /1,40/ , the data listed above also give evidence for the presence of such a rotation.

c) Chemical shift studies. In solutions of Al^{3+}, Ga^{3+}, I^{3+}, Be^{2+}, Mg^{2+}, Ni^{2+} at sufficiently low temperatures the nmr spectrum of the water nucleus (^1H or ^{17}O) consists of two lines. One line is due to the hydration water, the other one represents the bulk water - including the hydration water of the anions /46-49/. According to eq.(4) this allows the determination of a space integral of the water-ion distribution function. The space integral has to be extended over that range of ion-water distances within which the ion-water interaction suffices to cause a chemical shift which is larger than the exchange rate ofthe water molecule (or proton) in the corresponding position. This region will just be the first hydration sphere. In this way it was found that the hydration number of Be^{2+} is $n_h = 4$, the other ions quoted above have $n_h = 6$. It is interesting to note that for ions like Li^+ and Ca^{2+} even at lowest accessible temperatures and even in the solvent MeOH the situation of slow exchange cannot be achieved /50/47/.

B. The Relaxing Nucleus is the Ion Nucleus

If the ion nucleus relaxes by magnetic dipole-dipole interaction, then from the ion relaxation the same result regarding the water-ion distribution function must be obtainable as was derived by aid of the proton relaxation. The two examples which have been investigated, are $^7Li^+$ and $^{19}F^-$, and indeed, consistent results have been found /38,40/. The corresponding numbers are given in the fourth column in table 1.

In favourable cases not only information regarding the radial part of the ion-water distribution function is obtainable, the orientational part may also be studied. This can be done by a comparison of the central ion relaxation effect which is caused by the water protons with that which is prduced by the ^{17}O of the water. It is necessary that the ion nucleus relaxes by magnetic dipole-dipole interaction. If the proton causes very much more relaxation than ^{17}O, then one of the protons must point directly towards the centre of the ion. If the effect of ^{17}O is stronger, then the water protons point towards the bulk water. F^- is a suitable example and it has been shown that the asymmetric configuration as shown by model II in Fig.11 is the correct one /1/.

Most of the ions with noble gas structure have nuclei which relax by nuclear quadrupole interaction. ^7Li relaxes by magnetic dipole-dipole and quadrupole interaction /38,51/. Cl^-, Br^-, I^-, Cs^+, Rb^+, Na^+ are typical ions which should be mentioned here, they have been studied rather extensively in aqueous solution /51-54/. Recently, relaxation measurements in non-aqueous solvents have been performed to a fair amount /55,56/. So far these measurements show that the relaxation rate in water is the smallest when compared with those in other simple organic solvents like methanol, ethanol, formamide and ethylene glycol /55/.

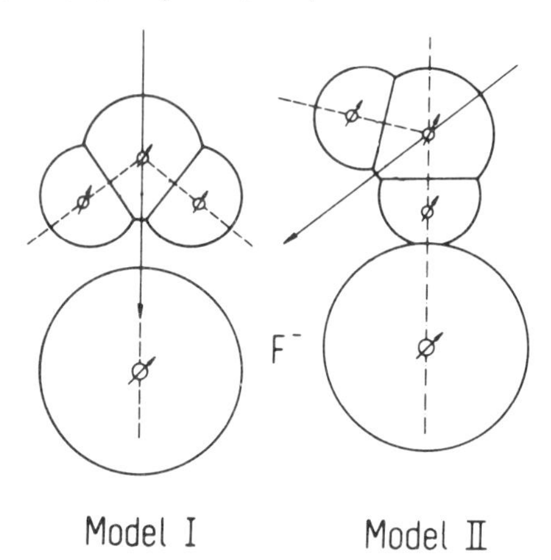

Fig. 11 Proposed orientation of water molecule in the hydration sphere of F^-

An electrostatic theory of the quadrupolar relaxation in electrolyte solutions has been developped (see ref. /57/ and the literature cited therein). The relaxation rates of structure breaking ions which are calculated by this theory are in good agreement with the observed ones /57/. The calculations imply an ion-water distribution function with a flat radial minimum which leads to a low first coordination number. Furthermore, the orientational distribution of the water dipoles should be fairly broad, or at least distinct deviations from the radial dipole orientation should occur. In non-aqueous solvents the packing of molecules becomes closer, the polar groups are orientated more in radial direction. Around Li^+ and Na^+ in H_2O a hydration complex of slightly distorted cubic symmetry is formed which contributes very little to the relaxation rate. Apparently the almost spherical H_2O molecules favour the formation of a symmetric arrangement, whereas in non-aqueous solvents the molecular geometry makes symmetric packing around the spherical ion more difficult, thus the relaxation rate is high.

VII.2.4. ION-ION DISTRIBUTION FUNCTIONS

We begin with paramagnetic electrolyte solutions. A simple model radial pair distribution function $p(r_0)$ which is a step function, may be introduced in eq.(11). The step function is defined by the mean ion concentration $c_p = N/V$ and by the location $r_0 = a$ of the step. At $r_0 = a$ $p(r_0)$ jumps from $p(r_0) = 0$ to $p(r_0) = 1/V$. Thus, a is the closest distance of approach between the ions. Probably, the simplest situation is given when the relaxing nucleus and the paramagnetic interaction partner are both spherical ions. In particular, if both ions are cations then the ions keep far apart from one another, the closest distance of approach, a, becomes large. Now for $P(\vec{r}_0, \vec{r}, t)$ in eq.(1) the solution of the diffusion equation should be reasonably valid. From the frequency dependence of $1/T_1$ the two parameters $\bar{D}$ and a may be obtained. For the systems Li^+ - Mn^{2+} and Na^+ - Mn^{2+} reasonable distances of closest approach have been derived /58/: a = 4.8 Å and a = 4.6 Å, respectively. Preliminary studies regarding the distribution Rb^+-Mn^{2+} and Rb^+-Cu^{2+} were less satisfactory, probably here scalar interaction contributes too much to the relaxation mechanism /59/. In diamagnetic electrolyte solutions, the anion-anion distribution functions in KF, RbF, and CsF solutions have been investigated /60/. It is remarkable that these measurements yielded a F^--F^- distribution function with a closest distance of approach between the anions of ≈ 3 Å and a first F^- coordination number ≈ 1. This means that we have a marked F^--F^- association. The salt concentration was about 3 molal. This particular behaviour was predicted from ion nuclear magnetic relaxation studies in concentrated aqueous alkali halide solutions /51/. The quadrupolar relaxation behaviour in these solutions can be understood in terms of higher distribution functions, i.e. at least three ion distribution functions /51,61/.

REFERENCES

1 H. G. Hertz and C. Rädle, Ber.Bunsenges.physik.Chem. 77 (1973) 521

2 A. Abragam, The Principles of Nuclear Magnetism, At the Clarendon Press, Oxford 1961

3 C. P. Slichter, Principles of Magnetic Resonance, Harper and Row, New York 1963

4 see also e.g. H.G. Hertz in: Progress in Nuclear Magnetic Resonance Spectroscopy, 3, p.159 (J.W.Emsley, J.Feeney, L.H.Sutcliffe, Eds.) Pergamon Press, Oxford 1967

5 T.L. Hill, Statistical Mechanics, McGraw-Hill, New York 1956

6 see e.g. H.G. Hertz in: Water, A Comprehensive Treatise, 3 Chapt.7 (F.Franks ed.) Plenum Press, New York 1973

7 L. Endom, H. G. Hertz, B. Thül, and M.D. Zeidler, Ber.Bunsenges. physik. Chem. 71 (1967) 1008

8 G. Engel and H. G. Hertz, Ber.Bunsenges.physik.Chem. 72 (1968) 808

9 H. G. Hertz, Ber.Bunsenges.physik.Chem. 74 (1970) 666

10 H. G. Hertz, Ber.Bunsenges.physik.Chem. 75 (1971) 572

11 H. G. Hertz, Ber.Bunsenges.physik.Chem. 75 (1971) 183

12 H. G. Hertz and M. D. Zeidler, Ber.Bunsenges.physik.Chem. 67(1963) 774

13 D. E. Woessner, B. S. Snowden jr. and A. G. Ostroff, J.chem. Physics 50 (1969) 4714

14 F. Fister and H. G. Hertz, Ber.Bunsenges.physik.Chem. 71(1967)1032

15 R. E. Connick and K. Wüthrich, J.chem.Physics 51(1969) 4506

16 J. N. Shoolery and B. J. Alder, J.chem.Physics 23 (1953) 805

17 H. G. Hertz and W. Spalthoff, Z.Elektrochem.Ber.Bunsenges.physik. Chem. 63 (1959) 1096

18 M. S. Bergqvist and E. Forslind, Acta chem.Scand. 16(1962) 2069

19 J. C. Hindman, J.chem.Physics 36 (1962) 1000

20 C. Franconi and F. Conti, Nuclear Magnetic Resonance in Chemistry, p. 349, New York 1965; C.Franconi, C. Dejak and F. Conti: Nuclear Magnetic Resonance in Chemistry, p.363, New York 1965; C. Dejak, C. Franconi and F. Conti, Ric.Sci. 35 (IIA) (1965) 3

21 R. E. Glick, W. E. Stewart and K.C. Tewari, J.chem.Physics 45 (1966) 4049

22 Z. Luz and G. Yagil, J. physic.Chem. 70 (1966) 554

23 H. G. Hertz, G. Stalidis and H. Versmold, J.Chim.Physique, Numéro Spéciale, Octobre 1969, p.177

24 L. O. Morgan and A. W. Nolle, J.chem.Physics 31 (1959) 365

25 N. Bloembergen and L. O. Morgan, J.chem.Physics 34 (1961) 842

26 R. Hausser and F. Noack, Z.Physik 182 (1964) 93

27 H. C. Torrey, Physic.Rev. 92 (1953) 962

28 H. Pfeifer, Ann.d.Physik 7.Flg. 8 (1961) 1

29 H. Pfeifer, Z.Naturforsch. 17a (1964) 279

30 P. S. Hubbard, Proc.Roy.Soc. A 291 (1966) 537

31 K. Günther and H. Pfeifer, Zhurn.strukt.Khimi 5(1964) 193
English translation, J.Struct.Chem. 5(1964) 177

32 H. G. Hertz, Ber.Bunsenges.physik.Chem. 71(1967) 999

33 L. P. Hwang, C. V. Krishnan, and H. L. Friedman, Chem.Physics Letters 20(1973) 391

34 T. J. Swift and R. E. Connick, J.chem.Physics 37 (1962) 307
Err. J.chem.Physics 41 (1964) 2553

35 A. M. Chmelnick and D. Fiat, J.chem.Physics 51 (1969) 4238

36 M. V. Olson, Y. Kanazawa and H. Taube, J.chem.Physics 51 (1969) 289

37 R. A. Bernheim, T. H. Brown, H. S. Gutowsky and D. E. Woessner, J.chem.Physics 30 (1959) 950

38 H. G. Hertz, R. Tutsch and H. Versmold, Ber.Bunsenges.physik.Chem. 75(1971) 1177

39 K. Hermann, unpublished work

40 H. G. Hertz, G. Keller and H. Versmold, Ber.Bunsenges.physik.Chem. 73 (1969) 549

41 H. S. Frank and W. Y. Wen, Disc. Faraday Soc. 24 (1957) 133

42 O. Y. Samoilov, Disc.Faraday Soc. 24 (1957) 141
O. Y. Samoilov, Structure of Aqueous Electrolyte Solutions and Hydration of Ions, Consultants Bureau, New York (1965)

43 R. W. Gurney, Ionic Processes in Solution, McGraw-Hill, New York (1953)

44 H. G. Hertz, Angew.Chemie 82 (1970) 91, Angew.Chemie Intern. Edition 9 (1970) 124

45 A. Geiger, Thesis Karlsruhe 1973

46 A. Fratiello, R.E. Lee, V. M. Nishida and R. E. Schuster, J.chem. Physics 48 (1968) 3705

47 N. A. Matwiyoff and H. Taube, J.Amer.Chem.Soc. 90(1968) 2796

47 a J. A. Jackson, J. F. Lemons and H. Taube, J.chem.Physics 32(1960) 553

48 R. E. Connick and D. N. Fiat, J.chem.Physics 39(1963) 1349; 44 (1966) 4103

49 T. J. Swift, O. G. Fritz jr. and F. A. Stephenson, J.chem.Physics 46 (1967) 406

50 S. Nakamura and S. Meiboom, J.Amer.chem.Soc. 89 (1967) 1765

51 H. G. Hertz, M. Holz, G. Keller, H. Versmold and Ch. Yoon, to be published in Ber.Bunsenges.physik.Chem.

52 H. G. Hertz, M. Holz, R. Klute, G. Stalidis and H. Versmold, Ber.Bunsenges.physik.Chem. in the press

53 C. Deverell, Progress in Nuclear Magnetic Resonance Spectroscopy 4 p. 235 (J.W.Emsley, J.Feeney, L.H.Sutcliffe, Eds.) Pergamon Press, Oxford 1969

54 C. Hall, Quarterly Reviews 25 (1971) 87

55 C. Melendres and H.G. Hertz, to be published

56 P. Neggia, M. Holz and H. G. Hertz, J.Chim.Physique, Physico-Chim.Biol. in the press

57 H. G. Hertz, Ber.Bunsenges.physik.Chem. 77(1973) 531

58 R. Göller, H. G. Hertz and R. Tutsch, Pure and Appl.Chem. 32 (1972) 149

59 W. Handschmann, Diplomarbeit Karlsruhe 1973

60 H. G. Hertz and C. Rädle, to be published in Ber.Bunsenges.physik. Chem.

61 H. G. Hertz, Ber.Bunsenges.physik.Chem. 77 (1973) 688

62 E. Wicke, Angew.Chemie 78(1966) 1 , Angew.Chemie internat.Edit. 5, (1966) 106

VII. 3. NMR METHODS :

AQUEOUS NON-ELECTROLYTE SOLUTIONS

M.D. Zeidler

Institut für Physikalische Chemie und Elektrochemie
der Universität Karlsruhe, Germany

ABSTRACT

In this article the modern developments of obtaining dynamical and structural information from nmr methods, i.e. through relaxation times and chemical shifts, are reviewed. After a short theoretical introduction several examples are given: complete relaxation rate analysis of binary aqueous mixtures, study of hydrophobic association from intermolecular relaxation, structure of the hydrophobic hydration layer from intermolecular relaxation. The effect of hydrophobic hydration can also be studied through chemical shifts of the water proton signals and the corresponding systematic analysis of such shifts is discussed.

VII.3.1. INTRODUCTION

The technique of nuclear magnetic resonance can be utilized to yield information on the dynamics and structure of aqueous solutions. The two experimental quantities of interest are nmr relaxation times and chemical shifts. The former quantity is the source for the large amount of dynamical data which is presently available, but recently it has been shown how also structural information can be obtained from this type of experimental data. Since different types of time-dependent perturbation determine the relaxation time, different dynamical aspects are involved. This fact requires clear experimental separations of the different contributions and these methods will be considered in some detail. Chemical shifts on the other hand provide information about the electronic environment of a nucleus, or rather changes thereof. The structural information therefore is much less direct and heavily depends on the interpretation of experimental data, thus there is not yet a common viewpoint reached by different authors.

The peculiar effects which are observed when nonpolar molecules are dissolved in water, are termed hydrophobic hydration and hydrophobic association. The former refers to solute-water distributions, the latter to solute-solute distributions. These effects can be studied with advantage by nmr methods and several examples will be presented.

VII.3.2. RELAXATION DATA

A. Fundamental Aspects /1-3/

The relaxation process, which very often can be characterized by a single time constant - T_1 for longitudinal and T_2 for transversal relaxation - may arise from different types of time-dependent interactions. Two of these are of major importance for our discussion:

a) Quadrupolar interaction. This occurs for nuclei having a quadrupole moment (spin > 1/2) which interacts with fluctuating electric field-gradients. Typical examples are ^{2}H, ^{14}N, ^{17}O. The interaction strength is usually very large, however, the range is short, thus confining this process to the intramolecular contribution.

b) Dipolar interaction. This occurs for all nuclei possessing a magnetic moment (spin ≥ 1/2), which interact with other magnetic nuclei in the surroundings. The interaction strength is usually much weaker than for quadrupolar interaction, therefore mainly spin-1/2 nuclei are of interest, ^{1}H being a typical example. The interaction range is longer than for quadrupolar interaction, comprising intramolecular as well as intermolecular contributions.

The time-dependence of the interactions is due to the molecular motion in the liquid. For intramolecular interaction only the molecular rotations can produce the time dependence, whereas in intermolecular contributions relative translations and rotations come into play.

We have the theoretical relation

$$\frac{1}{T_1} \sim \int_0^{\infty} e^{-i\omega t} \cdot \langle f(0)\, f(t) \rangle \, dt \qquad (1)$$

connecting the relaxation time with Fourier transforms of certain time-correlation functions. These are functions of parameters specifying the orientation and length of internuclear vectors, for example, within the laboratory coordinate system.

The time-correlation functions decay from $\langle f^2 \rangle$ at $t = 0$ to $\langle f \rangle^2$ at times where correlation is lost. Typical times after which the correlation functions have approached $\langle f \rangle = 0$ are of the order 10^{-12} to 10^{-11} sec for the liquid systems concerned in this paper. Since the nmr frequency $\omega = \gamma H_0$ is of the order 10^8 sec^{-1} or less we have $\exp(-i\omega t) \approx 1$ in the range where $\langle f(0)\, f(t) \rangle \neq 0$ and thus

$$\frac{1}{T_1} \sim \int_0^{\infty} \langle f(0)\, f(t) \rangle \, dt \qquad (2)$$

i.e. we measure the area under a certain time-correlation function. This fact is described in nmr terminology as the condition of "extreme narrowing", in which case $T_1 = T_2$ for the interaction types considered.

Time-correlation functions may be expressed as

$$\langle f(0) f(t) \rangle = \iint f(y_1) f(y_2)\, p(y_1, 0 \mid y_2, t)\, dy_1\, dy_2 \qquad (3)$$

where the joint probability density $p(y_1, 0 \mid y_2, t)$ for finding the variable y_1 at time zero and y_2 at time t, was introduced. Here y stands for the set of afore-mentioned variables that specify orientation and length of internuclear vectors.

The joint probability density may be expressed in terms of the conditional probability density $P(y_1 \mid y_2, t)$, giving the chance to find the variable y_2 at time t if it had the value y_1 at time zero:

$$p(y_1, 0 \mid y_2, t) = p(y_1) \cdot P(y_1 \mid y_2, t) \, . \qquad (4)$$

$p(y_1)$ is the a priori probability density for finding the variable y_1 at any time. This latter quantity contains the structural information about the liquid system, whereas the dynamics is described by the conditional probability (propagator).

We may consider two extreme cases for the dynamical process: Rotational motion, where the separation between the interaction partners remains fixed, and independent relative translational motion, where this separation varies. With the appropriate expressions for the propagators and

$$p(r) = \delta(r - r_0)$$
$$p(r) = \begin{cases} 0 & r < a \\ 1/V & r > a \end{cases} \quad (5)$$

for the two cases, where r_0 is the fixed dipole-dipole distance (for example), a the distance of closest approach and V the sample volume, we can find T_1 from eq.(2), (3) and (4). The results are:

$$\frac{1}{T_1} \sim \tilde{Q} \cdot \tau_c \quad (6)$$

$$\frac{1}{T_1} \sim r_0^{-6} \cdot \tau_c \quad (7)$$

for rotational motion, where the first equation is for quadrupolar, the second for dipolar interactions, and

$$\frac{1}{T_1} \sim \frac{N}{a\,D} \quad (8)$$

for translational motion and dipolar interaction. D is the translational self-diffusion constant, N the spin density, $\tilde{Q}$ the quadrupole coupling constant and

$$\tau_c = \int_0^\infty \frac{\langle f(0)\, f(t)\rangle}{\langle f(0)^2\rangle}\, dt$$

the correlation time.

Eqs.(6) and (7) of course are appropriate for intramolecular contributions to the relaxation process, but the intermolecular contribution may be describable by the extreme situations eq.(7) or (8) or any intermediate stage.

B. Experimental Results

In the aqueous solutions of non-electrolytes, we are concerned with, both components contain hydrogen, of which the two isotopes ^1H and ^2H are of special importance. As mentioned before their relaxation is governed by different types of interactions. For protons ^1H intra- and intermolecular contributions to the relaxation process exist, thus

$$\frac{1}{T_1} = \left(\frac{1}{T_1}\right)_{intra} + \left(\frac{1}{T_1}\right)_{inter} . \quad (9)$$

Experimental separations of these two contributions are possible by use of the isotopic dilution technique /4-7/. Here the intermolecular contribution is reduced by a factor of 24 by substituting the neighbouring proton-containing molecules by completely deuterated ones; i.e. the extrapolated value of $1/T_1$ for protons at infinite dilution in the deuterated analogue yields (after corrections for the remaining intermolecular amount) the intramolecular contribution. This procedure, however, does not work for water because isotopic exchange produces HDO.

A complete analysis of a binary aqueous mixture may, however, be performed in the following way /8/: Proton and deuteron relaxation rates $1/T_1$ are measured in mixtures of the light component 1 and the deuterated component 2 and also in mixtures of the deuterated component 1 and the light component 2 over the whole composition range. One of these components is water. Furthermore in the pure non-aqueous component the isotopic dilution method is applied to find the intra- and intermolecular contributions to the proton relaxation rate. Now the ratio $(1/T_1)^{Proton}_{intra}/(1/T_1)^{Deuteron}$ is known in the pure non-aqueous component. Assuming this ratio to remain constant over the total composition range, the intramolecular contribution to the proton relaxation rate is found through the measurement of the deuteron rate. If this procedure is justified, then at vanishing concentration of the non-aqueous component the total proton rate and the indirectly measured intramolecular proton rate must give identical results (or nearly identical results if isotopic effects are interferring) since at high dilution the intermolecular contribution disappears. Of course the same argument is valid at vanishing concentration of the aqueous component, but in the pure aqueous component we have no comparable reference or checking point. There is no a priori justification for assuming the ratio of intramolecular proton rate over deuteron rate to remain constant, since this implies the constancy of the deuteron quadrupole coupling constant and of the ratio of corresponding correlation times. However, experiments in three different binary systems indicate the correctness of the procedure. Thus one now knows the two contributions of the proton rate and consequently the correlation time over the total composition range for both components of the aqueous binary mixture.

The advantage of the described procedure is that from intramolecular proton rates through eq.(7) the correlation time can be calculated on a relatively safe basis since intramolecular distances are quite accurately known from molecular structure studies. In contrary, application of eq.(6) to obtain the correlation time is less safe since the quadrupole coupling constant must be estimated. Still in order to find correlation times of molecular groups which carry no hydrogen atoms one has to rely on quadrupole relaxing nuclei /9/.

By determining the rotational correlation times of various vectors inside the molecule with different positions relative to the principal axis system of the tensor describing the rotational motion one can get detailed information about the anisotropic motion. Acetonitrile for which in the pure liquid such anisotropic rotation had been known /10, 11/ was found to preserve this anisotropy also in aqueous mixtures/9/.

A summary of aqueous binary systems which have been analyzed in the described way is found in Figure 1.

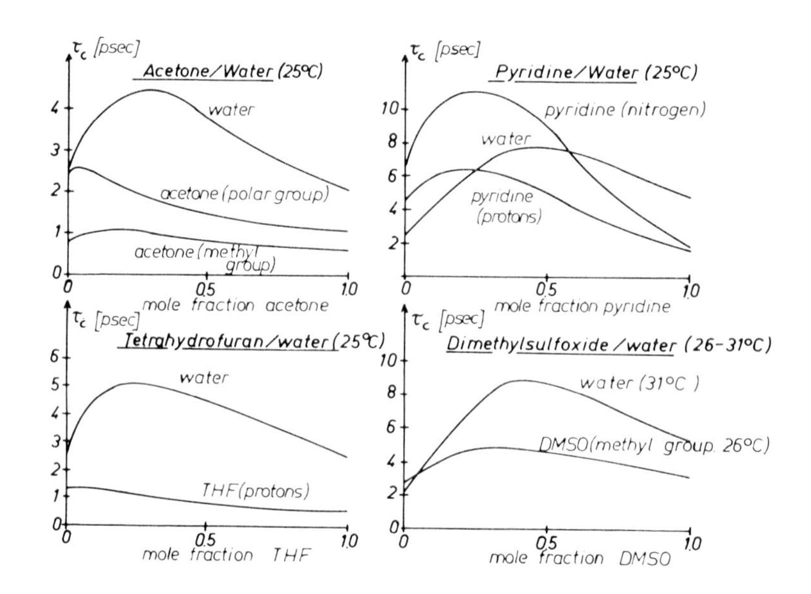

Fig.1 Rotational correlation times of both components in different mixtures of non-electrolytes with water. Relaxation rates from various nuclei as reported by different authors /8,9,12,13/ were combined. From Zeidler /14/.

On the left weakly interacting, on the right strongly interacting systems are shown, as from the crossing of the correlation time curves is evident. These results thus permit a qualitative interpretation of intermolecular interactions between the components in the mixture.

The limiting rotational correlation times of water at vanishing water concentration are a measure of solvent-water interaction just as are infrared shifts of the OH-stretching vibration of water (relative to water vapor). A satisfactory correlation between both quantities is indeed observed /14/.

In the aqueous systems of acetic acid, propionic acid, butyric acid and ethanol the intramolecular and intermolecular proton relaxation rates for the non-aqueous component have been determined /15-17/. This time we will focus our attention on the intermolecular rates. Since the self-diffusion constants of the non-aqueous components are also known the quantity $(1/T_1)_{inter} \cdot D/N_H$ is experimentally accessible, where N_H is the number of protons per cm^3. In view of eq.(8) this quantity corresponds to a^{-1}, the reciprocal distance of closest approach. Now it is observed experimentally that with decreasing concentration of the acids - the alcohol behaves completely different - the quantity $(1/T_1)_{inter} \cdot D/N_H$ increases. This increase is stronger for the non-polar group closest to the polar carboxylic group, i.e. it is stronger for intermolecular methylene-methylene interactions than for methyl-methyl interactions in propionic and butyric acid. Also mixed intermolecular interactions between methyl and methylene groups have been studied. The physical picture arising from the measurements is that of an association between acid molecules where the methylene groups are the most pronounced contact points. Since the effect is stronger for butyric than for propionic acid the non-polar groups are responsible for association (hydrophobic association). The conclusion follows from the fact that a decrease of the distance of closest approach can be reinterpreted, by taking a realistic pair distribution function instead of the step function eq.(5), as an increase of the first neighbour maximum in $\rho(r)$ which means increase of the local acid concentration around a given acid molecule with respect to the mean acid concentration in the solution. In contrary for ethanol no effect of hydrophobic association was found.

Next we will concentrate on the intermolecular dipolar interaction between solute and water nuclei. Such studies permit to extract structural information, i.e. relative orientations of the molecules, by comparing distances of closest approach with respect to different sites at the molecules. The aqueous systems of formic acid, propionic acid and methanol have been studied with the intention to investigate the orientation of water molecules in the hydrophobic hydration sphere /18/. Non-aqueous components containing just a single proton (the relaxation rate of which was measured), the other hydrogens being deuterium, had to be used in water D_2O^{16} for which in the first experiment D_2O^{17} and in the second experiment H_2O^{16} was substituted. Thus one determines either proton-oxygen^{17} or proton-proton interactions in the two experiments respectively. Depending on the choice of the propagator either eq.(7) or eq.(8) or any mixed expression may be used to extract the corresponding internuclear distances. The qualitative result is always the same: The proton-oxygen distance is smaller than the proton-proton distances meaning that the protons of the water molecules in the hydrophobic hydration sphere are directed away from the solute molecule. The same structure has been predicted to exist at the liquid-vapor interface /19/.

VII.3.3. CHEMICAL SHIFT DATA

A decrease of the temperature in pure water is accompanied by a shift of the proton resonance signal towards lower magnetic field which means that the hydrogen nucleus is deshielded. Thus we conclude that the formation of a hydrogen bond corresponds to a downfield shift. This empirical relation is supported by a large number of independent experimental checks observing the behaviour of hydrogen-bonding materials on dilution with inert solvents or on temperature change /20/.

We seek an answer to the following question: In which direction is the water proton resonance signal shifted by non-electrolytes in aqueous solution? Or more specifically: Does the hydrophobic part of the molecule produce a downfield shift as to be expected if the water hydrogen bonding increases? Many experiments have been performed at room temperature and the expected downfield shift did not show up, even when comparing relative shifts in homologous series of solutes where the size of the hydrophobic part increased (see for example /14/). Glew et al. /21/ and at the same time Symons et al./22/ were successful in obtaining downfield shifts when they performed their experiments at $0^{o}C$ and $6^{o}C$ respectively. This was the case for example for aqueous solutions of t-butanol, tetrahydrofurane and acetone at rather low solute concentrations, typically around 4 mole %. However, the first systematic procedure to decompose this downfield shift into contributions, in order to extract finally the downfield shift due to the hydrophobic hydration, was recently published by Wen and Hertz /23/. It should be anticipated that such a decomposition into various contributions cannot be performed rigorously and different approaches to this subject /23,24,25/ are of great value in order to estimate the reliability of the obtained results.

Consider for example the system t-butanol/water at $0^{o}C$. The composition dependence of the hydroxyl proton shifts is shown in Figure 2. The distinctly separate signals for the water and alcohol hydroxyl protons coalesce below 30 mole % alcohol. Here the observed shift is an

and the isothermal compressibility is

$$\kappa_T = -\frac{1}{V}\left[\sum n_k\left(\frac{\partial V_k}{\partial p}\right)_T + \sum V_k\left(\frac{\partial n_k}{\partial p}\right)_T\right]. \tag{13}$$

In this model the enthalpy is

$$H = \sum n_k H_k \tag{14}$$

and the specific heat at constant pressure is

$$C_p = \sum n_k\left(\frac{\partial H_k}{\partial T}\right)_p + \sum H_k\left(\frac{\partial n_k}{\partial T}\right)_p \ . \tag{15}$$

It can be seen from eqs. (12), (13), and (15) that the first terms of the derivatives thermal expansion coefficient, isothermal compressibility, and specific heat are the high-frequency or nonrelaxational contributions, the second terms are the relaxational contributions.

The equilibrium condition for the different structural forms may be written as follows:

$$\sum dn_k \mu_k = 0, \tag{16}$$

where the chemical potential of the component k in the mixture is

$$\mu_k = \mu_{k,o} + RT \ln \gamma_k n_k. \tag{17}$$

Here $\mu_{k,o}$ is the chemical potential of the pure component k and γ_k the activity coefficient of component k.

The rates for the transitions between the different structural forms are determined in the transition state theory by the free activation enthalpies $\Delta G^{\neq}$ for the transitions, which contain the activation volumes $\Delta V^{\neq}$, the activation enthalpies $\Delta H^{\neq}$, and the activation entropies $\Delta S^{\neq}$ /18,19/.

The excess ultrasonic absorption for n = 2 is for frequencies f[*)] well below the structural relaxation frequency ($2\pi f\tau \ll 1$) /2/:

$$\frac{2\alpha_s}{f^2} = \frac{4\pi^2 n_1 n_2 (\Delta V)^2 \tau}{c_o V R T \kappa_s N} \ . \tag{18}$$

*) $f=\nu$ in Chapt. VIII.1.

For n = 3 the states are

$$M_{i,1} \rightleftharpoons M_{i,2} \rightleftharpoons M_{i,3} \ . \tag{19}$$

The subscript i denotes the ionic environments as discussed below for ionic solutions. The corresponding formula for the ultrasonic absorption is for the two normal reactions j of eq. (19) /18,19/:

$$\frac{2\alpha_s}{f^2} = \frac{4\pi^2}{c_o} \sum_i \sum_j \frac{(\nu'_{i,j} V_{i,1} + \nu''_{i,j} V_{i,2} + \nu'''_{i,j} V_{i,3})^2 \tau_{i,j}}{VRT\ \kappa_o \left(\frac{\nu'^2_{i,j}}{n_{i,1}} + \frac{\nu''^2_{i,j}}{n_{i,2}} + \frac{\nu'''^2_{i,j}}{n_{i,3}}\right)} \ , \tag{20}$$

where $\nu_{i,j}$ are the pseudostoichiometric coefficients /22/ and κ_o the total compressibility. Equation (20) was derived assuming that only pressure induced relaxation processes occur and that no distinction need be made between adiabatic and isothermal loss processes.

In order to fit the thermodynamic and ultrasonic properties discussed above the following quantities of the n-state model have to be adjusted to the experimental results simultaneously: Partial molar volumes and their derivatives V_k, $(\partial V_k/\partial T)_p$, $(\partial V_k/\partial p)_T$ (3n parameters); nonrelaxational specific heats $(\partial H_k/\partial T)_p$ (n parameters); differences of free enthalpies ΔG between the states (2(n-1)parameters); free activation enthalpies $\Delta G^{\neq}$ for a linear chain of transition steps (3(n-1)parameters). Thus, in the n-state model a total of 9n-5 quantities have to be fitted simultaneously.

VIII.2.3. Water

A. Results and Discussion

Several attempts have been made /2,6,11,23-25/ to describe the ultrasonic absorption of water together with a small number of thermodynamic properties, e.g.,isothermal compressibility or molar volume, using a two-state model. For a free choice of a number of parameters, this always seems to be possible. However, in order to obtain a self-consistent set for the ultrasonic and thermodynamic properties all quantities discussed above should be fitted to as many reliable experimental data as possible.

If the temperature and pressure dependence of the ultrasonic absorption and the total compressibility as well as the molar volume, the thermal expansion coefficient, and the specific heat are fitted simultaneously with the least-squares method in a computer program, it turns out that no satisfactory description of all the experimental properties can be obtained with a two-state model. Furthermore, the variation of the activation energy for the viscosity in the temperature range covered cannot easily be incorporated into the two-state model.

Extension of the model to n = 3 gives satisfactory agreement with the experimental results listed above. Here M_1 of eq. (19) is a so-called open-packed structure, M_2 an intermediate or porous structure, and M_3 a close-packed structure. A set of thermodynamic parameters with which the experimentally determined properties of water are in good agreement with the calculated values is shown in Table 1.

Table 1

Temperature independent parameters for H_2O

ΔH_1	ΔH_2	$\Delta H_1^{\neq}$	$\Delta H_2^{\neq}$	ΔS_1	ΔS_2	$\Delta S_1^{\neq}$	$\Delta S_2^{\neq}$
kcal/mol				e.u.			
2.53	-0.91	5.07	1.88	9.36	-1.86	11.68	1.26

The difference of enthalpy ΔH_1 for water between states 2 and 1 corresponds roughly to the enthalpy required to break a hydrogen bond. The negative sign of ΔH_2 is a consequence of the definition of $\Delta H_2 = H_3 - H_2$. It indicates that the close-packed state (M_3) has the lower enthalpy. The activation enthalpy $\Delta H_1^{\neq}$ given for water in Table 1 corresponds roughly to the activation energy for the viscosity at lower temperatures, while $\Delta H_2^{\neq}$ corresponds to that at higher temperatures.

Mole fractions, relaxation times, partial molar volumes, and relaxation compressibilities are given in Fig. 1 For water two relaxation times, corresponding to the two reactions in normal coordinates, exist. However, the reaction with the longer relaxation time dominates the ultrasonic absorption completely. The absolute values in Fig. 1 are somewhat different from the values published previously /19,20/. This is due to the fact that pressure data /7,23,26/ up to 2 kbar have been

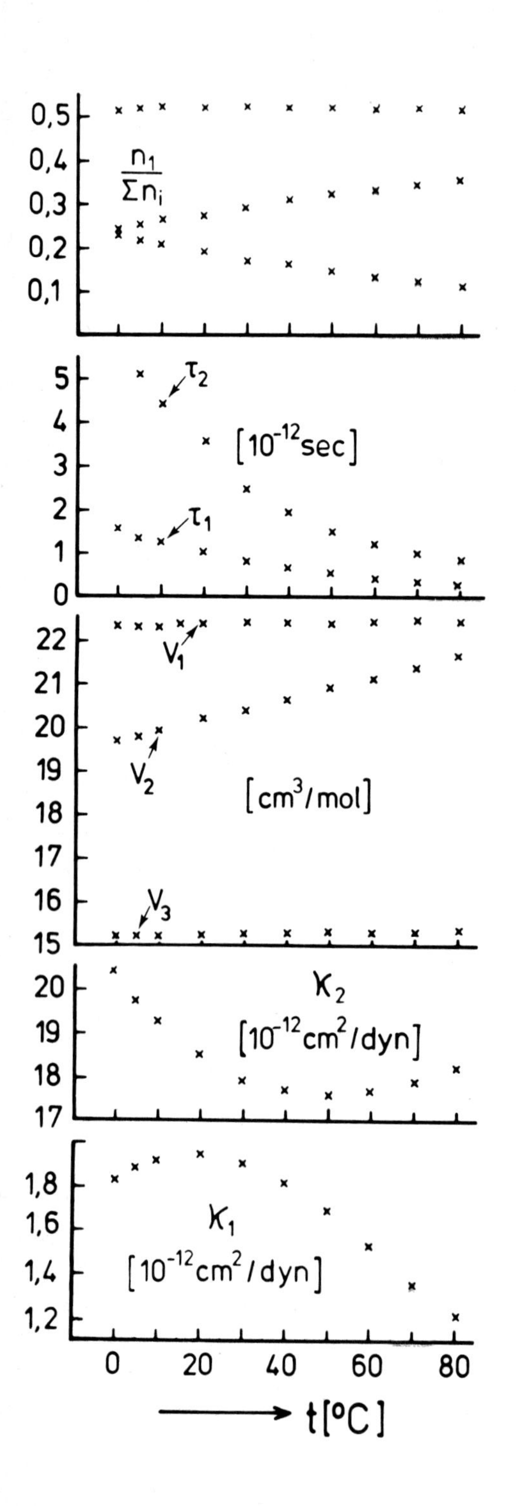

Fig. 1

Three-state model to describe the ultrasonic absorption of water needs the following parameter:

a) Mole fraction of structural states.

b) Relaxation times for the transition between the different structural forms.

c) Partial molar volumes

d) Relaxational Compressibilities

included in the numerical analysis.

In the evaluation of the thermodynamic and relaxation parameters for water presented here the nonideality /27/ of the mixture of different structural forms, expressed by the activity coefficients γ_k, has been included in the entropy term, i.e., the activity coefficients have been taken to be constant in order to keep the number of adjustable parameters as small as possible. However, variable activity coefficients can easily be introduced in the computations /28/.

With the parameters given in Tbl.1 and Fig.1 the calculated temperature dependence of the excess absorption in water is in good agreement with the experimental results (Fig. 2). The pressure dependence of the ultrasonic absorption and the compressibility can only be described within larger error limits (Fig. 3) since the experimental pressure data are not as reliable as the other data and therefore were weighted lower in the numerical analysis. For the same reason, the activation volumes $\Delta V_k^{\neq}$ were not fitted.

A comparison between theoretical and experimental values of the total molar volume, the thermal expansion coefficient, the total compressibility, and the total specific heat is given in Fig. 4 Generally it can be said, that with the exception of the pressure data the agreement between theoretical and experimental results is satisfactory.

Since the detailed structural arrangements are not known, the non-relaxational or high-frequency contributions of compressibility κ_∞, thermal expansion coefficient β_∞, and specific heat $c_{p\infty}$ cannot be theoretically determined. They are taken from the numerical analysis (Table 2). High-frequency compressibility κ_∞ and high-frequency thermal expansion coefficient β_∞ depend strongly on the structural form. They are more liquid-like for the porous structure and more solid-like for the other structures. The high-frequency specific heat $c_{p\infty}$ and the activation volume $\Delta V^{\neq}$ were taken to be the same for all structural forms since no information is available for a distinction. The activation volume was taken from the pressure dependence of the viscosity (Table 2).

B. Conclusion

A comparison between the theoretical results of the n-state model and the experimental data shows that the ultrasonic and thermodynamic properties of water can be described with the n-state model, using n = 3,

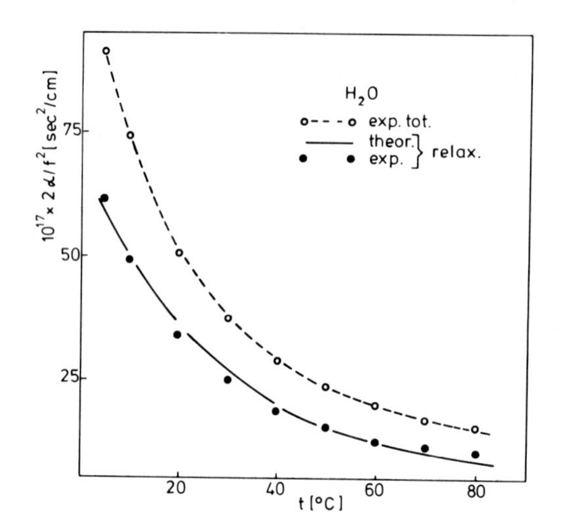

Figure 2 Comparison between theoretical and experimental temperature dependence of the ultrasonic absorption in water

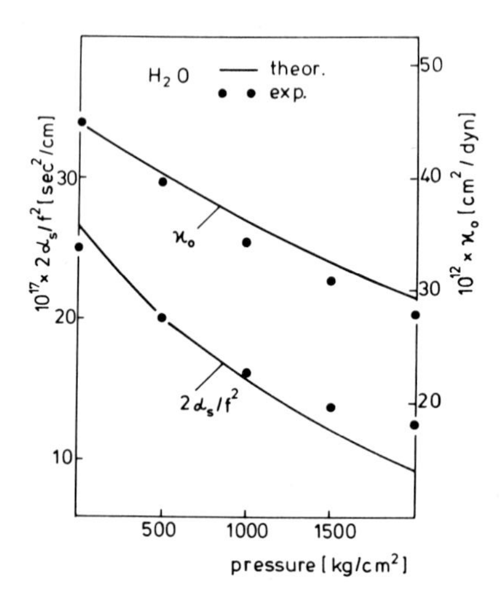

Figure 3 Comparison between theoretical and experimental pressure dependence of the ultrasonic absorption in water and of the total compressibility

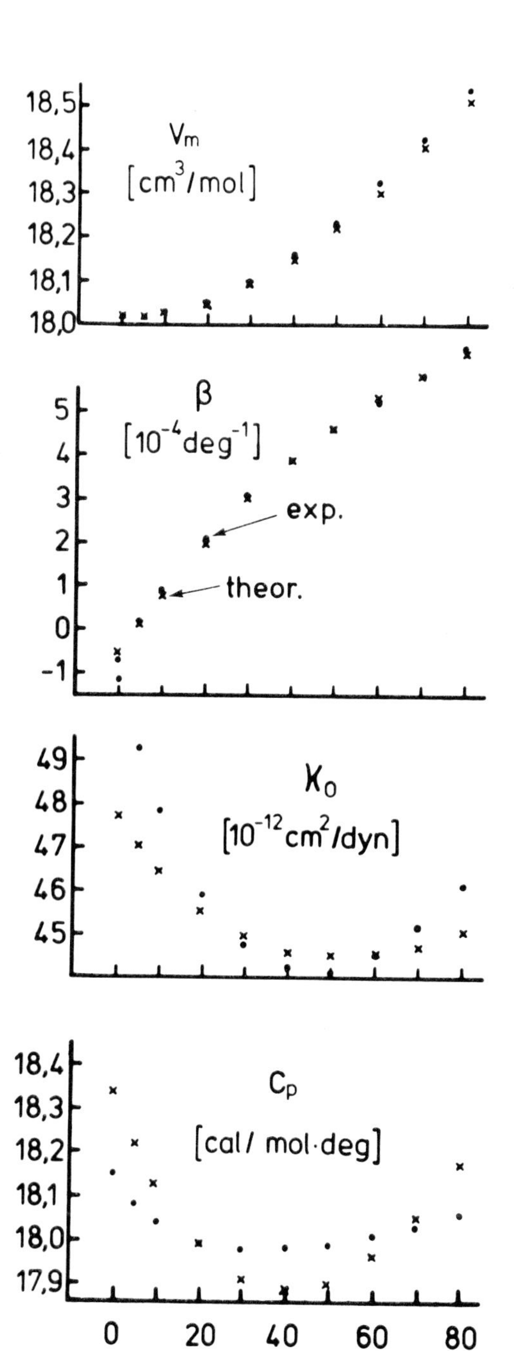

Fig. 4

Properties of water

- • experimental
- x calculated with the parameters of table 1 and fig. 1

a) Total molar volume V_m

b) Thermal expansion coefficient ß

c) Total compressibility K_0

d) Total specific heat C_p

Table 2

High-frequency parameters and activation volume for H_2O

$\beta_{\infty,1}$	$\beta_{\infty,2}$	$\beta_{\infty,3}$	$\kappa_{\infty,1}$	$\kappa_{\infty,2}$	$\kappa_{\infty,3}$	$C_{p\infty}(t)$	$\Delta V^{\neq}$
$10^{-5}deg^{-1}$			$10^{-12}cm^2\cdot dyn^{-1}$			$cal\cdot mol^{-1}\cdot deg^{-1}$	$cm^3\cdot mol^{-1}$
8	120	8	10	60	6	11.9(1+0.005t)	4.8

with reasonable accuracy. The relaxation times obtained are also in good agreement with the values derived from other experimental methods /7/. The distribution of water molecules by hydrogen-bond numbers obtained from molecular dynamics studies /29/ indicates that the assumption of essentially three states of comparable mole fractions with different binding characteristics is not unreasonable. The spatial extension of the different structural forms does not enter into the model. Thus, the structural forms between which thermal fluctuations occur may consist of a minimum number of molecules and may even penetrate into each other. X-ray data show that correlations between density fluctuations /30/ in adjacent regions are practically zero at all distances greater than a few molecular diameters. Considering these points, one may therefore conclude that the approximation of the distribution function of water molecules by three discrete states is not purely arbitrary.

VIII.1.4. Ionic Solutions

A. Results and Discussion

The ultrasonic absorption in aqueous solutions of alkali halides depends in a characteristic way on the concentrations and the types of ions present /19,20/. The solutions of LiCl (Fig. 5) and KBr (Fig. 6) represent the two types of ultrasonic absorption behavior found: solutions, such as LiCl solutions, which show after an initial decrease a strongly increasing ultrasonic absorption with concentration, and solutions, such as KBr solutions, in which the ultrasonic absorption essentially decreases with concentration in the total concentration range.

The total excess absorption is composed of contributions from different regions around anions and cations as well as from the region of

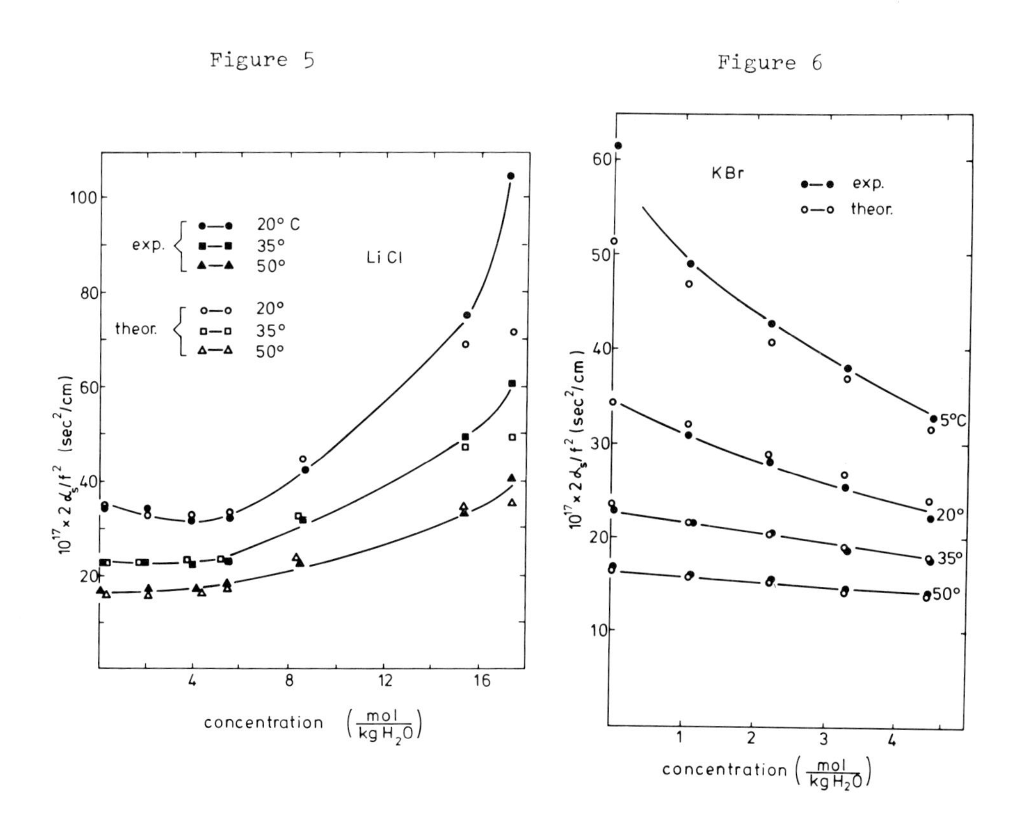

Figure5 Comparison between theoretical and experimental concentration dependence of the ultrasonic absorption in LiCl solutions

Figure 6 Comparison between theoretical and experimental concentration dependence of the ultrasonic absorption in KBr solutions

undisturbed water. Therefore the number of parameters in the n-state model increases considerably, and as many parameters as possible must be kept constant in the least-squares method. Again, as in the case of pure water, certain boundary conditions have to be taken into account in the evaluation of the parameters. In aqueous solutions at least two solvation regions have to be distinguished in order to describe the ultrasonic absorption in solutions of ions with high charge density. For ions with low charge density this distinction, although here somewhat arbitrary, is maintained.

The parameters of the region of undisturbed water and, with the exception of $\Delta H_1^{\neq}$, those of the outer regions of anions and cations were assumed to be the same as the parameters of pure water. Somewhat smaller $\Delta H_1^{\neq}$ values for the outer regions are necessary to explain the reduced ultrasonic absorption at low ion concentrations, i.e., the transition between the open-packed and intermediate structure is facilitated and the relaxation time of this process is somewhat shorter than in undisturbed water. Since the special open-packed structure M_1 of water is not likely to exist in the immediate vicinity of ions, it suffices to use a two-state model with k = 2 and 3 for the inner regions of the ions. A complete set of thermodynamic parameters is given in Table 3 for the different ions investigated.

The relaxation times calculated with the parameters of Table 3 are presented in Table 4. The single relaxation times for the inner regions are listed in the column of τ_2 since, as discussed above, these values belong to the normal reaction which practically determines the ultrasonic absorption. The same consideration applies for the ratio of the relaxation time τ_{ion} in the vicinity of an ion and that in the undisturbed liquid τ_{H_2O}. This ratio may be used to define the dynamic structure effects of the individual ions. It can be seen from Table 6 that the relaxation time of the water molecules in the inner region of ions with higher charge density (Li^+, Na^+) is longer than in the undisturbed liquid, it is shorter in the inner region of ions with lower charge density (K^+, Cl^-, Br^-, and I^-). The ratio τ_{ion}/τ_{H_2O} roughly agrees with values reported in the literature /31,32/ which have been obtained with other techniques.

The partial molar volumes of the different structural forms are also presented in Table 4. The corresponding molar volumes of pure water were taken for the region of undisturbed water and for the outer regions. For geometrical reasons the molar volume of the close-packed structure is somewhat different in the inner regions of anions and cations from that

Table 3

Temperature independent parameters for ion environments

	Region	ΔH_1	ΔH_2	$\Delta H_1^{\neq}$	$\Delta H_2^{\neq}$	ΔS_1	ΔS_2	$\Delta S_1^{\neq}$	$\Delta S_2^{\neq}$
		kcal/mol				e.u.			
Li^+	outer	2.53	-0.91	4.75	1.88	9.36	-1.86	11.68	1.26
	inner	-	-1.00	-	4.30	-	-1.86	-	6.70
Na^+	outer	2.53	-0.91	4.75	1.88	9.36	-1.86	11.68	1.26
	inner	-	-1.00	-	4.20	-	-1.86	-	5.30
K^+	outer	2.53	-0.91	4.75	1.88	9.36	-1.86	11.68	1.26
	inner	-	-1.00	-	3.30	-	-1.86	-	5.30
Cl^-	outer	2.53	-0.91	4.75	1.88	9.36	-1.86	11.68	1.26
	inner	-	-1.00	-	3.30	-	-1.86	-	5.30
Br^-	outer	2.53	-0.91	4.75	1.88	9.36	-1.86	11.68	1.26
	inner	-	-1.00	-	2.90	-	-1.86	-	4.50
I^-	outer	2.53	-0.91	4.75	1.88	9.36	-1.86	11.68	1.26
	inner	-	-1.00	-	2.70	-	-1.86	-	4.80

Table 4

Relaxation times, molar volumes and number of solvent molecules in the regions

	Region	τ_1	τ_2	τ_{ion}/τ_{H_2O}	$V_{1,i}$	$V_{2,i}$	$V_{3,i}$	Z_i
		10^{-12} sec			cm^3/mol			
H_2O	-	1.10	3.60	-	22.35	19.73	15.25	-
Li^+	outer	0.95	3.06	0.85	22.35	19.73	15.25	6
	inner	-	7.90	2.20	-	20.75	15.30	4
Na^+	outer	0.95	3.06	0.85	22.35	19.73	15.25	6
	inner	-	4.80	1.33	-	20.75	15.30	4
K^+	outer	0.95	3.06	0.85	22.35	19.73	15.25	6
	inner	-	2.90	0.80	-	21.55	15.30	6
Cl^-	outer	0.95	3.06	0.85	22.35	19.73	15.25	6
	inner	-	2.90	0.80	-	21.55	15.30	6
Br^-	outer	0.95	3.06	0.85	22.35	19.73	15.25	6
	inner	-	2.18	0.60	-	21.55	15.30	6
I^-	outer	0.95	3.06	0.85	22.35	19.73	15.25	6
	inner	-	1.33	0.37	-	21.55	15.30	6

in pure water. It is found, however, that the range in which the values of $V_{3,i}$ can be varied is rather limited /19/. Additional information concerning the dynamic structure around the various ions may be obtained from the number Z_i of water molecules in the different regions around the ions (Table 4). Although these numbers are defined on the basis of the dynamic properties of the liquid, there is a certain relationship to the number of nearest neighbors around the ions.

The experimental curves of the excess ultrasonic absorption in aqueous solutions may qualitatively be interpreted with the relaxation times of Table 4, if the changes introduced by the different values of V_k in different ion environments are neglected for a moment: At lower ion concentrations the excess ultrasonic absorption is essentially determined by the region of the undisturbed liquid and the outer regions of the ions. In the outer regions the relaxation time is reduced and the excess ultrasonic absorption decreases. At higher concentration, where the contribution of the inner regions becomes dominant, the ultrasonic absorption decreases further for ions with reduced relaxation time in the inner region, for ions with longer relaxation time in the inner region the ultrasonic absorption increases again /19,20/.

A comparison between the experimental values of the structural absorption and the theoretical values calculated with the parameters of Tables 3 and 4 is shown in Fig. 5 for LiCl solutions at three different temperatures over the total concentration range. The agreement is reasonably good not only as far as the absolute values of the structural absorption are concerned but also as far as the detailed ultrasonic behavior as a function of concentration and temperature is concerned. The decreasing absorption versus concentration and the absorption minimum which exists at lower temperatures is represented correctly by the theory. In order to explain the structural absorption for concentrations higher than about 6 mol/kg H_2O it has to be assumed that the influence of the inner region with the longer relaxation time dominates when the inner regions overlap. As one would expect, the model cannot describe the excess absorption at concentrations where the inner regions of the cations overlap.

For KBr solutions a comparison between experimental values of the structural absorption and the theoretical values calculated with the parameters of Tables 3 and 4 is given in Fig. 6 at four different temperatures in the total range of solubility. Here, too, a reasonable agreement of the absolute value of the structural absorption as well as of the detailed shapes of the absorption curves versus concentration at different temperatures is found. A comparison between theoretical and experimental results of the structural absorption for a number of solutions of alkali halides is shown in Fig. 7.

B. Conclusion

The ultrasonic absorption of aqueous solutions of alkali halides can be described quantitatively with the n-state model. Because of the large number of adjustable quantities for the ionic solutions a complete self-consistent computation of the parameters is not possible at present. However, if as many quantities as possible are kept constant with respect to pure water the rest of the parameters may be adjusted so that the experimental curves are reproduced within small error limits. The ultrasonic data indicate that in the immediate vicinity of ions with low charge density the relaxation times of the water molecules are reduced, while in the immediate vicinity of ions with high charge density the relaxation times are longer than in pure water. For the latter ions a second layer of water molecules with reduced relaxation times has to be introduced in order to explain the ultrasonic data. This behavior is different from that found for ions in liquid ammonia and methanol /21/.

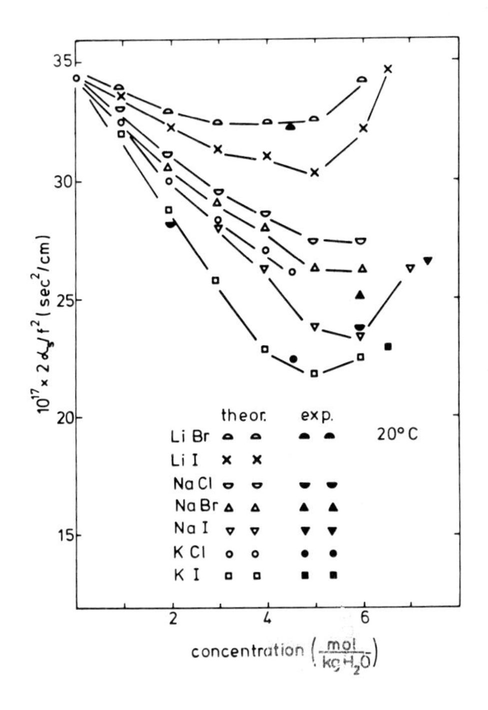

Fig.7

Comparison between theoretical and experimental concentration dependence of the ultrasonic absorption in different alkali halide solutions

REFERENCES

1 J.J.Markham, R.T.Beyer, and R.B.Lindsay, Rev.Modern Phys. 23, 353 (1951)

2 K.F.Herzfeld and T.A.Litovitz, Absorption and Dispersion of Ultrasonic Waves, Academic Press, New York 1959

3 D.Sette in Handbuch der Physik XI 1, 275, Springer Verlag, Berlin-Göttingen-Heidelberg 1961

4 A.B.Bhatia, Ultrasonic Absorption, Oxford 1967

5 A.J.Matheson, Molecular Acoustics, Wiley 1971

6 C.M.Davis, Jr. and J.Jarzynski, Advan.Mol.Relaxation Processes 1, 155 (1967-68)

7 C.M.Davis, Jr. and J.Jarzynski in Water, F.Franks, Ed., Vol.I, 443, Plenum Press, New York 1972

8 P.A.Egelstaff, An Introduction to the Liquid State, Academic Press, New York 1967

9 K.F.Herzfeld, J.Acoust.Soc.Am. 13, 33 (1941)

10 H.O.Kneser, Ann.Physik 32, 277 (1938)

11 L.Hall, Phys.Rev. 73, 775 (1948)

12 A.Gierer and K.Wirtz, Phys.Rev. 79, 906 (1950)

13 L.Tisza, Phys.Rev. 61, 531 (1942)

14 C.M.Davis, Jr. and T.A.Litovitz, J.Chem.Phys. 42, 2563 (1965)

15 G.Némethy and H.Scheraga, J.Chem.Phys. 43, 1869 (1965)

16 M.S.Jhon, T.Ree, and H.Eyring, J.Chem.Phys. 44, 1465 (1966)

17 S.K.Kor, G.Rai, and O.N.Awasthi, Phys.Rev. 186, A 105 (1969)

18 K.G.Breitschwerdt, Chem.Phys.Letters 6, 587 (1970)

19 K.G.Breitschwerdt, Ber.Bunsenges.physik.Chem. 75, 319 (1971)

20 K.G.Breitschwerdt and H.Kistenmacher, J.Chem.Phys. 56, 4800 (1972)

21 K.G.Breitschwerdt and H.Wolz, Ber.Bunsenges.physik.Chem. 77 (1973)

22 A.Bechtler, K.G.Breitschwerdt, and K.Tamm, J.Chem.Phys. 52, 2975 (1970)

23 T.A.Litovitz and E.H.Carnevale, J.Appl.Phys. 26, 816 (1955)

24 N.Hirai and H.Eyring, J.Appl.Phys. 29, 810 (1958)

25 G.Rai, B.K.Singh, and O.N.Awasthi, Phys.Rev. 5, A918 (1972)

26 J.Hawley, J.Allegra, and G.Holton, J.Acoust.Soc.Am. 47, 137 (1970)

27 A.Ben-Naim, J.Chem.Phys. 57, 3605 (1972)

28 K.G.Breitschwerdt and H.Wolz, to be published

29 F.H.Stillinger and A.Rahman, J.Chem.Phys. 57, 1281 (1972)

30 T.R.Chay and H.S.Frank, J.Chem.Phys. 57, 2910 (1972)

31 L.Endom, H.G.Hertz, B.Thül, and M.D.Zeidler, Ber.Bunsenges.physik. Chem. 71, 1008 (1967)

32 G.Engel and H.G.Hertz, Ber.Bunsenges.physik.Chem. 72, 808 (1968)

VIII.2. ULTRASONIC RELAXATION DUE TO HYDROGEN BOND DISSOCIATION

Friedrich Kohler
Institute of Physical Chemistry, University of Vienna, Austria

ABSTRACT

The principal relations between the experimentally observable parameters of an ultrasonic relaxation process and the corresponding molecular changes are briefly reviewed. As the molecular interpretation is never unique, it should be corroborated by additional thermodynamic and structural investigations. An important possibility is the study of ultrasonic relaxation upon dilution with different solvents. Among the homocomplexes formed by hydrogen bonds, the cyclic dimers of the carbonic acids are investigated best. For alcohols in non-polar solvents, we have a preliminary kinetic analysis. For some heterocomplexes, we have kinetic data in aqueous surroundings by a n.m.r. technique. Though substantial information has been accumulated on systems where complexes involving water molecules occur, a reliable molecular interpretation is still impossible. It is hoped that a better understanding can be gained with the development of a systematic knowledge on the kinetics of hydrogen bonded complexes.

VIII.2.1. INTRODUCTION

We begin this review by recalling the experimental and theoretical background of ultrasonic relaxation processes.

The absorption coefficient is connected to transport coefficients (for small amplitudes of the sound wave [1]) by

$$\frac{\alpha}{\nu^2} = \frac{8\pi^2}{3c^3\rho}\left[\eta_s + \frac{3}{4}\eta_v + \frac{3}{4}\,(M\lambda/C_p)(C_p/C_v - 1)\right], \qquad (1)$$

where α is the sound absorption per unit length, ν the frequency, c the sound velocity, ρ the mass density, η_s the shear viscosity, η_v the bulk viscosity, M the molecular weight, λ the thermal conductivity, and C_p and C_v the molar heat capacities at constant pressure and volume, respectively. A non-zero value of the bulk viscosity implies that in compressions equilibrium is not established instantaneously. If this lag becomes comparable to the inverse of the frequency, relaxation occurs. If part of the bulk viscosity relaxes at measureable frequencies (5.10^9 s^{-1} or lower), then it is usually this term which is dominant in α/ν^2 at low frequencies. Fig. 1 shows, using acetic acid as an example, the characteristics of a relaxation region which can be determined experimentally [2,3,4]. The relaxing part of the

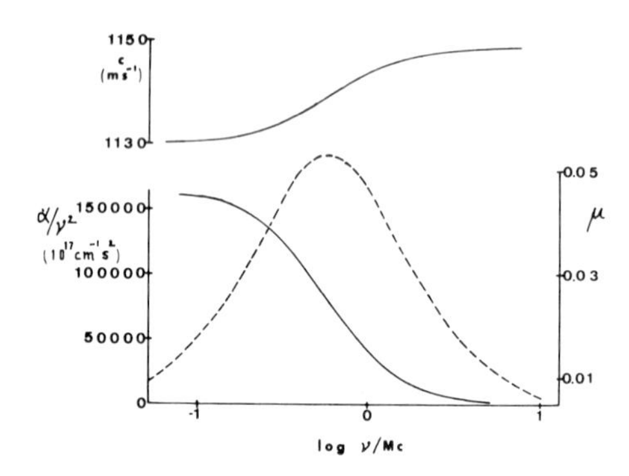

Fig. 1. Ultrasonic velocity (upper curve), absorption α/ν^2 (lower curve), and absorption per wavelength μ (dashed curve) for acetic acid at 20°C in the relaxation region.

absorption coefficient over the square of the frequency, α_1/ν^2, drops over two decades of frequency from its constant low frequency value to almost zero. Another usual plot for indicating the absorption in a relaxation region is the absorption coefficient per wavelength, $\mu = \alpha_1\lambda$, which is a dimensionless quantity and shows the characteristic bell shape. The velocity of sound exhibits a small dispersion in the relaxation region.

The individual characterisation of each relaxation region needs two parameters, relaxation strength A and relaxation time τ.[*)] Their relation to the quantities which are experimentally accessible are best understood by writing the magnitude of the wave vector $k = \omega/c$ (ω being the angular frequency) as a complex quantity $k = k - i\alpha$, or

$$\left(\frac{1}{c^*}\right)^2 = \left(\frac{1}{c} - i\frac{\alpha}{\omega}\right)^2 . \tag{2}$$

Here the asterisk is used to label complex quantities.

If the reaction rate for reestablishing equilibrium in a temperature or pressure wave is proportional to the deviation from equilibrium (exponential decay of the fluctuation), we have

$$\left(\frac{1}{c} - i\frac{\alpha_1}{\omega}\right)^2 = \frac{1}{c_o^2}\left(1 - A\,\frac{i\omega\tau}{1+i\omega\tau}\right) . \tag{3}$$

Here the notation α_1 indicates that only the contribution from a single relaxation process is considered. The maximum of the absorption coefficient per wavelength occurs at

$$\omega^2\tau^2 = 1/(1-A) \quad , \tag{4}$$

and has the value

$$\mu_{max} = (\pi/2)\ A/\sqrt{1-A} \quad . \tag{5}$$

The dispersion of sound velocity is also determined by the relaxation strength:

$$c_o/c_\infty = \sqrt{1-A} \tag{6}$$

where the indices indicate the limits on the low and high frequency side, respectively. The low frequency absorption is given by the product of relaxation strength and relaxation time:

$$(\alpha_1/\nu^2)_{\nu\to 0} = (2\pi^2/c_o)A\tau \quad . \tag{7}$$

---

*) $\tau = 1/\omega_V$ in the nomenclature of chapt.VIII.1.

VIII.2.2. MOLECULAR INTERPRETATION OF RELAXATION

The molecular interpretation of relaxation requires the assumption of an equilibrium between states of different energy and/or volume. This equilibrium is disturbed by the change of temperature and/or pressure caused by the sound wave. The equilibrium parameters of the excitation reaction determine the relaxation strength A, whereas τ is given by the kinetics of that reaction. The detailed relations are the following [4,5,6,7]:

$$A = \frac{c_o^2 M}{C_p(C_p - C_r)} \frac{n}{n_s} \frac{\phi}{RT} \left[\alpha_p \Delta H - \frac{C_p}{V} \Delta V\right]^2 , \quad (8)$$

$$\tau = \phi/(kxf) \quad . \quad (9)$$

In eq.(8), ΔH is the enthalpy change and ΔV the volume change connected with the excitation reaction, whereas ϕ is given by the stoichiometry of that reaction, defined by the degree of advancement of the reaction ξ and the equilibrium constant K,

$$\phi^{-1}\delta\xi = n\delta \ln K \quad . \quad (10)$$

Furthermore, C_r is the part of C_p due to the excitation reaction,

$$C_r = \frac{1}{n_s} \frac{\delta\xi}{\delta T} \Delta H \quad ,$$

α_p the coefficient of expansion, V the molar volume, n the mole number of species, and n_s the mole number of substance.

In eq.(9), k is the rate constant of the excitation reaction, whereas x and f are mole fraction and activity coefficient of the unexcited species, respectively.

An inconvenience of eq.(8) is the occurrence of both ΔH and ΔV, and the possibility that the corresponding terms may compensate each other. In water (pure liquid or solvent) $\alpha_p \simeq 0$, and only the volume term is important. In most other cases, including mixtures of water + organic substances, it is believed that the ΔH-term is dominant [7]. Some attempts have been made [8] to estimate ΔV of various reactions giving rise to ultrasonic absorption, but thorough experimental determinations of the pressure dependence of ultrasonic relaxation remain to be done. A classical study [9] on the pressure dependence of relaxation time furnished the activation volume of some reactions. In the case of $\Delta V^{\ddagger} = 0$, one might also assume $\Delta V = 0$ (as has been done for the relaxation process in acetic acid).

If the assumption $\Delta V = 0$ is justified, then eq.(8) contains the equilibrium constant (through the equilibrium concentrations contained in the stoichiometric coefficient ϕ) of the reaction and its temperature derivative (ΔH). In principle two determinations of A at different temperatures would suffice for the derivation of these two quantities. In the case of $\Delta V \neq 0$ at least one additional determination of A at elevated pressure would be necessary.

Having traced ultrasonic relaxation to the thermodynamic and kinetic characteristics of a reaction, we might ask about the detailed nature of that reaction. The details can be quite varied. As examples, we know of dissociation of associated species, vibrational excitation, and conformational change. The assignment of a certain reaction to a relaxation process is not always easy. The usual basis is a comparison between model assumptions and the thermodynamic and kinetic quantities of the relaxation process. A better way would be to check the postulated reaction by independent experimental investigations. An important possibility is dilution with "inert" solvents, since this should reveal the stoichiometry of the underlying reaction [10].

Though it is true that very different types of reactions can cause ultrasonic relaxation, the investigation of ultrasonic absorption and relaxation is very selective for those reactions which produce non-negligible values of A (eq.8), and have at the same time a sufficiently low value of the rate constant k. In general, k should be between 10^5 and 10^8 s^{-1} (writing the excitation reaction as monomolecular). Outside that interval, relaxation would be hard to observe experimentally. Moreover, for values of k higher than 10^{10} s^{-1} (corresponding to low values of τ, eq.9), only extreme values of A would make the product $A\tau$ big enough for a significant contribution to absorption (eq.7). Let us investigate the possibilities for higher A-values in the case of thermal excitation only ($\Delta V = 0$), and under the assumption that the excitation equilibrium $A_m \rightleftharpoons mA^*$ leads to a small mole fraction of excited species (m originating from one unexcited species). In this case $n\phi/n_s$ can be approximated by the mole fraction of the excited species and this is given approximately by $e^{-\Delta G/mRT}$:

$$A \approx R \frac{C_p - C_v}{C_v(C_p - C_r)} \left(\frac{\Delta H}{RT}\right)^2 e^{-\Delta G/mRT} \tag{11}$$

Denoting the ratio $\Delta H/\Delta G$ by c (usually about 2 for hydrogen bonding dissociation), we have

$$A = \frac{(C_p - C_v)C_r}{C_v(C_p - C_r)} \quad \text{with } C_r \approx R \left(\frac{\Delta H}{RT}\right)^2 e^{-\Delta H/mcRT} \tag{12}$$

It follows, that C_r has the maximum value for $\Delta H = 2mcRT$, which is (for m = 2, c = 2) about 9R. On the basis of $(C_p - C_v)/C_v \approx 0{,}4$ and $C_p \approx 25$ R, this gives $A \approx 0{,}2$. Most actual values of A are much lower. But it is seen that the usual enthalpy changes for hydrogen bonding dissociation have a favourable magnitude for creating significant values of A. The situation is not so favourable with respect to the rate constants. In most pure liquids where hydrogen bonding occurs, dissociation goes so fast that relaxation cannot be observed and the contribution to absorption is small. The carbonic acids are a noticeable exception. One might think that in general cyclic associates dissociate more slowly and are therefore more readily observed by ultrasonic investigations.

The other extreme is the case that the bulk viscosity does not relax at significantly lower frequencies than the shear viscosity (cf. eq. 1, the term with the thermal conductivity being negligible in most cases). This is known as structural relaxation [11]. Here it is only the disturbance of the molecular equilibrium positions (and/or orientation) by the compression wave which matters. A simple example are monoatomic liquids, but the ratio of bulk to shear viscosity is enhanced when the structure of the liquid is less random as in the case of water and simple alcohols. A typical characteristic of structural relaxation is the near constancy of the ratio η/Φ at different temperatures. Unfortunately, the direct measurement of the frequency dependence of η and Φ is beyond the experimental possibilities (except perhaps Brillouin scattering of slow neutrons). Indirect evidence has to be used to divide α/ν^2 into the factors A and τ (eq.7). Some aspects of the structural relaxation of aqueous systems will be dealt with in the contribution of Breitschwerdt to this volume.

VIII.2.3. DISSOCIATION OF HYDROGEN BONDS BETWEEN MOLECULES OF THE SAME KIND

Here the most interesting cases are the simple monocarbonic acids and some alcohols, the latter showing relaxational behaviour mainly in solutions. In spite of many investigations there is still some controversy about the detailed reaction mechanism. In the case of acetic acid and propionic acid, there seems to be some agreement that the cyclic dimer is the dominant species, and that the open dimer is less important than the monomer [4,5,12,13,14].

It has been pointed out that the concentration behaviour of α_1/ν^2, μ_{max}, and τ, in mixtures with an "inert" solvent might allow a judgement on the stoichiometry of the reaction. To illustrate this suggestion, the quantity $n\phi^2/n_s x$, which is mainly responsible for the concentration behaviour of $A\tau$ or α_1/ν^2, is plotted in Fig. 2 for the cases that (1) excitation leads to the same number of species (like conformational change, or dissociation of a cyclic into an open dimer), (2) to twice (dissociation of a dimer into a monomer), (3) to three times (dissociation of a trimer), or (4) to four times the number of species. The varying values of ϕ are calculated with a mole fraction equilibrium constant, i.e., neglecting activity coefficients. In order to calculate the real concentration dependence of α_1/ν^2, the activity coefficients of the species and the concentration dependence of expansion coefficient and heat capacity should be known, i.e., quite substantial information on the thermodynamic and structural behaviour is needed. For the case of acetic acid, an attempt has been made [4] to calculate the activity coefficients of the species so as to give agreement to the thermodynamic, ultrasonic and chemical shift (n.m.r.) data of acetic acid + carbon tetrachloride. In Fig. 3 it is indicated how much the activity coefficients can effect the quantity $n\phi^2/n_s x$.

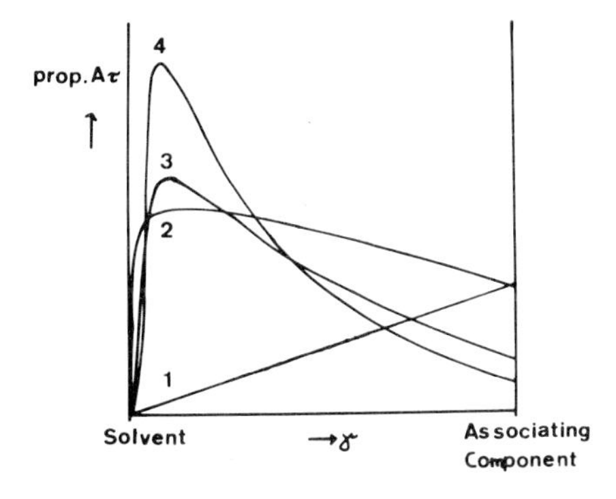

Fig. 2. The quantity $n\phi^2/n_s x$ (which determines α_1/ν^2) for the reaction $A_m \rightleftharpoons mA^*$, with m = 1 (mole fraction constant K = 1/5), m = 2 (K = 1/250), m = 3 (K = 1/100), m = 4 (K = 1/500) versus the total mole fraction γ of the associating component.

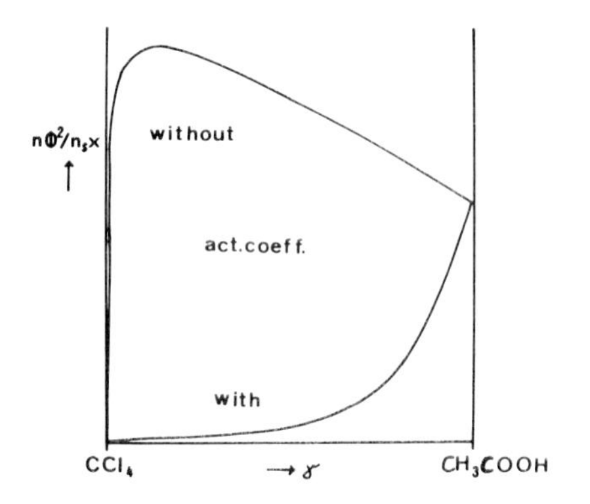

Fig. 3. The quantity $n\phi^2/n_s x$ in the system acetic acid + carbon tetrachloride, calculated with [4] and without activity coefficients of the reacting species. The curve without activity coefficients is identical with curve 2 of Fig. 2.

On the basis of that calculation, the overall rate constant for the dissociation of the cyclic dimer to the monomer was found to be 5.10^4 s^{-1} at 20°C (if the cyclic dimer reacts first to an open dimer, which reacts back faster than dissociating into the monomer, the dissociation rate from the cyclic to the open dimer might be faster than the overall rate constant given above).

In the case of phenol (Fig. 4) [15,16], benzyl alcohol [17], n-alcohols [18], and tert-butanol [19,20], maximum values of α/ν^2 have been found for low concentrations in non-polar solvents. Looking at Fig.2, it seems likely that dissociation of trimers or tetramers are

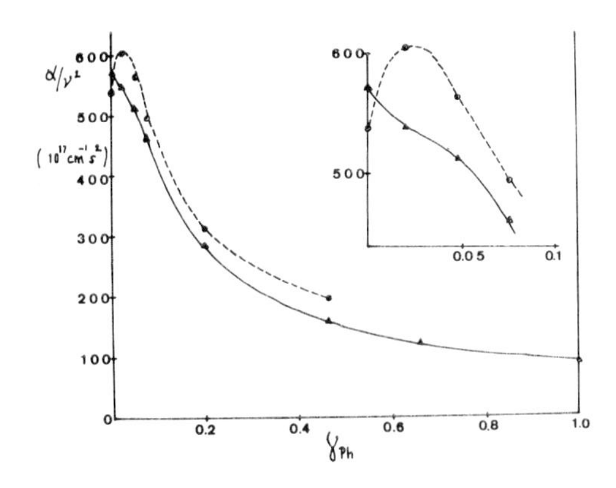

Fig. 4. The quantity α/ν^2 for the mixture phenol + carbon tetrachloride at 20°C (dashed curve) and 41°C versus the mole fraction of phenol. The low concentration region is shown separately on a larger scale.

responsible for the maxima. A kinetic analysis has been done [20] for tert-butanol in cyclohexane on the basis of a model which assumes continuous association, with the dimer being much less stable than the higher associates. Then for the dissociation of hydrogen bonds a rate constant of $1{,}2.10^{8}$ s^{-1} (15°C) has been found. However, dielectric relaxation of pure alcohols gives a higher rate for the dissociation of hydrogen bonds [21]. Furthermore, the small increase of dielectric constant by adding alcohols to non-polar solvents points to the formation of associates of small polarity [22,23]. From this point of view, the formation of cyclic trimers or tetramers in non-polar solutions seems more probable than a continuous association model.

As a whole we are at the very beginning of elucidating the reaction mechanisms of self-association by hydrogen bonding.

VIII.2.4. DISSOCIATION OF HYDROGEN BONDS BETWEEN DIFFERENT MOLECULES

Some dissociation rate constants obtained by Grunwald et al. [24] by a nuclear magnetic resonance technique (the rate of proton exchange was measured, the dissociation rate constant derived) are relevant and are shown in Table 1.

Table 1

Dissociation rate constants of hydrogen bonded complexes, obtained by Grunwald et al. by a n.m.r. technique

Complex	Dissociation rate constant [s^{-1}]	t [°C]
$H_3N.HOH$	$2{,}2.10^{11}$	25
$Me_3N.HOH$	$1{,}0.10^{10}$	25
$Et_3N.HOH$	$3{,}8.10^{9}$	25
$Bz_2MeN.HOH$	$2{,}7.10^{9}$	30
B.HOH	$6{,}6.10^{9}$	25
$B.HOCH_3$	$4{,}3.10^{8}$	30
B.HO(t-Bu)	$2{,}6.10^{5}$	25

B = (benzene ring with Me, bearing NEt₂ → written: B = $C_6H_4(Me)NEt_2$ as drawn: ring–N(Et)(Et), Me on ring)

t-Bu = tert-butyl
Bz = benzyl

The dissociation rate decreases with increasing size of the complexing molecules. Grunwald et al. interpret this in the sense of the hydrogen bond getting strengthened by van der Waals forces. Another way of looking at this phenomenon is to consider the difficulty of big organic molecules to reorientate in aqueous surroundings. A special case is the tert-butanol + m-diethyltoluidine complex. Here Grunwald et al. point to the possibility of a cyclization of this complex by van der Waals interactions (shown schematically in Fig. 5), and cyclic complexes seem to have an especially long life time.

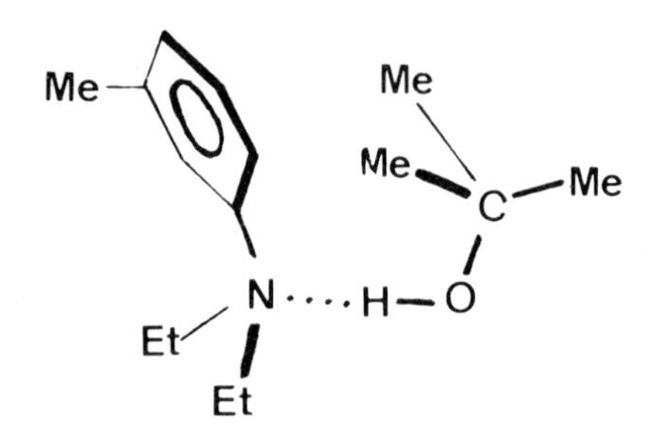

Fig. 5. A possibly stabilized structure of the tert-butanol+m-diethyltoluidine complex.

Ultrasonic properties of mixtures of water + organic components have been studied intensively, and these investigations have been reviewed [7,25]. But clear molecular models for the relaxation processes have not emerged. Some general statements can be made, however:

(1) Even for relatively high concentrations of water, the ΔH-term in eq. 8 seems to be more important than the ΔV-term [7] (a possible exception being the system t-butanol + water [25]). In other words, the expansion coefficient α_p increases very rapidly with increasing concentration of the organic component.
(2) The relaxation time becomes larger (relaxation frequency smaller) with increasing size of the organic molecule (e.g., it goes from 10^{-9} s in the case of acetone + water to 6.10^{-9} s in the case of n-octylamine + water [7]). In most cases, the experiments cannot be fitted to a single relaxation process [25,26,27].
(3) The maximum absorption occurs between pure water and the equimolar concentration, and is shifted more to pure water in the case of amines,

and with increasing size of the organic molecule [7,25].
(4) The absorption remains first almost constant when adding the organic component to pure water, and increases very sharply at mole fractions of order 0,01 [25].

Statement (3) agrees with the finding [28] that in the hydrogen bond between tertiary amines (strong proton acceptors) and water the hydrogen is somewhat shifted toward the nitrogen, thus leaving behind an extra negative charge at the OH group, which induces further association by water molecules (proton donors).

Statement (4) is paralleled with thermodynamic investigations [29] of the mixture water + triethylamine. There the idea has been put forward, that the first additions of the organic component to water will be accomodated in regions more rich on defects in the quasi-lattice structure, and that some abrupt changes can be noticed when the organic component is forced into the bulk of water structure.

Of the molecular models (all calculated without using activity coefficients) which have been fitted to the ultrasonic experiments, two examples [7] should be mentioned. The first model considers the reaction $B + (H_2O)_m \rightleftharpoons B.mH_2O$, the second combines the reaction $B + H_2O \rightleftharpoons B.H_2O$ with a change in water structure described by a two state-model $(H_2O)_{free} \rightleftharpoons (H_2O)_{bound}$ where the equilibrium constant depends on the total concentration of water, $K = K_o(\gamma_{H_2O})^u$. But in most cases values of the parameters m and u are required which are physically unreasonable. These difficulties and the lack of single relaxation frequencies has led to a model which uses concentration fluctuations [30]. Instead of the properties of a chemical reaction ($n\phi/n_s$, ΔH, ΔV), this model requires the second derivatives of the thermodynamic functions of mixing (Gibbs energy, enthalpy, and volume) with respect to concentration. A distribution of relaxation times corresponds to the time of diffusion out of the fluctuating regions. Though this theory looks rather interesting, the assumed size distribution of fluctuating regions seems to be inconsistent with our experience from scattering of electromagnetic waves, which would require a rather narrow distribution of relaxation times. Morever, the theory leads to a mean size of fluctuating regions which does not exceed the first shell of neighbours. This fact leads back to the necessity of constructing molecular models.

It would be interesting indeed to make some association mechanism probable where water molecules are involved. Similarly an understanding of the reaction mechanism for alcohol or phenol association in non-polar solvents would help much to improve our view on the pure hydrogen bonded liquids, though another association mechanism might be dominant there. In pure water the relaxation time is of order 10^{-12} s, and all evidence points to the impossibility for obtaining any resolution into detailed association mechanisms. Selfassociation of water in "inert" solvents has not yet been detected by ultrasonic absorption. A preliminary investigation in 1,2-dichloroethane, where trimerisation of water has been claimed [31] to take place, was not successful [32]. This increases our need for understanding the association mechanism of water complexes under the influence and the incorporation of a second partner (like, e.g., triethylamine).

VIII.2.5. CONCLUSION

It is believed that ultrasonic relaxation is a promising tool for studying those hydrogen bonded complexes, which have a comparable long lifetime. But it has not been really exploited. For assigning reaction mechanisms, combination with thermodynamic and structural data on the same system is necessary, including the use of different solvents. Only when a systematic knowledge on dissociation rates of hydrogen bonds in the various complexes has been accumulated, the interpretation of more complicated structures will be possible.

REFERENCES

1 L.D. Landau and E. M. Lifschitz, Lehrbuch der Theoretischen Physik, Bd. 6 Hydrodynamik, Kap. 8, Akademie-Verlag Berlin 1966

2 J. Lamb and J. M. M. Pinkerton, Proc. Roy. Soc. [London] A 199, (1949) 144

3 F. Kohler, The Liquid State, chapter 11.5 and 12.4, Verlag Chemie, Weinheim 1972

4 G. Becker und F. Kohler, Monatsh. Chem. 103 (1972) 556 (Erratum: Monatsh. Chem. 104, 1138 (1973))

5 K. F. Herzfeld and T. A. Litovitz, Absorption and Dispersion of Ultrasonic Waves, Academic Press 1959

6 J. Lamb in W.P. Mason (ed.), Physical Acoustics, Vol. II A, Academic Press 1965

7 J. H. Andreae, P. D. Edmonds, and J. F. McKellar, Acustica 15 (1965) 74

8 E. Wyn-Jones and W. J. Orville-Thomas, Advan. Molec. Relaxation

Processes 2 (1972) 201

9 T. A. Litovitz and E. Carnevale, J. Acoust. Soc. Amer. 30 (1958) 134

10 J. E. Piercy and J. Lamb, Trans. Farad. Soc. 52 (1956) 930

11 T. A. Litovitz and C. M. Davis in W.P. Mason, ref. 6.

12 U. Jentschura und E. Lippert, Z. phys. Chem. NF 77 (1972) 64

13 N. Tatsumoto, T. Sano, and T. Yasunaga, Bull. Chem. Soc. Japan 45 (1972) 3096

14 L. V. Lanshina, M. I. Lupina, and P. K. Khabibullaev, Soviet Physics-Acoustics 16 (1971) 343

15 W. Maier und A. Mez, Z. Naturf. 7a (1952) 300; 10a (1955) 167; A. Mez und W. Maier, Z. Naturf. 10a (1955) 997; K. Eppler, Z. Naturf. 10a (1955) 744

16 H. Mang und F. Kohler, unpublished

17 J. Rassing and B. N. Jensen, Acta Chem. Scand. 24 (1970) 855

18 Z. Lang and R. Zana, Trans. Farad. Soc. 66 (1970) 597

19 R. S. Musa and M. Eisner, J. Chem. Phys. 30 (1959) 227

20 F. Garland, J. Rassing, and G. Atkinson, J. Phys. Chem. 75 (1971) 3182; cf. also J. Rassing, Ber. Bunsenges. 75 (1971) 334

21 S. K. Garg and C. P. Smyth, J. Phys. Chem. 69 (1965) 1294

22 R. Mecke, A. Reuter and R. L. Schupp, Z. Naturf. 4a (1949) 182

23 F. Kohler, E. Liebermann, R. Schano, H. E. Affsprung, J. K. Morrow, H. Kehiaian, and K. Sosnkowska-Kehiaian, to be published

24 E. Grunwald, R. L. Lipnick, and E. K. Ralph, J. Amer. Chem. Soc. 91 (1969) 4333; E. Grunwald and E. K. Ralph, J. Amer. Chem. Soc. 89 (1967) 4405

25 M. J. Blandamer and D. Waddington, Advan. Molec. Relax. Proc. 2 (1970) 1; cf. also M. J. Blandamer in F. Franks (ed.), Water, Vol. 2, Plenum Press 1973

26 J. M. Davenport, J. F. Dill, V. A. Solovyev and K. Fritsch, Akust. Zh (USSR) 14 (1968) 288

27 M. J. Blandamer, M. J. Foster, and D. Waddington, Trans. Farad. Soc. 66 (1970) 1369

28 R. Schano, H. E. Affsprung und F. Kohler, Monatsh. Chem. 104 (1973) 389

29 F. Kohler, H. Arnold und R. J. Munn, Monatsh. Chem. 92 (1961) 876

30 V. P. Romanov and V. A. Solovyev, Soviet Physics-Acoustics 11 (1965) 68 und 219

31 J. R. Johnson, S. D. Christian, and H. E. Affsprung, J. Chem. Soc. [London] A 1966, 77; T. F. Lin, S. D. Christian, and H. E. Affsprung, J. Phys. Chem. 69 (1965) 2980

32 H. Mang und F. Kohler, unpublished

IX SEA-WATER DESALINATION

IX.1. ECONOMIC ASPECTS OF LARGE-SCALE DESALINATION

B.C. Drude

8520 Erlangen, Germany

ABSTRACT

Desalination processes are handicapped by the low cost of the product. For seawater, the lowest cost-limit possible is reckoned as equivalent to 7 kWh/m^3. Today's best figure is at least four times as much. Some guiding lines for eventual further cost reductions are given.

IX.1.1. INTRODUCTION

While many parts of the world are suffering today from a serious water shortage the full import of the problem will become apparent only in the next few decades, with the huge additional quantities of water needed for a higher standard of life, for a rapidly progressing industrialization, and, above all, for the production of the food needed for an exploding world population. Paradoxically, the quantity of water existing at the surface of the earth in liquid or solid form is abundant for all times to come. But it is very often not available when it is needed, nor where it is needed.

In addition to its many physical and chemical peculiarities water has a unique economic property: its dilute value, i.e. its minute worth per pound. The price commonly conceded to water is 2 or 3 orders of magnitude below that of the most common mass products such as coal, oil, cement or pig-iron. While these may be shipped around the world without excessive cost, penalty water transport costs over long distances are, whatever the means, prohibitive. Likewise the artificial storage of water is as a rule unusually expensive. At 2 ¢ per gallon, which is near the lowest limit /1/, the container is 20 to 50 times as dear as its contents. Water storage over some length of time is hence a rather costly expedient.

The dilute value of water is also the main problem in the production of fresh from saline water. The capital and operating costs of desalination plants have to be borne by a uniquely cheap commodity. The production costs per gallon must therefore be excessively low. For any economic desalting project the basic facts to be considered are the minimum energy requirements, depending on the salinity and temperature of the feed and the percentage of water recovery. With increasing salt concentration they mount at the end so steeply as to make a complete separation and production of a dry residue impracticable. The disposal of the highly enriched brine poses a major problem at inland locations, open evaporation ponds being no satisfactory solution. Thus large-scale desalination is practically restricted to coastal areas where the brine is returned to the sea, though the salinity of ocean water is at least ten times that of many inland brackish water sources.

IX.1.2. COST OF DESALINATION

A rough idea of the lowest principally attainable cost of desalinated seawater can be deduced from the minimum energy requirements as represented in Fig.1 /2/. At 25 °C and 0% recovery, i.e. water withdrawal from an unlimited supply, they are 0,71 kWh/m^3 (2.68 kWh/1000 gals.).

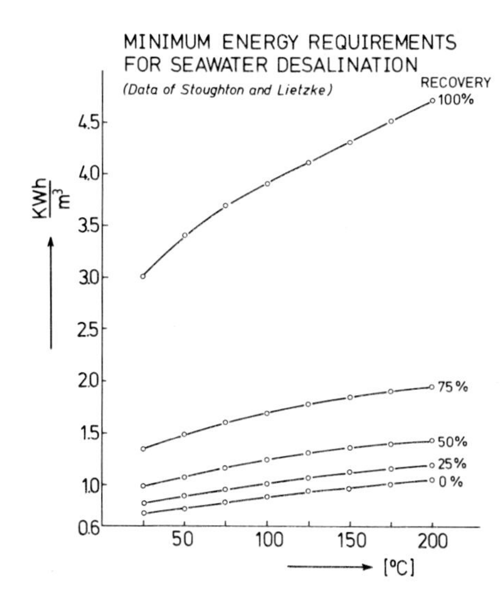

Assuming for a technical process four times the minimum amount of the reversible case the energy consumption will be about 2.8 kWh/m^3. In most large desalting plants the ratio of energy to all other costs is not far from 1 : 1.5. The lowest limit of water costs may therefore be assumed equivalent to 7 kWh/m^3 or at 21 Pfg/m^3 with a power price of 3 Pfg/kWh, resp. 20¢/1000 gals. with a power price of 7.5 mils/kWh.

This minimum figure for the unsubsidized water production exclusive of storage and distribution, low as it is, is still too high for the wasteful modes of conventional irrigation. New methods are called for. Compared to water of conventional supply 3 valuable advantages can, however, be claimed for desalted water: it is free from undesirable components; it can be supplied in any needed quantities; it is always available and not subject to the vagaries of the weather.

Though water production costs have been brought down sharply over the last 15 years today's figure of 80¢/1000 gals. (under favourable conditions) is still about four times the minimum costs. In spite of extensive scientific research and important technical development the desalting industry has not yet succeeded in reducing the equipment to the bare minimum of principally essentials, nor in approaching somewhat closely ideal operating conditions.

While the problems vary greatly with the type of process adopted there are some general points common to all. The biggest shares in the water costs are normally the financial burden due to the capital invested, and the power consumption. With investment costs averaging (within wide limits) 1 $/gallon or 1000 DM/m^3 of daily production (including water intake, pretreatment equipment, the plant proper and perhaps some storage and water posttreatment) and an amortization period of 20 or at the utmost 30 years the interest rate on capital is a decisive factor and particułarly prejudicial at today's high levels.

Whereas the annual "fixed costs" (including insurance) would amount to 9.4% of the capital investment at an interest rate of 4% they exceed 13.7% at 10% interest. This heavy burden has to be borne by the year's water production. A load factor of 90%, corresponding to 330 full stream days, is the practical limit; if mere 60% are reached the fixed costs per unit produced will be 50% higher.

The "variable" or "operating costs" on the other hand are nearly proportional to the load factor. Their main item, besides labour and supervision, repair and maintenance, chemicals and other materials are the costs of steam and/or electric power. Since energy consumption per gallon of water can mostly be reduced at the expense of a more elaborate and costly design, and vice versa, an optimization of design parameters is called for in each particular case.

As a rule the energy contributes to the water costs from 33 to 40%. Solar radiation, though gratis, is unsuitable for any but minor plants as the investment per daily gallon is 10 to 20 times higher than in other types of plants. Where fuel is very cheap the plant design can be somewhat simpler, and the water costs considerably lower. Kuwait, the country with the world's largest total capacity, is an example.

IX.1.3. FEASIBLE PROCESSES

Out of numerous technically feasible desalination processes only a few have gained so far practical importance. They may be grouped under membrane and phase-changing processes. In the former, consisting in transport of solute or solvent across a physical membrane, the energy requirements are strongly dependent on the salinity of the feed. Electro-dialysis and Reverse Osmosis are therefore particularly suited for brackish waters. In case of phase-changing processes on the other hand the salinity is of little importance as far as the energy demand is concerned. The highly disadvantageous ratio of latent to sensible heat in both distillation and crystallization should favour the latter, but other factors outweigh the theoretical advantage. Thus the field of seawater desalination is left at present entirely to distillation processes. The salient points of their economy are dealt with in another section of this book. Though recuperation of heat or multiple effect flowsheets alleviate the problem the actual energy demand of circa 50 kWh/m^3 remains very unsatisfactory.

Valuable improvements may be attained by reducing the floor space required, by combining different processes such as Vertical Tube and Multistage-Flash evaporation, by the preceding removal of bi-valent ions and in other ways, provided that the savings obtained are not balanced by higher investment costs. Other efforts are aimed at the reduction of operating, in particular energy costs, be it by increasing the efficiency of the plant or by drawing on cheaper energy sources. Distillation processes are affected by a peculiar dilemma. Low temperature heat, as required for those processes, is admittedly of little commercial value but (except by tapping geothermal reservoirs) not available from independent sources. Hence it is necessary either to leave the valuable high-temperature range of fossil or nuclear fuels almost unused - a serious sacrifice - or to resort to waste steam from power plants, linking the evaporator closely and permanently to their locations, operative programs and economies /3/. A limited margin of freedom may be secured at the expense of higher investment and/or energy costs. As in other cases of coupled production the rules for the allotment of costs will remain controversial between the partners.

While it is true that all large installations are perforce of the dual-purpose type complete independence combined with (relatively) good economy may be expected from autonomous single-purpose plants utilizing all the energy generated within their own precincts /4/ , either by adding to the evaporator a power-consuming process such as vapour compression, crystallization or, particularly, reverse osmosis (now widely considered the most promising contestant even for seawater), or by employing exclusively power-driven processes. No special steam raising boiler or diesel generator plant is required for the latter if a reliable network is available.

IX.1.4. GENERAL LINES

Without going into technical details some general guiding lines for cost reduction will emerge from this brief exposition:

- The lowest energy requirements may be expected from cold seawater and with minimal water withdrawal. Electric processes allow a high degree of utilization of the energy applied, phase-changing processes being handicapped by irretrievable thermal losses.
- Contrivances requiring little or no purification or pretreatment bring about savings in both investment and operating costs. Designs approaching a three-dimensional disposition of desalting interfaces result in higher compactness of the desalting cells. Space and energy requirements are reduced if the work is performed rather on the fresh water than the seawater side. Pressure losses along the water paths should be kept as low as possible.
- A process demanding for its proper performance the maintenance of a minimum number of parameters will simplify instrumentation and facilitate automation.
- Processes allowing to be started and interrupted easily and quickly may profit from cheap energy in off-peak periods.
- A design reaching optimum economy at low unit capacity will facilitate adaptation to a fluctuating, and to a gradually growing demand, as well as a spreading of production centres.

- Plants are best located within the immediate vicinity of the most suitable quality of seawater and as close as possible to the centres of consumption, without being tied to the site of the power plant.
- Pollution by effluents is minimized by processes involving no chemical and thermal pollution, and in places from where they can spread freely in all directions.

REFERENCES

1 United Nations: The Design of Water Supply Systems based on Desalination, New York 1968, p. 24

2 Taken from K. S. Spiegler: Principles of Desalination, Academic Press, New York 1966

3 B. C. Drude, Techn. Mitt., Hs. d. Technik 63, 81 (1970)

4 O. Pugh and M. C. Tanner, Proc. 4th Intern. Symp. on Fresh Water from the Sea, Athens 1973, 2, p. 387

IX.2. THE PHYSICAL CHEMISTRY AND STRUCTURE OF SEA WATER*

Frank J. Millero

Rosenstiel School of Marine and Atmospheric Science

University of Miami

Miami, Florida 33143

ABSTRACT

To obtain an understanding of how the physical chemical properties of sea water vary with ionic composition and how sea water affects chemical processes that occur in the oceans, it is necessary to have an understanding of the structure of sea water. In the present paper, the structure of the ionic medium of sea water is briefly discussed. An ion pairing model is used to predict the speciation of the major components of sea water. The model is also used to predict stoichiometric activity coefficients of the major ionic components. Since the major components of sea water are in nearly constant relative composition, it is possible to treat the physical properties in terms of a "sea salt" which is proportional to all of the individual ionic components. The properties of sea water can be related to: $P = P^{\circ} + \Sigma$ ion-water and Σ ion-ion; where P° is the property of pure water, Σ ion-water is a term related to the equivalent weighted sum of the ion-water interactions of the major components, and Σ ion-ion is a term related to the equivalent weighted sum of the ion-ion interactions of the major components. The Σ ion-water term can be estimated from the infinite dilution terms for the individual components. The Σ ion-ion term can be divided into two terms: a Debye-Hückel limiting law term (Σ Debye-Hückel) and term related to the weighted deviations from the Debye-Hückel equation (Σ Dev. D.H.). The Σ ion-ion term can be estimated by using Young's Rule as reformulated by Wood and Anderson. This equation has been used to estimate the heat capacity of sea water. The estimates are nearly within the experimental error of the measured values.

*Contribution Number 1683 from the Rosenstiel School of Marine and Atmospheric Sciences, University of Miami, Miami, Florida.

IX.2.1. INTRODUCTION

In recent years there has been a number of major advances made on the interpretation of interactions in multicomponent electrolyte solutions (1-5). In this chapter we shall briefly review the application of these concepts to sea water. A more complete description of the work done on the physical chemistry of sea water is given elsewhere (4, 5).

Since the pressure and temperature range of the oceans is quite extensive (-2 to 40°C; 0 to 1000 bars applied pressure), reliable physical chemical data for sea water is needed over a large range of conditions. Surprisingly, such data is available and most of the physical chemical data (5) is of very high quality. The data, also covers a wider range of P-T space than that available for the simple electrolytes, like NaCl and KCl. The reason, of course, is due to the recent interest in desalinization and oceanography.

Before we examine the physical chemistry of sea water, we will, in the next section, briefly examine the composition of sea water.

IX.2.2. COMPOSITION OF SEA WATER

As has been adequately demonstrated by numerous workers (6), the relative composition of the major (greater than 1 ppm by weight) components of sea water is nearly constant. Thus, by measuring one constituent, the composition of the other components can be characterized. The constituent normally selected to characterize a given sea water sample is the chlorinity, Cl(°/oo). The chlorinity is determined by the titration of sea water with $AgNO_3$ and is defined as the mass in grams of Ag necessary to precipitate the halogens (Cl and Br) in 328.5233 gms of sea water. Since this definition is based on the 1930 molecular weights, the Cl(°/oo) is not exactly equal to the gms of Cl plus the gms of Br as if it were Cl (the difference is 1.00045). The ratios of the mass of the major constituents of natural sea water (g /kg of solution) to the chlorinity (gms/kgm of solution) are given in column two of Table I. The results given for HCO_3^-, CO_3^{2-}, $B(OH)_3$ and $B(OH)_4^-$ were calculated by using the equilibrium constants of Lyman (7) for average sea water at 25°C and pH = 8.1. The value given for Na^+ was determined by difference by making the total cation equivalents equal to the total anion equivalents ($\Sigma e_+ = \Sigma e_- = 0.62778$eq/1000 g H_2O).

The total salt content of sea water is also frequently used to characterize the composition of sea water. The salinity, S(°/oo), of sea water was originally defined as the mass in grams of dissolved inorganic matter in one kgm of sea water after all the Br and I have been replaced with an equivalent amount of chloride and all the carbonate is converted to oxide. Due to the difficulty of determining the salinity by heating to dryness (i.e., the loss of HCℓ, CO_2, etc.), the salinity is normally determined by indirect methods such as measuring the Cℓ(°/oo) or using a conductance salinometer. The salinity is related to the chlorinity by the relation

$$S(^{o}/oo) = 1.80655\ C\ell(^{o}/oo) \quad [1]$$

The "average" sea water in the world oceans has a salinity of $\sim 35^{o}/oo$ or a chlorinity of $19.374^{o}/oo$. The total grams (g_i) and equivalents (e_i) for "average" sea water ($S = 35^{o}/oo$ and pH = 8.1 at 25°C) are also given in Table 1. As is quite apparent from these results, the total salt content for sea water of $S(^{o}/oo) = 35.000$ is 35.1696 gms/kgm of sea water. Thus, the total grams of sea salt is related to $S(^{o}/oo)$ and $C\ell(^{o}/oo)$ by

$$g_T = 1.81578\ C\ell(^{o}/oo) = 1.005109\ S(^{o}/oo) \quad [2]$$

Table 1

The Composition of Natural Sea Water[a]

Species	$g_i/C\ell(^{o}/oo)$[b]	$C\ell(^{o}/oo) = 19.374$		
		g_i	e_i[c]	E_i[d]
Na^+	0.55566	10.7653	0.48534	0.77270
Mg^{2+}	0.06680	1.2942	0.11038	0.17573
Ca^{2+}	0.02125	0.4117	0.02129	0.03895
K^+	0.02060	0.3991	0.01058	0.01684
Sr^{2+}	0.00041	0.0079	0.00019	0.00030
$C\ell^-$	0.99894	19.3534	0.56579	0.90078
SO_4^{2-}	0.14000	2.7124	0.05853	0.09318
HCO_3^-	0.00608	0.1178	0.00200	0.00318
Br^-	0.00348	0.0674	0.00087	0.00138
$B(OH)_3$	0.00105	0.0203	0.00034	0.00054
CO_3^{2-}	0.00063	0.0122	0.00042	0.00067
$B(OH)_4^-$	0.00034	0.0066	0.00009	0.00014
F^-	0.00006_7	0.0013	0.00007	0.00011
		35.1696		
H_2O		964.8304		

a) Taken from the compilation made by Millero (4, 5).

b) g_i is the gms of constituent/kgm of sea water for "average" sea water, pH = 8.1.

c) e_i is the equivalents/1000 gms of water.

d) $E_i = e_i/e_T$ where $e_T = 1/2\ \Sigma e_i + e_B$, where e_B is the moles of boric acid.

To calculate the molality, molarity, normality and ionic strength of sea water, it is convenient to determine the mean molecular weight (M_T) and mean equivalent weight (M_T') for "sea salt". Using the composition given in Table 1, we obtain $M_T = 1/2\ \Sigma(m_i/m_T)\ M_i = 62.808$ and $M_T' = 1/2\ \Sigma(e_i/e_T)\ M_i' = 58.049$, respectively, for the mean molecular and equivalent weight of "sea salt". These values can be used to calculate the molality, molarity and normality of sea water at any chlorinity. For example, the molality and molarity are given by

$$m_T = \frac{g_T}{M_T\ (1000 - g_T)} = \frac{28.9047\ C\ell(^o/oo)}{1000 - 1.81531\ C\ell(^o/oo)} \qquad [3]$$

$$C_T = \frac{g_T}{M_T}\ d = 0.0289047\ C\ell(^o/oo) \times d \qquad [4]$$

where d is the density of the solution (for $S = 35.000^o/oo$, $m_T = 0.5804$ and $C_T = 0.5731$). As will be demonstrated in later calculations, it is frequently convenient to define $C\ell(^o/oo) \times d$ by $C\ell_V$, which is the volume chlorinity (frequently called the chlorosity at 20°C).

The molal and molar ionic strength of sea water are given by

$$I_m = 1/2\ \Sigma(g_i/M_i) = \frac{35.9997\ C\ell(^o/oo)}{1000 - 1.81578\ C\ell(^o/oo)} \qquad [5]$$

$$I_V = 1/2\ \Sigma(g_i/M_i)\ Z_i^2 d = 0.0359997\ C\ell(^o/oo) \times d \qquad [6]$$

for $35^o/oo$ salinity sea water, $I_m = 0.7229$ and $I_V = 0.7137$. The densities of sea water solutions can be calculated at various $C\ell(^o/oo)$'s from equations of the form

$$d = d^o + A_V\ C\ell_V + B_V\ C\ell_V^{3/2} \qquad [7]$$

where d^o is the density of water, A_V and B_V are empirical constants given elsewhere (4).

The Debye-Hückel limiting law slope for various thermodynamic functions contains a valence factor that must be considered when comparing the properties of sea salt to other electrolytes. The valence factor w is defined as

$$w = 1/2\ \Sigma\ \gamma_i\ Z_i^2 \qquad [8]$$

where γ_i is the number of ions i of charge Z_i. For a monovalent ion, w = 1/2 and for a divalent ion, w = 1. When the property of interest is expressed in moles, $w = \Sigma(n_i/n_T)w_i = 1.081$ and in equivalents, $w = (e_i/e_T)w_i = 1.151$ for sea salt. These results are very important since they indicate that according to the Debye-Hückel theory, sea salt should behave in dilute solutions as either 1.081 or 1.151 times the behavior of a 1-1 electrolyte.

For many calculations[4], it is frequently convenient to assume that all the carbonate exists as HCO_3^- and all the boron exists as $B(OH)_3$. This assumption will only have a small effect when calculating the physical, chemical property of sea water, however it will not be valid when discussing the effect of the medium of sea water on chemical processes occurring in the oceans.

It should be pointed out that all of the concentration equations derived in this section are only valid for average sea water and dilutions made with pure water (which is true for most of the world oceans). When sea water is diluted with river waters of various compositions, the composition will vary slightly (especially for HCO_3^-, Mg^{2+}, Ca^{2+} and SO_4^{2-}).

IX.2.3. IONIC INTERACTIONS IN SEA WATER

Classically, the ionic interactions in sea water have been treated by using an ion pairing model (8). The basic assumption of this approach is that short range interactions in sea water can be represented by the formation of cation-anion ion pairs

$$M^+ + A^- \rightarrow MA^o \qquad [9]$$

The extent of ion pair formation is characterized by an association constant

$$K_A = \frac{a_{MA^o}}{a_{M^+}\, a_{A^-}} = K_A{}^* \frac{\gamma_{MA}}{\gamma_{M^+}\, \gamma_{A^-}} \qquad [10]$$

It is quite apparent from the data given in Table 2 that the measured stoichiometric activity coefficients of some of the major ions (HCO_3^-, CO_3^{2-}, SO_4^{2-}, Mg^{2+}, Ca^{2+}, etc.) are quite a lot lower than values calculated using the ionic strength principle. This difference can be interpreted using the ion pairing model. Some of the major ion pairs, thought to exist are: $NaSO_4^-$, KSO_4^-, $MgSO_4^o$, $CaSO_4^o$, $SrSO_4^o$, $NaHCO_3^o$, $MgHCO_3^+$, $CaHCO_3^+$, HCO_3^+, MgF^+, CaF^+ and $MgOH^+$. The relative concentration or speciation of these various ion pairs obtained by various workers are given in Table 3 (5).

Table 2

The Measured and Calculated Activity Coefficients of the Major Ions in Sea Water at 25°C[a]

Ion	Measured	Calculated	
		Ionic Strength	Ion Pairing
H^+	0.74	0.85	0.74
Na^+	0.68	0.71	0.70
Mg^{2+}	0.23	0.29	0.25
Ca^{2+}	0.21	0.26	0.22
K^+	0.64	0.63	0.62
Sr^{2+}	----	0.25	0.22
Cl^-	0.68	0.63	0.63
SO_4^{2-}	0.11	0.22	0.10
HCO_3^-	0.55	0.68	0.43
CO_3^{2-}	0.02	0.21	0.02
F^-	----	0.68	0.31
OH^-	----	0.65	0.11
$B(OH)_4^-$	0.35	0.68	0.38

a) Taken from Reference (5).

Table 3

The Speciation of the Major Cations and Anions in Sea Water[a]

Cations						
Ion	% Free	% MSO_4	% $MHCO_3$	% MCO_3	% MF	% MOH
Na^+	97.7 - 99	1.2 - 2.2	0.01 - 0.03	0.002	----------	0.002
Mg^{2+}	87 - 89.2	9.2 - 11	0.3 - 1	0.1 - 0.3	0.06 - 0.1	0.02
Ca^{2+}	85.4 - 91	8 - 13.2	0.5 - 1	0.2 - 0.9	0.01	0.001
K^+	97.6 - 99	1 - 2.4	-----------	---------	----------	-----
Sr^{2+}	86.4	12.3	0.4	0.9	0.01	0.001

Anions						
Ion	% Free	% NaA	% MgA	% CaA	% KA	% SrA
SO_4^{2-}	39.0 - 54	21 - 37.2	19.4 - 22.5	2.6 - 4.8	0.4 - 0.5	0.04
HCO_3^-	62.9 - 81.3	7.9 - 20.5	6.5 - 19.0	1.5 - 4.0	---------	0.02
CO_3^{2-}	8 - 10.6	3.4 - 17.5	43.9 - 67.3	4.8 - 37.5	---------	0.3
F^-	45.4 - 51.8	4.4	46.8 - 47.0	1.3 - 2.0	---------	0.01
OH^-	14.6	0.6	83.9	0.9	---------	0.002
$B(OH)_4^-$	56.1 - 76.0	15.1	21 - 24	7.0	---------	-----
$C\ell^-$	100	----------	-----------	----------	---------	-----
Br^-	100	----------	-----------	----------	---------	-----

a) Taken from reference (5).

The wide spread of values for some of the ion pairs is caused by the large selection of association constants used in the calculation (5). The stoichiometric activity coefficients calculated from the speciation using

$$\gamma_T = \frac{[M^+]_F}{[M^+]_T} \gamma_F \qquad [11]$$

(where γ_T and γ_F are the total and free activity coefficient; $[M^+]_T$ and $[M^+]_F$ are the total and free concentration of M^+) are given in Table 2. The values calculated using the ion pairing model are in good agreement with the measured values. Although these results might be used to support the ion pairing model, many other non-ion pairing methods (9, 10) also predict γ's that are in good agreement with the measured values. For example, Leyendekkers (9) has used the Guggenheim methods to predict the γ's of ions, while Robinson and Wood (10) have used the equation of Wood and co-workers (2) to predict the γ's for sea salts. The predicted γ's by both are in excellent agreement with the measured values.

The effect of pressure and temperature on the speciation of sea water has been discussed elsewhere (5).

IX.2.4. Physical Properties of Sea Water

In recent years a number of studies of the physical, chemical properties of sea water have been made by various workers. A complete review of these results is given elsewhere (5). In this section we will examine how a typical thermodynamic and transport property of sea water can be treated in terms of its major components. A more thorough review is given elsewhere (4).

As a first approximation (i.e., by neglecting the excess properties) the apparent equivalent properties ($\bar{\Phi}$) of sea water, defined by

$$\bar{\Phi} = (P - P^{o})/e_{T} \qquad [12]$$

where P and P^{o} are, respectively, the properties (such as volume, compressibility, expansibility, enthalpy and heat capacity) for sea water and pure water, and e_T is the total equivalents, can be estimated by using Young's rule (11)

$$\bar{\Phi} = {}_{MX}\, E_M\, E_X\, \phi_{MX} + E_B\, \phi_B \qquad [13]$$

where E_M and E_X are, respectively, the equivalent fraction of cation M and anion X, E_B is the equivalent fraction of boric acid; ϕ_{MX} and ϕ_B are the apparent equivalent property of MX and boric acid at the total ionic strength of the mixture. The extension of Young's rule to electrolyte and nonelectrolyte solutions (i.e., the addition of the $E_B\, \phi_B$ term to the equation) has recently been demonstrated to be reliable for the volumes of boric acid-NaCl solutions over a wide concentration range (12). By making the summation for each cation over all its possible salts, the total electrolyte term is

$$E_M\, E_X\, \phi_{MX} = \phi(Na\Sigma X_i) + \phi(Mg\Sigma X_i) + \phi(Ca\Sigma X_i) + \phi(K\Sigma X_i) + \phi(Sr\Sigma X_i) \qquad [14]$$

where each cation contribution $\phi(M\Sigma X_i)$ is given by

$$\phi(M\Sigma X_i) = E_M\, E_{C\ell}\, \phi(MC\ell) + E_M\, E_{SO_4}\, \phi(MSO_4) + E_M\, E_{HCO_3}\, \phi(MHCO_3) + E_M\, E_{CO_3}\, \phi(MCO_3) + E_M\, E_F\, \phi(MF) + E_M\, E_{B(OH)_4}\, \phi(B[OH]_4) \qquad [15]$$

It should be pointed out that by determining the $\bar{\Phi}$ by this method, one essentially eliminates excess cation-anion interaction (due to ion pair formation).

These equations are difficult to use in their present form since there is a paucity of reliable physical, chemical data for all the major sea salts over the concentration (0.1 to 0.8m ionic strength), temperature (0 to 40°C) and pressure (0 to 1000 bars, applied pressure) ranges of interest. Since salts such as $CaSO_4$ are not soluble in pure water at high ionic strengths, one must estimate its properties by some additivity method, e.g.

$$\phi(CaSO_4) = \phi(CaC\ell_2) + \phi(Na_2SO_4) - 2\phi(NaC\ell) \qquad [16]$$

or

$$= \phi(CaC\ell_2) + \phi(MgSO_4) - \phi(MgC\ell_2) \qquad [17]$$

By dividing the apparent property into an infinite dilution term (Φ^o) and one or more concentration terms (S' or b)

$$\Phi = \Phi^o + S' I_V^{1/2} \quad [18]$$

$$\Phi = \Phi^o + S I_V^{1/2} + b I_V \quad [19]$$

(where I_V is the volume ionic strength and S is the Debye-Hückel limiting law slope), it is possible to simplify the use of Young's rule. Since the infinite dilution properties are always additive, the $\Phi^o = \Sigma E_i \phi_i^o$, can be determined from data on soluble salts. Thus, Young's rule need only be applied to the concentration dependent terms, S' and b, (S is the theoretical Debye-Hückel term and can be estimated for any electrolyte system).

By combining equations [10], [16] and [17] and noting that $e_T \alpha Cl_V$, we have

$$P = P^o + A' Cl_V + B' Cl_V^{3/2} \quad [20]$$

$$P = P^o + A Cl_V + B Cl_V^{3/2} + C Cl_V^2 \quad [21]$$

The constants, A' and A, are related to infinite dilution ion-water interactions and the constants, B, B' and C, are related to ion-ion interactions. The B term is related to the theoretical Debye-Hückel contribution. At a given concentration, any physical property of sea water can be attributed to

$$P = P^o + \Sigma \text{ ion-water interactions} + \Sigma \text{ ion-ion interactions} \quad [22]$$

where the second term is due to the weighted ion-water interactions of the major sea salts (at infinite dilutions) and the third term is due to the weighted ion-ion interactions of the major sea salts. The ion-ion interaction term can be divided into two terms

$$\Sigma \text{ ion-ion interactions} = \text{Debye-Hückel Term} + \text{deviations from Debye-Hückel} \quad [23]$$

the first being a theoretical Debye-Hückel limiting law term and a term due to deviations from the limiting law. This general approach of examining the physical, chemical properties of sea water serves two purposes: 1) it provides the theoretical concentration dependence, and 2) it emphasizes the importance of ion-water and ion-ion interactions of the major components of sea water.

In the next section we will examine the use of Young's rule to estimate the heat capacity (13) of sea water. A more complete discussion of the application of Young's rule to sea water is given elsewhere (4).

IX.2.5. THE HEAT CAPACITY OF SEA WATER

The apparent equivalent heat capacity of sea salt can be calculated from

$$\Phi_{cp} = [1000(cp - cp^o)/e_T] + M_T' \, cp \quad [24]$$

where cp and cp^o are, respectively, the specific heat of sea water and pure water and e_T is the equivalent molality. The apparent equivalent heat capacity of sea salt at 25°C calculated from the specific heat data of Millero, Perron and Desnoyers (13), are shown plotted vs $I_m^{1/2}$ in Figure 1. In dilute solutions the $\bar{\Phi}_{cp}$'s (j deg^{-1} eq^{-1}) approach the limiting law and can be represented by the equation

$$\bar{\Phi}_{cp} = -91.4 + 33.3\ I_m^{1/2} + 5.6\ I_m \qquad [25]$$

Using ϕ_{cp} data of the major sea salts, Millero, Perron and Desnoyers (13), calculated $\bar{\Phi}_{cp}$ for sea water using Young's rule (equation [12])

$$\bar{\Phi}_{cp} = -95.3 + 42.6\ I_m^{1/2} \qquad [26]$$

A comparison of the calculated and measured $\bar{\Phi}_{cp}$ and cp for sea water is shown in Table 4. The agreement is excellent and indicates that Young's rule holds very well for the heat capacity of sea water.

IX.2.6. SUMMARY

The recent developments that have been made in understanding multicomponent electrolyte solutions can be very useful in interpreting the physical chemistry and structure of sea water. The physical, chemical properties of sea water can be examined in terms of the weighted ion-water and ion-ion interactions of the major components.

Table 5

Comparison of the Measured and Estimated ϕ_{cp} and cp for Seawater Solutions at 25°C

	$-\bar{\Phi}_{cp}$ (j deg^{-1} eq^{-1})			cp (j deg^{-1} g^{-1})		
I_m	Meas.	Calc.	Δ	Meas.	Calc.	Δ
0	91.4	95.3	3.9	4.1793	4.1793	------
0.1	80.3	81.8	1.5	4.1514	4.1513	0.0001
0.2	75.4	76.2	0.8	4.1246	4.1245	0.0001
0.3	71.5	72.0	0.5	4.0986	4.0985	0.0001
0.4	68.1	68.4	0.3	4.0734	4.0733	0.0001
0.5	65.1	65.2	0.1	4.0489	0.0489	0.0000
0.6	62.2	62.3	0.1	4.0251	4.0250	0.0001
0.7	59.6	59.7	0.1	4.0018	4.0017	0.0001
0.8	57.1	57.2	0.1	3.9790	3.9790	0.0001
0.9	54.8	54.9	0.1	3.9568	3.9568	0.0000
1.0	52.5	52.7	0.2	3.9352	3.9350	0.0002
		Ave.	0.7		Ave.	0.0001

FIGURE 1

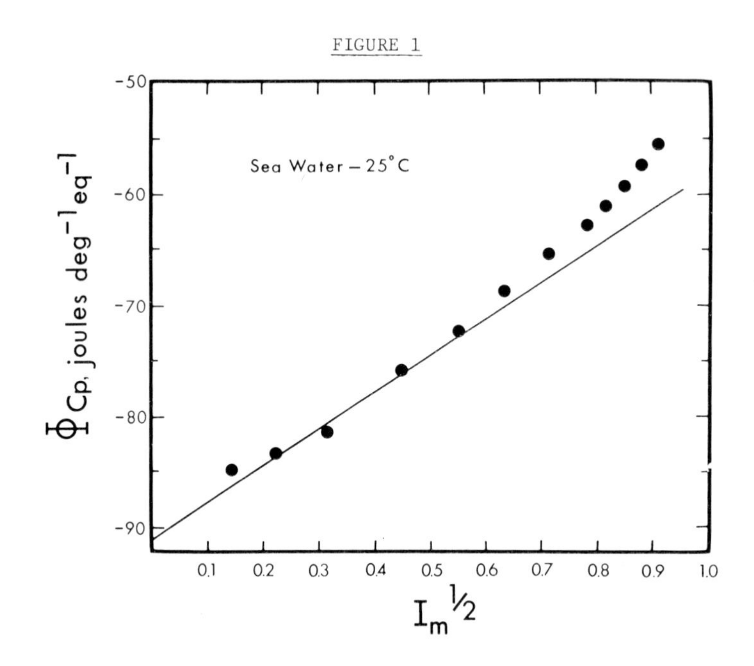

ACKNOWLEDGEMENT

The author wishes to acknowledge the support of the Office of Naval Research (Grant #N00014-67-A-0201-0013) and the Oceanographic Branch of the National Science Foundation (Grant #GA-17386) for this study.

REFERENCES

1. H. S. Harned and B. B. Owen, The Physical Chemistry of Electrolytic Solutions, ACS Monograph No. 137 Reinhold, New York (1958).
2. R. H. Wood and R. J. Reilly, Ann. Rev. Phys. Chem., 21: 287 (1970).
3. F. J. Millero, In: Biophysical Properties of Skin (Chapter 9), Ed., H. R. Elden, Wiley & Sons, New York, 329 pp (1971).
4. F. J. Millero, In: The Sea (Chapter 1), Ed., E. D. Goldberg, Wiley Interscience, New York (1973).
5. F. J. Millero, Ann. Rev. Earth & Planet. Sci., 2 (in press).
6. J. P. Riley and R. Chester, Introduction to Marine Chemistry, Academic Press, London (1971).
7. J. Lyman, Buffer mechanism of sea water, Ph.D. Thesis, University of California Los Angeles, California (1957).
8. R. M. Garrels and M. E. Thompson, Amer. J. Sci., 260: 57 (1962).
9. J. V. Leyendekkers, Mar. Chem (in press).
10. R. A. Robinson and R. H. Wood, J. Soln. Chem., 1: 481 (1972).
11. T. F. Young, Rec. Chem. Progr., 12: 81 (1951).
12. G. Ward and F. J. Millero, J. Soln. Chem. (submitted).
13. F. J. Millero, G. Perron and J. E. Desnoyers, J. Geophys. Res., 78: 4499 (1973).

IX.3. DESTILLATION PROCESSES

F. Mayinger, H. Voos
Institut für Verfahrenstechnik, Hannover, Germany

ABSTRACT

Following three destillation processes for technical use are described: vertical tube evaporator, multi-stage flash evaporators and the vapour compression method. As an example for a new European destillation plant the Terneuzen (Netherlands) plant is described. For the further development of large destillation plants some of the main problems like material difficulties, thermo- and fluiddynamic questions and plant design and economic problems are discussed.

IX.3.1. INTRODUCTION

The world fresh-water reserves are calculated on 20 - 70 x 10^{15} m^3/ year /1/. This would mean that even on a water consumption of 3.000 m^3 per person and year, just like at present in the U.S.A. for instance, 7 - 24 · 10^{12} people could be provided. The world population nowadays increases to about 2 % per year, this means that in the year 2000 the world population can be estimated on 6 Billion and in the year 2100 on 20 Billion people. At present most of our population have a much lower waterconsumption than it is described above by the example of the U.S. It is supposed that it augments slowly due to the increasing industrialisation and the improving living standard.

Through these comparisons an obviously high necessity of seawater desalination plants can not be derived. One has yet to consider that - like nearly all of our good in the world - the fresh water is most irregularly distributed. A big part of our globe is covered by dry and half-dry zones and though only 200 Mio people live in these areas. It will become necessary before long to build up the agriculture of these countries, in order to guarantee the nutrition of the entire population. This is only possible by desalting sea-water and brackish water.

But also in water-rich countries like Germany for instance a local shortage of water can occur due to the increasing industrialisation. The industry uses water for cooling purposes, as carrierfluid for waste-products, but also for the manufacturing of products. This resulted in a heavy pollution of our rivers not only on a chemical point of view - our rivers having the function of a sewage canal system - but also thermodynamically as a heat exchanger in power plants. For cooling power plants there is already a need of installing cooling towers which operate with water evaporation which is the reason for an increasing water consumption considering the fact that our society's need for energy is permanently growing. So the problem of preparing fresh-water for the future on acceptable costs comes up.

Aereas rich in water where only a locally limited shortage of water is recorded will of course consider to carry water through pipelines, concreted tunnels and open runnings. As shown in picture 1 /1/ acceptable specific transport costs can occur if the distances are not too long and if the water consumption is not to high. The costs of the transport make about 0,2 - 0,4 DM/m^3 water on a fresh-water source-distance of 40 - 50 km. The desalting of sea- or brackish-water and sewage for providing the population not only in dry areas with fresh water is a necessity for living. For areas where natural water reserves are polluted by industrial and municipal sewage re-use of water becomes of great economic interest. The industrial desalting of sea water or sewage is therefore an essential main task for engineers in our days. Industrially produced potable water is still expensive. The production-costs of water in operating plants with a capacity of 1000-5000 t/d amount to about 1-1,5 DM/m^3. Expenses for the distribution are not included. Now plants with 20-50000 t/d capacity which diminish the fresh-water costs per m^3 are planned for the future.

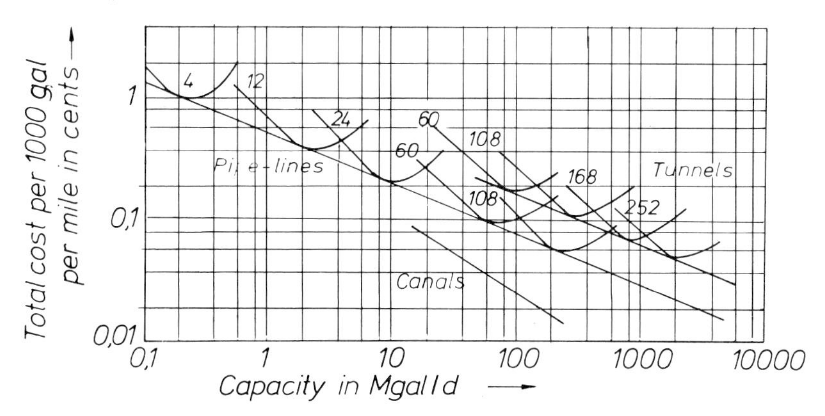

Fig.1: Total cost of transmission for a given water supply

IX.3.2. PRINCIPLES OF DESTILLATION PLANTS

Huge desalination plants with a daily out-put of some thousand tons are built according to the destillation method.

This method is as old as mankind itself and had already been used in a most simple form by the ancient egypts. Seamen in antiquity used to boil seawater during their long journeys on the oceans; they collected the steam by means of sponges and thus they gained pure fresh-water. This single-stage evaporation or destillation means a high expense of energy. Therefore nowadays multi-stage plants are used in technical plants diminishing the energy-expense by recuperating of the evaporation heat. For large plants three methods for the destillation process are of technical importance:

a) vertical tube evaporator (VTE)

b) mult-stage flash evaporator (MSF)

c) evaporator with vapor compression

Of these 3 methods the flash evaporation is the most developed. These three methods - the flash evaporation being technically the best known - shall be outlined as follows: In the tube evaporator salt water flows vertically down through rod tubes which are heated by saturated steam from the outside as it is shown on picture 2 /2/. The rod bundle is installed in a pressure vessel which circuits the heating steam. Due to the heat supply by heating steam sea water partly evaporates. Because of the extremely low steam-pressure of the salts the up-coming steam consists in pure H_2O. The brine steam mixture is led to a flash-chamber where the now more concentrated brine can settle. The steam is led to another evaporation unit and can now be used as heating steam being condensed on the tubes outside and distilled as fresh water. Thus it is possible to recuperate the latent heat of evaporation transferred to this steam primarly. In order to gain high efficiency that means a heat consumption as low as possible this process is repeated in a cascade of evaporators. The steam for the first evaporator is produced by a steam generator, oil-gas- or coal-fired, and flows as a condensate from this first evaporator back again. The brine flowing from the rod-bundle of the first evaporator is led to the second evaporator for the next destillation step.

The temperature of the brine becomes lower on its way through the evaporators. In the pilot plant in Freeport, Texas, the brine enters the first stage with a temperature of $120^{\circ}C$ and is cooled down to a temperature of about $30-35^{\circ}C$ reaching the 12th and last chamber. The pressure drops correspondingly from chamber to chamber in order to make an evaporation possible. The brine flowing out of the last chamber is rejected to the sea.

Similar to the multiple effect vertical tube evaporation the multi stage flash evaporation makes also use of the physical fact that water is boiling for decreasing temperatures at lower pressures. The principal scheme of a flash evaporation plant is shown in a simple way on picture 3 /3/. Seawater has been stage-wise preheated in condense and is superheated to a certain maximum temperature in the final pre-heater. This system's pressure has to be fixed so high that an evaporation in the heat exchangers does not occur. The heated brine now enters the first flash chamber through an orifice. The pressure in this chamber is lower than the saturation pressure of the entering brine. Part of the water evaporates. The steam is streaming up into the condense which is installed in the upper part of the chamber. There the steam condensates on metallic surfaces. The condensate is collected in a tank and led to the next stage just like the rest-brine.

On this stage the pressure is again lower than on the preceding stage, that means another part of the brine and a part of the entering destillate flashes too. The process is stage wise repeated until reaching the last stage. Here the brine is pumped off. The destillate and the non-condensable gases are pumped off too. The available temperature gradient and the circulating quantity of brine are thermodynamically related. This will not be discussed here. It should be mentioned, however, that with the increasing temperature gradient the circulating quantity can be decreased for the same quantity of destillate. The temperature gradient per stage is a characteristic factor for the consumption of energy.

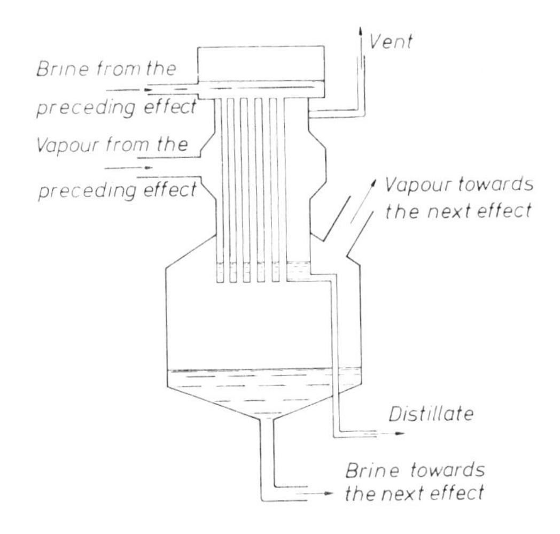

Fig. 2: Falling film L.T.V. effect [2]

Because of economical purposes one will try to operate with possibly high entering temperatures and a thus wide-spread total temperature difference. The hydrogen carbonates of calcium and magnesium, soluted by sea-water, form - due to thermal dissociation at higher temperatures - hard soluable hydrogencarbonates of magnesium or calcium

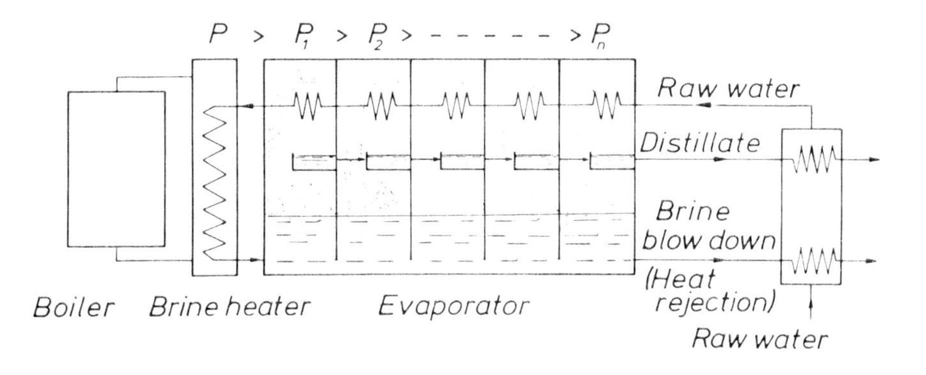

Fig. 3: Multi-stage flash distillation [2]

which easily fall out and occur as a coating mainly on the condenser tubes. For the above discussed tube evaporator this coating grows inside in the evaporation tubes. In order to reduce the coating or even to prevent it hydrochloric or sulfuric acid is added to the brine in the stochiometric quantity in order to destroy the carbonate-hardness. These chemicals have to be added to the total seawater quantity if an arrangement of the plant layout is used as shown in figure 3.

Since in technical plants the relation of destillate quantity and brine quantity is about 1 to 8 till 1 to 10, relatively high chemical costs occur. That is why the multi stage flash evaporator with circulating brine has been developed. Its principle is shown on picture 4 /2/. Plants which are constructed according to this system have nowadays been installed all over the world. The brine quantity which is

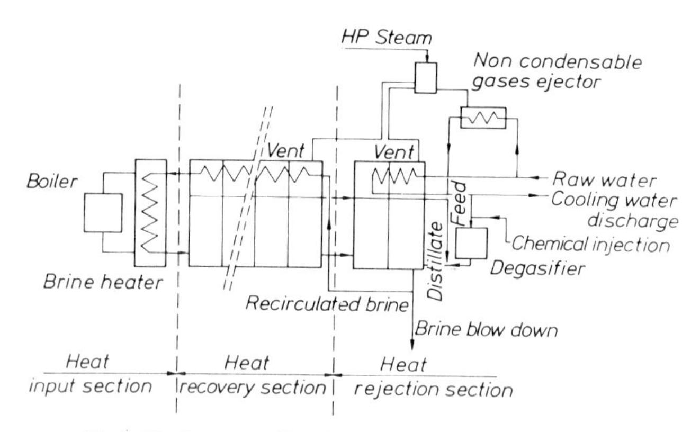

Fig. 4: Flash evaporation plant with recirculation (single-effect multi-stage) [2]

necessary for a certain destillate quantity is circulated in a closed loop. The circulating water flows through all the flash chambers but not through all the condensers. Fresh sea water, serving as cooling water flows through these condensers which belong to the so called heat removal part of the plant, removing the amount of heat primarly added in the final preheater. A certain quantity of this fresh sea water is given into the loop to avoid that the concentration of the circulating brine cannot overcome a pretended maximum value. Due to this principle the chemicals must only be added to this part of the fresh seawater which is to be given into the loop. That is how the costs for the additivs decrease to 25-35 % of the value for a plant of the kind above described.

The principal of the vapor compression is finally shown on picture 5 /2/. The steam of the flash evaporator is exhausted by a compressor and after the compression it is led as a heating steam with a higher temperature into a heat exchanger. There the steam is heating up the uprising brine, is condensed and is gained as fresh water.

The essential difference between this procedure and the above discussed destillation-process is the way of feeding the system with energy which is necessary for the desalination. In the two other processes energy was added in form of heat, while the process of vapor compression claims the compression energy as energy supply. A supplementary heat supply is only necessary when the plant is set into operation. This method demands an amount of energy of 15 kWh per ton of produced fresh water. While the two other processes - the tube evaporator and the flash evaporator need 50 kWh for the same quantity of water.

For giving an example for a huge modern European desalination plant some facts of the flash evaporation plant Terneuzen /3/ in Holland shall be mentioned. Because of the wide-spread industrial settlements in this area the demand for fresh water and electric energy had become

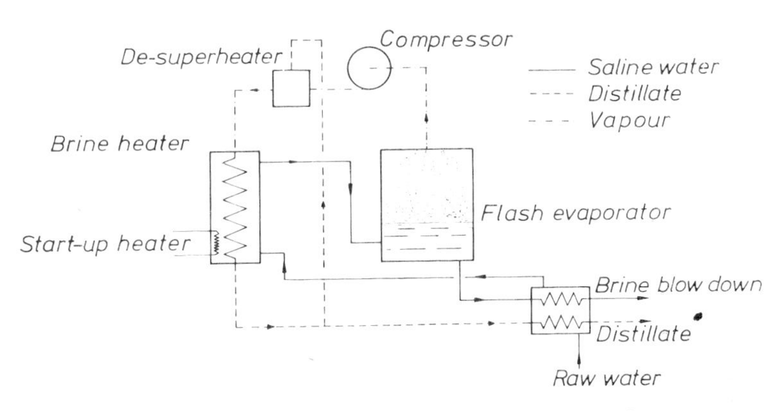

Fig.5: Vapour compression distillation [2]

stronger and stronger. Considering the fresh water consumption of Antwerpen the natural sources could not be exploited without taking a risk and even a fresh water transfer from far away aereas, from Westerschelde i.g., would have meant too high costs. Because of these reasons one decided in October 1966 to build up an combined energy supply - and sea-water desalination plant in Terneuzen with expenses of 232 Mio Dutch guilders, where the expenses of the desalination plant amounted to 18 Mio Dutch guilders. The essential part of this plant consists in a natural gas heated boiler with a production of 136 t/h steam, a turbogenerator and the desalination plant. In the turbogenerator the high-pressure steam is expanded to low-pressure-steam and 24 MW of electric energy can be exploited. The waste steam of the turbogenerator serves as a heat source for the desalination plant; the evaporation process occurrs in two parallelly operating evaporation plants each of these having a production capacity of 14.000 m^3 daily and consists in 38 evaporation stages. The maximum brine temperature is 120 ^oC, the destillate flows off having a temperature of 37 ^oC. The cost for the extracted fresh water is 0,85 DM/m^3 destillate, plus the transfer costs of 0,25 DM/m^3.

Recently proposals /4/ for a vertical construction of the multi stage flash evaporator have been made. In the sense of chemical engineering its way of operation is identical with the above discussed multi stage flash evaporator with horizontally constructed chambers. If the chambers are constructed vertically as it is shown on picture 6 /4/, the brine flashes in vertical venturi jets and is lifted up to the higher situated chamber by means of the so-called "gas-lifting-effect" so that the potential energy increases. This way of construction makes possible the recuperate of part of the pressure drop in form of potential energy. Thus the energy consumption for the circulation of the massflows is diminished. It has to be mentioned moreover that the vertical construction-method claims a far smaller base than the horizontal one. Because of the "higher water level" the pumps do not tend so much to cavitation. The vertical construction method can also be used combined with rising and falling stages as it is shown on picture 7 /4/.

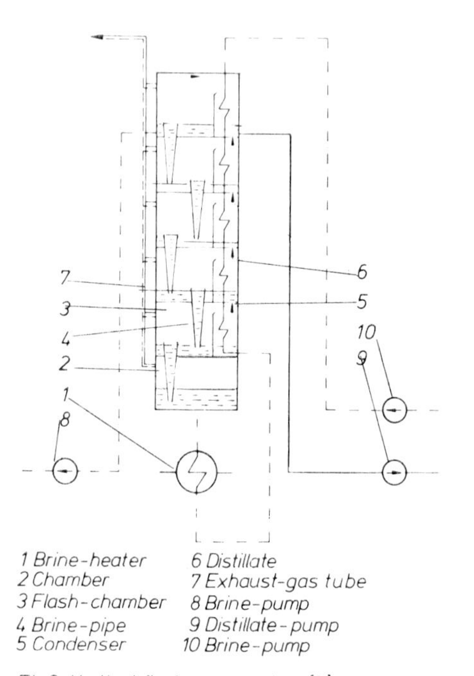

Fig6 Vertical flash evaporation [4]

IX.3.3. PROBLEMS OF DEVELOPMENT

The fact, that large destillation plants for desalinating sea water are already in technical operation could easily give the impression that the development of that kind of plants has nearly come to an end and that only inferior detail problems might be of some importance. This fact becomes most evident, if one considers that there are operating more than 700 desalination plants with a minimum capacity of 100 tons daily per unit and that the whole capacity of these plants has already exceeded a daily out-put of one Mio tons. Most of these plants are in the Middle East and the United States with an out-put of about 250.000 t/d followed by Europe and Russia with a production of 130.000 t/h fresh water. It is estimated that the production of desalinated water will reach an out-put of 4 Mio t/h by 1975. More than 90 % of this quantity will be produced by destillation plants. The increasing technical use of these plants intensifies the demand with regard to their efficiency and reliability. The product costs for the fresh water consist of the capital interest for the investments, the costs for energy and chemicals and the operating costs. An allotment as follows can be made:

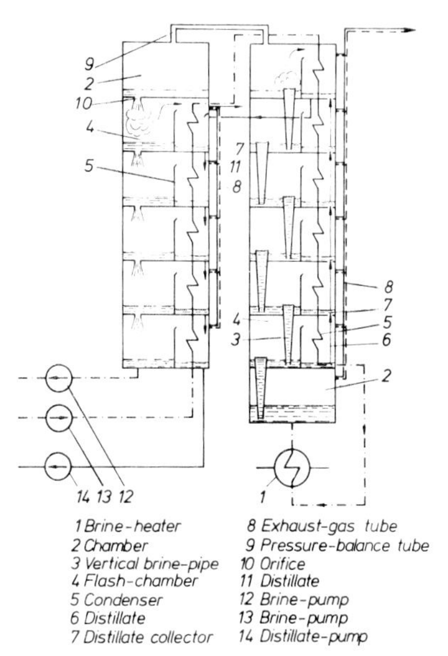

Fig.7: Combined vertical flash evaporation with vertical lifting plant [4]

capital interest for investments	35 %
energy expenses	37 %
chemicals	3 %
other expenses for the operation	25 %

The cost of the condensors in case of the flash evaporator or the evaporator in the VTE-plant makes with 40-50 % the biggest part of the investment expenses. A reduction of the product cost is only possible if the specific investment expenses can be reduced i.g. by means of operating with bigger construction units or a reduction of the energy expenses by means of increasing the temperature gradient. As far as I know the biggest plant projected has a capacity of about 100.000 t/h fresh water. Even operating with units as big as this which works according to the principle of the flash evaporation the investment expenses can hardly be reduced more than 1000 DM/t·h. According to the liquidation and bank rate the charge of just these expenses exploiting the plant to 90 % will be 0,30-0,40 DM/m^3 destillate. The energy expenses of these enormous plant probably amount to 0,12-0,15 DM/m^3 destillate, so that already the sum of these two different kind of expenses makes a water price of at least 0,35 DM/m^3. Altogether one has to expect a water price of 0,50 DM/m^3 for large plants.

A question of economy but to a high extent also of the reliability is the life-time of the plant, problems of corrosion being of great importance. In our days the plants are projected with an availibility of 90 % and a life of 15-20 years. The demands for a most economical way of operating, high availibility a long durability and low product costs lead to a variety of technical problems which shall be partly discussed as follows.

IX.3.4. MATERIAL PROBLEMS

As mentioned above acid has to be added to the sea water or to the circulating brine - in order to prevent a crusting of the tubes caused by hydrogen carbonates. This additive increases the corrosion in the plant and might lead to a reduction of the operating security and a less of economic operating of the plant. Therefore it is important to know all kinds of corrosion and corrosion velocities for the materials which occur in seawater desalination plants. This question has been discussed recently e.g. by Wangnik /5/. Because of the low cost one aspires to use cast iron and steel for sea water desalination plants like it had been made e.g. constructing the first huge American testplant in Freeport/Texas. These materials proved there due to the fact that the velocities of water are low and neither turbulences nor water bounding did occur. Todd /6/ reports that the velocity of corrosion depends on the water velocity, further for carbon steel the p H value and the oxygen-content has an essential influence on the corrosion. Stainless steel with the quality X2CrNiMo 1810 and X5CrNi 189 only show a very unimportant material attrition in calm sea water. But perforation corrosion by overgrowing of sea organismes and sediments have, however, been observed. High velocities of the brine prevent the growing of sediments but only for stainless steel there is than no loss of metal by erosion.

Because of their good heat transfer qualities the copper compositions are the most important materials used for seawater desalination plants, especially referring to aluminium bronzes, brass compositions CuZn 20 Al, special 76, CuNi 10 Fe and CuNi 30 F. Using those materials turbulences with increased velocities might cause an erosion - corrosion process and dying micro organismes can form sulfides and ammoniac, which cause heavy corrosions. It has also to be considered, since these materials are especially used for the tubes of condensers, the corrosion on steam-side, especially if the steam contents, non-condensable gases like carbon dyoxid, sulfor hydrogen and ammoniac.

Nickel-copper-composition generally have a good corrosion constancy, perforation corrosion can, however, occur if some substances or micro organismes settle down. Because of their higher expenses as against the copper compositions they are generally used only for extremely strained segments.

Investigation for titan and aluminium as tube materials have been made and also for concrete and synthetics as a material for chamber parts.Titan in seawater desalination plants can be generally considered as corrosionproof it is, however, susceptible to perforation corrosion if there is unsufficient ventilation together with sedimentation. Considering the due to its good physical qualities possible low wall-thickness and the low specific weight titan can not yet be used for sea water desalination plants because of its high expenses. An estimation of the influence of different materials on the capital cost of the condensating tubes was given by Wangnik /5/ and comes, as it is shown on picture 8, to a strong dependence of the planned life durability of the plant, with special brass 76 (Co Zn 20 Al) as the best solution by the economic point of view.

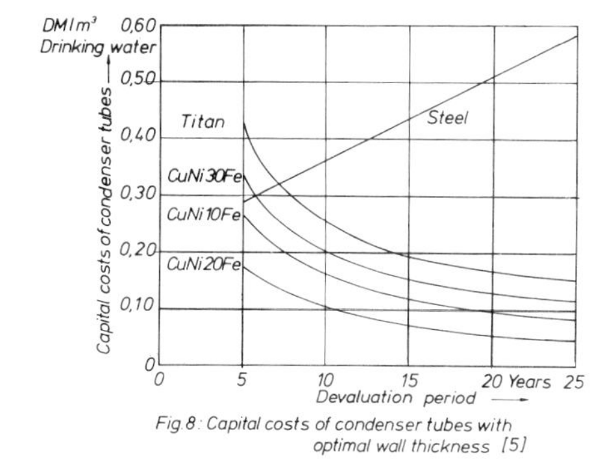

Fig.8: Capital costs of condenser tubes with optimal wall thickness [5]

The latest development tends to use condensers with longitudinal welded tubes with possibly thin walls. There still exist an amount of problems referring to the corrosion proof fabricating of these tubes and to its clamping to the tube plates.

Nearly no experiences have been made concerning the behaviour of concrete as a construction element for sea water desalination plants. At first a convenient method for sealing has to be found; it is generally planned to provide the concrete with a protective layer from the inside.

IX.3.5. THERMO- AND FLUIDDYNAMIC PROBLEMS

The lowest energy quantity for the desalination is needed if there is a thermodynamically reversible processing. Theoretically the required energy would be equal to the equilibrium potential difference of the two differently concentrated waters and that is brine and freshwater. It comes up to 0,77 kWh/m^3 if the temperature is 25^oC. As we realized in the beginning an energy quantity of 50 kWh is necessary for the production of one m^3 freshwater in flash evaporation plants. The actual process is thus far away from the reversible process. The reason is that for the heat- and mass-transfer processes of the desalination temperature- and concentration differences are necessary. The temperature-differences are mainly determined by the heat transfer resistance between steam- and brine part of the condenser which results from the sum of the single resistances in condensation, heat conduction through the tube wall and the forced convection on the brine part. Efforts have been made to increase the heat transfer coefficient by fins artificial roughnesses or twist producers. Profiled tubes are generally more expensive than smooth tubes and therefore the thermodynamic profit caused by a possible decrease of the temperature gradient and the material savings compared to the higher production expenses for profiled tubes resulting from the smaller heat exchange aerea have to be considered carefully.

For the engineer there a broad field for the investigation of the heattransfer coefficients with different profile forms is open. Best results can only be achieved by a team work with a fabrication engineer. For an investigation of the efficiency of different finforms on the heat transfer in forced convection, that means on the brine part, optical measuring methods are of great importance. Theoretical observations of finefficiency and roughness-influences, can systematize the expensive complicated experimental investigations. Since the condenser causes about 40 % of the entire investment costs of a seawater desalination plant, the developing operations for the best result are surely advantageous.

Another irreversibility concerning the heat- and masstransfer in flash evaporation occurs because of the unsufficient mixing of the brine and an imperfect evaporation the thermodynamic equilibrium in the flash chambers cannot be attained. The difference of the saturation temperature due to the pressure differences between two chambers caused by the flash is generally 3 K that means, the brine flows to the next chamber with a 3 K higher temperature than the belonging saturation temperature and the stored heat which becomes free due to the reduction of the temperature difference is available for the evaporation of fresh water partial quantity. Generally this temperature difference is only reduced to 2/3, that means the thermodynamic equilibrium cannot be reached, the brine is in an unstable thermodynamic state with 1 K overtemperature. A better reversibility and therefore smaller exergy losses might be caused by a more intensive mixing of the brine in the flash chamber, an artificial adding of seeds or promotion of the evaporation process in general. Of course longer chambers would cause a better approximation to the thermodynamic equilibrium too; this would, however, mean higher investment expenses.

A forced mixing of the brine in the flash chamber might cause a foam formation at the brine surface and promotes the carrying out of brine droplets into the steamflow. Both processes lead undesired salt-particles into the condensate. The carry over of the brine drops into the condensate is to be prevented by demisters which are disposed in the steam chamber in front of the condenser tubes. To get the droplets out of the steam one generally uses the inertia and adhasic forces of the brinedrops, by bounce and "turpentine" separation. With increasing separation efficiency the pressure drop in these demisters however increases too. This can lead to the result that mainly in vacuum-stages the pressure losses in the demister influence the total pressure drop between the chambers and that there will hardly be left a drifting pressure gradient for the evaporation and condensation process, that means for the actual masstransfer process. Because of the low density of the steam in the vacuum stages the demisters have to overcome relatively big volum flows. The aim of this development is to develop hydrodynamicly favourable demister units which still guarantee good seperation for small dropsizes with a minimum of pressure loss. Its production costs should not be too high.

Another problem which is presently researched on all over the world is the specific mass flow rate in the orifice between the flash chambers. The nowadays usual values of 200-700 t/h water per meter chamber width are to small for large plants. Specific weir rates of 1.500 - 2.000 t/hm are aspired. Some intensive experimental and theoretical investigations of the hydrodynamic processes in orifices of different design have become necessary. In the past it was frequently assumed that the steam formation happens only in form of evaporation at the water surface and that the flow in and behind the orifice had to be considered as merely single phase flow. Other investigations have proved however, that there is a sufficient amount of smallest seeds and small particles in the brine to cause boiling and bubble formation in and behind the orifices. The flow through the orifice is thus to be considered as a two-phase-flow, which complicates especially the exact calculation of the pressure drop in the orifice.

Thus the scaling of this part of the plant has become vague and generally in practice one has been forced to start some restauration works during the operating of the plant and the orifices have been constructed in a adjustable way. An exact knowledge of these two-phase-flow processes requires a number of theoretical and experimental investigations, especially because of the dependence of the evaporation process during the flash on a series of hydrodynamic and thermodynamic parameters.

Experimental observations have proved moreover, that there is some danger of oscillations and pulsations in case of a certain kind of orifice construction. These pulsations cause a strong surface waves on the brine level which anew causes considerable dropemission deteriorating the purity of the condensate. On one hand these pulsations occur due to rising vortexes of diametral motion in the brine, on the other hand they can also be caused by the characteristic pressure loss behaviour of the two-phase flow, since the pressure loss depends to a higher extent on the steam content than on the massflow.

Pressure loss, evaporation process, fluxconditions are the counting factors for the dynamic behaviour of a multistage flash evaporation plant in case of sudden massflow changes. It is true that calculations on the propagation of thermodynamic disturbances in the flash plant stages are available, the influences of hydrodynamic disturbances, their growth or damping on the way through the cascade and an eventual hydrodynamic oscillation behaviour of the plant are however, hardly examined. For a better understanding of these problems the knowledge of the pressure loss behaviour and the hydrodynamic processes in the orifices is a prior condition.

Losses - not only in a thermodynamic sense, in form of exergy losses - are the unavoidable temperature differences for the heat transfer, e.g. between condensing steam and the heated brine. An essential part forms the structure material of the heatexchanger, that means the tubewall, which separates both fluxes and acts as a heat-flux resistance. There is primarly of course the chance to diminish the thus caused difference of temperature by choosing a material with a high thermal conductivity. Further considerations tend to improve the heat transfer process in a way that a fluid can be used operating as heat carrier instead of the structuring material of the heatexchanger. But there any solubility between the condensate or the brine and this fluid must not exist. The heat transfer process is there divided in mixing and demixing processes where at the heat carrying fluid e.g. oil or low melting metal is sprayed dropwise into the rising steam from the flash chamber, condensing this steam by absorbing heat. The now heated fluid is mechanically separated, added to the brine flow and there losing its heat. It is true that thus the temperature differences between condensate and brine flow can be diminished and the energy expenses can also be decreased, but it has to be proved how far the necessarily increased construction expenditure really improves the entire economic situation.

IX.3.6. PLANT TECHNIQUE AND COST OPTIMISATION

One of the engineer's essential task is to offer his products in an economical low price way. The price of a product is not only influenced by energy- and investment expenses but also by the reliability and durability of the production plant. A number of specific engineering problems like e.g. attrition of material by corrosion, cavitation and erosion, the tightening of axels and tubelines and the dosage of small portions emerge concerning different parts of seawaterdesalination plants, especially brine pumps but also heatexchangers and acid-dosage-constructions; these problems are probably of no importance for the physicist interested in more fundamental problems.

The energy which is necessary for the flash evaporation desalination is even nowadays generally produced by fossil fuels, usually oil, but also coal and natural gas. During the last years a number of studies have been made how the energy expenses can be decreased by using nuclear energy that means exploiting a nuclear reactor as a heat source for the seawater desalination. In this connection as well so called combined systems, in which electric energy and freshwater are produced as pure "one purpose plants" were discussed in which the nuclear reactor serves as heat source. As to the preceding example the nuclear plant could be managed in a less complicated way because of the low pressures and temperatures. Concerning all these considerations one must not forget that an essential part of the investment costs for nuclear reactors is caused by the extensive emergency installations which are necessary for the operation. A considerable energy price could only occur if units have a thermal capacity of more than 2 to 3000 MW. Seawater desalination plants, which operate with these heat quantities producing accordingly high water quantities are not at all the todays technical knowhow and such a high water consumption only occurs in few areas such as South California. Even combined reactor-desalination plants can normally not be set according to thermodynamic aspects, they have to adapt to the local consumption of electric energy and water.

According to the increasing extension of the plant one has to try to use other structuring materials such as concrete instead of steel for the construction of the flash chambers which are mostly under pressure. A number of material investigations, sealing problems and the solution of constructive details such as the fixing and designing of the liner called inner diaphragm have still to be discussed.

IX.3.7. RESUMÉ

I can imagine that the above mentioned technical scientific problems concerning further developments of seawater desalination plants did not always find your physicist's interest from a fundamental point of view. An engineer according to his task is, however, forced to think in anapplied manner and therefore he has always to consider carefully the problems of the economy and reliability of the production plants and the quality of his products. Therefore interesting physical fundamental investigations with no important influence on the product have to be inferior at times. Therefore I would like to beg your pardon if I have been setting to high a value on the engineer's task.

REFERENCES

1 Linaweaver, F.P., and C. Scott Clark: Costs of water transmission
Water Works, December 1964

2 D'Orival, M.: Water Desalting and Nuclear Energy, München 1967

3 Anonimous: Meerwasserentsalzung in Holland, Wasser, Luft und
Betrieb 15 (1971), Heft 12, 452 - 454

4 Klaren, IR.D.G.: Onderzoek naar de werkwijze en uitvoeringsvormen
van verticale ontspanningsverdampers
Technische Hogeschool Delft, Laboratorium voor Energievoorziening
en Kernreactoren

5 Wangnik, K.: Die Auswahl kostengünstiger Werkstoffe für Meerwasser-
entsalzungsanlagen, Techn. Mitt. Krupp, Werksberichte Bd. 30,
(1972) H.4

6 Todd, B.: The corrosion of materials in desalination plants.
In: Proc. 2nd Europ. symposium on fresh water from the sea,
Athens, May, 9-12th, 1967, paper 49

IX.4. PHYSICO-CHEMICAL PROCESSES IN ASYMMETRIC CELLULOSE ACETATE MEMBRANES

W. Pusch
Max-Planck-Institut für Biophysik, Frankfurt am Main, Kennedy-Allee 70, Germany

ABSTRACT

In reverse osmosis or hyperfiltration, membranes are used for the desalination of brackish or sea water by pressing the salt solution through cellulose acetate membranes using pressures up to 100 at. The cellulose acetate membranes used are asymmetric in structure as a result of the method of preparation. Since until recently they were the only synthetic membranes which possess an asymmetric structure, the influence of the asymmetry on membrane transport properties has been of interest. Therefore, the transport coefficients of these membranes have been measured using the linear relations of thermodynamics of irreversible processes. The experimental results manifest a strong dependence of the transport coefficients on solute concentration. This strong dependence on solute concentration can be attributed to a concentration gradient within the porous sublayer of the asymmetric membrane. Since the magnitude of the concentration profile depends on the direction of volume flow and, thereby, on the boundary conditions, the measured transport coefficients also depend on the boundary conditions. The concentration profile within the porous sublayer can be calculated using certain assumptions, and it is therefore possible to find the intrinsic transport coefficients. These transport coefficients can also be obtained from hyperfiltration experiments since there exists no concentration profile within the porous sublayer due to the different boundary condition. For the comparison of dialysis experiments and hyperfiltration experiments, a relation between salt rejection r and transport coefficients is derived. Furthermore, the relation between the volume flux q and the pressure difference ΔP for hyperfiltration experiments is derived, taking into account the fact that the osmotic difference $\Delta\Pi$ across the membrane is not an independent variable in hyperfiltration.

IX.4.1. INTRODUCTION

During the years 1952-1958 a new technique for the production of drinking water from brackish and sea water was developed within a research program in the United States. This new technique is attractive in its simplicity. In the early work, based on the principles of semipermeable membranes salt solutions were pressed through suitable membranes. These membranes should be permeable to water but impermeable to salt. Looking for suitable membranes, Reid and Breton /1/ demonstrated that membranes prepared from cellulose acetate reject salt in "reverse osmosis" or "hyperfiltration" very well. They used homogeneous cellulose acetate membranes which had a thickness of about 6 µm. Due to their very low water permeability (water flux $q \approx 3 \cdot 10^{-5}$ cm/sec at 40 at) these membranes could not be used for the economic production of drinking water from brackish or sea water. An obvious development of much thinner membranes was not pursued because Loeb and Sourirajan /2/ developed so-called modified membranes. Stimulated by the work of Reid and Breton, Loeb and Sourirajan studied the salt rejection of commercially available cellulose acetate membranes produced by the Schleicher & Schuell Co., because these membranes were known to have a much higher water permeability than the membranes used by Reid and Breton. First, Loeb and Sourirajan found that these commercially available membranes do not reject salt. Having in mind the pore model of membranes they tried to diminish the supposed large pores of these membranes by annealing the membranes in hot water at temperatures between 75°C and 95°C. Thereby, they proved that membranes annealed at temperatures between 80°C and 90°C exhibit high salt rejection. It was found that salt rejection incresed with increasing annealing temperature while the water permeability decreased with increasing annealing temperature. Figure 1 shows the

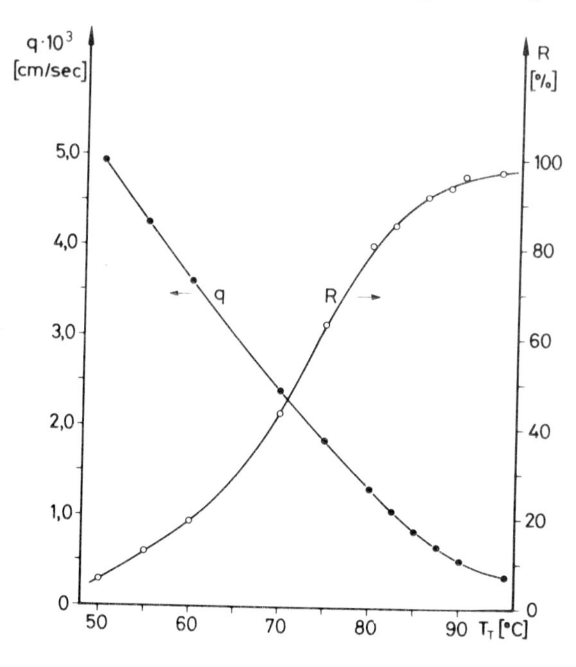

Figure 1
Volume flux, q, and salt rejection, R, of an asymmetric, paper-reinforced flat cellulose acetate membrane (produced by KALLE AG, Wiesbaden, Germany) as a function of annealing temperature T_T at ΔP = 100 at, using NaCl solution of c_s' = 0,1 m NaCl

salt rejection, R, and the volume flux, q, as a function of the annealing temperature at 100 at. The salt rejection is defined as $R = \{(c_s' - c_s'')/c_s'\} \cdot 100$ and is given in %. Here, c_s' is the salt concentration of the brine solution and c_s'' the salt concentration of the product (filtrate). These membranes exhibit the same salt rejection as the membranes used by Reid and Breton, but the water permeability of the annealed membranes was much higher than that of the homogeneous membranes of the same thickness. Loeb and Sourirajan then developed a method of preparation of their so-called "modified" membranes. The method is as follows: From a casting solution of cellulose acetate, magnesium perchlorate, water, and acetone (ratio 22.5:1.1:10:66.4 weight-%) the membranes are casted on a glass plate at $-10^{\circ}C$. After an evaporation period for the solvent of about 3 minutes at the same temperature, the glass plate with the membrane is immersed into an ice water bath. It is kept there for about one hour during which the ice water is renewed several times. Thereafter, the membrane is removed from the plate and heated by immersion in hot water for about 5 minutes at $65^{\circ}C$ to $85^{\circ}C$.

IX.4.2. Structure of Modified Membrane

In hyperfiltration experiments with modified membranes two unexpected experimental results were first observed. First, the water permeability turned out to be independent of the membrane thickness. Second, the membranes were effective in rejecting salt only if the membrane surface which was adjacent to the air during casting was juxtaposed with the brine solution. If the membrane was turned over, so that the membrane surface which was adjacent to the glass plate during the casting procedure was juxtaposed with the brine solution, the salt rejection of the membrane was very low. These experimental findings may be explained in terms of the membrane structure, first investigated by Merten et al. /3/. Under the electron microscope the membranes proved to consist of a fine-pored matrix with a very thin, dense layer of cellulose acetate on the air-dried suface. The dense surface layer was estimated to be about 500-3000 Å thick, the total membrane thickness being 100 µm. The porous substructure was estimated to have a pore size in the order of 0.1 µm. Figure 2 reproduces an electron microphotograph of the two surfaces of an asymmetric cellulose acetate membrane. It may be concluded that the asymmetric membrane consists essentially of two single membranes. The active layer at the air-dried surface is equivalent to a homogeneous membrane of about 500 - 3000 Å thickness and is responsible for the salt rejection. The porous sublayer is a very good natural support for the active layer. Using this membrane model of Merten et al., it is possible to explain the experimental results with regard to the asymmetric salt rejection and the independence of volume flow on membrane thickness. Since the water permeability is determined mainly by the thickness of the active layer, which is nearly independent of the thickness of the whole membrane, the volume flux is nearly independent of the overall thickness of the membrane. Furthermore, if the porous membrane surface is juxtaposed with the brine solution, the salt penetrates into the porous sublayer, where it is rejected by the active layer and accumulates there. A concentration profile is therefore built up within the porous sublayer /4/

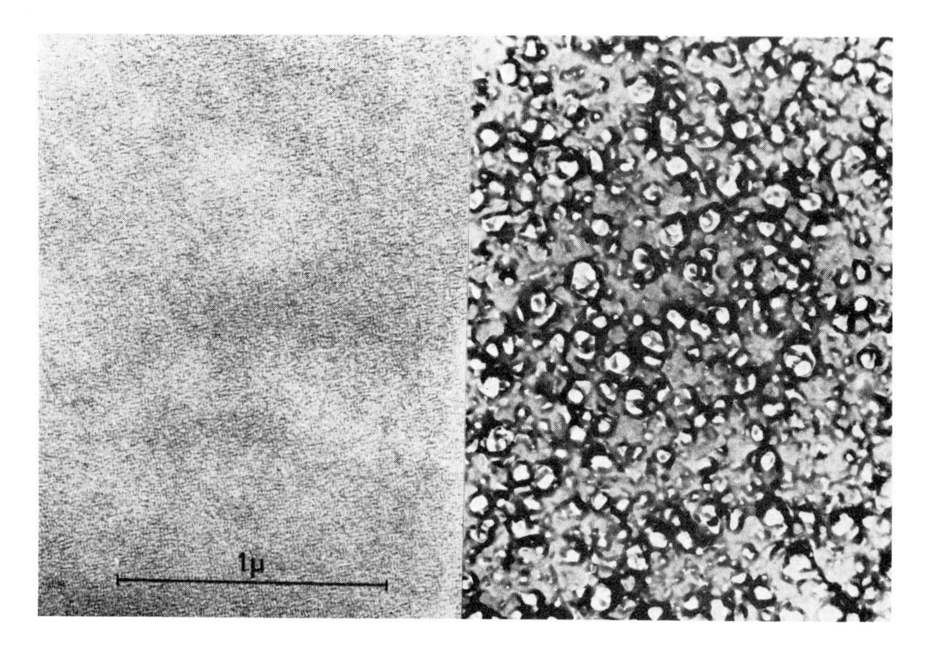

Figure 2: Electron micrograph of a preshadowed carbon replica of the air-dried surface (left) and the bottom surface (right) of an asymmetric cellulose acetate membrane (carbon produced by decomposition of benzene vapour; original magnification 35000 x)

which leads to a higher salt concentration at the active layer. Since this salt concentration is responsible for the salt concentration of the permeate, the salt concentration of the product increases and may become equal to the salt concentration of the brine solution.

IX.4.3. TRANSPORT EQUATIONS

Parallel to the development and improvement of modified cellulose acetate membranes as well as their application in hyperfiltration, there has also been an interest in the description and explanation of salt and water transport through these membranes. For a general description of transport through artificial membranes one may use the phenomenological relations of thermodynamics of irreversible processes. Thereby, one gets linear relations between fluxes, Φ_i, and conjugated general forces, X_i, like $\Phi_i = \sum_k \ell_{ik} \cdot X_k$. Since in this case the membrane is treated as a black box, it is impossible to get any theoretical information about the transport coefficients, ℓ_{ik}. They have to be determined by experiment. The ℓ_{ik} depend on the physico-chemical state of the system, i.e. on temperature, T, pressure, P, and mean salt concentration, $\bar{c}_s$, for instance, in a binary system. If one would like to have any insight into the physico-chemical reasons for water and salt transport, one would need a membrane model. Then it would be possible to get relations between the ℓ_{ik} and membrane parameters like pore radius, water content of the membrane, diffusion coefficients of salt and water within the membrane, distribution coefficient of salt, and so on. That is, the physics of the system is contained in the ℓ_{ik}. The linear relations are very good guidelines for relating fluxes and forces as is done with Ohm's or Fick's law. With these linear relations it is possible to give an overall description of transport through artificial membranes, taking into account the transport parameters of the membrane, ℓ_{ik}. In the description of transport through artificial membranes, especially modified cellulose acetate membranes, the general transport relations and model theories have often been mixed up. In this section, the transport phenomena in asymmetric cellulose acetate membranes will be described by using the linear relations of thermodynamics of irreversible processes.

Given an isothermal system, with a membrane separating two salt solutions of different salt concentrations c_s' and c_s'' (Figure 3) which are kept under different hydrostatic pressures P' and P", one will have a water flux Φ_w {mol/cm^{2}sec} and a salt flux Φ_s {mol/cm^{2}sec} through the membrane. Since these fluxes cannot be determined directly by experiment, it is useful to introduce other fluxes. These are the volume flux q{cm/sec} and the so-called chemical flux χ{cm/sec}. Between the original fluxes, Φ_w and Φ_s, and the new fluxes, q and χ, the following relations exist:

(1) $$q = V_s\Phi_s + V_w\Phi_w$$ (2) $$\chi = \Phi_s/c_s - \Phi_w/c_w$$

where V_w and V_s are the partial molar volumes of water and salt, respectively. As Schlögl /5/ has shown, the following linear relations exist between fluxes and conjugated forces for a system near equilibrium, if no electrical current passes through the membrane:

(3) $$q = \ell_p\Delta P + \ell_{\pi p}\Delta\Pi$$ (4) $$\chi = \ell_{\pi p}\Delta P + \ell_\pi\Delta\Pi$$

In these expressions, the Onsager relation, $\ell_{\pi p} = \ell_{p\pi}$, was used. The chemical flux χ, a measure of the variation of concentration with time within the two phases (') and ("), is defined by:

(5) $$\chi \equiv \frac{1}{A}\frac{V'}{\bar{c}_s}\cdot\frac{dc_s'}{dt} = -\frac{1}{A}\frac{V''}{\bar{c}_s}\cdot\frac{dc_s''}{dt}$$

where V' and V" are the volumes of phase (') and ("), respectively; A is the membrane area {cm^2}; and $\bar{c}_s = \frac{1}{2}(c_s' + c_s'')$ is the mean salt concentration. With the definition of the reflection coefficient, $\sigma \equiv -\ell_{\pi p}/\ell_p$, given by Staverman /6/, Equations (3) and (4) yield:

(6) $$q = \ell_p(\Delta P - \sigma\Delta\Pi)$$ (7) $$\chi = -\sigma\ell_p\Delta P + \ell_\pi\Delta\Pi$$

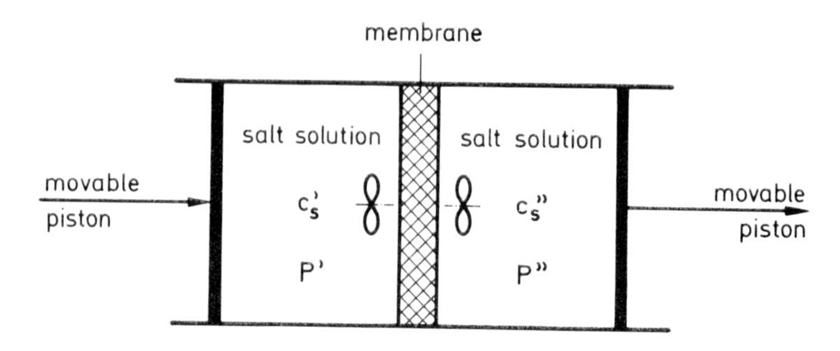

Figure 3: Diagrammatic presentation of an isothermal membrane system

Using these linear relations, the transport coefficients can be determined by measuring the dependence of volume flux, q, and chemical flux, χ, on pressure difference, ΔP, and osmotic difference, $\Delta \Pi$. This was done with homogeneous ion exchange membranes by Schlögl et al. /7/, Schmid et al. /8/, and Spiegler et al. /9/. Among other things, it was found that the transport coefficients depend on the mean salt concentration $\bar{c}_s$ in binary electrolyte solutions. For a long time, it was not clear if the transport across modified cellulose acetate membranes could be treated in the same way because these membranes have an asymmetric structure. Many investigators found contradictory results. On the other hand, it should be possible to describe the transport across asymmetric membranes in the same way as for homogeneous ones since no special membrane model is taken as a basis.

IX.4.4. Experimental Determination of Transport Coefficients

With the cell shown in Figure 4 the transport coefficients of a membrane can be measured. With two conductivity cells (L) immersed in the two solutions, the variation of salt concentration with time is followed by measuring the conductivity using a Wayne Kerr Bridge (accuracy ±0,1%). The volume flux, q, is determined from the increase in volume, ΔV, of the corresponding phase using a calibrated capillary (K, 1/1000 ml), or by determining the volume of solution passing through a small capillary into an Erlenmeyer flask by weighing. To reduce evaporation the Erlenmeyer flask is sealed with a silicon rubber stopper (St) containing a perforation for the capillary. Pressure equalization is attained by a canula through the stopper. The partition wall (MT) contains the membrane (M) which is fixed eccentrically. The membrane is supported by a CONIDUR stainless steel screen (Hein, Lehmann & Co., Düsseldorf, Germany). The screen is 0.35 mm thick and has holes of 0.15 mm diameter. With different partition walls, membranes with an effective area of $\sim 1\ cm^2$, $\sim 4\ cm^2$, and $\sim 16\ cm^2$ can be used. The membrane holder is shown separately in Figure 5. The solutions are well stirred by means of magnetic stirrers. All surfaces of contact are sealed by O-rings. Air is used for pressurizing one phase. The air pressure is maintained constant to $\pm 5 \cdot 10^{-4}$ at by means of KENNDAL

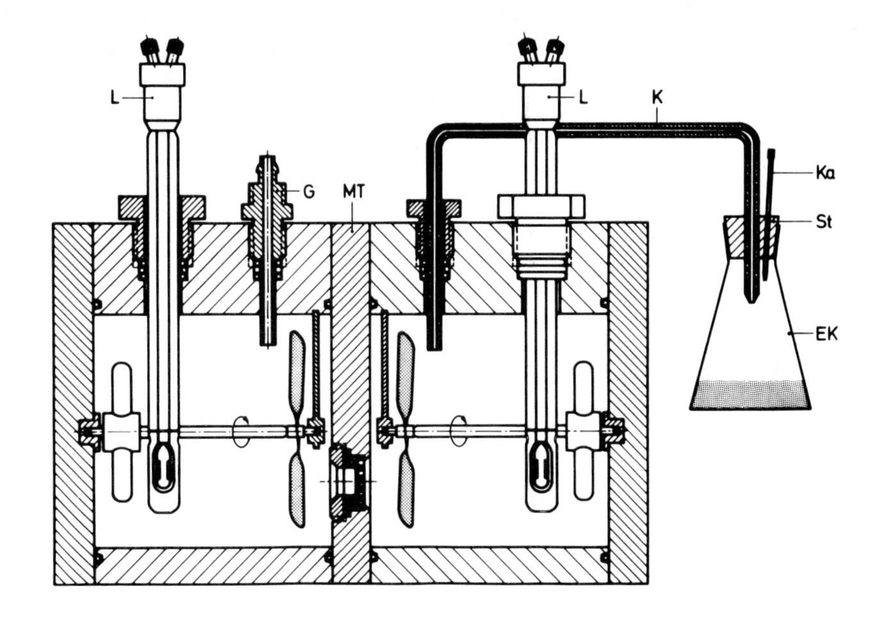

Figure 4: Cross-sectional view of dialysis cell

valves (VIA, Düsseldorf, Germany). The pressure is measured with high accuracy pressure gauges (±0,1%) produced by WIKA (Klingenberg am Main, Germany). - The four transport coefficients - three of which are independent - were measured with the following three different experimental setups.

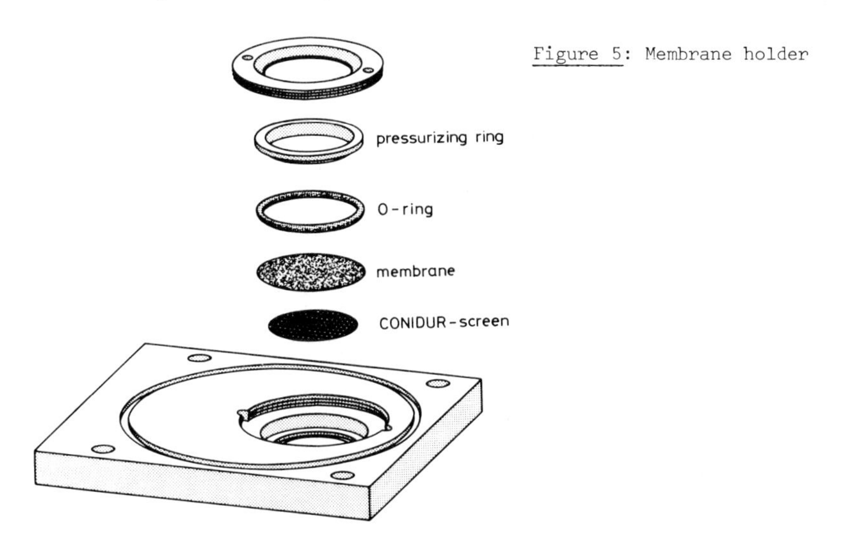

Figure 5: Membrane holder

1. Determination of ℓ_p at $\Delta\Pi = 0$

Both cell compartments - phase (') and (") - contain salt solutions of equal concentration. Measuring the volume flow $Q = A \cdot q$, as a function of ΔP using four different values of ΔP, one gets ℓ_p from the slope of the straight line in a Q-ΔP diagram if the effective membrane area, A, is known.

2. Determination of ℓ_p and σ at $\Delta\Pi$ = constant $\neq 0$

An osmotic difference $\Delta\Pi$ between the two cell compartments is established ($c_s' > c_s''$) and the volume flow Q is measured as a function of ΔP. From the slope of the corresponding straight line in a Q-ΔP diagram, one again gets ℓ_p. In addition, from the intersection of the straight line with the abscissa, one gets $\sigma = (\Delta P/\Delta\Pi)_{q=0}$. Thereby, a mean value $\overline{\Delta\Pi}$ of $\Delta\Pi$ is determined from $\Delta\Pi_a$ at the beginning and $\Delta\Pi_e$ at the end of the measurement (arithmetical mean). The osmotic difference $\Delta\Pi_e$ is obtained by conductivity titration of the two salt solutions after the experiment is finished.

3. Determination of ℓ_π and $\ell_{\pi p}$ at $\Delta P = 0$

At time t = 0 a concentration difference between the two cell compartments is set up ($c_s' > c_s''$). The variation of concentration with time is then followed by measuring the conductivity of both phases at a time interval of about 24 hours. After the measurements are completed (about ten days) the cell constants of the conductivity cells used are determined once more by conductivity titrations of the salt solutions from both compartments. Assuming a linear relation between the salt concentration and conductivity for small concentration differences, one gets the salt concentration from a calibration curve. The volume flow Q is obtained by determining the volume of solution which flowed into an Erlenmeyer flask during about 24 hours. From the curves of concentrations, c_s' and c_s'', and the volume increase, ΔV, with time, a mean value of the chemical flux, χ, and the volume flux, q, as well as a mean value $\overline{\Delta\Pi}$ is estimated by averaging over 24 hours. The mean values are obtained from the measured values for the time interval t_1 to t_2 as follows:

$$(8) \quad \Delta\Pi(t_1) = \Pi'(t_1) - \Pi''(t_1) \qquad (9) \quad \Delta\Pi(t_2) = \Pi'(t_2) - \Pi''(t_2)$$

$$(10) \quad \overline{\Delta\Pi} = \frac{1}{2}\{\Delta\Pi(t_1) + \Delta\Pi(t_2)\}$$

$$(11) \quad \overline{\chi}' = \frac{1}{A}\frac{V'}{\overline{c}_s}\{c_s'(t_2) - c_s'(t_1)\}/\Delta t \qquad (12) \quad -\overline{\chi}'' = \frac{1}{A}\frac{V''}{\overline{c}_s}\{c_s''(t_2) - c_s''(t_1)\}/\Delta t$$

$$(13) \quad \overline{q} = \Delta V/A\cdot\Delta t \qquad (14) \quad \Delta t = t_2 - t_1 \qquad (15) \quad \overline{c}_s = \frac{1}{4}\{c_s'(t_1) + c_s'(t_2) + c_s''(t_1) + c_s''(t_2)\}$$

Plotting $\overline{\chi}$ and $\overline{q}$ as functions of $\overline{\Delta\Pi}$, one gets straight lines. The slopes of the calculated regression lines yield the coefficients ℓ_π and $\ell_{\pi p}$, respectively. If ℓ_p is known from a different measurement, one gets once more $\sigma = -\ell_{\pi p}/\ell_p$.

IX.4.5. Transport Coefficients for Asymmetric Cellulose Acetate Membranes

Using solutions of NaCl, Na_2SO_4, NaF, $CaCl_2$, and saccharose the mechanical permeability ℓ_p was measured as a function of solute concentration ($0 < c_s < 1$ mol/l) at 20°C. Using solutions of NaCl, Na_2SO_4, NaF, and $CaCl_2$ the reflection coefficient σ was also measured. The results are shown graphically in Figures 6, 7, and 8. In addition, ℓ_π and $\ell_{\pi p}$ were determined as a function of concentration ($0 < c_s < 0,5$ mol/l) at 25°C using NaCl and Na_2SO_4 solutions. The corresponding results for ℓ_π are also graphically represented in Figure 9. The curve $\ell_{\pi p}$ as a function of c_s has the same shape.

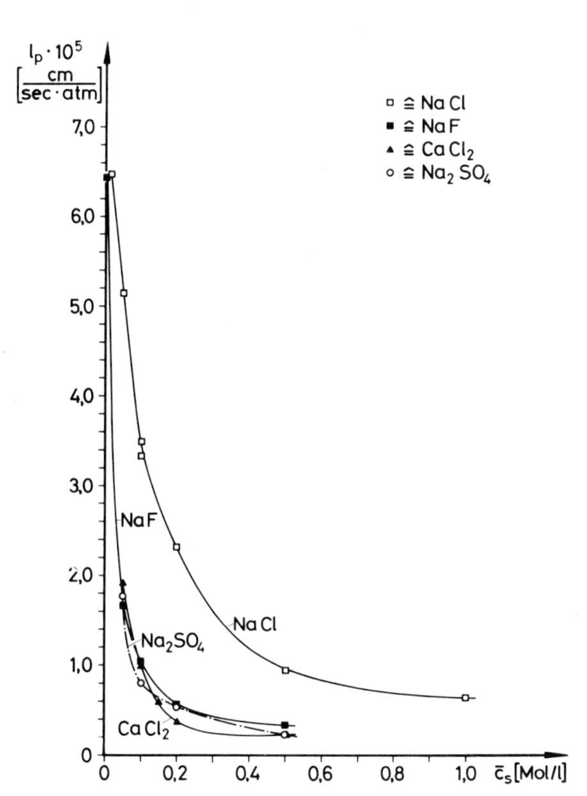

Figure 6: Mechanical permeability ℓ_p as a function of salt concentration $c_s = \bar{c}_s$ at 20°C using different salts

As one can see from these figures, there exists a strong dependence of ℓ_p, ℓ_π, and $\ell_{\pi p}$ on salt concentrations c_s or $\bar{c}_s$, respectively. On the other hand, the reflection coefficient σ varies only slightly with salt concentration. If one calculates σ from ℓ_p and $\ell_{\pi p}$ ($\sigma = -\ell_{\pi p}/\ell_p$) measured by methods (1) and (3), respectively, it is found that these values agree with the σ values measured directly using method (2). If $\sigma = 1$, there exists the following relation between ℓ_p, ℓ_π, and $\ell_{\pi p}$ /10/: $|\ell_p| = |\ell_\pi| = |\ell_{\pi p}|$. As can be seen from Figure 10, that relation is satisfied for Na_2SO_4 solutions since the straight lines $\overline{q}$ as a function of $\overline{\Delta\Pi}$ and $\overline{\chi}$ as a function of $\overline{\Delta\Pi}$ coincide ($\ell_\pi = -\ell_{\pi p}$) and $\sigma = 1$. Therefore, it is also $\ell_\pi = \ell_p$. The experimental results show, above all, that a consistent description of the transport through asymmetric cellulose acetate membranes is possible for dialysis experiments. In the next section, we examine the validity of this statement for hyperfiltration experiments.

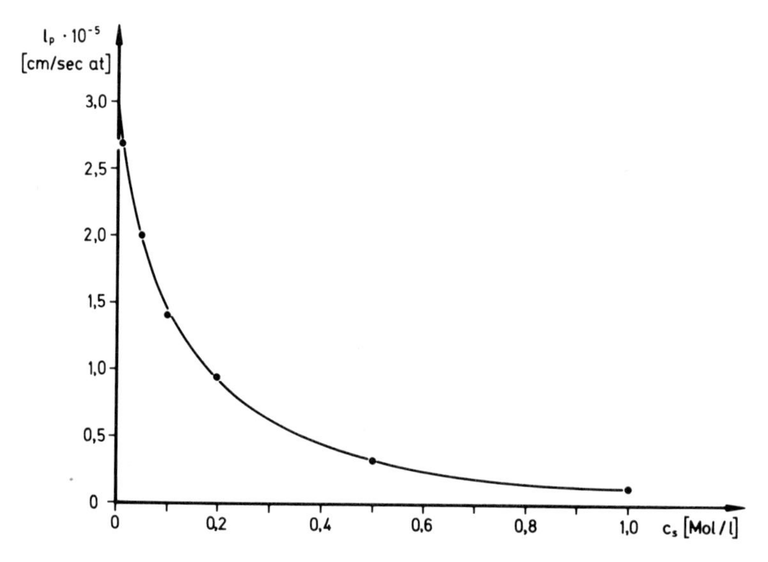

Figure 7: Mechanical permeability ℓ_p as a function of concentration c_s at 25°C using saccharose solutions. With regard to the concentration dependence of ℓ_p in this case, the increase of viscosity of saccharose solutions with concentration has to be taken into account.

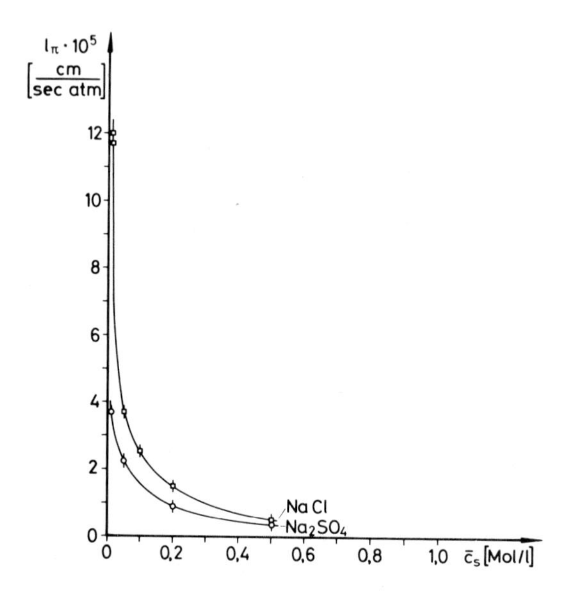

Figure 8: Osmotic coefficient ℓ_π as a function of mean salt concentration $\bar{c}_s$ at 25°C for solutions of NaCl and Na_2SO_4

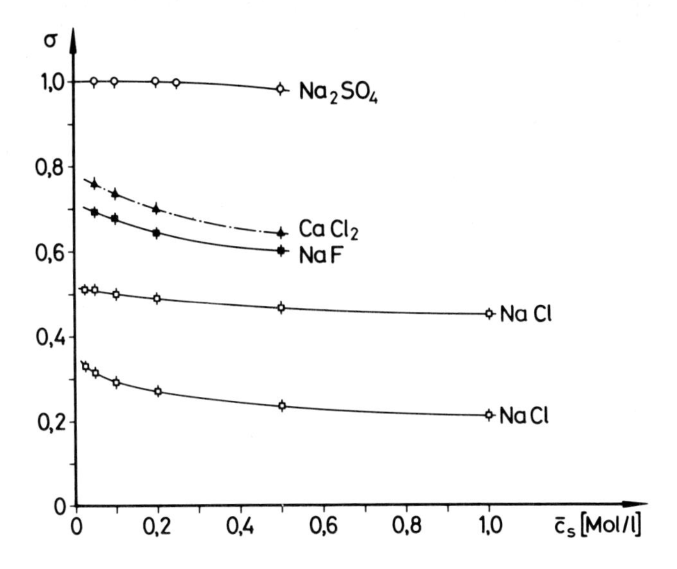

Figure 9: Reflection coefficient σ as a function of salt concentration for different salts at 20°C (NaCl-(I) curve measured with a virgin membrane; NaCl-(II) curve measured with a membrane which was used in hyperfiltration at 50 at

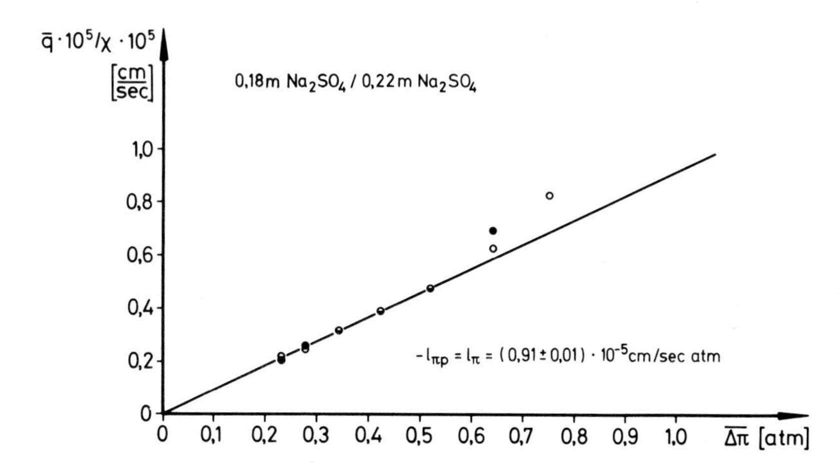

Figure 10: Mean volume flux, $\bar{q}$, and mean chemical flux, $\bar{\chi}$ as a function of $\overline{\Delta\Pi}$ at 25° for a solution with mean concentration of 0.2 m Na_2SO_4

IX.4.6. Relation Between Salt Rejection and Transport Coefficients

In hyperfiltration experiments the volume flux, q, and the salt rejection, r, are determined. Here, r is defined as follows:

$$r = (c_s' - c_s'')/c_s' \qquad (16)$$

Figure 11 shows an experimental setup for hyperfiltration measurements. If one would like to compare hyperfiltration with dialysis experiments, one has to relate r to the transport coefficients. This can be done /11/ if the following boundary condition is used:

$$c_s''/c_w'' = \Phi_s/\Phi_w \qquad (17)$$

Here, c_w'' is the water concentration of the product. This boundary condition states that the composition of the product is determined by the composition of the solution which passes through the membrane. With relation (16), which can be rearranged into $r = 1 - c_s''/c_s'$, one receives the following relation using (17):

$$r = 1 - (c_w''\Phi_s/c_s'\Phi_w) \qquad (18)$$

Assuming $c_w'' \approx c_w'$ and rearranging Equation (18) one gets:

$$r = (c_w'/\Phi_w)(\Phi_w/c_w' - \Phi_s/c_s') \qquad (19)$$

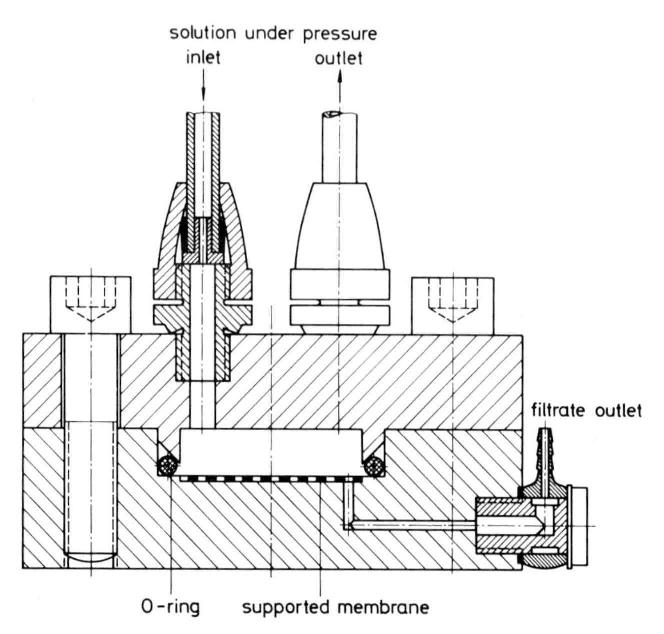

Figure 11: Cross-sectional view of hyperfiltration apparatus

Using Equation (2) and the approximations $c'_w V'_w \approx 1$ as well as $q \approx V_w \Phi_w$ one gets:

$$r \approx -\chi/q \tag{20}$$

If Equations (6) and (7) are substituted in Equation (20) and if the approximation $r \approx \Delta\Pi/\Pi'$ is used, one obtains a quadratic equation for r which can be solved to give:

$$r = (1/2\sigma)\{(\Delta P/\Pi' + \ell_\pi/\ell_p) - \sqrt{(\Delta P/\Pi' + \ell_\pi/\ell_p)^2 - 4\sigma^2(\Delta P/\Pi')}\} \tag{21}$$

Figures 12, 13, and 14 present r as a function of some interesting variables using different parameters such as σ, Π', and ℓ_π/ℓ_p. From Figure 12 one can see that $r \to \sigma$ if $\Delta P \to \infty$. Figure 13 shows the dependence of r on pressure difference ΔP for different salt concentrations (Π'). As can be seen from this figure, r approaches its limiting value much more slowly with increasing salt concentration. This theoretical prediction is in agreement with experimental results which are shown graphically in Figure 15. Another interesting conclusion from Equation (21) is the dependence of r on ℓ_π/ℓ_p if σ, Π', and ΔP are kept constant (Figure 14). As ℓ_π/ℓ_p increases, r decreases. This result is of interest in connection with the different salt rejection of ion exchange membranes and asymmetric cellulose acetate membranes which have nearly the same reflection coefficient. From experiments it is well known /12/ that ion exchange membranes having $\sigma \approx 0.9$ will exhibit a salt rejection of only 0.4 - 0.5 at 100 at whereas asymmetric cellulose acetate membranes with the same reflection coefficient will give about 90% salt rejection (r = 0.9). Equation (21) shows that this difference is due to the different values of ℓ_π/ℓ_p for ion exchange membranes and cellulose acetate membranes. For asymmetric cellulose acetate membranes $\ell_\pi/\ell_p \approx 1$ whereas for ion exchange membranes $5 < \ell_\pi/\ell_p < 20$.

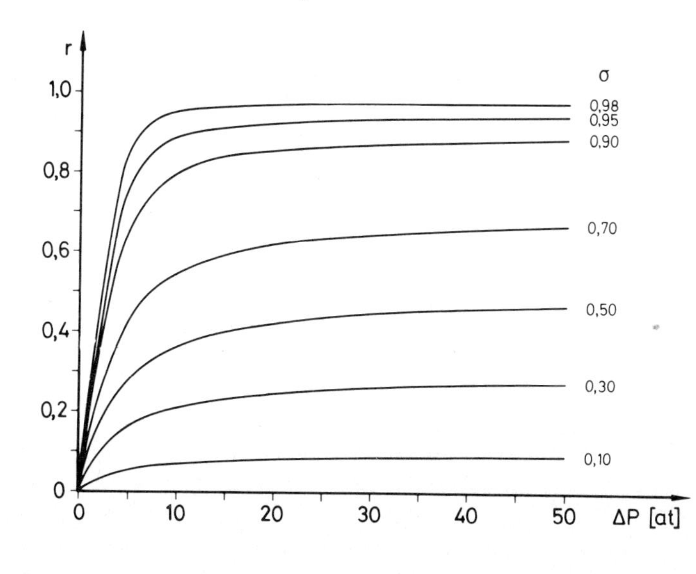

Figure 12: Salt rejection as a function of ΔP using different parameter values of σ (c'_s = 0.1 m NaCl; ℓ_π/ℓ_p = 1; Π' = 4.56 at)

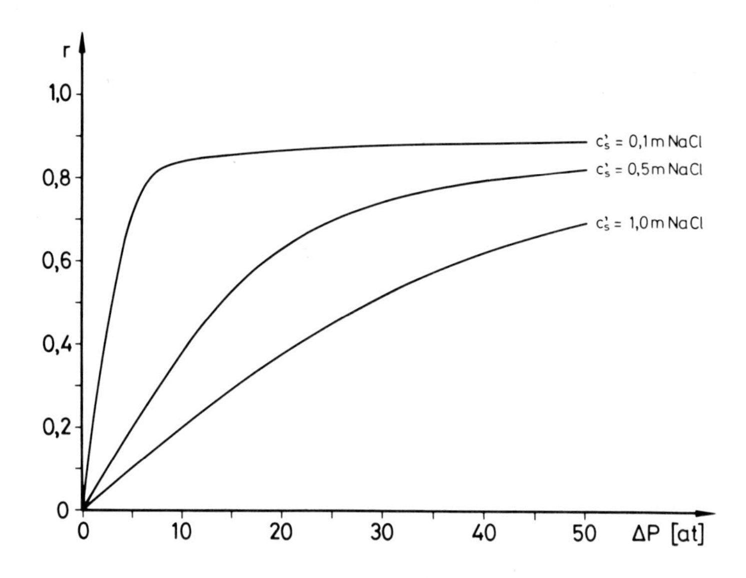

Figure 13: Salt rejection, r, as a function of pressure difference, ΔP, using different salt concentrations c_s' (σ = 0.9; ℓ_π/ℓ_p = 0.95)

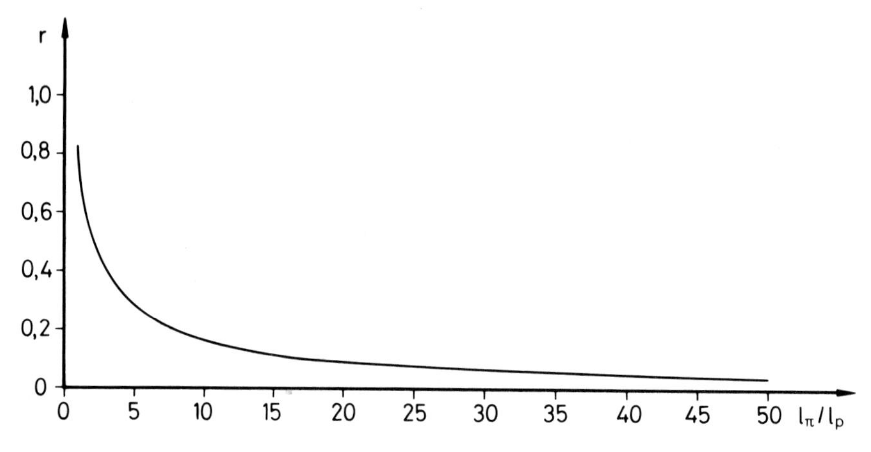

Figure 14: Salt rejection, r, as a function of ℓ_π/ℓ_p (c_s' = 0.1 m NaCl; ΔP = 10 at; σ = 0.9)

One can also obtain /11/ a relation between volume flux, q, and pressure difference, ΔP. Since in hyperfiltration $\Delta\Pi$ is itself a function of ΔP, it is not an independent variable as it would be in dialysis measurements. Using the boundary condition (20), the linear relation (7) for χ, and taking into account that $r \simeq \Delta\Pi/\Pi'$, one arrives at:

$$q = \{(\ell_\pi/\ell_p - \sigma^2)/(\ell_\pi/\ell_p - r\cdot\sigma)\}\cdot\ell_p\cdot\Delta P \tag{22}$$

Figure 16 shows this relation graphically. As can be seen, there exists no linear relation between q and ΔP in hyperfiltration experiments at low pressures as has been

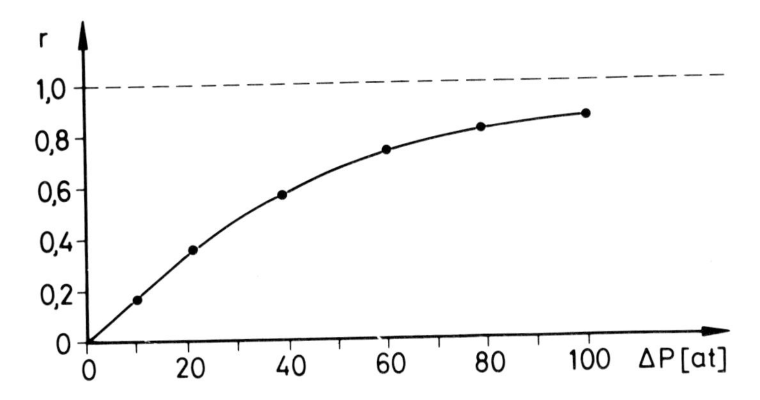

Figure 15: Salt rejection, r, as a function of ΔP using 1 m NaCl solution (asymmetric CA membrane annealed at 82.5°C)

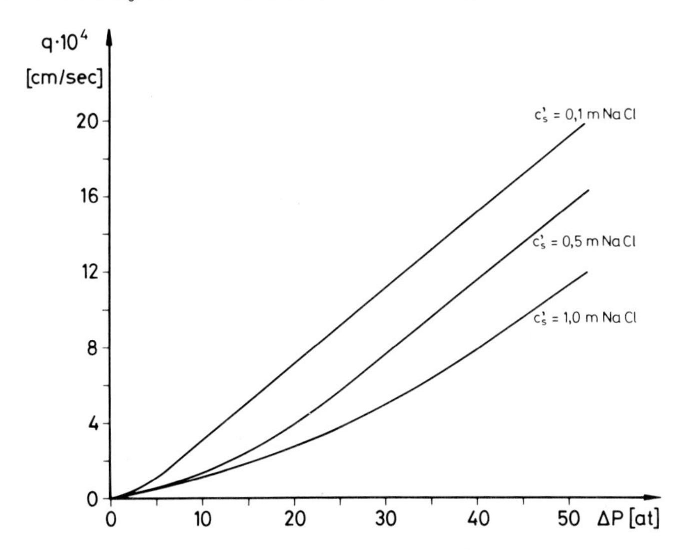

Figure 16: Volume flux, q, as a function of pressure difference, ΔP, in hyperfiltration using different salt concentrations c_s ($\sigma \doteq 0.9$; $\ell_\pi/\ell_p = 0.95$)

frequently assumed in the past. Arriving at the asymptotic value of r, one gets:

$$q = \ell_p \cdot \Delta P \tag{22a}$$

That means, at sufficiently high pressures there is a linear relation between q and ΔP. From the slope of the corresponding asymptotic line one obtains the mechanical permeability, ℓ_p. On the other hand, one can again obtain the reflection coefficient, σ, from the intersection of that asymptotic line with the abscissa. Using $r \simeq \Delta\Pi/\Pi'$ and Equation (6), one gets:

$$q = \ell_p(\Delta P - r \cdot \sigma \cdot \Pi') \tag{23}$$

For the asymptotic case, that yields /13/:

$$q = \ell_p(\Delta P - \sigma^2 \Pi') \tag{23a}$$

From the intersection of the asymptotic straight line of a plot of q against ΔP one thus arrives at the following relation:

$$\sigma^2 = (\Delta P/\Pi')_{q=0} \tag{24}$$

Measuring ℓ_p in hyperfiltration experiments by the use of Equation (22a), one finds that there is nearly no dependence of ℓ_p on salt concentration. At first glance, that seems to contradict the experimental results obtained with dialysis experiments since there one observes a very strong dependence of ℓ_p on solute concentration. The reason for this difference can be found in the different boundary conditions. Whereas, in hyperfiltration the boundary condition of free flow exists at the low pressure side, in dialysis experiments the salt concentration on both sides of the membrane is maintained constant. Therefore, one should check the influence of different boundary conditions in dialysis measurements on the value of the mechanical permeability, ℓ_p. To investigate that, measurements with the following boundary conditions have been made.

1. The salt concentration c_s' was varied between 0 and 0.2 m NaCl keeping $c_s'' = 0$ (pure water in the corresponding cell compartment). Thereby, the active layer of the membrane was juxtaposed with the salt solution. The volume flow was directed from phase (') to phase (").

2. The same arrangement as in case 1 but the membrane was turned over so that the active layer was juxtaposed with pure water. Thereby, the volume flow was directed from phase (") to phase (').

3. The salt concentration c_s'' was varied between 0 and 0.2 m NaCl keeping $c_s' = 0.2$ m NaCl. The active layer was again juxtaposed with phase ('), the solution of constant salt concentration. The volume flow was directed from phase (') to phase (").

The experimental results of these measurements are represented graphically in Figure 17. As can be seen from this figure, the experimentally determined values of ℓ_p depend strongly on the salt concentration of that solution which is adjacent to the porous surface of the asymmetric cellulose acetate membrane.

The experimental results demonstrate that the transport coefficients of an asymmetric cellulose acetate membrane depend strongly on the concentrations of both solutions, c_s' and c_s'', and on the direction of the volume flow. This statement agrees with conclusions drawn by Jagur-Grodzinski and Kedem /14/ treating theoretically a double-layer membrane. Therefore, the transport coefficients do not depend on the mean salt concentration $\bar{c}_s$ as is found with homogeneous membranes. This is a consequence of the asymmetric structure of the modified cellulose acetate membrane. With a volume flow through the membrane, a concentration profile within the porous sublayer is developed by the interaction of salt rejection at the active layer, volume flow, and diffusion within the porous sublayer. Therefore, a concentration profile develops as is shown in Figure 18. Thereby, an osmotic difference $\Delta\Pi^m$ across the active layer is developed. This osmotic difference is responsible for the fluxes through the active layer and thereby, for the fluxes through the entire membrane if one assumes that the reflection coefficient of the porous sublayer is nearly zero.

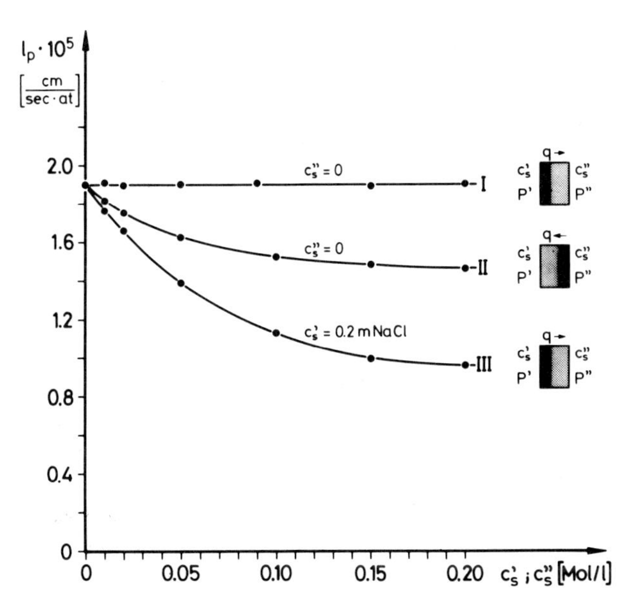

Figure 17: Volume flux, q, as a function of concentration c_s' or c_s'', respectively. Thereby, different boundary conditions were choosen with curves I, II, and III.

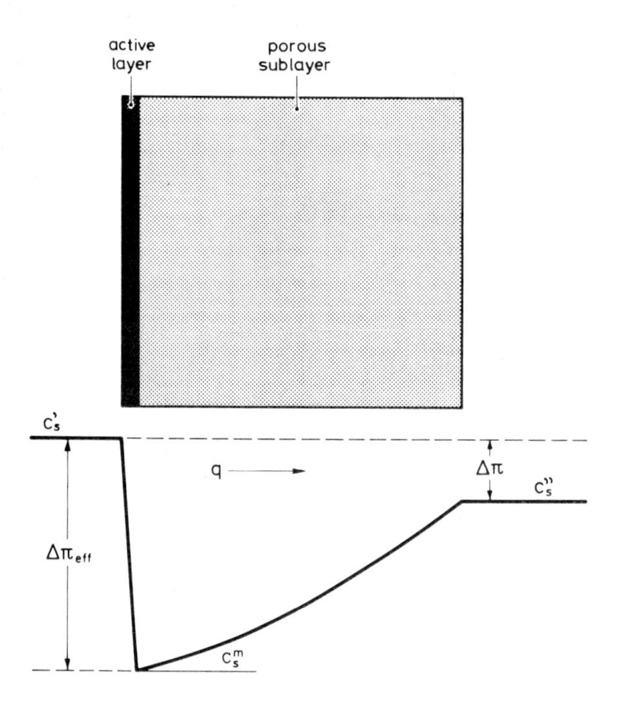

Figure 18: Concentration profile within the porous sublayer of an asymmetric cellulose acetate membrane.

IX.4.7. Theoretical Treatment of the Influence of the Porous Sublayer on the Mechanical Permeability in Dialysis Experiments

Consider a membrane which is composed of two single membranes (Figure 18). One of these is assumed to be a homogeneous cellulose acetate membrane (active layer) of a thickness of about 2000 Å. The other single membrane is assumed to be porous (sublayer) with pore radii $r > 100$ Å and a thickness of $\delta \simeq 100$ µm. This double layer model was proposed by Merten et al. /15/ as a consequence of electron microscopic investigations. With this model it should be possible to estimate the influence of the concentration profile on the transport properties of the membrane by solving the corresponding differential equation for the mass transport in the porous sublayer. Now, the following assumptions are made:

a) The transport properties of the active layer can be described using the linear relations of thermodynamics of irreversible processes. The following relations are used.

$$q = \ell_p'(\Delta P^m - \sigma\Delta\Pi^m) \tag{25}$$

$$\chi = -\sigma\ell_p'\Delta P^m + \ell_\pi'\Delta\Pi^m \tag{26}$$

where ℓ_p' = mechanical permeability of the active layer (cm/sec·at)
ℓ_π' = osmotic coefficient of the active layer (cm/sec·at)
ΔP^m = pressure difference across active layer (at)

$\Delta\Pi^m = \Pi' - \Pi^m$ = osmotic difference across active layer (at)

b) The reflection coefficient of the porous sublayer is assumed to be zero ($\sigma = 0$). Therefore, the following linear relation between volume flux, q, and pressure difference, $\Delta P''$, exists:

$$q = \ell_p''\Delta P'' \tag{27}$$

where ℓ_p'' = mechanical permeability of porous sublayer (cm/sec·at)
$\Delta P'' = P^m - P''$ = pressure difference across porous sublayer (at)

The chemical flux, χ, through the porous sublayer can be determined by the salt and water flux using Equations (1) and (2).

c) The main influence of the porous sublayer is on the salt flux and thereby, on the concentration profile within the supporting layer. Due to this concentration gradient the effective osmotic difference $\Delta\Pi^m$ which determines the transport behaviour of the active layer deviates from the osmotic difference $\Delta\Pi = \Pi' - \Pi''$ across the entire membrane. With the simplified Nernst-Planck Equation the following relation for the salt flux is obtained:

$$\Phi_s = -D_s(dc_s/dx) + c_s q \tag{28}$$

where Φ_s = salt flux across the entire membrane (mol/cm^{2}sec)
c_s = local salt concentration in the porous sublayer of the membrane (mol/cm^3)
D_s = diffusion coefficient of salt within porous sublayer (cm^2/sec)
x = coordinate perpendicular to membrane surface (cm)

In the stationary state the salt flux, Φ_s, is constant through the entire membrane. Therefore, one can integrate the differential equation with the following boundary conditions: $c_s = c_s''$ at $x = \delta$ and $c_s = c_s^m$ at $x = 0$ (δ = thickness of porous sublayer). Here, the origin ($x = 0$) is choosen to be at the interface active layer/porous sublayer. With these boundary conditions, one arrives at the following relation:

$$c_s^m q - c_s'' q\exp(-q\delta/D_s) = \Phi_s\{1 - \exp(-q\delta/D_s)\} \tag{29}$$

Now, if $q\delta/D_s < 1$ one can use a series expansion of the exponential function, terminating after one term, obtaining:

$$c_s^m q - c_s'' q(1 - q\delta/D_s) = \Phi_s q\delta/D_s \tag{30}$$

After some rarrangment as well as adding and subtracting c_s' one arrives at:

$$\Delta c_s^m = \Delta c_s + c_s'' q\delta/D_s - \Phi_s\delta/D_s \tag{31}$$

where $\Delta c_s^m = c_s' - c_s^m$ and $\Delta c_s = c_s' - c_s''$

If one multiplies this relation by $RT \cdot f_o$, taking a mean value for the osmotic coefficient f_o, the following relation is obtained:

(32) $$\Delta\Pi^m = \Delta\Pi + \Pi'' q\delta/D_s - RTf_o\Phi_s\delta/D_s$$

As can be seen from Equation (32) the decisive osmotic difference $\Delta\Pi^m$ depends on the volume flux, q. This relation shows, furthermore, that the measured transport coefficients depend on the concentrations of the external phases and, thereby, on $\Delta\Pi$.

If one combines Equations (25), (26), and (27), after a short calculation one arrives at:

(33) $$q = \ell_p(\Delta P - \sigma\Delta\Pi^m)$$

(34) $$\chi = -\sigma\ell_p\Delta P + \ell_\pi\Delta\Pi^m$$

where $\ell_p = \ell_p' \cdot \ell_p''/(\ell_p' + \ell_p'')$ and $\ell_\pi = \ell_\pi' + (\ell_{p\pi}')^2/(\ell_p' + \ell_p'')$

If one substitutes, now, $\Delta\Pi^m$ in Equation (33) and solves for q, one arrives at:

(35) $$q(1 + \sigma\ell_p\Pi''\delta/D_s) = \ell_p(\Delta P - \sigma\Delta\Pi) + \sigma\ell_p\Pi_o(\delta/D_s)\Phi_s$$

For small salt fluxes the second term in Equation (35) is of the order of $10^{-3}\sigma\ell_p$ and may, therefore, be neglected. For simplicity only this case will be discussed here, since the same approximate result is obtained even with high salt fluxes. As can be seen then, one gets a relation between q and ΔP and $\Delta\Pi$ which gives a linear dependence of q on ΔP but the slope of the corresponding straight lines depend on the salt concentration c_s'' (Π''). Therefore, the measured mechanical permeability depends on the salt concentration c_s''. From Equation (35) the following relation between measured permeability, $\bar{\ell}_p$, and intrinsic permeability can be made:

(36) $$\bar{\ell}_p = \ell_p/(1 + \sigma\ell_p\Pi''\delta/D_s)$$

If the term $\sigma\ell_p\Pi''\delta/D_s << 1$, one can use an series expansion for $\bar{\ell}_p$:

(36a) $$\bar{\ell}_p \simeq \ell_p(1 - \sigma\ell_p\Pi''\delta/D_s)$$

As one can see, a linear relation between the measured mechanical permeability, $\bar{\ell}_p$, and Π'' exists in these cases if one assumes a constant value for σ and D_s. Now, if σ is very small, the above mentioned inequality is fulfilled. Therefore, the linear relation should hold with membranes of very small reflection coefficients. These are membranes of low annealing temperatues, $50^\circ C$ - $70^\circ C$. In Figure 19 the measured values, $\overline{\ell}_p$, are plotted as a function of $\Pi'' = \Pi$. As one can see, there is very good agreement between experimental results and the theoretical relation (36a) for membranes with reflection coefficients up to about 0.01 ($70^\circ C$ annealing temperature). Thus, the theoretical analysis confirms that the concentration gradient in the porous sublayer is responsible for the strong concentration dependence of the mechanical permeability. It should be pointed out that the relation (36a) was obtained with special boundary conditions and a volume flow directed from phase (') to phase ("). If the boundary conditions and the direction of the volume flow are changed, one will obtain somewhat different results but the main result will be the same.

The author is indebted to Professor R. Schlögl for his interest in this work and many valuable discussions as well as to Dr. H. Lonsdale for reading the proofs. Furthermore, the author would like to thank Mrs. R. Lachmann and Mrs. U. Schaffner for carrying out the time-consuming and difficult measurements and Mr. R. Gröpl for his help in designing and producing the dialysis cell. The work was supported by the "Bundesminister für Forschung und Technologie", Bonn, BRD.

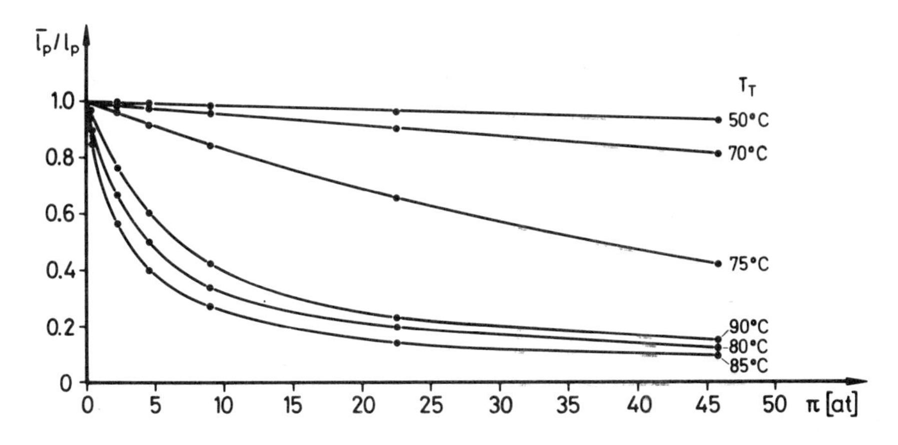

Figure 19: Measured mechanical permeability, $\overline{\ell}_p/\ell_p$, as a function of $\Pi = \Pi''$

REFERENCES

1 C. E. Reid and E. J. Breton, J. Appl. Polymer Sci. 1 (1959) 133

2 S. Loeb and S. Sourirajan, Advan. Chem. Ser. 38 (1962) 117

3 R. L. Riley, U. Merten and J. O. Gardner, Desalination 1 (1966) 30

4 R. Gröpl and W. Pusch, Desalination 8 (1970) 277

5 R. Schlögl, *Stofftransport durch Membranen*, Dr.-Dietrich-Steinkopff-Verlag, Darmstadt, 1964

6 A. J. Staverman, Rec. Trav. chim. Pays-Bas 70 (1951) 344

7 R. Schlögl, Z. phys. Chem. N.F. 1 (1954) 305; 3 (1955) 73

8 G. Schmid, Z. Elektrochem. 54 (1950) 424; 55 (1951) 229; 56 (1952) 181

9 K. S. Spiegler, J. electrochem Soc. 100 (1953) 303 C; also in: Ion Exchange Technology, F. C. Nachod, J. Schubert, New York 1956, 118

10 N. Lakshminarayanaiah, *Transport Phenomena in Membranes*, Academic Press, New York 1969, page 314

11 W. Pusch, unpublished

12 W. Pusch and D. Woermann, Ber. Bunsenges. physik. Chem. 74 (1970) 444

13 M. Minning and K. S. Spiegler, unpublished

14 J. Jagur-Grodzinski and O. Kedem, Desalination 1 (1966) 327

15 H. K. Lonsdale, *Desalination by Reverse Osmosis*, U. Merten, ed., M.I.T. Press, Cambridge, Mass., 1966

16 H. K. Lonsdale, *Reverse Osmosis Membrane Research*, H. K. Lonsdale and H. E. Podall, ed., Plenum Publishing Corporation, New York, 1971

HYPERFILTRATION AND ULTRAFILTRATION WITH CELLULOSE ACETATE MEMBRANES

IX.5.

E. Staude

Technische Entwicklung, Kalle AG, 6202 Wiesbaden-Biebrich, Germany

ABSTRACT

Hyperfiltration, commonly called reverse osmosis, and ultrafiltration are new techniques of separation. These pressure driven processes work with membranes, made of cellulose acetate or polyamide, as barriers which are produced by phase inversion using different solutions. The separation process of hyperfiltration is applied to produce fresh water from sea water or brackish water as well as the reclamation of waste waters. Ultrafiltration, employing a different transport process through the membrane as in hyperfiltration, is applied to concentration and separation of macromolecules. The availibility of hyperfiltration and ultrafiltration is limited by various restrictions imposed by the membrane material itself as well as by process peculiarities.

IX.5.1. INTRODUCTION

Processes for separation of gases by passage through membranes were already of scientific interest towards the middle of the last century. Similarly, the possibilities of desalination of salt solutions based on membranes were known and examined fifty years ago. The term reverse osmosis originates from this work, yet until the last decade reverse osmosis did not become a practical process. On the one hand new sources such as brackish water or seawater were required for water supplies, and on the other hand environmental protection made it necessary to introduce new processes to clean up sewage.

The basic advantage of the separation processes with membranes is that they allow separation of dissolved materials from one another or from a solvent with no phase change, in contrast with evaporation or crystallization processes. Thus the total operating cost for the membrane process is the cost of the pumping energy necessary to produce an operating pressure in excess of the osmotic pressure of the system.

Naturally the economy of any process is not only a matter of the costs of energy, in this case the pump energy, but also the costs of installation which are determined by the technical application. Therefore efficient and economical desalination membranes must be extremely thin so that high filtration rates can be obtained in order to reduce the required membrane area. On the other hand the membranes must be sufficiently thick so that they will be mechanically stable at the operating pressure. These two contradictory requirements may be satisfied by the utilization of the asymmetric or skinned membranes first developed by Loeb and Sourirajan /1/. They employed cellulose acetate as the membrane material.

IX.5.2. MEMBRANE FORMATION PROCESS

Membranes may be considered as a special type of film. To prepare membranes, therefore, one might hope to adapt the traditional methods of film formation: extrusion, calendering, or solvent casting. However, these procedures do not allow or only in a limited degree the preparation of asymmetric membranes. By using a casting solution, homogeneous films are generally formed with a thickness not less than 1 µm. These films exhibit a very low permeability unfit for economic purposes. Assuming a linear relationship between volume flux and reciprocal film thickness, it is possible to calculate from the permeability of homogeneous films the thickness required for economic separation processes. The thickness calculated in this way is usually less than 5000 Å. There are two different ways to produce such thin films. First, following the method of Loeb and Sourirajan asymmetric membranes can be formed by using a solvent system which consists of a good solvent and a poor solvent or a nonsolvent for the polymer. In the case of cellulose acetate (CA) as the film forming material, appropriate casting solutions may have the following composition, in which acetone (Ac) acts as the good solvent. CA:dimethyl formamide:Ac (ratio 14,3:21,4:64.3 wt-%); CA:dimethyl sulfoxide:Ac (ratio 25.0:37,5:37,5 wt-%); or CA:acetic acid:Ac (ratio 22,5:9,1:68,4 wt-%). But there are also quaternary or binary systems used as casting solutions, the former often containing electrolytes. An optimum composition is as follows: CA:formamide:Ac (ratio 25:30:45 wt-%) /2/.

This so-called modified Loeb-Sourirajan solution is cast onto a solid substrate through an appropriate shaped die to a wet film thickness between 100 and 200 µm. The solvent is then allowed to evaporate from the surface of the cast film at ambient temperature for a controlled period of time - 30 seconds maximum. After that time the membrane is immersed in water.

This phase-inversion method of formation of porous membranes is the result of a series of steps. After casting the solution on an

Fig. 1 Phase diagramm for CA-formamide-Ac at various temperatures.

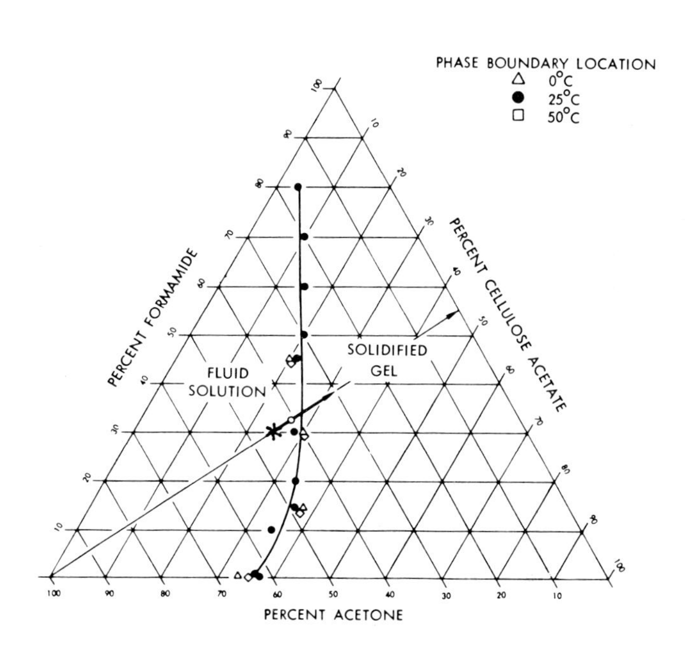

appropriate support the acetone evaporation begins. During this evaporation a change in composition of the ternary cellulose acetate - formamide - acetone system is induced which has an important influence on the structure of the final membrane. Figure 1 shows the phase diagram of the ternary system at various temperatures /3/.

The decrease in content of the volatile solvent causes the separation of the initially homogeneous solution into two interspersed phases. This so-called coacervate can be considered as consisting of droplets of one liquid phase embedded in a matrix of the other. It is assumed that these droplets represents the incipient voids or vacuoles within the finished membrane. It is evident that they must be gelled and the gel stabilized before complete evaporation leads to their disappearance.

Frommer et al. /4/ have shown that the salt-selective portion of the membrane is only part of the upper layer - the upper - most portion. This permselective skin can be removed by slight abrasion of the membrane surface. When the wet film consisting of the upper layer of solidified gel ahd the supporting lower layer of fluid solution is immersed in water the structure of the membrane is fixed. The solvents are exchanged with the nonsolvent water and the polymer is precipitated. This leaching causes the porous substructure; the two lower layers are the support for the uppermost dense skin responsible for desalination.

The membranes immersed in water have a negligible desalination efficiency and high troughputs, however. These primary gel membranes may be utilized as such particularly for low-pressure applications, e.g. the ultrafiltration processes. To achieve a desalting membrane it must be subject to a physical alteration of the primary gel structure to affect decreases in porosity and water content. The most important means to this end is thermal annealing in a water bath. Annealing a porous membrane results in a diminishing of the permeability and an increase in permselectivity.

In figure 2 may be seen the volume flow density q and the salt rejection R as a function of annealing temperature using 0,085 m sodium chloride solution at an operating pressure of 40 atmos. The brine for the curves at 5 atmos. was municipal water of 30° hardness. It is possible to obtain membranes with different desalination efficiency and volume flow density, as may be seen in the figure.

The other method for making extremely thin permeable membranes is the concept of Riley et al. /5/ for producing composite membranes. Whereas the structure of the membrane is similar to the above mentioned membrane, namely a thin semipermeable layer supported by a porous support, the preparation is quite different.

Fig. 2 Volume flow density q and salt rejection R as function of annealing temperature. (Kalle CA membranes, reinforced; solutions see text).

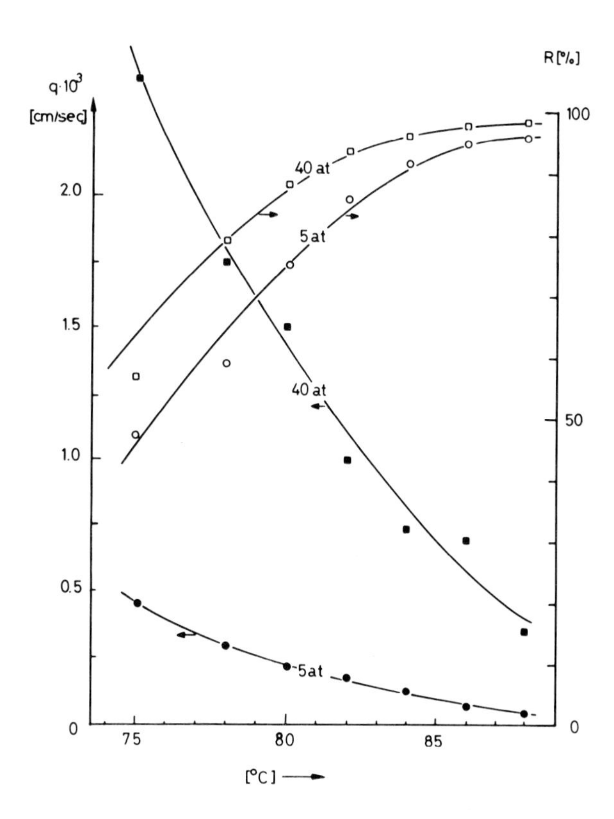

A homogeneous thin film is formed from a 0,5 % solution of cellulose triacetate in chloroform by dipping an appropriate porous support into this solution and removing the solvent by an air stream. The membrane thickness can be as low as 350 Å. This membrane formation can be done continuously also.

IX.5.3. TECHNICAL MEMBRANES AND MODULES

The technical production of membranes on a commercial scale is similar to casting procedures used for most solvent-cast films, such as continuous belt or drum casting. The difference is in the membrane post-treatment. Current production techniques yield membranes in flat sheet or tubular form, the latter with the active dense layer either on the inside or the outside the tube, and also in the form of hollow fibers.

The membranes thus formed are mounted either in plate-and-frame permeators or in tubes which combine two functions in one, in that the surface of the tube is used as a support for the membrane and the tube wall serves as a pressure vessel. These units, combining membranes and pressure supporting equipment, are termed modules. There is considerable engineering interest in constructing modules with high packing density to reduce apparatus volume. This is best achieved, at present, by the concept of spiral wound modules and hollow fibers. Besides the modules a reverse osmosis plant requires pumps, storage vessels, and measuring and control devices such as those used for pH adjustment or conductivity measurement.

IX.5.4. HYPERFILTRATION OF SALT WATER AND WASTE WATER

CA membranes used for hyperfiltration must be annealed after immersion in water. The transport through the membrane allows diffusive passage of water or hydrogen-bonding low-molecular weight substances but hinders the passage of salts. The driving force for this process is a pressure gradient across the membrane. Pusch gives an extensive description of the transport phenomena in the preceding chapter.

The use of membrane separation processes has created great interest for the production of fresh water from the sea. But there is an important point to note here. A minimum salt rejection of 98,6 % is required to convert seawater with a content of 35000 ppm TDS to potable product water (500 ppm TDS) in one passage or cycle. To obtain a significant fractional water recovery from the feed, 50 % for example, an average salt rejection of no less than 99 % is needed. But this may be too much to ask of membranes fabricated on a production scale. Only the composite membranes and the new polyamide-hydrazide membranes of duPont /6/ are able to desalt seawater in one stage.

Generally the main application of hyperfiltration is in industrial waste water purification and reclamation of municipal water. The principal municipal applications for this separation process, with the present state of membrane technology, could be in the desalting of brackish waters with total dissolved solutes between 1000 and 15000 ppm. Fortunately the pressure requirements for demineralization of brackish water are not as high as for seawater desalination, so pressures of 40 atmos. guarantee an economical volume flow density at the desired salt rejection, and the flux decline or "compaction" effect is not as serious as in the case of seawater desalination.

Today there are over 1000 communities in the U.S. and Canada consuming brackish waters which are several times in excess of recommended health standards. In Germany there are only a few communities with the same problem. In the next few decades the reclamation of municipal water by hyperfiltration seems to be a good possibility, especially here in Germany. A reduction of water hardness can be affected even at the low pressures existing in municipal water pipes if the TDS is lower than 1000 ppm. Volume flow density and salt rejection at 5 atmos. are shown in figure 2. The ever increasing pollution loads on streams, lakes, and coastal waters, which has led to growing public concern, may be effectively treated by hyperfiltration. Though industry is making continuous efforts it is difficult to make rapid advances because of the diversity of the wastes produced in different manufacturing plants. The hyperfiltration process may be advantageously used on individual waste water streams, since the quality and quantity of wastes are fairly well defined. But a characteristic of the membrane process is that, as with other separation processes, a concentrate is produced, and while the concentrate may be of very small bulk in a distillation process, in hyperfiltration it is not likely to be less than 10 % of the feed volume and this concentrate has to be disposed of.

Hyperfiltration is thus used for treating effluents produced by separate industrial plants. Some problems which result should, however, be pointed out. There are limits to the concentration of solutions of inorganic salts, e.g. electroplating effluents, that can be achieved because of the increasing osmotic pressure which must be overcome. In the case of solutions containing both inorganic and organic substances it must be noted that low molecular weight organic substances will readily diffuse through the membrane. Figure 3 demonstrates this process for an effluent containing monoethylene glycol and salts. Whereas the retention is independent of brine concentration with different desalting membranes (number 1, 2 and 3 in the figure) for the inorganic constituents (a in the figure), the retention for organic substances (b) is very much less than that for the salts and, moreover, it becomes even less as the concentration of the brine is increased during the separation process. These differences in retention could possibly be utilised for selective separation. The effect can be reinforced taking effects of temperature into consideration: at higher temperatures the retention of inorganic substances remains constant while that for organic substances decreases, although the working life of the membranes is shortened.

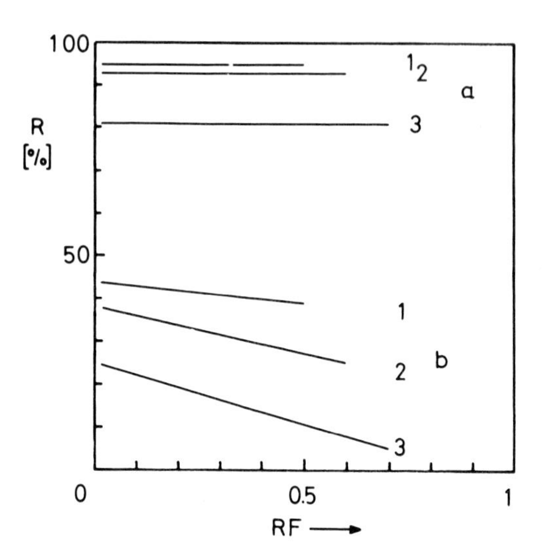

Fig. 3 Retention R as function of brine concentration RF, using Kalle CA membranes, reinforced. (Operating pressure 40 atm; further explanations see text).

A different effluent system demonstrates yet another effect of organic substances on membranes, as shown in figure 4.

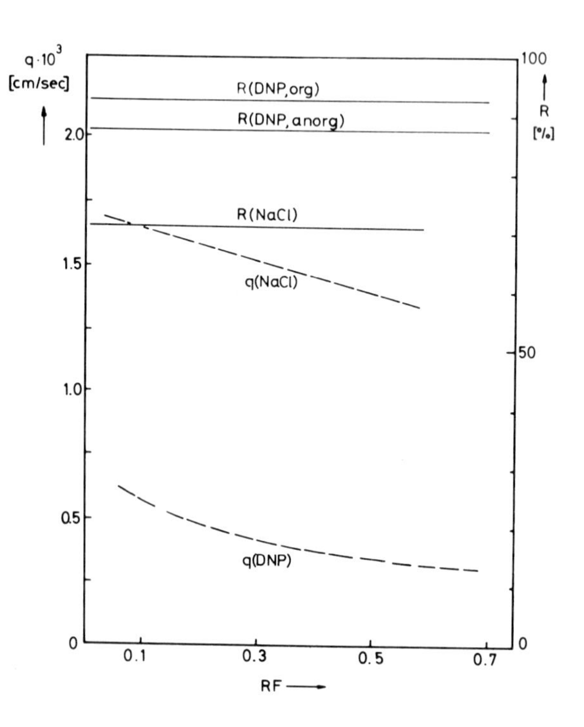

Fig. 4 Volume flow density q and retention R as function of brine concentration for pure sodium chloride (0,085 m) and waste water, containing substituted dinitrophenole (DNP). (Kalle CA membranes, reinforced; operating pressure 40 atm.).

The considerable flow reduction which is apparent on concentrating a solution of a substituted dinitrophenol and inorganic salts, as compared with the concentration of a pure sodium chloride solution, is due to the solubility of organic species in cellulose acetate.

In the interest of environmental protection it is important to reduce oxygen requirements of effluents. For waste waters produced by factories which handle natural products, which have a great BOD, it is sometimes possible to separate a large proportion of the organic constituents by using hyperfiltration as tests on the effluent produced by a fermentation works have proved. At an initial loading of 78 g O_2/litre, a retention of 98 % was achieved by using suitable membranes. The effluent containing 1,3 g O_2/litre does, however, have to be biologically purified prior to being passed into a main drainage channel.

IX.5.5. ULTRAFILTRATION

In contrast with hyperfiltration, after immersion in water the CA membranes are ready for ultrafiltration. A pressure gradient across the membrane is the driving force in ultrafiltration as in hyperfiltration, but the type of transport is quite different. In ultrafiltration the passage of large molecules is impeded by pores in the membranes, but these pores allow the passage of solvents and salts. The ultrafiltration membranes can be operated at low pressures of 3 to 10 atmos.

The dairy industry is making concerted efforts to use ultrafiltration for obtaining purified proteins and lactose from whey. This technique is also being studied for the athermal concentration of milk. These efforts are not only appropriate, but essential, in view of the fact that the highest proportion of lactose, most vitamins, and 20 % of milk proteins, that is, 70 % of the nutritive value of milk contained in whey, goes largely to waste in rivers, streams, and lakes.

Generally speaking, ultrafiltration is the ideal method of concentrating and purifying proteins, since these very delicate substances are not harmed by this process.

Another almost ideal use for ultrafiltration is in electrophoretic paint deposition in which wash water is separated from the paint bath, into which it is returned after washing has been completed.

IX.5.6. EFFECTS LIMITING PRACTICAL MEMBRANE APPLICATION

For any given modular design, tubular or flat, certain compromises in the processing conditions may be required. One of the factors which may impose certain limitations on the performance characteristics of the membrane is concentration polarization. In hyperfiltration concentration polarization can result in osmotic pressures considerably higher than those represented by the bulk stream concentration. Therefore higher pressures are required to overcome the osmotic pressure in the adjacent boundary layer and produce reasonable flow rates. To restrict the degree of concentration polarization to an acceptable level, the geometry of the feed brine channel should allow for adequate mixing of rejected salt with the bulk stream at reasonable inlet feed velocities. For ultrafiltration the macromolecular solutes and colloidal species usually have insignificant osmotic pressures. In this case the concentration polarization at the membrane surface can rise to the point of incipient gel precipitation, forming a dynamic secondary membrane on top of the primary membrane. This secondary membrane can offer the major resistance to flow.

Furthermore, the polymeric membrane material does have several drawbacks. The asymmetric structure is subject to compaction under the pressures generally used. The higher the crosslinking achieved by the annealing process, the less the reduction of volume flow density by compaction. Nevertheless the rapid loss of flux at 100 atmos. necessary for direct seawater desalination, is considerable even with highly annealed membranes. From this successfull attempts of Riley et al. in making composite membranes can be appreciated. These membranes do not exhibit substantial loss of flux by compaction.

Moreover, cellulose acetate is subject to microbiological attack as well as hydolysis at pH's far removed from neutrality. While hyperfiltration membranes for salt rejection lose their permselectivity quickly outside a pH range of 3 to 8, ultrafiltration membranes may be cleaned with alkaline solutions at pH 10.

Finally the application of cellulose acetate membranes is limited to separation processes at temperatures below 30° C, for at that temperature a slow but continuous shrinkage of the membrane takes place. It is also necessary to avoid membrane drying, which results in an irreversible loss of membrane performance.

An approach to circumventing these characteristics of cellulose acetate is to provide another polymeric membrane material. This can be done by using aromatic polyamides such as duPont nomex which is used in hollow fibers.

ACKNOWLEDGMENT

I would like to thank Dr. H. K. Lonsdale for reading the proofs, and the "Bundesminister für Forschung und Technologie" for financial support of the work on hyperfiltration.

REFERENCES

1. S. Loeb and S. Sourirajan, University of California, Los Angeles, Department of Engineering, Report No. 60-60 (1960)

2. S. Manjikian, Ind. Eng. Chem. Prod. Res. Devel. 6 (1967) 23

3. Saline Water Conversion Report, OSW, US. Dept. Interior 1969-1970, p. 449

4. M. A. Frommer and D. Lancet, in "Reverse Osmosis Membrane Research" ed. by H. K. Lonsdale and H. E. Podall, Plenum Press, New York 1972 p. 103

5. R. L. Riley, H. K. Lonsdale and C. R. Lyons, J. Appl. Polymer Sci. 15 (1971) 1267

6. N. W. Rosenblatt, Proceed. of the 4th Int. Symp. on Fresh Water from the Sea, Vol. IV, p. 349, ed. by A. and E. Delyannis, Athen 1973.

X. Titles and References of Short Communications

II. Theory

1. "Some General Questions Concerning the Theory of the Structure of Aqueous Electrolyte Solutions"
   O.J.Samoilov, Moskau, UdSSR
   "General Problems in the Theory of Ionic Hydration in Aqueous Solutions" in "Water in Biological Systems" Vol.1, p.20, Consultans Bureau, New York 1969

   "Strucutre of aqueous electrolyte solutions and the hydration of ions." Consultans bureau , New York 1965 ; Teubner Verlag ,Leipzig 1961.

2. "Problems of Characterization of Water Structure"
   J.Finney, London, U.K.
   Acta Crystallographica, A28 (S4), S133 (1972); Proc.Roy.Soc. A280, 299 (1964); Nature 212, 1353 (1966); Nature 212, 1355 (1966).

3. "Calculated Spectroscopic Properties of the Hydrated Proton"
   R.Janoschek, Stuttgart, Germany
   J.Amer.Chem.Soc. 94, 2387 (1972); Faraday Transactions II 69, 505 (1973)

III. Infrared Methods

1. "Fundamental Region"
   A.P.Zukovskij, Leningrad, UdSSR
   Main lecture without manuscript
   "Hydrophobic Bonding in Dilute Aqueous Nonelectrolyte Solutions" in "Water in Biological Systems" Vol.3, p.1, Consultans Bureau, New York 1971; "On Aqueous Electrolyte Solutions" in "Water in Biological Systems" Vol.4, in press.

2. "Normal Coordinate Analysis of H-bond Water and Structure"
   J.Bandekar, Manhattan, U.S.A.
   J.Mol.Spec. 41, 500 (1972); Spectry.Letters 5, 345 (1972); to be published in J.Mol.Spec. under title: "Incomplete Hydrogen-bonding and Structure of Liquid Water"; to be published in J.Mol.Spec. under title: "A Monte Carlo Normal Co-ordinate Analysis Treatment of Intermolecular Vibrations in Liquid Water".

3. "Structure of Electrolyte Solutions as Reflected in the XH Vibrational Band Splitting"
Z. Kecki, Warszawa, Poland
"Effect of Electrolytes on the Intensity of the Infrared Bands of Water" Roczniki Chem. 42, 1749 (1968); "Effect of Perchlorates on the Infrared Bands of Water. Part.I. The Cationic Effect" Roczniki Chem. 43, 1053 (1969); "Splitting of Infrared Absorption Bands of Water by Anions" Roczniki Chem. 44, 1141 (1970); "Assignment of High-frequency Band in Aqueous Perchlorates Solutions - A Reply to the Argument of S.Subramanian and H.F.Fischer" Roczniki Chem. 45, 937 (1971); "Effects of Perchlorates on the Infrared Bands of Water. Part II. The Origin of Band Splitting" Roczniki Chem. 46, 671 (1972); "Effect of Perchlorates on the Infrared Bands of Water. III. The Structural Model" J.Mol.Structure 12, 219 (1972); "The Structure of Electrolyte Solutions as Reflected in the OH and NH Band Splitting in Vibrational Spectra" Adv.Mol.Relax.Proces. 5, 261 (1973).

4. "I.R. Spectroscopic Study of Water Association in Aqueous Non-Electrolyte Solutions"
P.Saumagne, E.Gentric, A.Le Narvor, Brest France
"Etude par spectroscopie infrarouge de l'eau dissoute dans quelques solvants aromatiques" C.R.Acad.Sci.Paris 270, C, 1053-1056 (1970); "Etude par spectroscopie infrarouge de l'eau dissoute dans des mélanges de solvants peu polaires; mise en évidence de complexes de contact" C.R.Acad.Sci.Paris 270, B, 1332-1334 (1970); "Comparison Between Two Models of Complexed Water: Open Water-dimer and 1-1 Complexed Water" J.Chem.Phys. 53, 3768 (1970).

5. "Near I.R. Spectra of Water Solutions, as an Approach of Liquid Water Spectra"
A.Burneau and J.Corset, Paris, France
"Vibrational Spectra and Anharmonicity of H_2O, D_2O and HOD in Diluted Solution" Chem.Phys.Letters 9, 99 (1971); "Spectres d'absorption entre 1000 - 1100 cm^{-1} des molécules H_2O, D_2O et HOD libres et an interaction avec des bases en solutions diluées" J.Chim.Phys. 69, 142 (1972); "Calcul des perturbations du spectre de vibration des molécules H_2O, D_2O et HOD, sous l'effect de la dissolution" J.Chim.Phys. 69, 153 (1972); "Interprétation des variation de coefficients d'anharmonicité et de fréquences harmoniques observées par dissolution des molécules H_2O, D_2O et HOD" J.Chim.Phys. 69, 171 (1972); "Near Infrared Spectroscopic Study of the Interactions Between Water and Acetone" J.Phys.Chem. 76, 449

(1972); "On the Near Infrared Studies of the Structure of Water and the Use of Mixed Solvent Systems" J.Chem.Phys. 58, 5188 (1973); "Transitions vibrationelles simultanées impliquant des molécules de donneurs de proton en solution" Can.J.Chem. 51, 2059 (1973); "Etude comparée des spectres d'absorption dans le proche infrarouge des complexes 1-1 entre l'eau et les bases, en mélanges ternaires" Can.J.Chem. (soumis pour publication).

6. "I.R. Evidence of "Cooperative Effects" in Complexes where the Water Molecule Acts Both as a Donor and an Acceptor"
M.-H. Baron, J.Corset, C.de Loze and L.Schriver, Paris, France
"Infrared Evidence of Cooperative Effects in Complexes where the Water Molecule Acts Both as a Donor and an Acceptor: Etude par spectroscopie infrarouge des interactions entre l'eau et les sels. I - Interactions eau-sel en solution diluée dans des solvants organiques" J.Chim.Phys. 9, 1293 (1971); "Etude par spectroscopie infrarouge des interactions entre l'eau et les sels. II - Structures des solutions aqueuses saturées en sel" J.Chim.Phys. 9, 1299 (1971); "Infrared Evidence for the Structure of Salt Saturated Aqueous Solutions" Chem.Phys.Letters 9, 103 (1971); "Etude par spectroscopie infrarouge des associations moléculaires entre l'acide thiocyanique, l'eau et la méthylisobutylcétine. Application à l'extraction de l'acide thiocyanique" J.Chim.Phys. (sous presse).

7. "Fermi Resonance of Water in Solutions with Organic Solvents"
H.Wolff, D.Mathias and E.Millermann, Heidelberg, Germany
Spectrochim.Acta 27A, 21 + 9 (1971); "Hydrogen Bonding and Fermi Resonance of Aniline" J.Phys.Chem. 77, 2081 (1973).

8. "Anomalies in the Temperature-Dependence of the 1,2 nm Absorption of Liquid Water"
G.Andoloro, M.B.Palma-Vittorelli and M.U.Palma, Palermo, Italy

9. "I.R. Investigations of Amorphous Ice"
U.Buontempo, Roma, Italy

IV. Raman Methods

1. "Structure of Aqueous Magnesium Sulfate Solutions"
   W.A.Adams and A.R.Davis, Ottawa, Canada
   "A High Pressure Laser Raman Spectroscopic Investigation of Aqueous Magnesium Sulphate Solutions" J.Phys.Chem. 78 (March) 000 (1974); Spectrochim. Acta, Part A 27, 2401 (1971); J.Chem.Phys. (Feb. 15 issue), (1974).

2. "Raman Studies on the HSO_4-Ion in Aqueous Solution"
   D.J.Turner, Leatherhead, U.K.
   J.Chem.Soc.Faraday Transactions 1974, in press; J.Chem.Soc.Faraday Transactions II, 68 643 (1972).

V. Scattering Methods

1. "Structure Study of Amorphous Ice by Neutron Scattering"
   C.U.Linderstrom-Lang and J.Wenzel, Roskilde, Denmark

2. "X-Ray Scattering and Radial Distribution Function of Liquid Water"
   F.Hajdu, S.Lengyel and G.Pâlinkás, Budapest, Hungaria

3. "The Structure of Water by Slow Neutron Scattering"
   J.G.Powles
   (main lecture without manuscript)

4. "Light Scattering and Molecular Motion in Water"
   T.A.Litovitz
   (main lecture without manuscript)

VI. Dielectric Methods

1. "Essay About Interpretation of the Dielectric Relaxation of Water in Concentrated Na-DNA Solutions"
   D.Bourgoin, Paris, France

2. "Kinetics of Micelle Formation in Aqueous Solutions"
   H.Hoffmann and T.Janjic, Erlangen, Germany

VII. NMR Methods

1. "Relations Between NMR-Solvent Shifts and Thermodynamic Properties"
W.Schroer and E.Lippert, Berlin, Germany

2. "Molecular Structure from NMR in Liquid Crystal Solvents"
L.C.Snyder, Murray Hill, U.S.A.
Acc.Chem.Res. 4, 81 (1971); "Molecular Wave Functions and Properties" Wiley-Interscience, New York 1972.

3. "NMR Investigations of Less-Common Nuclei in Aqueous Solutions"
O.Lutz, Tübingen, Germany
"Atomic Reference Scale for Chemical Shifts of Rubidium" Z.Naturforschg. 27a, 1577 (1972); "The Magnetic Moment of ^{207}Pb and the Shielding of Lead Ions by Water" Phys.Letters 35A, 397 (1971); "The Magnetic Moment of ^{67}Zn and the Shielding of Zinc Ions by Water" Phys.Letters 45A, 255 (1973); "Nuclear Magnetic Resonance Studies of ^{43}Ca" Z.Naturforschg. 28a, 1534 (1973); "Nuclear Magnetic Resonance Studies of ^{111}Cd" Z.Pysik, accepted for publication; "NMR Investigations and Nuclear Magnetic Moment of ^{33}S" Z.Naturforschg. 28a, 1370 (1973)

4. "Aqueous Solution of Triethylenediamine"
J.W.-Y. Wen, Worcester, U.S.A.
"Viscosity and Apparant Molal Volumes of Aqueous Triethylenediamine" Journal of Solution Chemistry, Vol.3, 103-118 (1974); "Proton Chemical Shifts of Water in Aqueous Triethylenediamine: Cosphere Model and Temperature Dependence" Journal of Solution Chemistry, Number 5, Vol.3, 1974.

IX. Sea-Water

1. "The Refraction Index of H_2O, D_2O and Seawater between 0-25° and Pressures up to 2 kbar"
W.A.Adams, Ottawa, Canada
"A Digitized Laser Interferometer for High-Pressure Refractive Index Studies of Liquids (MS)" submitted to Applied Optics, 1973; Proc.Am.Electrochem.Soc., Miami Beach, October 1972, Marine Electrochemistry Symposium.

2. "Equilibrium Thermodynamics of Desalination by Freezing"
J.Bilgram, Zürich, Swiss

X. Water-Alcohol Mixtures

1. "Hydrophobic Interactions in Water Ethanol Mixtures"
   A.Ben-Naim and M.Yaacobi, Jerusalem, Israel

2. "PMR Investigation on Structural Problems in Water-Ethyl Alcohol Mixtures"
   A.Coccia, P.L.Indorina, F.Podo and V.Vitti, Roma, Italy
   "Studi strutturali mediante Risonanza Megnetica Nucleare in miscele aqua-etanolo." Rapporti dei Laboratori di Fisica, Instituto Superiore di Sanita', ISS 71/6, ISS 71/7, ISS 71/8 (1971); "PMR Studies on the Structures of Water-Ethyl Alcohol Mixtures" (to be submitted to Journal of Solution Chemistry)

3. "NMR-Chemical Shift and Relaxation Study of Alcohol-Water Mixtures"
   J.Oakes, Port Sunlight, U.K.
   Chem.Soc.Farad.Transact.II 69, 1311 (1973)

4. "Selected Solvatation in Highly Diluted Electrolyte Solutions"
   G.Kortüm, Tübingen, Germany

5. "Nichtrelaxierende Volumenviskosität in Wasser-Äthylalkohol-Mischungen"
   M.Sedlacek, Wien, Austria

XI. Thermodynamics and General Properties

1. "Heat Capacity and Thermal Expansivity of Supercooled Water"
   C.A.Angell, Lafayette, U.S.A.
   "Anomalous Properties of Supercooled Water: Heat Capacity Expansivity, and PMR Chemical Shift from 0 to -38°C"; Anomalous Heat Capacities of Supercooled Water and Heavy Water" Science 181, 342 (1973)

2. "The Vapour Pressure Isotope Effects"
   H.Wolff
   "The Vapour Pressure Isotope Effect of Ice" Physics of Ice, Plenum Press 1969, p.305; "Über den Dampfdruck-Isotopie Effekt von Wasser und Eis" Ber.Bunsenges. 72, 393 (1969); "Hydrogen Bonding and Vapour Pressure Isotope Effect of Dimethylamins" J.Phys.Chem. 74, 1600 (1970)

3. "Isotope Effects and Hydrate Structure of Aqueous Electrolyte Solutions"
   K.Heinzinger, Mainz, Germany

4. "Rate of Liquid Vapour Transitions of Water and other Polar Liquids and its Relation to Structure"
H.K.Cammenga, Braunschweig, Germany
"Rate of the Liquid Vapour Transition of Water and other Polar Liquids and its Relation to Structure" - To be published in Ber. Bunsenges.; "Untersuchungen über die Verdampfungsgeschwindigkeit von Flüssigkeiten" Fortschr.Kolloide and Polymere 55, 118-123 (1971); "Die Bestimmung der Verdampfungsgeschwindigkeit von Flüssigkeiten und Festkörpern im Bereich kleiner Dampfdrucke" Z.Physik. Chem. N.F. 63, 280-296 (1969).

5. "Thermodynamic Properties of Aqueous Solutions of Non-Electrolytes at High Pressure"
G.Götze, P.Engels and G.M.Schneider, Bochum, Germany
Ber.Bunsenges.physik.Chem. 76, 1239 (1972); P.Engels, thesis, Karlsruhe 1970; Proceedings of the Third International Conference on Chemical Thermodynamics, 3-7 September 1973, Baden, Austria (Section 5, page 145).

6. "Some Experimental Results on the Transport of Inorganic Ions from the Dissolved State into the Gas Phase"
P.Koske, Kiel, Germany
"Ion Fraction at the Surface of Aqueous Inorganic Salt Solutions by Means of a Film Centrifuge" J.Geophysical Res. 77 (27), 5201 - 5203 (1972); "Über eine Methode zur Messung von Oberflächenkonzentration flüssiger Mischphasen" Z.Physikal.Chem. NF 82, 287-294 (1972); "Surface Structure of Aqueous Salt Solutions ans Ion Fractionation" J.Recherches Atmosphériques (in print).

7. "Equation of State and Effective Potential for Liquid Water"
O.Singh, Rurkee, India

8. "Structure-Function of Antifreeze Glycoprotein from the Antarctic"
R.E.Feeney, Santa Barbara, U.S.A
"Structure and Role of Carbohydrate in Freezing-point Depressing Glycoproteins from an Antarctic Fish" J.Biol.Chem. 247, 7885 - 7889 (1972); "Macromolecules from Cold-adapted Antarctic Fishes" Naturwissenschaften 59, 22-29 (1972); "Inhibition of Lectins by Antifreeze Glycoproteins from an Antarctic Fish" Nature 242, 342-343 (1973); "Depressions of Freezing Point by Glycoproteins from an Antarctic Fish" Nature 243, 357-359 (1973); "Effects of Chemical modifications of carbohydrate residues on antifreeze and antilectin activities" J.Biol.Chem. 248 (in press, 1973); "A Biological Antifreeze" Amer.Sci (in press, 1974).

9. "Anomalous Disjoining Processes in Thin Aqueous Electrolyte Layers"
   G.Peschel, K.H.Adlfinger and G.Schwarz
   "Viscosity Anomalies in Liquid Surface Zones. IV. The Apparent Viscosity of Water in Thin Layers Adjacent to Hydroxylated Fused Silica Surfaces" J.Colloid Interface Sci. 34 (4), 505 (1970); "Thermodynamic Investigations of Thin Liquid Layers Between Solid Surfaces. II. Water Between Entirely Hydroxylated Fused Silica Surfaces" Z.Naturforsch. 26a (4), 707 (1971).

10. "Ion Diffusion and Momentum Correlation Functions in Dilute Electrolytes"
    S.Harris, Stony Brook, U.S.A.
    Molecular Physics, Vol. 2b, 953 (1973).

SUBJECT INDEX